AF566766

Dynamic Aspects of Explosion Phenomena

Edited by
A. L. Kuhl
Lawrence Livermore National Laboratory
El Segundo, California

J.-C. Leyer
Université de Poitiers
Poitiers, France

A. A. Borisov
Russian Academy of Sciences
Moscow, Russia

W. A. Sirignano
University of California
Irvine, California

Volume 154
PROGRESS IN ASTRONAUTICS AND AERONAUTICS

A. Richard Seebass, Editor-in-Chief
University of Colorado at Boulder
Boulder, Colorado

Technical papers from the Thirteenth International Colloquium on Dynamics of Explosions and Reactive Systems, Nagoya, Japan, July 1991, and subsequently revised for this volume.

Published by the American Institute of Aeronautics and Astronautics, Inc.,
370 L'Enfant Promenade SW, Washington, DC 20024-2518

ISSN 0079-6050

Progress in Astronautics and Aeronautics

chapter includes articles on flash x-ray visualization of steam explosions created by molten metal drops in water by *Frost and co-workers*; models of rapid vaporization of nonequilibrium mixtures of tin and water by *McCahan and Shepherd*; a steady, one-dimensional thermal detonation model for molten Sn-water suspensions by *Gelfand and co-workers*; and an equilibrium Hugoniot analysis of thermal detonations by *Frost and Ciccarelli.*

Chapter IV, Nonsteady Flows, presents a number of articles on nonsteady processes occurring in explosions. For example, *Kuhl et al.* describe numerical simulations of dusty boundary layers by various shock wave structures. They find that the dusty boundary layer grows as a power law function of the distance behind the shock, similar to turbulent boundary layers on clean flat plates. *Matsuo and Fujiwara* present numerical predictions of oscillatory instabilities of shock-induced combustion, while *Merzhanov and Gordopolov* report on the use of shock waves in self-propagating, high-temperature synthesis (SHS).

The companion volumes, *Dynamics of Gaseous Combustion* (Volume 151), *Dynamics of Heterogeneous Combustion and Reacting Systems* (Volume 152), and *Dynamic Aspects of Detonations* (Volume 153) include papers on the chronology of research on detonation waves during the period 1920–1950, gaseous detonations, initiation of detonation waves, and nonideal detonations and boundary effects; papers on the behavior of propagating premixed flames, ignition dynamics, diffusion flames and their structure, nonsteady flames, and combustion in shear layers; and papers on the dynamics of turbulent combustion, combustion in dust-air mixtures, droplet combustion, pulsed jet combustion, and internal combustion engines.

These four volumes will, we trust, help satisfy the need first articulated in 1966 and will continue the tradition of augmenting our understanding of the dynamics of explosions and reactive systems begun the following year in Brussels with the first colloquium. Subsequent colloquia have been held on a biennial basis: 1969 in Novosibirsk, 1971 in Marseilles, 1973 in La Jolla, 1975 in Bourges, 1977 in Stockholm, 1979 in Göttingen, 1981 in Minsk, 1983 in Poitiers, 1985 in Berkeley, 1987 in Warsaw, 1989 in Ann Arbor, and 1991 in Nagoya. The Colloquium has now achieved the status of a principal international meeting on these topics, and attracts contributions from scientists and engineers throughout the world.

To provide an enduring focal point for the administrative aspects of the ICDERS, the organization was formally incorporated in the state of Washington under the name Institute for Dynamics of Explosions and Reactive Systems (IDERS). Professor J. R. Bowen is serving as the current president. Communications may be sent to:

Dean J. R. Bowen
President, IDERS
College of Engineering FH-10
University of Washington
Seattle, Washington 98195
USA

Papers from the first six colloquia have appeared as a part of the journal *Acta Astronautica*, or its predecessor, *Astronautica Acta*. With the publication of the Seventh Colloquium, selected papers have appeared as part of the Progress in Astronautics and Aeronautics series published by the American Institute of Aeronautics and Astronautics (AIAA). These are the last Dynamics of Explosions and Reactive Systems Colloquium papers to appear in the Progress in Astronautics and Aeronautics series.

Acknowledgments

The Thirteenth Colloquium was held under the auspices of Nagoya University from July 28 to August 2, 1991. Local arrangements were organized by Professors T. Fujiwara and A. K. Hayashi. Publication of selected papers from the Colloquium was made possible by grants from the National Science Foundation and the Defense Nuclear Agency of the United States. Generous financial support for the meeting was received from the following organizations: Aichi Machine Industry Company, Aichi Prefecture, Aishin AW, Canon Sales Company, Central Japan Nagoya Airport, Central Japan Nagoya Station, Chubu Aeronautics and Space Technology Development Association, Chubu Electric Power Company, Daikin Industry, DAIKO Foundation, ENGAKU, Haruki (Mr.), ET Planning, FUJIMA Sohke School of Kabuki Dances, Gifu Auto Body Industry Company, Hitachi, Honda Motor Company, IBM Japan, Ishikawajima-Harima Heavy Industries, Isuzu Motor Company, Japan Gas Association, KATO Ryutaro Foundation, Kawasaki Heavy Industries, Kobe Steel, Matsushita Graphic Communication Systems, Mazda Motor Company, Meitec Corporation, Mitsubushi Heavy Industries, Nagoya City, Nippon Denso, Nippon Oil and Fats Company, Nippon Sanso, Nippon Steel Corporation, Nissan Motor Company, Rinnai Corporation, Science Research Fundings from the Ministry of Education, Science, and Culture (Profs. K. Abe, T. Fujiwara, and K. Takayama), Shachihata Industrial Company, Sogo Solvent Company, Takashimaya-Nippatsu Kogyo Company, Toho Gas, Tokai Bank, Toshiba Corporation (Chubu Branch), Toyoda Automatic Loom Works, Toyoda Gosei Company, Toyoda Machine Tools, Toyota Central Research and Development Laboratory, Toyota Motor Company, and Toyota Techno Service Company.

A. L. Kuhl
J.-C. Leyer
A. A. Borisov
W. A. Sirignano
May 1993

Table of Contents

Chapter II. Dust Explosions

Chapter III. Vapor Explosions

Chapter IV. Nonsteady Flows

Table of Contents for Companion Volume 151

Preface

Chapter I

Chapter II. Ignition Dynamics

Chapter III. Diffusion Flames and Their Structure

Chapter IV. Nonsteady Flames

Chapter V. Combustion in Shear Layers

Table of Contents for Companion Volume 152

Preface

Chapter I. Dynamics of Turbulent Combustion

Chapter II. Combustion in Dust-Air Mixtures

Chapter III. Droplet Combustion

Chapter IV. Pulsed Jet Combustion

Chapter V. Internal Combustion Engines

Table of Contents for Companion Volume 153

Modeling of Turbulent Unvented Gas-Air Explosions

Francesco Tamanini*
Factory Mutual Research Corporation, Norwood, Massachusetts 02062

Abstract

A phenomenological model of turbulent confined gas explosions is presented and its predictions are compared with experimental results obtained in a 1.35-m^3 spherical vessel. The novel contribution of the model is in the approach taken to account for the effect of turbulence on the combustion process. This is assumed to be characterized by engulfment of unburnt mixture by large-scale eddies and subsequent consumption of the engulfed material by laminar flame propagation through eddies of a size equal to the Taylor microscale. The laminar burning velocity of the combustion system being considered is a free parameter that must be supplied to the model. Calculated values for the peak explosion pressure, p_m, an equivalent turbulent burning velocity, $u_{t,eq}$, and the normalized maximum rate of pressure rise, K_G, have been obtained using thermodynamic properties based on the assumption of chemical equilibrium. The predictions have been compared with experimental data for three CH_4/air (5.5, 7.5, and 9.5%) and two C_3H_8/air (4.0 and 4.8%) mixtures, with good overall agreement. This result is quite encouraging, given the range of reactivities of the five mixtures, as expressed by laminar burning velocities ranging from 0.08 to 0.48 m/s. The paper also discusses improvements that are required in order for the model to be extended to the case of dust/air mixtures and vented explosions.

---

*Manager, Explosion Section.

Introduction

Because of the difficulties associated with the complex processes that control the rapid combustion of flammable mixtures, improvements in explosion protection technology have so far been mostly driven by experimentation. In the case where emergency venting is used as a means to mitigate the severity of explosions in enclosures, the flow dynamics associated with the gas outflow leads to interactions with the combustion which usually complicate direct extrapolation of test results. As in other areas of engineering, however, theoretical modeling offers an attractive complementary approach to testing, if not yet a complete alternative, since it holds the promise of developing a deeper understanding of the interactions among the various aspects of the phenomenon being modeled. An additional, shorter-term benefit of modeling is in the form of tools for extending the applicability of experimental results to conditions differing from those of the available test data.

The contribution offered by this work focuses on the modeling of the effects of turbulence on the process of flame propagation in an explosion. Turbulence can be present in the flow independently of the combustion, or else it can be generated by the flame propagation process itself due to the interaction of the flow induced by volume expansion with internal obstacles and/or openings in the enclosure boundary. Flame instabilities provide another mechanism for enhancement of the rate of combustion of a flame propagating in an initially quiescent mixture. These phenomena represent important contributing factors in determining the severity of accidental (vented or unvented) gas explosions and they eventually need to be considered. A similar assessment applies to dust explosions, with the possible exception that, in this latter case, flame instabilities have not been observed and may in fact not occur.

Whereas the ultimate goal of this modeling effort is to include the treatment of gas and dust explosions under both unvented and vented conditions, the focus of the present contribution has been limited to unvented gas explosions, for which new data have recently been obtained as part of an ongoing experimental effort[1]. These results have provided information on the interaction of background (i.e., existing independently of the explosion) turbulence with the flame propagation process. A simplified treatment has been used for the fuel mixture reactivity by assuming the fundamental flame propagation to be characterized by a laminar burning velocity and the chemical reaction to be fast. This latter assumption implies a thin reaction front and is generally appropriate to gas explosions. In the case of dust explosions, the same formulation may be applicable in selected situations.

Background

Several studies have appeared in the literature describing models of unvented and vented explosions, with most of the efforts to date directed to the problem of gas explosions[2-6]. In these models, the flame is assumed to propagate as a thin front separating unburnt fuel from completely reacted material. The propagation speed is then adjusted to a multiple of the laminar flame speed to approximate the effect of turbulence.

While proposals have been made[7,8] on possible empirical approaches to deal with the vent flow-induced flame acceleration phenomena observed in several gas explosion experiments[9-12], the modeling of the effect of the turbulence itself on the flame has remained rather primitive. This is somewhat surprising, since careful measurements of these effects have been made under controlled experimental conditions[13-15] and for combustion in practical systems (mostly internal combustion engines[16-17]).

From these studies[14-17], the process by which fuel is consumed by an advancing turbulent flame front has been found to be dependent on the characteristics of the turbulence field and on the chemical reactivity of the mixture. More specifically, if the integral scale of the turbulent eddies is at least a few times greater than the laminar flame thickness, then the flame propagates as a convoluted laminar front, which can break up and be partially quenched if the intensity of the fluctuations is sufficiently high. Under these conditions, the flame propagation process can be described as initial engulfment of fresh mixture by large-scale turbulent eddies. The unburned mass behind the entrainment front is subsequently ignited as the reaction spreads rapidly along the boundaries of small eddies of characteristic size equal to the Taylor microscale. Fuel consumption is finally brought about by flames propagating through these eddies at the fundamental laminar burning velocity.

The main contribution of the model described in the following sections is in the adaptation of this physical concept to the problem of flame propagation in an unvented explosion. The actual implementation is done within the framework of a simple, lumped-parameters representation of the flowfield since a more detailed numerical description does not appear to be justified at this stage of the development effort. Comparisons with experimental data are used to identify areas where model improvements are needed and to give an indication of the extent to which this approach can provide reliable engineering predictions.

Theoretical Model

Overall Conservation Equations

The selected representation of the flowfield is shown schematically in Fig. 1. For simplicity, central ignition and spherical symmetry of the enclosed volume are assumed, which implies a macroscopically spherical flame propagation process. The mass involved in the modeled process can be in one of three states, depending on its location with respect to the advancing flame front and on whether it has been affected by the reaction:

1) Unburnt/unentrained (*uu*) mass: material ahead of the flame front which is totally unreacted.

2) Unburnt/entrained (*ue*) mass: unreacted material which has been engulfed by the advancing turbulent flame front and is, therefore, mixed with combustion products within the macroscopic flame envelope.

3) Burnt (*b*) mass: completely reacted material within the flame envelope.

In this formulation, the burnt and unburnt/entrained portion of the total mass contribute to the flame volume ($f=b+ue$), whereas the total unburnt material is given by the sum of the unburnt/unentrained and unburnt/entrained fractions ($u=uu+ue$). Spatially uniform properties are assumed for the burnt (b) and unburnt (u) material, respectively, and pressure is constant throughout the enclosure volume. Since the unburnt mass (u) is divided between two "zones", it is

$$m_u = m_{uu} + m_{ue} \,, \tag{1}$$

and its net rate of change is given by:

$$\frac{dm_u}{dt} = \frac{dm_{uu}}{dt} + \frac{dm_{ue}}{dt} \tag{2}$$

In accordance with the lumped-parameter approach used by the model, overall conservation of energy for the unburnt (u) and burnt (b) mass is written as:

$$\frac{d}{dt}(m_u\, u_u) + p\,\frac{dV_u}{dt} = \frac{dm_u}{dt}\, h_u + \dot{Q}_{ex,u} \tag{3}$$

and

$$\frac{d}{dt}(m_b\, u_b) + p\,\frac{dV_b}{dt} = \frac{dm_b}{dt}\, h_u + \dot{Q}_{ex,b} \tag{4}$$

where p is the total pressure, V is the volume, u_i is the internal energy ($= u_i^0 + c_{v,i}\,(T_i - T_0)$), h_i is the thermal enthalpy ($= h_i^0 + c_{p,i}\,(T_i - T_0)$), and $Q_{ex,i}$ is the total energy input from the surroundings.

The subscript (u or b) indicates whether the property refers to the unburnt or to the burnt gas. Constant specific heats at constant volume and pressure (c_v and

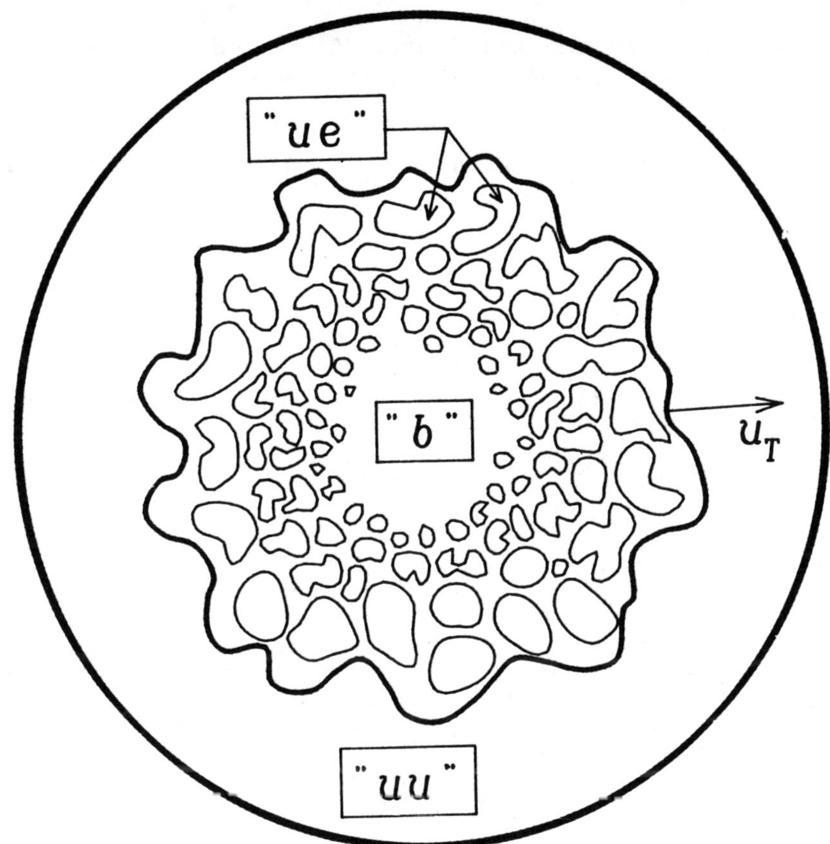

Fig. 1 Schematic representation of the flame propagation process.

c_p) have been assumed in the definitions for the internal energy and the enthalpy of the gas. In addition, the reference temperature, T_0, the enthalpies, h_u^0 and h_b^0, and the internal energies, u_u^0 and u_b^0, have been introduced.

In Eqs. (3) and (4), the two terms on the left represent the increase in internal energy of the gas and the external work done by the gas. The terms on the right represent the enthalpy flux contributed by the reacting mass and the net energy addition due to heat transfer to the gas from outside sources. The thermal energy liberated by the chemical reaction ($\Delta h_c^0 = h_u^0 - h_b^0$) would appear explicitly in Eq. (4) if the equation were to be written in terms of the sensible internal energies and enthalpies, rather than using the form for these variables which includes the chemical terms.

The volume of the enclosure, V, can be expressed as the sum of the volumes of the burnt and unburnt masses:

$$V = V_u + V_b \tag{5}$$

where the latter two volumes can be calculated from the equation of state as:

$$V_u = m_u \frac{R\, T_u}{p\, M_u} \quad ; \quad V_b = m_b \frac{R\, T_b}{p\, M_b} \tag{6}$$

with R the perfect gas constant, and M_u and M_b the molecular weights of the unburnt and burnt gases. Since the volume of the enclosure is constant, the rates of change of the unburnt and burnt volumes are equal and of opposite sign, namely:

$$\frac{dV_u}{dt} = - \frac{dV_b}{dt} \tag{7}$$

Turbulence Model

The turbulence in the flow is characterized through the introduction of two variables: the turbulence kinetic energy, k, and its dissipation, ε. For consistency with the handling of the other aspects of the model, a different set of values for k and ε should be calculated for the unburnt and burnt mass, respectively. However, since the combustion is assumed to be affected only by the turbulence in the unburnt mass, the present version of the model limits itself to calculating k and ε for the unburnt fraction of the flow and neglects the evolution of the turbulence properties in the reacted portion of the flowfield.

A more general treatment of turbulence should include ways to account for production by some external action, or due to the interaction between the flow induced by the explosion and obstructions or openings in the enclosure volume. At this point, the model simply provides a means to simulate the decay of background turbulence in a way that reproduces experimental measurements obtained under the same dynamic injection conditions of the explosion tests[1].

In accordance with the prescription of the k-ε model of turbulence[18,19], equations for the evolution of the two properties of the turbulence in the flow are written as:

$$\frac{dk}{dt} = P_k - \epsilon \tag{8}$$

and

$$\frac{d\epsilon}{dt} = P_\epsilon - c_{\epsilon 2} \frac{\epsilon^2}{k} \tag{9}$$

where the constant $c_{\varepsilon 2}$, which is usually set in the range 1.74-1.92, was found in Ref. 1 to require a value of 1.62 in order to simulate the natural decay in a 1.35-m^3 vessel of the turbulence induced by the injection of a compressed air charge in the volume. Since the model does not simulate the active injection portion of the transient, there is no need to account for the sources of k and ε. The corresponding terms in the equations are, therefore, set to zero ($P_k = P_\varepsilon = 0$).

Characteristic scales of the turbulence can be derived from k and ε. More specifically, the integral scale, L, is obtained from[20]:

$$L = c_\mu^{3/4} \frac{k^{3/2}}{\epsilon} \qquad (10)$$

For a value of the constant $c_\mu = 0.09$, as normally used in the k-ε model, the coefficient on the right becomes equal to 0.162. This value is consistent with experimental data on integral length scale. The Taylor microscale can be calculated from[21]:

$$\lambda = \sqrt{\frac{10\, \nu\, k}{\epsilon}} \qquad (11)$$

The turbulence intensity, u', which normally appears in Eq. (11), has been replaced by the turbulence kinetic energy, k, by taking advantage of the fact that, for isotropic turbulence, it is:

$$k = \frac{3}{2} u'^2 \qquad (12)$$

The expression for the Taylor microscale given as Eq. (11) is used by the model to calculate the characteristic time scale for consumption of the eddies of unburnt fuel existing behind the flame front. In this process, the dependence of the kinematic viscosity on temperature and pressure is taken into account through the following expression:

$$\nu = \nu_0 \left(\frac{T}{T_0}\right)^{1.75} \frac{p_0}{p} \qquad (13)$$

which is based on the assumption that the dynamic viscosity is proportional to the 3/4 power of temperature and independent of pressure. The value for air at ambient conditions (15.5 10^{-6} m^2/s) is used for the reference dynamic viscosity, υ_0.

Combustion Model

As indicated, the combustion model postulates initial entrainment of unburnt material as a result of the advancement of the turbulent flame front, and subsequent burning of the entrained mass in smaller scale eddies behind the flame front. This formulation has been implemented through the following simple set of equations.

The rate of depletion of unburnt/unentrained (uu) mixture ahead of the flame front is given by:

$$\frac{dm_{uu}}{dt} = - A_f \rho_u u_T \qquad (14)$$

where A_f is the projected flame area,
ρ_u is the density of the unburnt mixture, and
u_T is the turbulent burning velocity.

The turbulent burning velocity is calculated from:

$$u_T = u_\ell + u' \qquad (15)$$

where u_ℓ is the laminar burning velocity, and u' is the turbulence intensity obtained from the value of turbulence kinetic energy, k, by using Eq. (12). In calculating the laminar burning velocity, its variation with temperature and pressure of the unburnt gas is taken into account according to:

$$u_\ell = u_{\ell 0} \left(\frac{T}{T_o}\right)^\alpha \left(\frac{p}{p_0}\right)^\beta \qquad (16)$$

where $u_{\ell 0}$ is the laminar burning velocity at reference conditions (typically at ambient temperature and pressure). The exponents α and β assume values which depend on the nature of the combustible mixture[22]. For the calculations presented in this report, these two variables will be set equal to 2 and -0.25, respectively.

Substitution of Eq. (15) into Eq. (14) yields:

$$\frac{dm_{uu}}{dt} = - A_f \rho_u u_\ell - A_f \rho_u u' \qquad (17)$$

In this equation, the first term on the right represents the combustion rate associated with laminar effects. The second term represents the rate of consumption (i.e., entrainment) of fresh mixture associated with the turbulent fluctuations and, therefore, it describes the transfer of mixture from the uu to the ue state. Accordingly, the rate of change of the unburnt/entrained (ue) mass fraction is written as:

$$\frac{dm_{ue}}{dt} = A_f \rho_u u' - \frac{m_{ue}}{t_c} \qquad (18)$$

In this expression, the second term on the right represents the depletion of unburnt/entrained mixture by eddy consumption. The combustion of the mixture in the eddies is assumed to be due to laminar flame propagation and is, therefore, characterized by an eddy burnout time, t_c, which is calculated from:

$$t_c = \frac{\lambda}{u_l} \tag{19}$$

where the Taylor microscale, λ, is obtained from Eq. (11). The expression given as Eq. (19) implies a quite simple idealized situation: if the variation in the eddy size during burning is neglected, this formula yields the time for complete burnout of eddies of average size λ with a distribution of sizes in the range $0-2\lambda$.

The effect of combustion on the rate of change of burnt (b) mass is finally given by:

$$\frac{dm_b}{dt} \quad A_f \, \rho_u \, u_l + \frac{m_{ue}}{t_c} \tag{20}$$

There are two sources of production of burnt gas represented by the two terms on the right of the equation: direct transfer from uu material by laminar combustion at the macroscopic projected flame front and transfer from the ue material by eddy combustion.

The area of the macroscopic flame front is calculated by assuming that the flame maintains a spherical shape. Due to the presence within the flame envelope of both burnt (b) and unburnt/entrained (ue) material, the flame area is calculated from:

$$A_f = (36\,\pi)^{1/3} (V_{ue} + V_b)^{2/3} \tag{21}$$

The split between a "laminar" and a "turbulent" contribution of the total rate of mixture consumption shown in Eqs. (17) and (20) is somewhat arbitrary. Refinements to this aspect of the formulation may be necessary since, at low turbulence intensities, a continuous wrinkled laminar flame front is probably more realistic than a broken one with unburnt pockets behind it[21]. However, comparisons between model results and experimental data seem to indicate that this refinement is not necessary for the range of conditions considered here.

Heat Losses

While the modeling of heat losses was not a major goal of this effort, heat transfer from the flame can account for significant departures from adiabatic conditions, particularly when dealing with confined explosions. For this reason, a provision has been made to include both radiative and convective heat transfer effects in the model. Full details of the models used are given in Ref. 23.

In general, the net heat transfer to the unburnt and to the burnt material is written as the sum of a convective and a radiative component, namely:

$$\dot{Q}_{ex,i} = \dot{Q}_{c,i} + \dot{Q}_{r,i} \tag{22}$$

where the subscript, i, refers to the unburnt (*u*) or the burnt (*b*) material. The heat transfer by convection is calculated by postulating a constant heat transfer coefficient, and by assuming that only unburnt material is in physical contact with the enclosure wall up to the point when the flame front reaches it. Afterwards, convective heat transfer takes place between the burnt material and the wall. Given the fact that convective heat transfer is relatively unimportant, at least for the cases discussed in this paper, refinements to this part of the model have not been deemed necessary.

For the radiative heat transfer, the energy exchanges between the unburnt and burnt material and the walls of the enclosure are calculated using a mean beam length approach[24]. Two mean beam lengths are calculated, one for the unburnt and one for the burnt material assuming that the two types of mass do not intermix. This is not entirely consistent with the picture of a propagating flame front that engulfs what has been referred to as unburnt/entrained (*ue*) mass, but this simplification is not expected to introduce errors that are large when compared to the intrinsic accuracy of the mean beam length approach. Approximate values for the emissivities of the gas volumes involved in the process are obtained from published correlations for the products of combustion of hydrocarbon flames[24].

Results From Experiments

Experimental Conditions

Most of the experimental data which will compared with the theoretical predictions from the model have already been presented in Ref. 1. They are briefly summarized here for the purpose of providing documentation of the conditions of those tests and to describe the methodology used to process the data, since this latter aspect is also relevant to the presentation of the model predictions.

The main objective of these experiments was to determine the effect of turbulence on the development of unvented gas explosions. Tests were carried out in a 1.35-m^3 spherical vessel for five fuel/air mixtures. The vessel was equipped with a high-pressure air injection system which was used to introduce the turbulence in the test volume. The evolution of the transient turbulence field was determined through time-resolved velocity measurements made with a bidirectional probe specifically developed for the application. After averaging the measurements made at three different locations in the sphere, the variation with time of the turbulence intensity was expressed by the following power fit:

$$u' = 1.286(t - 0.34)^{-0.803} \tag{23}$$

where u' [m/s] is the rms of the instantaneous velocity, and t [sec] is the time from the beginning of compressed air injection in the test volume. By assuming the relationship between turbulence intensity, u', and turbulence kinetic energy, k, given by Eq. (12), the dissipation of k can be written in terms of u' as:

$$\epsilon = -\frac{dk}{dt} = -3\,u'\,\frac{du'}{dt} \tag{24}$$

From the knowledge of k and ε, the integral length scale, L, and the Taylor microscale, λ, can be obtained using Eqs. (10) and (11). The power fit to the experimental measurements of u' and the calculated values for L and λ are shown in Fig. 2 along with the corresponding values for the turbulence Reynolds number, Re_L, defined as:

$$Re_L = \frac{u'\,L}{\nu} \tag{25}$$

As can be seen, when the turbulence intensity is at the highest level for which explosion data were obtained (u' = 13 m/s), the Taylor and the integral scale are calculated to be 2.3 and 91 mm, respectively, corresponding to a turbulence Reynolds number of 76,600. After a tenfold drop in turbulence intensity, when this quantity is down to u' = 1.3 m/s, the above three properties of the field are calculated to become equal to 9.8 mm, 161 mm, and 13,500.

For the case of a flammable mixture with a laminar burning velocity of $u_{\ell 0}$ = 0.43–0.48 m/s, within the range of the two conditions defined above, the classification of combustion regimes for premixed flames given by Ref. 21 would imply an expected burning behavior starting with wrinkled laminar flame sheets at the low end of the turbulence intensity scale, and extending into the fragmented reaction zone regime at the high turbulence end. This is shown in Fig. 3, where the regimes identified in Ref. 21 are illustrated and where the dashed line represents the range of conditions for the tests in the 1.35-m^3 sphere reported in Ref. 1 (laminar burning velocities, $u_{\ell 0}$, of about 0.45 m/s). Some of the additional data for lean methane/air mixtures presented in this paper correspond to conditions approaching the limits where flame quenching is expected (laminar burning velocities of 0.08 and 0.31 m/s for the two lean mixtures tested).

The combustion regimes identified in Fig. 3 are shown in a plot of the normalized turbulence intensity, u'/u_ℓ,

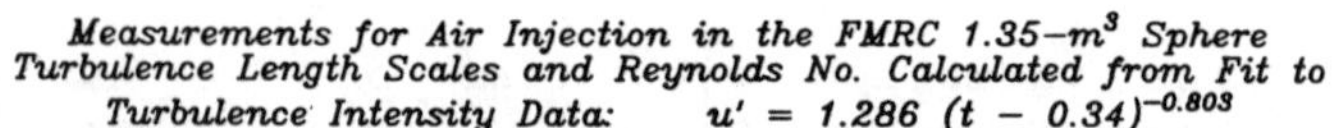

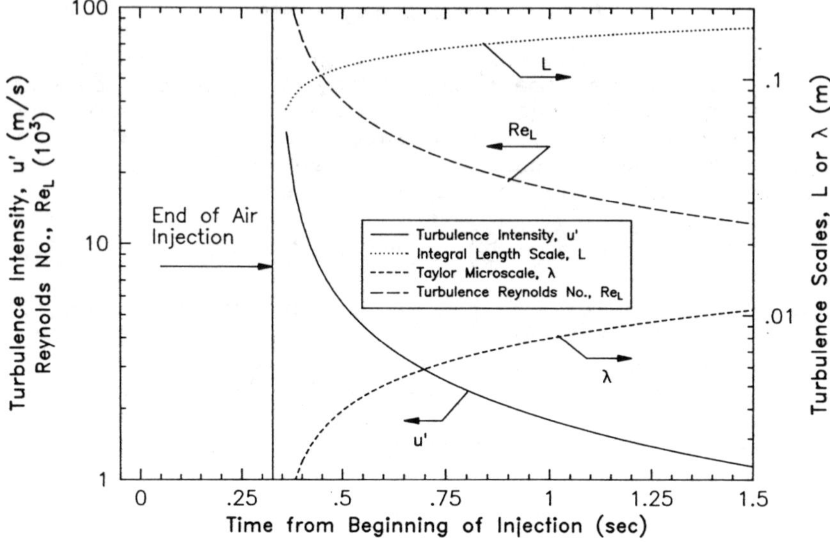

Fig. 2 Time variation of turbulence intensity data (u') obtained in the FMRC 1.35-m³ sphere and calculated integral length scale, Taylor microscale and turbulence Reynolds number.

versus the ratio of the integral scale, L, and the laminar flame thickness, δ_ℓ. An estimate for this latter quantity is obtained from the ratio u'/u_ℓ. The boundaries between the various regimes are defined by the values of the Karlovitz parameter, K, where:

$$K = 0.157 \left(\frac{u'}{u_\ell}\right)^2 Re_L^{-0.5} \tag{26}$$

The beginning of flame quenching is associated with a value of 1.5 for the parameter K. According to the slightly different classification given in Ref. 25, the region corresponding to $u'/u_\ell > 1$, up to the point where $K = 1$, is characterized by corrugated laminar flames, also described as wrinkled flames with pockets. On the basis of these correlations, the combustion model developed for this work does appear to embody a reasonable representation of the expected burning behavior typical of the flame data with which the model predictions will be compared.

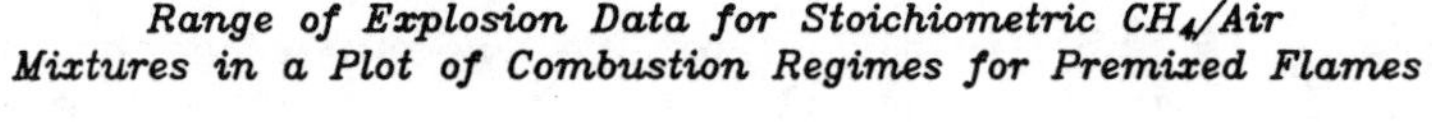

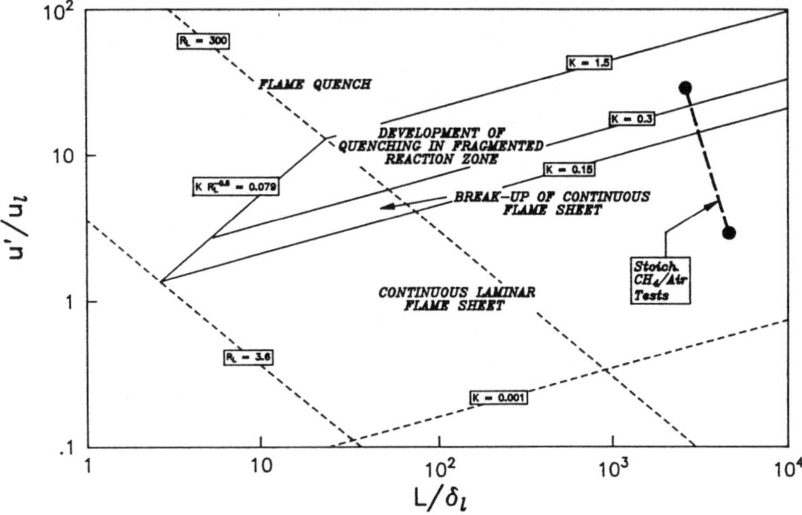

Fig. 3 Combustion regimes of premixed flames as a function of normalized integral length scale and turbulence intensity (according to Ref. 21). Also shown is the range of the 1.35-m^3 sphere data for the 9.5% CH_4/air mixture.

Data Processing

Through numerical processing of the pressure-time histories measured during unvented explosion tests, information can be obtained for quantities that are descriptive of the combustion. One such quantity is the peak explosion pressure. This was obtained from the data by calculating the maximum value of the smoothed pressure curve, to eliminate the influence of fluctuations in the pressure record on the reported value.

Two other quantities were calculated which provide a measure of the reactivity of the mixture. The first is the equivalent turbulent burning velocity, $u_{t,eq}$, defined as the burning velocity of a thin, undistorted flame front that would produce the same initial rate of pressure rise as displayed by the data. A simple analytical result, which is applicable to the case of constant burning velocity and isothermal gas behavior[2], provides a formula that relates the variation in pressure rise, Δp, to elapsed time:

$$\Delta p = p_0 \frac{4\pi}{3V} \frac{p_m - p_0}{p_0} \left(\frac{p_m}{p_0}\right)^2 (u_{t,eq}\, t)^3 \qquad (27)$$

where p_m is the maximum absolute pressure reached by the explosion, p_0 is initial pressure, and V is the volume of the explosion vessel. Equation (27) approximates the exact solution with an accuracy of 5% for a pressure variation from p_0 to $2p_0$. This formula, which was found to provide an excellent representation of the early rise of the pressure measured in the tests, was used by calculating the linear best fit to data of $(\Delta p)^{1/3}$ versus time, over the pressure range Δp = 1 to 8 psi (0.069 to 0.55 bar). Values for the equivalent burning velocity, $u_{t,eq}$, obtained by this method will be used in the comparisons of model predictions with experimental results.

A second quantity, which is descriptive of the reactivity of the mixture in the later stages of the combustion process, is the normalized maximum rate of pressure rise, K_G, defined as:

$$K_G = V^{1/3} \left(\frac{dp}{dt}\right)_{\max} \tag{28}$$

where V is the volume of the vessel in which the rate of pressure rise has been measured. The choice of the quantity K_G for the presentation of the data is based on the fact that this parameter is used in standardized procedures for the sizing of explosion vents[26].

While the evaluation of the equivalent turbulent burning velocity, $u_{t,eq}$, is relatively straight forward, care must be used in calculating the rate of rise, K_G, particularly when dealing with rapidly increasing and noisy pressure signals. In the experiments, data were acquired at the rate of 2000 readings/sec. In the processing of the raw data, smoothing was first introduced by applying a five-point (2.5 msec window) running average. Slopes were then calculated from a linear best fit to every five points of the smoothed curve. The maximum of the set of slope values obtained in this fashion was taken as $(dp/dt)_{\max}$. In order to make meaningful comparisons, the exact same processing steps have been applied to predicted pressure-time curves, after interpolation of the calculated values to force a 0.5-msec time interval between successive points of the numerically predicted curve.

Numerical Predictions

Selection of Model Parameters

The model requires the selection of a number of parameters, defining the nature of the mixture, the characteristics of the environment, the geometry of the enclosure, and the controls to be applied to the time step size in the numerical solution of the problem.

The physical properties of the burnt mixture were obtained from calculations performed using the NASA equilibrium program documented in Ref. 27, for the three methane (5.5, 7.5, and 9.5%) and the two propane (4.0 and 4.8%) mixtures in air studied. Mixture compositions and flame temperatures were calculated for adiabatic constant-volume combustion, for an initial temperature of 298 K. In this approach, the heat of combustion is derived from the enthalpy of formation of the species participating in the reaction going to the calculated equilibrium compositions.

The average specific heat at constant volume of the burnt mass is calculated from the heat of reaction and the adiabatic flame temperature rise. The specific heat at constant pressure is then derived from the relationship:

$$R = M_i(c_{p,i} - c_{v,i}) \tag{29}$$

The corresponding values for the unburnt mixture are calculated on the basis of handbook data[28] for the specific heats of the mixture components. The laminar burning velocity is calculated from the experimental data for the quiescent tests reported in Ref. 1 and from the tests for the lean methane mixtures performed after publication of that work.

It should be noted that the choice of equilibrium values for the mixture properties and the selection of the combustion and radiation submodels have limited the parameters that remain to be chosen to two: the convective heat transfer coefficient to the wall, h, and the laminar burning velocity, u_{l0}. The first was set equal to 50 W/m^2 K; its exact value, within a range of reasonable choices, has little effect on the results. The second quantity is obtained from experiment. As a result, the model in its present form is relatively free of adjustable parameters.

Comparisons With Experiment

Comparisons of model predictions with the experimental data obtained for the five fuel/air mixtures tested are shown in Figs. 4-8. The comparisons are made for the three variables: peak explosion pressure, p_m; equivalent burning velocity, $u_{t,eq}$; and normalized maximum rate of pressure rise, K_G.

The calculated curves were obtained by running the theoretical model for the laminar case ($u'=0$, quiescent conditions) and for nine levels of turbulence intensity. The highest turbulence was obtained by simulating an ignition delay time of 0.375 sec (from the beginning of mixture injection). The results from this run are represented by the point to the far right of the calculated curves. As can be seen, the curves for the slow burning mixtures (the 5.5% methane, in particular)

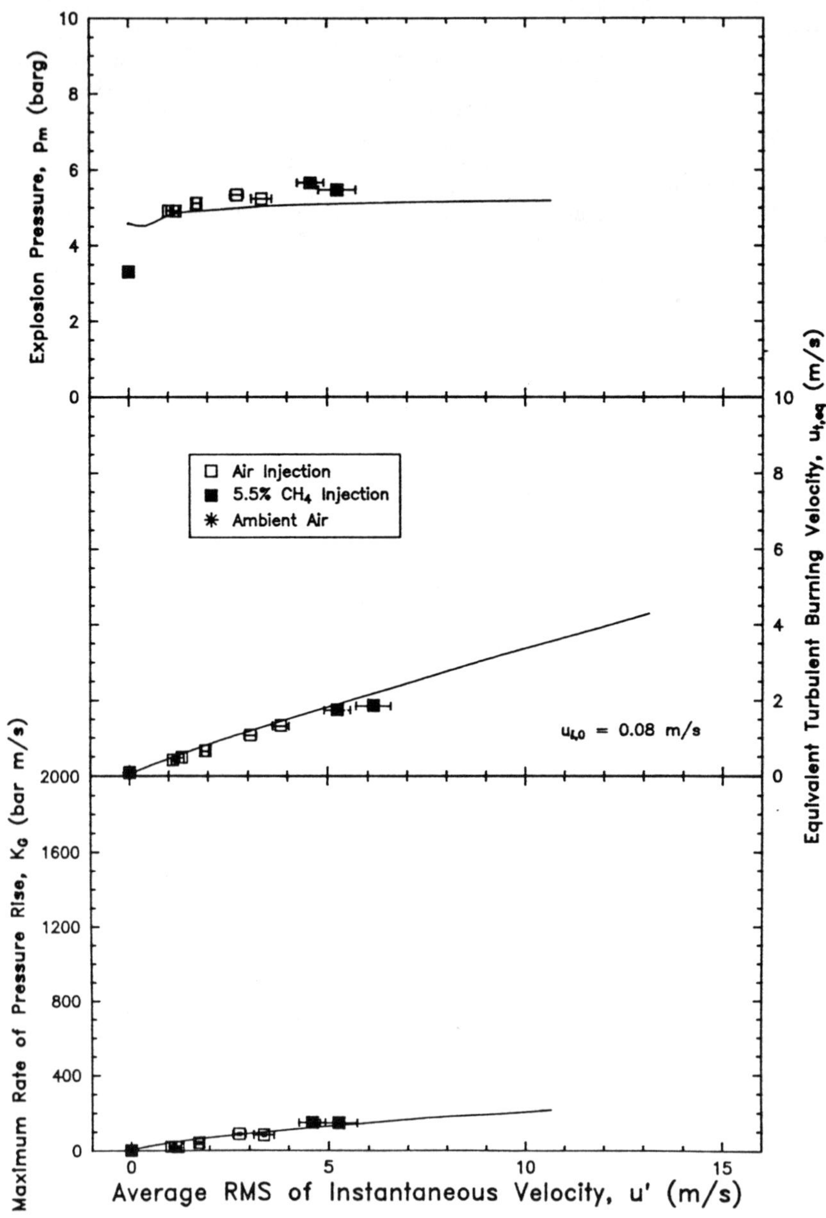

Fig. 4 Calculated explosion parameters as a function of turbulence intensity for the case of a 5.5% CH_4/air mixture ($u_{\ell 0}$ = 0.08 m/s).

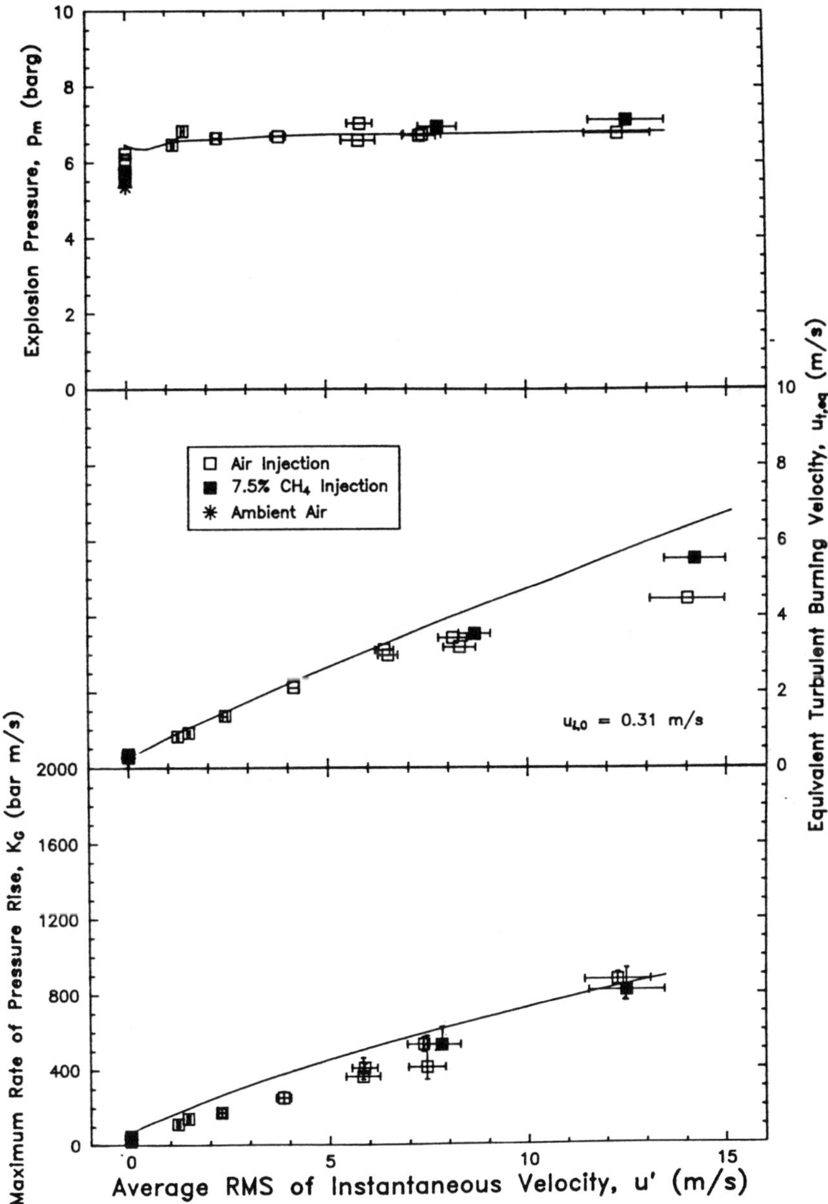

Fig. 5 Calculated explosion parameters as a function of turbulence intensity for the case of a 7.5% CH_4/Air mixture ($u_{\ell 0}$ = 0.31 m/s).

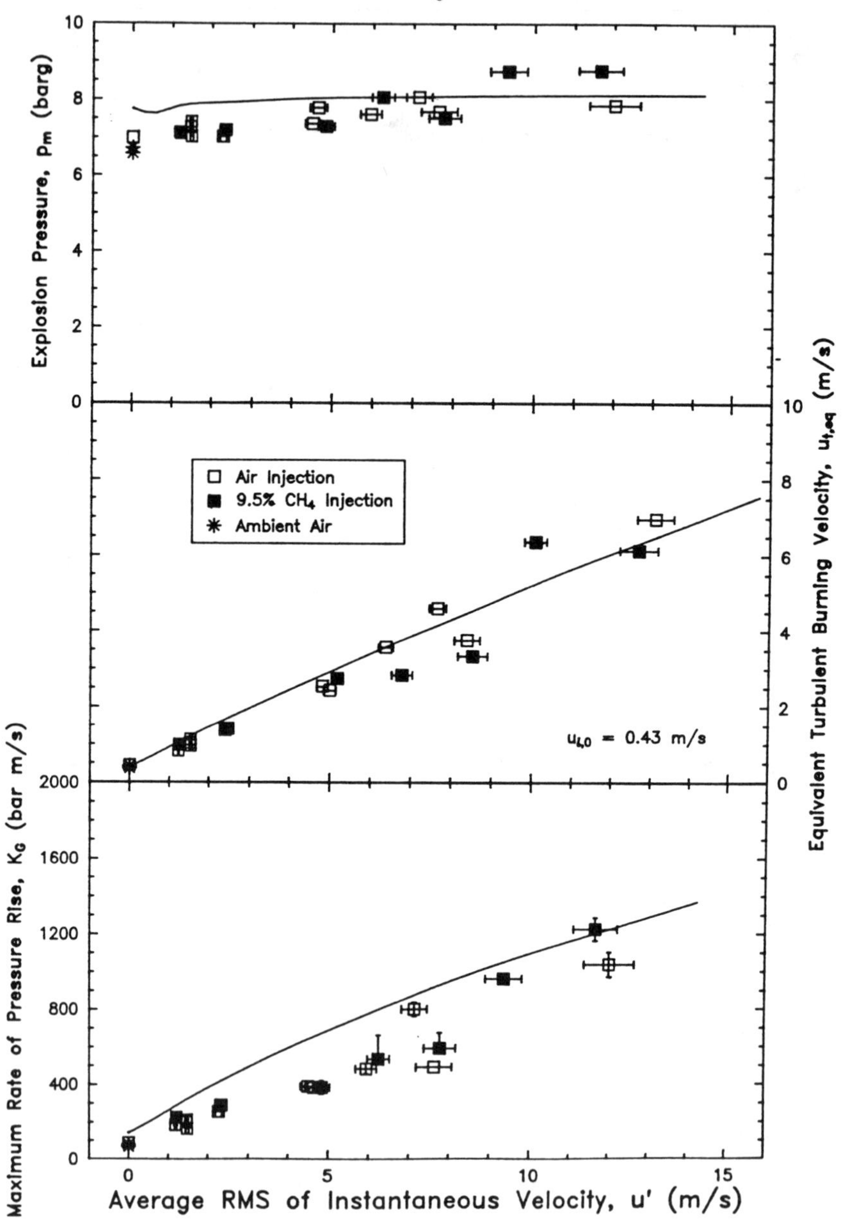

Fig. 6 Calculate explosion parameters as a function of turbulence intensity for the case of a 9.5% CH_4/air mixture (u_{t0} = 0.43 m/s).

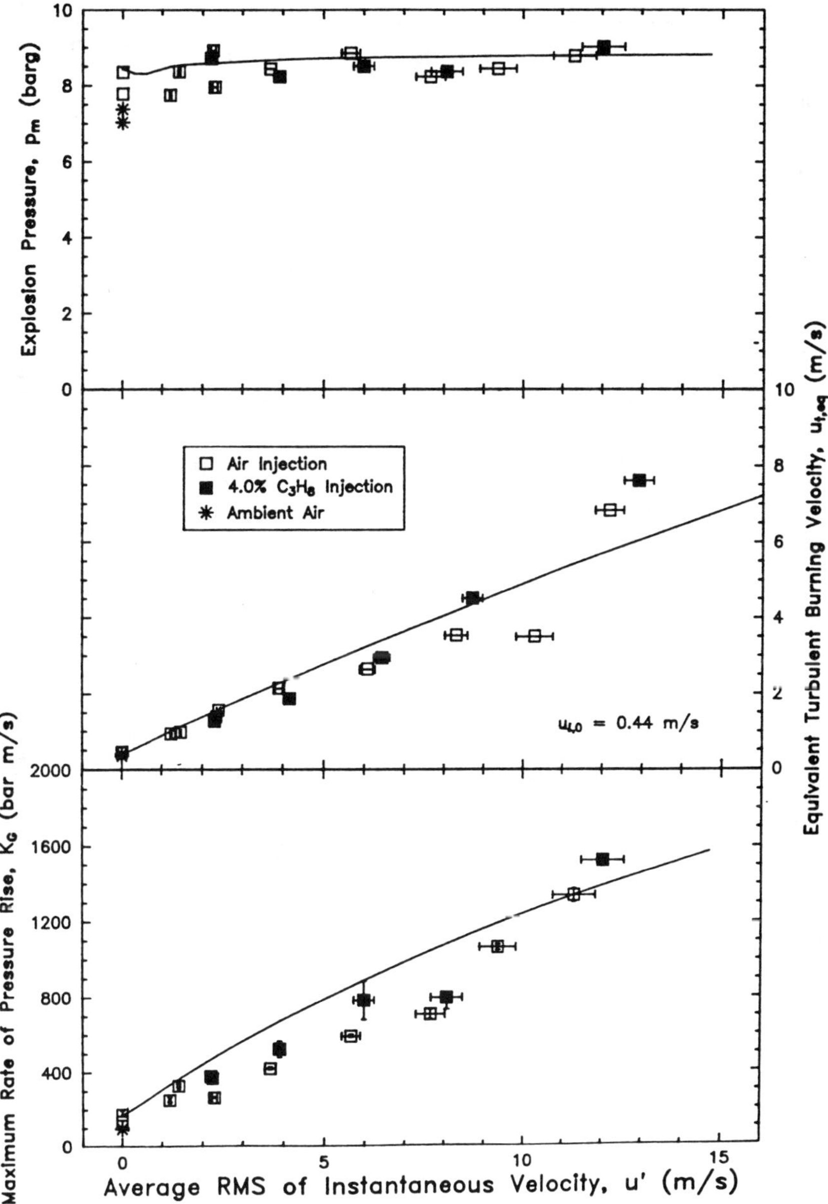

Fig. 7 Calculated explosion parameters as a function of turbulence intensity for the case of a 4.0% C_3H_8/air mixture (u_{l0} = 0.44 m/s).

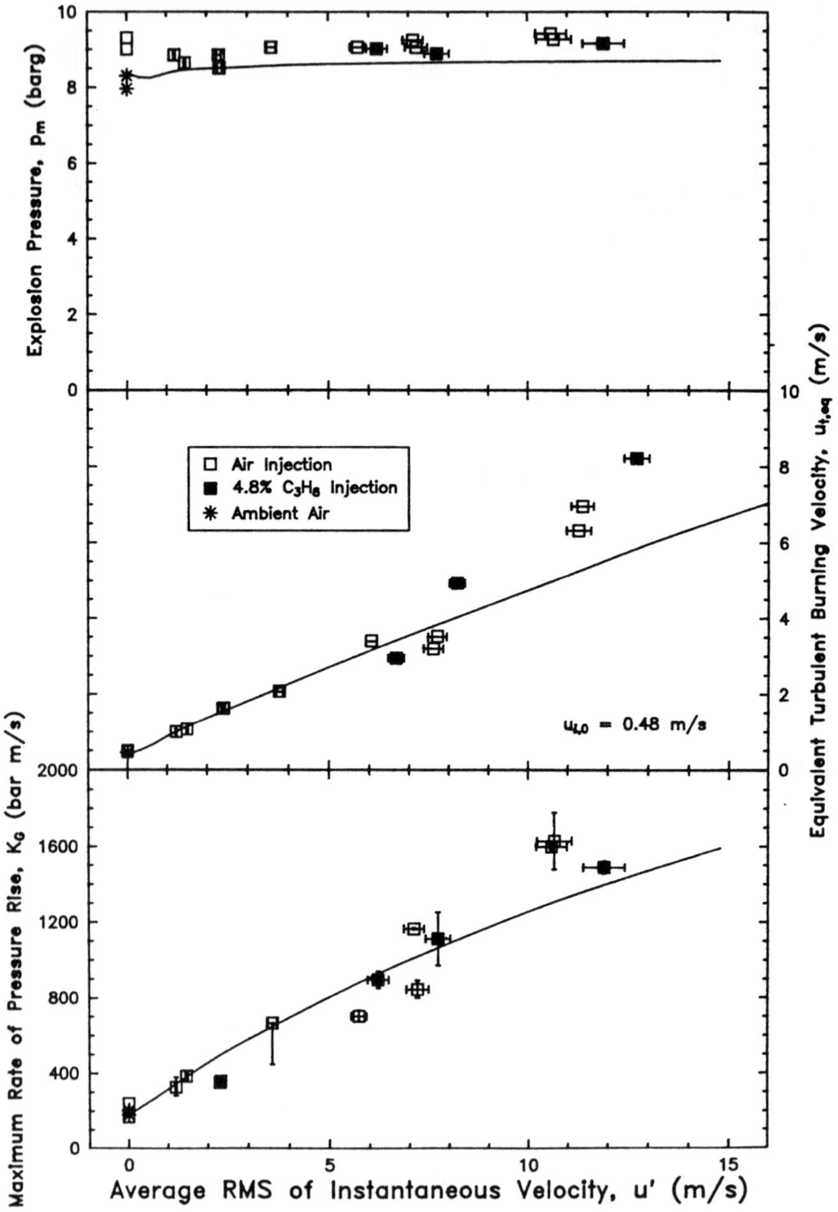

Fig. 8 Calculated explosion parameters as a function of turbulence intensity for the case of a 4.8% C_3H_8/air mixture ($u_{\ell 0}$ = 0.48 m/s).

extend to lower values of turbulence than those for the more reactive mixtures, due to the greater extent of turbulence decay taking place in this case during the flame propagation process. Both for the calculated results and for the experimental data, abscissa values for plotting were obtained by taking the average of the turbulence intensity during the portion of the pressure curve relevant to the variable being plotted. No account is given by the model for the fact that quenching effects may be important for the less reactive mixtures at high turbulence levels. In fact, during the experiments, ignition problems were encountered for these mixtures at the shortest ignition delays.

In general, agreement between theory and experiment is quite good, maybe surprisingly so, in view of the fact that there was no fine tuning of the model to optimize it. In the case of the peak explosion pressure, the agreement simply reflects the fact that the assumption of equilibrium compositions was reasonable in calculating the overall energy release. The turning up to the curve as the turbulence intensity goes down to zero (laminar conditions) is an artifact of the assumed perfect symmetry of the flame propagation process. The predicted slight increase in explosion pressure with turbulence intensity is attributable to the heat losses becoming progressively less important as the time scale of the flame propagation process is shortened. The general agreement between the predictions and the experiment cannot be used to validate the heat loss submodel due to the scatter in the data and to the limited influence of heat losses on the peak explosion pressure.

Despite the simplicity of the model, the agreement found between experiment and predictions for the two variables $u_{t,eq}$ and K_G is quite gratifying. Since the two quantities are obtained from different portions of the pressure vs. time curve, this agreement implies that the entire shape of the pressure curve is reproduced well by the model. A fairly good test of this is given by the results in Figs. 4-6 for methane, where the variation of laminar burning velocity of the mixture is fairly wide (from 0.08 to 0.43 m/s).

There are differences in the magnitude of the effect of the turbulence on these two quantities. For discussion purposes, selected numerical results are summarized in **Table** 1. The same numbers are listed twice in the table: first, to highlight the decrease in reactivity with mixture composition (from stoichiometric to 5.5%) at three levels of turbulence (u' = 0, 1, and 8 m/s), and second, to focus on the effect of increasing turbulence on the reactivity of each of the three mixtures. From the first set of results, the variable $u_{t,eq}$ is seen to be less sensitive to changes in mixture composition than the normalized maximum rate of pressure

Table 1 Calculated effect of turbulence on mixture reactivity (methane/air mixtures)

	$u_{t,eq}$ [m/s]	K_G [bar m/s]
1. Reactivity vs composition at constant turbulence intensity		
Laminar		
9.5% CH_4	.42	137
7.5% CH_4	.29 (69%)	68 (50%)
5.5% CH_4	.07 (17%)	8.4 (6%)
u′ = 1 m/s		
9.5% CH_4	.94	258
7.5% CH_4	.80 (85%)	156 (60%)
5.5% CH_4	.47 (50%)	43 (17%)
u′ = 8 m/s		
9.5% CH_4	4.31	1024
7.5% CH_4	3.95 (92%)	657 (64%)
5.5% CH_4	2.78 (65%)	172 (17%)
2. Reactivity vs turbulence intensity at constant composition		
9.5% CH_4		
Laminar	.42	137
u′ = 1 m/s	.94 (x2.2)	258 (x1.9)
u′ = 8 m/s	4.31 (x10.3)	1024 (x7.5)
7.5% CH_4		
Laminar	.29	68
u′ = 1 m/s	.80 (x2.8)	156 (x2.3)
u′ = 8 m/s	3.95 (x13.6)	657 (x9.7)
5.5% CH_4		
Laminar	.07	8.4
u′ = 1 m/s	.47 (x6.7)	43 (x5.1)
u′ = 8 m/s	2.78 (x39.7)	172 (x20.5)

rise, K_G. Furthermore, the decrease in K_G for decreasing methane concentration appears to be relatively independent of turbulence intensity.

The data grouped by constant composition (second part of the table) confirm the expectation that the effect of turbulence on mixture reactivity would be more pronounced the less reactive the mixture. For example, the weakest mixture (5.5% CH_4) is predicted to have an increase in reactivity of about 5-7 times (depending on whether one considers K_G or $u_{t,eq}$) as the turbulence increases from 0 to 1 m/s. Since the model is likely to overestimate the K_G for quiescent conditions, experiments would probably show this increase to be even greater than predicted. This result has practical implications because mixtures that show a very low reactivity and, therefore, low venting requirements based on K_G, in tests at quiescent conditions, may in fact present an unexpectedly high challenge if some turbulence is present. In this regard, it should be noted that strong forced ventilation flows can produce turbulence intensities of the order of $u' = 1$-2 m/s.

As already indicated, the model's ability to reproduce the experimental data is good, perhaps more so in the case of the equivalent turbulent burning velocity, $u_{t,eq}$, than in that of the maximum normalized rate of pressure rise, K_G. At high turbulence intensities, experimental uncertainties and scatter start to affect the data, particularly in the case of K_G. In addition, data acquisition limitations are believed to have caused an underestimate of the values of K_G in the range $u' =$ 3-8 m/s for some of the mixtures tested (cf. Figs. 6-7). However, the apparent rise of the high turbulence data for the 4.0 and 4.8% propane above the predicted curve may be real and may be the manifestation of flame instability effects which are not taken into account by the model. A possible alternative explanation for the departure is a different dependence of laminar burning velocity on pressure than the one used in the calculations [cf. Eq. (16)].

A discordant note in the picture of overall good agreement between theory and experiment is given by the fact that the model systematically overpredicts the initial rate of flame propagation immediately following ignition. The experimental curves display a slower rate of pressure rise in the early portion of the transient than that predicted by the calculation. This is probably due to the fact that, when the flame kernel is small, only eddies of a scale smaller than the flame envelope actually contribute to flame propagation; the larger eddies simply translate the entire flame about as a whole. As the flame envelope grows larger, then eddies from the entire spectrum of the turbulence contribute to its propagation. This effect, which has been quantified in at least one study of turbulent combustion[14], makes the flame respond in its early stages to a lower level of turbulence intensity than is present in the flow. While at some point it may be necessary to include this refinement in the model, its neglect is not believed to have influenced the comparisons for $u_{t,eq}$ and K_G presented in this paper, since these two quantities are related to portions of the pressure curve that occur late with respect to the initial development of the flame kernel. In addition, the calculated results and the experimental data have been related to the average turbulence intensity (calculated or measured) prevailing during the portion of the pressure/time history used to calculate the variable.

A graphical representation of combustion regimes has already been provided in Fig. 3 with the case of the stoichiometric (9.5%) methane/air tests used as an example. The beginning of breakup of continuous flame sheets, the development of quenching in a fragmented reaction zone, and the onset of total flame quenching are associated with values of the Karlovitz parameter, K (cf. Eq. (26)), of 0.15, 0.3, and 1.5, respectively. For the case of the three methane mixtures tested, **Table 2** summarizes the values of turbulence intensity at which these three boundaries are estimated to occur. As indicated in the table, the stoichiometric mixture was not exposed to conditions where total flame quenching should have been expected. The 7.5% mixture, on the

Table 2 Estimated boundaries of combustion regimes for the three methane/air mixtures

Methane concentration,	[Vol%]	5.5	7.5	9.5
Laminar burning vel., $u_{\ell 0}$	[m/s]	0.08	0.31	0.43
Turbulence intensity, u'	[m/s]			
@ K = 0.15 (Flame sheet break-up)		0.8	4.1	6.1
@ K = 0.3 (Quenching in fragmented reaction zone)		1.2	6.2	9.3
@ K = 1.5 (Flame quenching)		3.2	16.7	25.0

NOTE: Estimates for the Karlovitz parameter, K, are based on the values of integral scale, L, pertaining to the turbulence field produced in the 1.35-m^3 sphere as calculated from the correlations and formulas given in Eqs. (10), (12), and (23-26).

other hand, did approach quenching conditions for tests run with the shortest ignition delay and, in fact, ignition could not be obtained reliably in those cases. The weakest mixture (5.5%) is estimated to have been close to quenching already at the relatively low turbulence intensity of $u' = 3.2$ m/s. This estimate is not consistent with the fact that tests were successfully performed for this mixture at significantly higher levels of turbulence (cf Fig. 4).

Conclusions

The paper has presented the first results from a model development effort whose ultimate goal is the prediction of vented and unvented gas and dust explosions in cubical enclosures. In addition, plans call for the model to be extended to account for the effect of panel inertia and vent ducts.

In its current version, the model has demonstrated the ability to correctly simulate the effect of turbulence on unvented gas explosions. While some questions remain on the limits of applicability of the combustion model used, the comparisons with experiment carried out to date have covered a fairly wide range of mixture reactivities (laminar burning velocities, u_{l0}, from 0.08 to 0.48 m/s). This result is encouraging and tends to confirm the fundamental soundness of the modeling approach.

The model predictions have brought out effects of the turbulence that have implications on the behavior of practical systems. For example, as may have been expected, the relative impact of turbulence on mixture reactivity was found to be greater the less reactive the mixture. Providing a quantitative dimension to this result is important, because mixtures that show a very low reactivity, i.e., low K_G, when tested under quiescent conditions, may present a significantly higher challenge if turbulence is present. For the case of 1 m/s turbulence, which can be produced by a ventilation flow, effective flame speed increases of a factor of 5-10 would be possible. This result also hints at the order of magnitude of the combustion enhancements that could be expected in the presence of obstacles and/or during venting.

Several potential areas for model improvements have been identified, with the objectives of future model developments suggesting that certain deficiencies or limitations be addressed first. The intent to simulate dust explosions requires that the combustion model be modified to allow for a thick reaction zone. In addition, appropriate changes to the radiative heat transfer model will be needed to account for the presence of dust particles. One final question to be tackled in dealing with dusts is the choice of the thermodynamic properties of the combustion system, particularly with regard to the choice of the heat of combustion. Since the fuel concentration for maximum reactivity of dust/air mixtures is well on the rich side of stoichiometry (by a factor of 2-3 typically), a methodology will be needed to handle the heat sink effect associated with the portion of the fuel loading that remains unreacted.

For the model to be extended to vented explosions, issues of flame instabilities and vent-induced turbulence will need to be addressed. Proper accounting for these effects may present a hard challenge, however. A phenomenological treatment within the framework of the present lumped-parameter approach will be attempted first, with strong guidance from experimental data. If this approach cannot be made to work, since little would be gained from a one-dimensional model, a two- or three-dimensional description of the geometry may need to be implemented. The level of complexity of the formulation and the computational requirements would then greatly exceed those of the current approach.

In conclusion, the model presented in this report provides a useful tool for making predictions of a limited set of explosion problems. Its planned extensions to handle dusts and the presence of vents will require, at a minimum, the changes outlined above. After completion of the model extension to vented explosions, further development work will be undertaken to include the effects of vent panel inertia and the presence of ducts.

Acknowledgments

The author is grateful to Jeff Chaffee and Richard Jambor for generating the experimental data used for comparison with the predictions of the theoretical model. Erdem Ural provided significant help through useful discussions and by verifying the correctness of parts of the model. This work, which was made possible by the support and encouragement of Bob Zalosh and Cheng Yao, was carried out as part of a long-range, internally sponsored research effort in the explosion area.

References

[1]Tamanini, F., and Chaffee, J.L., "Turbulent Unvented Gas Explosions Under Dynamic Mixture Injection Conditions," Twenty-Third Symposium (International) on Combustion, The Combustion Institute, 1990, pp. 851-858.

[2]Nagy, J., Conn, J.W., and Verakis, H.C., "Explosion Development in a Spherical Vessel," U.S. Bureau of Mines, R.I. No. 7279, 1969.

[3]Tufano, V., Crescitelli, S., and Russo, G., "On the Design of Venting Systems Against Gaseous Explosions," *Journal of Occupational Accidents*, Vol. 3, 1981.

[4]Yao, C., "Explosion Venting of Low Strength Equipment and Structures," AIChE Loss Prevention Symposium, Vol. 8, 1974.

[5]Fairweather, M., and Vasey, M.W., "A Mathematical Model for the Prediction of Overpressures Generated in Totally Confined and Vented Explosions," Nineteenth Symposium (International) on Combustion, The Combustion Institute, 1982, pp. 645-653.

[6]Ural, E.A., and Zalosh, R.G., "A Mathematical Model for Lean Hydrogen-Air-Steam Mixture Combustion in Closed Vessels," Twentieth Symposium (International) on Combustion, The Combustion Institute, 1984, pp. 1727-1734.

[7]Solberg, D.M., and Pappas, J.A., "Modeling of Vented Gas Deflagrations," First Specialists Meeting (International) of the Combustion Institute, Bordeaux, July 1981, pp. 25-30.

[8]Chippett, S., "Modeling of Vented Deflagrations," *Combustion and Flame*, Vol. 55, 1984, pp. 127-140.

[9]Zalosh, R.G., "Gas Explosion Tests in Room-Size Vented Enclosures," *Loss Prevention*, Vol. 13, 1980, pp. 98-108.

[10]Zeeuwen, J.P., "Review of Current Research at TNO into Gas and Dust Explosions," pp. 687-702 in Fuel-Air Explosions, SM Study No. 16, University of Waterloo Press, 1982.

[11]Solberg, D.M., Pappas, J.A., and Skramstad, E., "Observations of Flame Instabilities in Large-Scale Vented Gas Explosions," Eighteenth Symposium (International) on Combustion, The Combustion Institute, 1981, pp. 1607-1614.

[12]Cooper, M.G., Fairweather, M., and Tite, J.P., "On the Mechanisms of Pressure Generation in Vented Explosions," *Combustion and Flame*, Vol. 65, 1986, pp. 1-14.

[13]Abdel-Gayed, R.G., Bradley, D., and McMahon, M., "Turbulent Flame Propagation in Premixed Gases: Theory and Experiment," Seventeenth Symposium (International) on Combustion, The Combustion Institute, 1979, pp. 245-254.

[14]Abdel-Gayed, R.G., Bradley, D., Lawes, M., and Lung, F. K.K., "Premixed Turbulent Burning During Explosions," Twenty-First Symposium (International) on Combustion, The Combustion Institute, 1986, pp. 497-503.

[15]Groff, E.G., "An Experimental Evaluation of an Entrainment Flame-Propagation Model," *Combustion and Flame*, Vol. 67, 1987, pp. 153-162.

[16]Beretta, G.P., Rashidi, M., and Keck, J.C., "Turbulent Flame Propagation and Combustion in Spark Ignition Engines," *Combustion and Flame*, Vol. 52, 1983, pp. 217-245.

[17]Daneshyar, H., and Hill, P.G., "The Structure of Small-Scale Turbulence and Its Effect on Combustion in Spark Ignition Engines," *Progress in Energy and Combustion Science*, Vol. 13, 1987, pp. 47-73.

[18]Launder, B.E., Morse, A., Rodi, W., and Spalding, D.B., "Prediction of Free Shear Flows - A Comparison in the Performance of Six Turbulence Models," Proceedings of the Conference on Turbulent Shear Flows, NASA-Langley Research Center, NASA SP-321 (N73-28154-173), July 1972, pp. 361-426.

[19]Launder, B.E., and Spalding, D.B., *Mathematical Models of Turbulence*, Academic, New York, 1972.

[20]Jeng, S.M., Lai, M.C., and Faeth, G.M., "Nonluminous Radiation in Turbulent Buoyant Flames," *Combustion Science and Technology*, Vol. 40, 1984, pp. 41-53.

[21]Abdel-Gayed, R.G., and Bradley, D., "Combustion Regimes and the Straining of Turbulent Premixed Flames," *Combustion and Flame*, Vol. 76, 1989, pp. 213-218.

[22]Methgalchi, M., and Keck, J.C., "Burning Velocities of Mixtures of Air with Methanol, Iso-octane, and Indolene at High Pressure and Temperature," *Combustion and Flame*, Vol. 48, 1982, pp. 191-210.

[23]Tamanini, F., "A Theoretical Model of Unvented Explosions Including a Treatment of Turbulence Effects," FMRC Technical Report J.I. 0Q2E4.RK, October 1990.

[24]Hottel, H.C., and Sarofim, A.F., *Radiative Transfer*, McGraw-Hill, New York, 1967.

[25]Peters, N., "Laminar Flamelet Concepts in Turbulent Combustion," Twenty-First Symposium (International) on Combustion, The Combustion Institute, 1986, pp. 1231-1250.

[26]NFPA68, "Guide for Venting of Deflagrations," National Fire Protection Association, Quincy, MA, 1988.

[27]McBride, B., and Gordon, S., "Computer Program for Calculation of Complex Chemical Equilibrium Compositions, Rocket Performance, Incident and Reflected Shocks, and Chapman-Jouguet Detonations," NASA SP-273, 1971.

[28]Weast, R.C. (ed.) *Handbook of Chemistry and Physics*, 59th Ed., The Chemical Rubber Company, 1979.

Dynamics of Flame Propagation in Multichamber Systems

R. H. Abdullin,* A. V. Borisenko,† and V. S. Babkin‡
Institute of Chemical Kinetics and Combustion, Novosibirsk, Russia

Introduction

Combustion of gases in systems of connected vessels is often realized for practical purposes. The problem of prechamber ignition in internal combustion engines has long been studied. It is assumed that the presence of a small prechamber in which the combustible mixture is ignited and which is connected through a narrow channel with the main volume of the engine cylinder can improve the working parameters and the emissive characteristics of the engine.[1-3]

Of particular interest for combustion in connected vessels is the problem of explosion prevention in industry and transport.[4-8] Many apparatuses of chemical technology, mine equipment, industrial compartments, fuel storehouses where combustible mixture can exist or form can be considered, from the physical point of view, as systems of connected vessels. The deflagration-detonation processes excited there in an emergency present difficulty in predicting the dynamics of their development and the especially destructive character of their consequences.

---

*Scientific Researcher.
†Researching Engineer.
‡Head, Laboratory of Physics and Chemistry of Gas Combustion.

From the scientific point of view, the interest for combustion in connected vessels is due to a number to interesting phenomena and specific regularities observed in these systems. A dramatic acceleration of flame, the development of anomalously high pressures, the possibility of a sudden transition from slow combustion to detonation or vice versa, the extinction of combustion, all these and some other unusual phenomena are rather typical for connected vessels and reflect the existence of a strong interaction of hydrodynamic, thermal, and chemical elementary processes and the presence of positive and negative feedbacks between them.

The best studied is the simplest dual-chamber system with one narrow connecting channel. The presence of narrowness causes, after the ignition of a combustible mixture in one of the vessels, the difference in the pressures in vessels and fast gas motion inside channel. This flow, in turn, leads to vortex formation and turbulization of the mixture in the second vessel and, consequently, a high-speed combustion of the mixture after flame transition to the second vessel. In this case, a higher rate of combustion and the increase of pressure in the second vessel can lead to changes in flow direction, with subsequent burning acceleration in ignition vessel.[9]

However, explosion transmission from one vessel to another is impossible in some cases. Thibault et al.[10] established that the absence of chemical reaction transmission can be observed when the diameters of the orifice are much larger than the critical diameters. Thus, it is concluded that the jet flow near the narrow chennel has not only a positive effect (accelerating combustion) but also a negative one, which leads to flame extinction. In the intermediate regime, the combustion in the second vessel starts after a certain period of induction.[3] Abdullin et al.[11] have shown that, in the dual-chamber systems, the main phenomena and regularities are determined by the interaction of two factors, i.e., combustion and outflow. Taking into account these factors in terms of dimensionless parameters allowed us to distinguish the basic combustion regimes, to find the optimal conditions for the development of anomalously high pressures, and to account for the other peculiarities of the process.

On the one hand, the study of combustion processes in three, four, and more connected vessels is a logical continuation of the above-mentioned series of papers. On the other hand, it is evident that the increase in the number of connecting channels in an individual vessel not only makes the problem more difficult because of the complication of dynamic heat and mass exchange but favors the manifestation of new unknown effects.

Our first experiments were performed using three-chamber and five-chamber linear systems consisting of successively connected same vessels. It was experimentally verified that in these systems the process displays an unsteady character: flame propagation velocities and maximum pressure differ in different vessels. It is not clear whether they can reach the asymptotic constant values with an increasing number of vessels.

To answer this question, the present report described not only a few-chamber system (the number of successively connected chambers $n = 7$) but also the multichamber linear systems ($n = 15$ and $n = 19$). In the last variant, the systems become similar (in terms of methods) to the experimental setups described in other of papers on gas combustion in tubes with regular obstacles.[12-18] In this case, the difference lies in the characteristic values of the blockage ratio $BR = 1-d^2/D^2$, where d is the diameter of a connecting channel and D is the vessel (tube) diameter. For our case, BR = 0.91–0.98 whereas, in the above-mentioned papers, BR = 0.28–0.6. Thus, in our experimental conditions, there is higher hydraulic resistance. Hence, it is possible to compare the concepts of gas combustion in connected vessels and in the tubes with regularly spaced obstacles.

Experimental Setup and Procedure

Multichamber systems consist of a number of successively connected cylindrical chambers 130 mm in length and 115 mm in diameter (Fig. 1). Two neighboring chambers are separated by a 30-mm baffle with a replaceable insertion, with orifice d situated on the system axis. The insertions with d = 17, 24, and 34 mm (BR = 0.982, 0.956, 0.913) were used. Each vessel was provi-

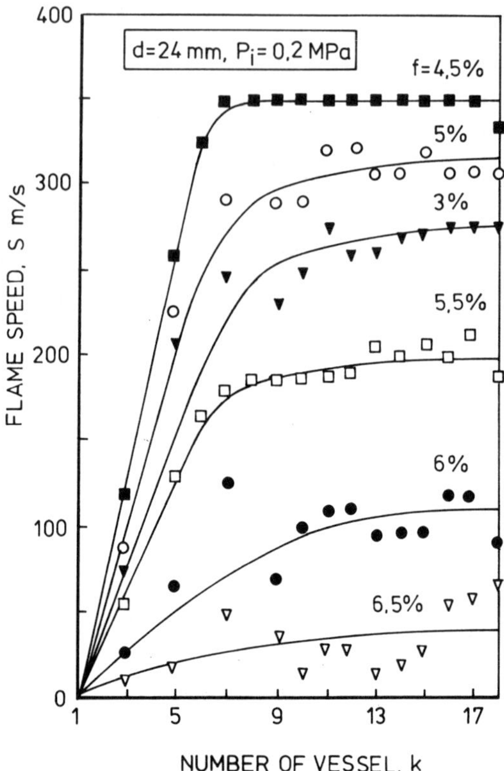

Fig. 3. The mean velocity of flame propagation in a 19-chamber system in propane-air mixtures.

The presence of a steady-state combustion process allows us to consider the question of time-independent velocity and the structure characteristics of the combustion wave: the parameters and the way they determine wave velocity, the structure and extension of the combustion zone, and the limiting conditions for the steady-state regime.

Figure 4 gives the dependencies of front propagation velocity on propane concentration in mixture f. For the system of vessels under study, the dependence

Table 1 Mean values of scatter in velocities

% C_3H_8	3.0	4.5	5.0	5.5	6.0	6.5
$\langle \delta \rangle$	0.058	≈ 0	0.03	0.055	0.10	0.50

Table 2 Velocity of steady-state flame propagation in 4% propane-air mixture at various initial pressures[a]

p_i, (MPa)	0.02	0.03	0.05	0.1	0.2	0.3	0.4
S^i, (m/s)	150	230	266	306	325	335	354

[a] 19-Chamber system, d = 24 mm, BR = 0.956.

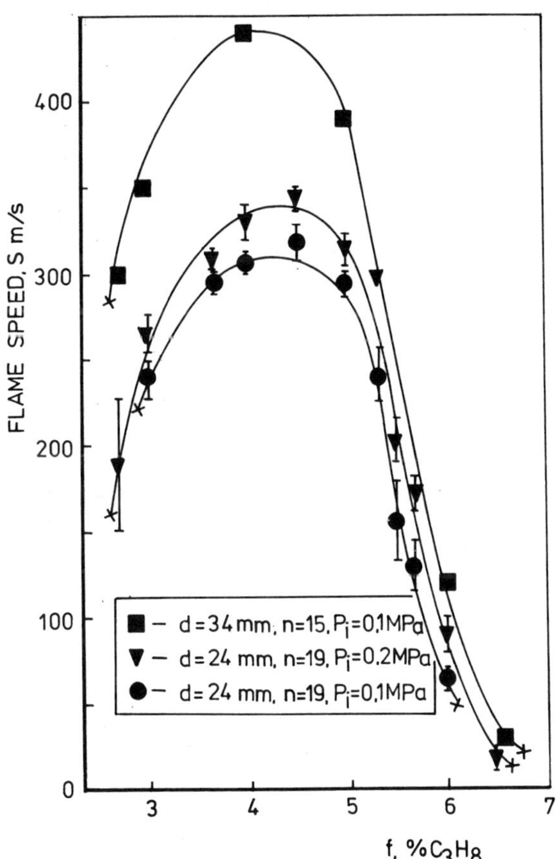

Fig. 4. The dependencies of the mean velocity of flame propagation on the composition of propane-air mixture. The limits of flame propagation in the system are designated with a cross.

S(f) is observed to be dome-shaped, which is typical of laminar and turbulent flames. When the mixture is rich or lean, the velocity decreases and, at some critical concentration of C_3H_8, combustion wave propagation in the system is impossible. In this case, propagation blows off in one of the system vessels.

The front propagation velocity is determined not only by mixture composition but also by the value of the orifice between vessels d and the initial mixture pressure p_i. The increase in the values of these parameters increases S (Fig. 4, Table 2). Table 2 lists the values of S for a 4% propane-air mixture for various pressures. As p_i increases from 0.03 to 0.4 MPa, the front propagation velocity increases as $S \sim p_i^{0.1-0.15}$.

Pressure Dynamics

Experiments show that the pressure in studied systems is highly unisotropic. The pressure differential is great not only over the system length but also in two contiguous chambers. This causes great variety in the dynamics of ressure development, depending on the number of chambers in the system, the mixture composi-

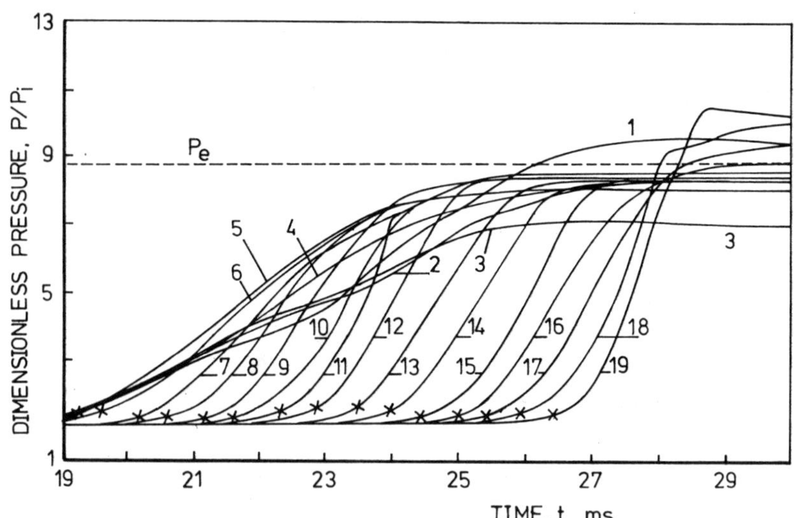

Fig. 5. The dependencies of p(t). The figures on the curves designate vessel numbers. A dotted line denotes the maximum pressure upon mixture combustion in a single closed vessel,[22] crosses indicate the flame emergence moments.

tion, the value of the orifice, and the initial pressure. However, some general features can be determined for pressure dynamics.

In strong mixtures and long systems (Fig. 5), the patterns of pressure p(t) in the first six to seven vessels vary considerably from vessel to vessel. In the next vessels (k = 7–16), the patterns are similar and invariant. The level and time of achieving maximum pressure remain the same. These pressure patterns correspond to a steady-state phase of combustion front propagation. Finally, in the last two to three vessels, the pressure patterns deform, passing to a relatively high maximum pressure that exceeds the maximum pressure in the single closed vessel p_e.

In weak mixtures and short systems (Fig. 6), the pressure patterns continuously vary in different vessels. The maximum pressure p_m increases in the course of the process. A steady state is not reached.

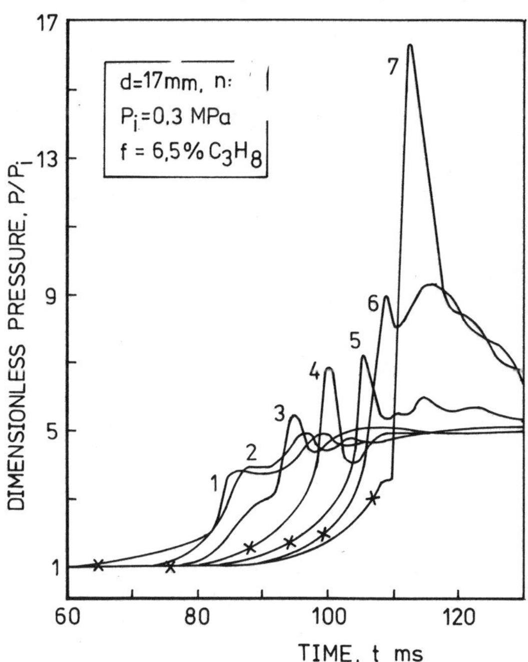

Fig. 6. The dependencies of p(t). The figures on the curves designate vessel numbers. Crosses indicate the flame emergence moments.

It is interesting to consider the spatial distribution of pressure in the system p(k) at a certain moment of time. Figure 7 shows pressure distribution upon combustion of the 4% propane-air mixture in a 19-chamber system. Flame propagation is followed by a pressure wave. First, this wave forms (t < 24 ms). The maximum profile pressure p(k) increases, and the front steepness of the pressure wave continuously grows. Behind the front, the pressure drops slightly. Thereafter, the front profile shape and the pressure behind the front are stabilized (t = 25–27 ms). At the final stage, the front profile deforms, and the maximum pressure in the wave increases, achieving values exceeding p_e (t = 28–30 ms). In weak mixtures and short systems, the process fails to reach a steady state (Fig. 8). The maximum pressure and the front steepness of the pressure wave continuously increase.

Figure 9 lists the values of the maximum pressure available in the different vessels upon combustion of

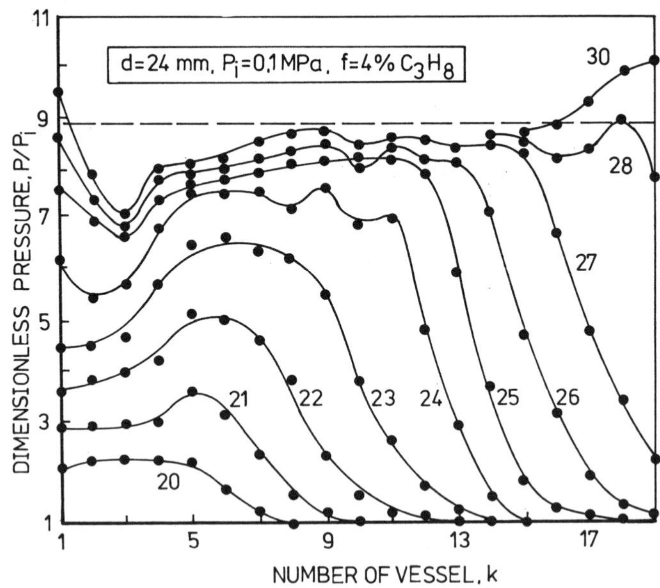

Fig. 7. The dependencies of p(k). The figures on the curves designate time in milliseconds. A dotted line indicate the maximum pressure upon mixture combustion in a single closed vessel.[22]

propane–air mixtures in a long system. The increased pressures are, as a rule, observed in the first and last system vessels. The highest pressures are recorded on combustion of rich mixtures in the last vessels.

Discussion of Results

Steady-State Propagation Regime

One of the main results of this paper is the establishment of the existence of steady-state regimes of combustion wave propagation in a multichamber system of connected vessels with high values of blockage ratio (BR > 0.9). This allows one to consider the velocity and structure characteristics of a combustion wave as the fundamental invariant values that are time-independent and determined not only by the properties of combustible gas but also by the geometrical system parameters.

The velocity of steady-state combustion wave propagation is the most important value of the process. It depends substantially on mixture composition, initial pressure, and system geometry. Figure 4 depicts the dome-shaped S(t) dependence typical of laminar and turbulent flames whose maximum is shifted slightly toward rich mixtures. The region of the steady-state regime, limited by lean and rich boundaries, is narrower than that of laminar flames. According to Ref.19, 2.3% and 8.8% C_3H_8 correspond to the usual flammability limits of propane–air mixtures, respectively, at lean and rich branches of the flammability region. Assuming, for our case, that the flame quenching is of a conductive nature, we find that the empiric relation Pe = 65, where the Peclet number is constructed by the burning velocity and orifice diameter d, must hold at the limits.[20] However, according to the estimates of the Peclet numbers at the flammability limits given in Fig. 4, their values are a few times higher. This is probably due to additional participation of gasdynamic factors in quenching.[10]

Note that, in the present report, relatively low values have been obtained for flame propagation speeds. Indeed, for a stoichiometric propane–air mixture at atmospheric pressure, we obtained 306–440 m/s (D = 11.5 cm, BR = 0.96–0.91) whereas, in the tubes with regular obstacles, we had 650–750 m/s (D = 5 cm, BR = 0.43–0.44),[13–15] 1100 m/s (D = 15 cm, BR = 0.5),[12]

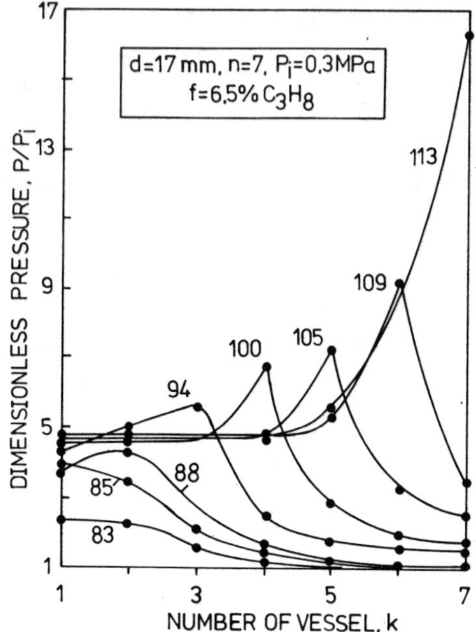

Fig. 8. The dependencies of p(k). The figures on the curves designate time in milliseconds.

700m/s (D = 30 cm, BR = 0.28),[13] 1400 m/s (D = 30 cm, BR = 0.43).[15] This effect might be assigned to a relatively small diameter of our vessels, D, and a high blockage ratio, BR. Figure 4 shows that, with BR = 0.91, the flame speed is 440 m/s and, for BR = 0.96, the value is lower, 306 m/s. Note that, during gas combustion in an inert porous medium, the flame propagation speed also decreases with the decreasing diameter of a pore channel.[21]

Another important characteristic of the steady-state combustion process is the wave propagation structure. As has been mentioned, flame front propagation is followed by a pressure wave with a time-independent profile. The front part of the profile smoothly increases as a thermal wave (Fig. 7). The maximum pressure p_m is close to the value of the final pressure upon gas combustion in a closed vessel with a constant volume. After p_m, some pressure drop is observed. Unlike a single closed vessel, the pressure dynamics in a single chamber of a multichamber system is determined not only by the combustion process but also by a heat and mass exchange between the neighboring chambers. The heat and mass exchange is provided by rather large pressure differences between the neighboring chambers (up to 0.2 MPa). In this case, according to the value and direction of the pressure gradient, the relation between inflow and outflow of gaseous masses continuously varies in an individual vessel in the course of combustion. From Fig. 5 it follows that, in the steady-state regime at the moment of flame emergence to the next vessel, there is some precompression of unburned gas due to its overflowing under the action of the pressure gradient. The value of precompression in different vessels is approximately the same and amounts to about 0.02–0.04 MPa. Since the precompression is small, it might be concluded that there is practically no gas motion before the gas front, and the intense motion toward wave propagation is observed in the zone of combustion and fast pressure growth.

The extension of the combustion zone can be estimated from the front part of the pressure p(k) profile. For the experimental conditions depicted in Fig. 7, this estimate gives the zone length of about 1 m. This means that combustion simultaneously proceeds in five to six chambers. Such a long zone reflects the axial-radial mechanism of mixture combustion, where a fast axial jet provides the advance of a chemical reaction and forms a radial flame front. In this front, in conditions of strong turbulence, the main mixture mass burns out. A turbulent character of combustion is confirmed by a positive dependence of flame speed on the initial pressure typical of turbulent flames (Table 2). In hydrocarbon-air mixtures, the turbulent combustion velocity is known to increase with growing pressure and decreasing laminar burning velocity. This situation holds for turbulent gas combustion in an inert porous medium.[21]

In the framework of an axial-radial mechanism, the flame speed must be determined by that of the central jet as the leader of flame propagation. Consequently, flame speed stabilization upon passage to a steady-state regime means the stabilization of the jet flow inside the narrow area under the action of limiting factors, such as the achievement of the critical conditions of outflow and the limitation of the value of the pressure differential between two neighboring chambers. Of importance is the manifestation of the kinetic mechanism of speed stabilization, especially in near-limiting conditions when flame propagation is of a periodic character and consists of the following: flame quenching upon passage through the narrow region, the induction period, self-ignition, flame propagation, flame quenching in the next chamber, etc. For this mechanism of reaction transfer, the speed-limiting stage could be represented by the induction period. A mechanism of speed stabilization by flame quenching in rather quick pulsations is realized upon turbulent gas combustion in a porous medium.[21]

Formation of a Steady-State Combustion Regime

Any steady-state front process of deflagration is preceded by the formation stage in a starting part. Since the formation process is of an asymptotic character, there is some uncertainty concerning the length of the starting region. The starting region is usually assumed to comprise a few lengths of the combustion wave zone in the steady regime. Hence, for a stoichiometric propane-air mixture, the region of the transition to a steady state must exceed 1 m. Indeed, the data on the pressure record (Fig. 5) and flame speed (Fig. 3) testify that the value of the starting region of 1.1 m (seven chambers) and the total length of the system are sufficient for passage to the steady-state combustion regime. On the contrary, if the system length is less than the starting region, it is impossible to reach a steady state. This situation is observed in a seven-chamber system (Fig. 6). For the same reason, even in the 19-chamber system, in the case of a near-limiting mixture, the steady state is not achieved to the full extent, and the maximum pressure in vessels increases slowly but steadily.

Influence of Boundary Conditions

A rigorous study of the steady-state wave process is possible only in infinitely long systems. In our case, the systems are limited. Therefore, one can expect the manifestation of end effects at distances about $x/l \sim 1$, where x is the distance to one of the ends and l is the combustion zone length. Indeed, such effects are recorded at both the fore (k = 1–3) and rear (k = 17–

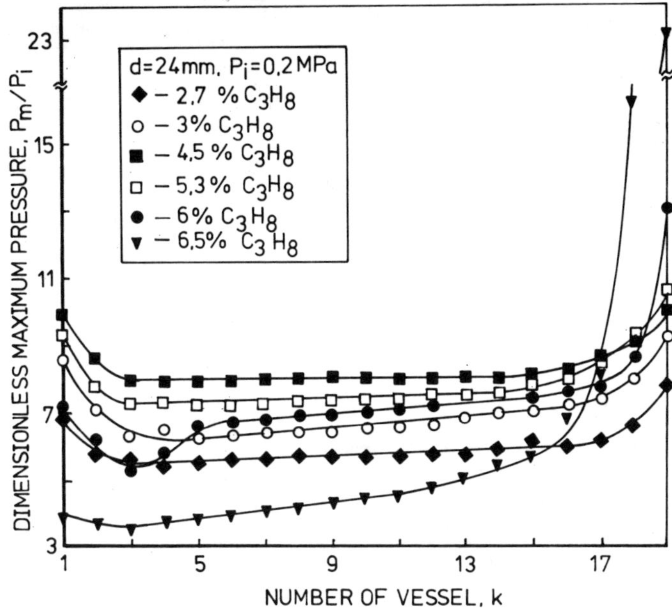

Fig. 9. The dependencies of the maximum pressures on vessel number for various propane–air mixtures.

19) system ends. As follows from Fig. 5, in the last two to three vessels, the pressure patterns deform, and the maximum pressures display anomalously high values. The observed phenomena are, in their nature, similar to those in dual-chamber systems. The rear wall stops the axial motion of gas. However, because of the pressure gradient, the inflowing of unburned mixture and combustion products to the last two to three chambers goes on. As there is no exit for gas, it is precompressed, which leads to the local increase in chemical and thermal energy. Since the gas flow stops, the kinetic energy of the central jet transforms into the energy of turbulent pulsations, providing a high burning rate at the final stage. In these conditions, the leveling of pressure with reverse gas flows cannot be effective. All this leads to the development of anomalously high pressures in the last two to three chambers. The effect is strengthened by a positive dependence of burning velocity on the initial pressure in a turbulent regime.

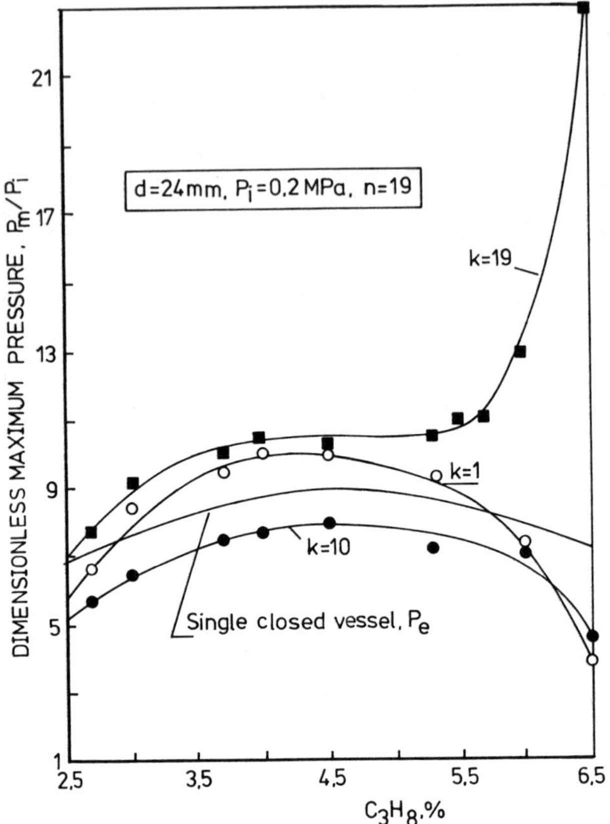

Fig. 10. The dependencies of the maximum pressures on the composition of propane-air mixture. The figures designate vessel numbers.

Figure 10 gives the maximum pressures for the last vessel (k = n) for different mixtures. For comparison, the figure shows the final pressures during combustion in the conditions[22] of the constant volume p_e. It is seen that, for all mixtures, $p_m > p_e$. However, in rich mixtures, this effect is especially strong and has no relation to the mixture reactivity characterized by the burning velocity of a laminar flame. Hence, the above interpretation of the anomalously high pressure in the last vessels needs further study. It might be assumed that the concentration anomaly is determined by the Lewis number controlling the tendency of the combustible mixture to form cellular flames, as well as the tendency to laminar flame instability. It is in rich propane-air mixtures that the flame instability[22] and the maximum rates of turbulent combustion[21] are observed. The increase in burning rate, as already mentioned, increases the effect of anomalously high pressures, thus decreasing the efficiency of reverse gas overflowing.

The end effects are also observed in the first two to three chambers. Figures 9 and 10 show that $p_m > p_e$ in the region of strong mixtures ($f = 3$–5.5% C_3H_8). Surely, this effect, as in the last chambers, is related to the concentrations of chemical and thermal energies due to gas overflowing. However, according to the scheme, the realization of excess energy and the development of excess pressure (relative to p_e) is somewhat different. Thus, in the last vessel, before achieving p_m, a two-stage scheme of mass transfer is realized: inflow and outflow (backflow). In the first vessel, before p_m, at least three stages are realized: outflow (direct flow), inflow (backflow), and again outflow (Figs. 5, 7). The appearance of a non-steady-state jet during either outflow or inflow generates turbulence and affects the combustion process. It is the interaction of two principal processes, of combustion and outflow (inflow) upon regular change of flow direction, that leads at last to the filling of the excess combustion product mass and the development of excess maximum pressure in the first vessel. It is only natural that the pressures observed in the third, fourth, and other vessels have some influence on the dynamics in the first one.

Conclusion

In long linear multichamber systems with a large hydraulic resistance, the steady-state regimes of combustion wave propagation are quite possible. In this regime, the dynamics of the pressure, the velocity flows, and combustion are reproduced in each vessel to form a united wave process of combustion with constant velocity and wave structure characteristics.

The steady-state wave velocity is comparatively low (about 300–400 m/s) and depends not only on the physicochemical properties of combustible gas but also on system geometry. The combustion wave possesses a particular feature; namely, it is followed by a pressure wave with a smoothly increasing leading front and the pressure maximum, which does not exceed the pressure upon gas combustion at constant volume. A positive pressure gradient at the front causes gas motion and flame transfer along the central line toward the wave motion. The lengths of the pressure wave and combustion zone are large enough, and they achieve, in the stoichiometric propane–air mixture, 1 m. This is important for understanding the non-steady-state processes occurring during the period of flame acceleration, processes upon gas combustion in short few-chamber systems, and combustion near the closed end of the system.

In a long system, the establishment of a steady-state regime is preceded by a period of formation. The distance of the passage to the steady state is of the order of magnitude of the combustion zone. When system length is comparable with, or shorter than, the typical size of the combustion wave, the process of flame propagation displays a non-steady-state character.

The fact that the system is limited in its length adds the effects of nonsteadiness to the combustion process. These are observed at both the fore and rear ends of the system and manifest themselves in the development of increased pressures in the first and last two to three chambers. The end effects are more prominent in short systems.

Upon combustion of propane–air mixtures in multichamber systems of connected vessels, the observed phenomena are determined by two main elementary processes, combustion and outflow, valid for each individual vessel, as well as by the interchamber interaction of these processes.

Acknowledgments

The authors are grateful to Dr. V.A.Bunev and Dr. P.K.Senachin for fruitful discussions.

References

[1] Gussak, L.A., Karpov, V.P., and Gussak, D.A., "Prechamber–Fuel Spray Ignition of Injected Fuels in Piston Engines," *Archiwum Termodynamiki i Spalania*, Vol. 7, 1976, pp. 507–527.

[2] Rychter, T.J., and Teodorczyk, A., "An Evaluation of Effectiveness of the Combustion Jet in a Dual-

Chamber Configuration," *Archivum combustionis*, Vol. 4, No. 3, 1984, pp. 255–266.

[3] Yamaguchi, S., Ohiwa, N., and Hasegawa, T., "Ignition and Burning Process in a Divided Chamber Bomb," *Combustion and Flame*, Vol. 59, No. 2, 1985, pp. 177–187.

[4] Heinrich, H.J., "Estimation of Relief Vents Using for Protection of Equipment in Chemical Industry," *Chemie-Ingenieur-Technik*, 38 Jahr, Nov, 1966, ss. 1125–1133 (in German).

[5] Bartknecht, W., *Explosions-Course, Prevention, Protection*, Springer-Verlag, Berlin/Heidelberg/NewYork, 1981.

[6] Harris, R.J., *The Investigation and Control of Gas Explosions in Buildings and Heating Plant*, E & F N Spon, London, 1983.

[7] Strel'chuk, N.A., Mishuev, A.V., Nikitin, A.G., and Ovakhelashvili, N.A., "Gas Dynamics of Combustion in a Gas-Air Mixture in a Semiclosed Volume with Pressure Release into a Adjacent Volume," *Combustion, Explosion, and Shock Waves*, Vol. 20, No. 1, 1984, pp. 59–62.

[8] Singh, J., "Gas Explosions in Compartmented Vessels: Pressure Piling," *Chemical Engineering Research and Design*, Vol. 62, No. 6, 1984, pp. 351–366.

[9] Heinrich,H.J., "Possibilities of Pressure Relief Vents," *Schadenprisma*, No. 4, 1975, ss. 69–75 (in German).

[10] Thibault, P., Liu, Y.K., Chan, C., Lee, J.H., Knystautas, R., Guirao, C., Hjertager, B., and Fuhre, K., "Transmission of an Explosion through an Orifice," 19*th Symposium (International) on Combustion*, Combustion Institute, Pittsburgh,1982, pp. 599–606.

[11] Abdullin, R.H., Babkin, V.S., and Senachin, P.K., "Combustion of Gas in Connected Vessels," *Combustion, Explosion, and Shock Waves*, Vol. 24, 1988, pp. 123–129.

[12] Gu, L.S., Knystautas, R., and Lee, J.H., "Influence of Obstacle Spacing on the Propagation of Quasi-Detonation," *Proceedings* 11*th International Colloquium on Dynamics of Explosions and Reactive Systems*, Warsaw, Aug. 1987.

[13] Lee, J.H., Knystautas, R., and Chan, C.K., "Turbulent Flame Propagation in Obstacle-Filled Tubes,"

20th *Symposium (International) on Combustion*, Combustion Institute, Pittsburgh, 1984, pp. 1663–1672.

[14] Lee, J.H.S., "The Propagation of Turbulent Flames and Detonations in Tubes," *Advances in Chemical Reaction Dynamics*, 1986, pp. 345–378.

[15] Peraldi, O., Knystautas, R., and Lee, J.H., "Criteria for Transition to Detonation in Tubes," 21*st Symposium (International) on Combustion*, Combustion Institute, Pittsburgh, 1986, pp. 1629–1637.

[16] Moen, I.O., Sulmistras, A., Hjertager, B.H., and Bakke, J.R., "Turbulent Flame Propagation and Transition to Detonation in Large Fuel-Air Clouds, "21*st Symposium (International) on Combustion*, Combustion Institute, Pittsburgh, 1986, pp. 1617–1627.

[17] Subbotin, V.A., and Kuznetsova, A.Ya., "Combustion Regimes of Explosive Gaseous Mixtures in Channels with Variable Cross-Section," *Dynamics of Multiphase Media*, Institute of Hydrodynamics SB USSR AS, Vol. 68, 1984, pp. 124–131 (in Russian).

[18] Taylor, P.H., "Fast Flames in a Vented Duct, " 21*st Symposium (International) on Combustion*, Combustion Institute, Pittsburgh, 1986, pp. 1601–1608.

[19] Babkin, V.S., Zamashchikov, V.V., Badalyan, A.M., Krivulin,V.N., Kudryavtsev, E.A., and Baratov, A.N., "Effect of Tube Diameter on Homogeneous Gas Flame Propagation Limits," *Combustion, Explosion, and Shock Waves*, Vol. 18, No. 2, 1982, pp. 164–171.

[20] Rozlowskii, A.I., *Technical Principles of Explosion-Proof upon Work with Gaseous Fuels and Vapors*, Khimiya, Moscow, 1980, (in Russian).

[21] Babkin, V.S., Korzhavin, A.A., and Bunev, V.A. "Propagation of Premixed Gaseous Explosion Flames in Porous Media," *Combustion and Flame*, Vol. 87 (2), 1991, pp. 182–190.

[22] Babkin, V.S., Bukharov, V.N., and Mol'kov,V.V., "Burning Velocity of Propane-Air Mixtures at High Pressures and Temperatures," *Combustion, Explosion, and Shock Waves*, Vol. 25, No. 1, 1989, pp. 52–57.

Fuel and Obstacle Dependence in Premixed Transient Deflagrations

A. T. Cates* and S. J. Bimson†
Shell Research Limited, Chester, United Kingdom

Abstract

Understanding the progress of a deflagration in an obstacle field is important for quantifying the hazard posed by an accidental vapour cloud explosion in both onshore and offshore environments. A series of deflagration experiments has been carried out in a vented duct with repeated obstacles. The roof of the duct was partly open, and the flame reached a steady flame speed along the duct, with a balance achieved between volume production and venting. These experiments were designed to examine details of the interaction of flow with obstacles. Although the geometry is complicated, for a given flame speed and roof venting, the flowfield away from the combustion region is specified, and very accurate comparisons between combustion rates can be made. The experiments establish two things. First, the comparative severity of two simple geometries (i.e., which geometry gives rise to faster flame speeds) is shown to be fuel dependent. These geometries are obstructed by grids with quite similar drag, but different blockage and obstacle shape. The gross turbulent properties behind the grids are similar, but the detail of the turbulence is different and can differentiate between fuels. This demonstrates that simulating the geometry using a single figure for the level of turbulence behind a grid has a limited accuracy. This is important in the treatment of subgrid obstacles in large-scale numerical simulators for real geometries. Second, under certain conditions, the influence of an obstacle grid on the course of an explosion can be better characterized by drag than by blockage. For a range of flame speeds, drag is shown to characterize grids to a similar level of accuracy as that to which it can be estimated.

* Scientist, Combustion & Fuels, Thornton Research Centre.
† Senior Technician, Combustion & Fuels, Thornton Research Centre.

Fuel and Obstacle Dependence in Premixed Transient Deflagrations

Background

Our knowledge of premixed deflagrations in obstacle fields has improved dramatically in recent years. We now have a good understanding of most of the principal underlying mechanisms of flame speed and pressure generation.

Obstacles have several roles in deflagrations. For example, they reduce the total volume of flammable gas; they can inhibit the flow of burnt gas away from the flame front, causing pressure buildup; and they distort the flame front and increase the burning rate.

Of these roles, the first two can be characterized fairly well. Porosity (or "volume blockage ratio") is clearly the correct characterization of obstacles as far as allowing for "dead" volumes goes. Drag is clearly the appropriate characterization for the inhibition of flow away from the flame front. Hence, numerical schemes that use a continuum model of subgrid obstacles, such as that described in Ref. 1, use porosity to form a modified density in the mass conservation equation and incorporate drag as a force term in the momentum equation.

However, flame-front distortion (which leads to an increased burning rate) is more complicated. A variety of physical mechanisms contributes to the distortion of the flame front. These include area enhancement by turbulence downstream of the obstacles, flame folding around obstacle wakes, and flame stretch in the shear layers in the immediate vicinity of an obstacle.

It has been recognized for a considerable time[2] that the level of turbulence some distance behind an obstacle or grid depends predominantly on mean flow and drag. Nearer an obstacle, the primary mechanism for distortion is flame folding around the obstacle wakes. Wake size is more closely related to drag than, say, to blockage. Very near an obstacle, the flow dynamics distorting the flame front depend strongly on obstacle shape, and drag alone cannot fully account for the distortion.

For many situations, drag-related mechanisms are dominant in distorting the flame front; making an exception, for example, of regimes where shocks are significant.[3] Therefore, we might expect obstacles with similiar drag (and similiar lengthscale) to have a similiar effect on a deflagration. The effect of the fuel mixture on the course of a deflagration may seem simpler. Inevitably, mixtures with higher laminar burning rates generally give rise to faster flame speeds and higher pressures in a given obstacle field. It is also to be expected that, in general, flame speeds will increase faster than linearly with laminar burning rates, since the degree of turbulence and flame-front distortion increase coincidently. A simple argument assuming a fractal flame front and a turbulence level proportional to flame speeds suggests that for flame propagation through a given obstacle field, flame speeds should increase proportionally to the product of the laminar burning rate with the expansion ratio raised to a power of around 1.36 (Ref.4). This suggests that ethene (ethylene) flame speeds should be around twice those with propane in the same geometry.

In a variety of deflagration geometries, particularly in a series of 4-m^3 wedge-shaped confinements obstructed by different configurations of pipe racks,[4] and in scaled models of some real onshore and offshore structures, the flame-speed ratio between propane and ethene has previously been found to be close to 2.3. This accords reasonably well with the ratio suggested in Ref. 4.

However, for deflagrations in the duct in the experimental program described in Ref. 5, the propane/ethene flame-speed ratios for a range of roof venting and flame speeds were consistently around 2.6 instead of 2.3. In itself this is unremarkable, since the processes involved in a deflagration are very complicated, and many explanations could be offered to explain the anomaly. However, it was apparent that all of the geometries in which the ethene/propane flame-speed ratio was close to 2.3 had predominantly round obstacles. It was conjectured that ratio was considerably above this figure because of the obstacle corners.

Below, we present the results of a series of experiments carried out to investigate this further, and also to investigate the accuracy of drag characterization. The geometry chosen is one where obstacles have a very small mean porosity (at most 6%), and where they are not expected to sigificantly inhibit vent flow.

Experimental Details

The experiments were carried out in an explosion duct described in Ref. 6. The duct is 2 m long, with obstacle grids at 0.1 m intervals, and has a centrally vented roof. Flame arrival was recorded with ion gaps between obstacles. This means of quantifying flame progress has been checked against high-speed photographs. The experiments were conducted with ethene (CP grade 99.9%) and propane (CP grade 99%). Roof venting was varied between 20 and 80%. The measured gas concentrations were 6.85% (by volume) for ethene and 4.24% for propane. Gas

Table 1 Average propane flame speeds (m/s)

% roof venting	40% rectangular	40% circular	28% rectangular
80	16.9	6.7	7.8
70	20.7	7.2	9.9
60	22.2	13.3	13.0
50	27.8	18.1	17.2
40	30.4	22.9	18.8
30	32.3	23.8	22.2
20	36.6	24.2	25.9

mixture was adjusted using a recirculation system. We estimate the measured gas concentration was accurate to within 0.05%. For both mixtures the concentrations were therefore between 1.04 and 1.06 times stoichiometric.

The flame reached a steady flame speed after the first half-dozen obstacles, and the single parameter of the steady flame speed is used here to compare geometries. This steady flame speed occurs when a balance is achieved between side venting and turbulent enhanced burning.

Three different sets of obstacle grids were placed in the duct. The grids had four cylindrical bars in them. The first set of grids were made of 10 x 15 mm rectangular bars, giving a blockage ratio of 40%; the second of 7 x 8 mm rectangular bars, giving a blockage ratio of 28%; and the third of 10-mm diameter circular bars, giving a blockage ratio of 40%. These grids were chosen so that the 28% square bars had a drag as close as possible to the 40% round bars, where the formula

$$\text{Grid Drag} = (\text{Grid Element Drag})/(1 - \text{Blockage Ratio})^2$$

has been used, with the element drag coefficient for the round bars taken as 1.0 at relevant Reynolds numbers, and the element drag coefficient taken as 2.0 for both rectangles. In fact, there is considerable uncertainty in estimating drag, particularly for circular cylinders at varying Reynolds number,[7] with upstream turbulence,[5] and also for grids of a given blockage ratio.[8] It is impossible, therefore, to guarantee that the drag of these two configurations is closer than approximately 10%. It would thus be impossible to detect the dependence on the detail of turbulence purely by varying geometry. For this reason, we emphasize the comparison between two different fuels.

Each experiment was repeated once, and if the results of the two tests were significantly different, subsequent tests were run to investigate this. However, the results quoted are the results of the first two (for 40% blockage) or three (otherwise) tests run. These results were chosen to ensure a statistically unbiased compilation, but have the disadvantage that several errant test runs are included. This is why the flame speed for round obstacles with ethene is slightly

Table 2 Average ethene flame speeds (m/s)

% roof venting	40% rectangular	40% circular	28% rectangular
80	45.0	14.3	20.1
70	52.5	15.9	25.2
60	59.0	32.0	32.4
50	67.0	40.3	41.0
40	72.5	50.3	49.2
30	84.5	58.5	60.7
20	92.7	55.9	71.4

less with 20% venting than with 30% venting, for example. Repetition of these 20% venting tests gave higher answers, but they have not been included here. All of the results had some scatter between repetitions, which is consistent with the scatter in Table 3.

The surfaces of the obstacles were smooth on the lengthscale of laminar flame thicknesses. The roughness (PRa: the arithmetic mean of the departure of the roughness from the mean line) was measured as 13 micrometres for the 10 x 15 mm rectangular obstacles, 2 micrometres for the 7 x 8 mm rectangular obstacles, and 0.2 micrometres for the circular obstacles. Since the Reynolds numbers were generally too low for turbulent boundary layers to occur, structure on a lengthscale less than the laminar flame thickness is unlikely to affect the flow much. The corners on the rectangular obstacles had radii of curvature of 0.25 mm (for 15 x 10 mm) and 0.15 mm (for 8 x 7 mm). Again this was intended to be sufficiently sharp so as to make no difference.

Experimental Results and Analysis

The experimental results for flame speeds with different fuels and geometries are summarized in Tables 1 and 2. Table 3 compares the flame-speed ratio between propane and ethene with the various grids. There is no statistically significant trend within the columns. There is, however, a highly significant difference between the second and other columns, which we now analyze.

We initially take each column to be part of a separate population, with the same variance. This gives a figure of 0.0149 for the variance. Taking a null hypothesis that the first and third columns (40 and 28% rectangular geometries, respectively) have the same mean, we compare with a student's T distribution with eighteen degrees of freedom, using the statistic

$$|m_i - m_j|(n/2V)^{1/2}$$

Table 3 Ethene/propane flame-speed ratios

% roof venting	40% rectangular	40% circular	28% rectangular
80	2.67	2.13	2.57
70	2.54	2.19	2.54
60	2.66	2.40	2.49
50	2.41	2.22	2.39
40	2.38	2.20	2.62
30	2.61	2.46	2.73
20	2.53	2.31	2.76

Table 4 Obstacle ratios

	Same drag		Same blockage	
% roof venting	Propane	Ethene	Propane	Ethene
80	1.29	1.41	2.52	3.15
70	1.37	1.58	2.89	3.30
60	0.98	0.99	1.70	1.82
50	0.95	1.02	1.53	1.66
40	0.82	0.98	1.33	1.44
30	0.93	1.04	1.36	1.39
20	1.07	1.28	1.51	1.67

where m_i is the mean of the ith column, n is the number of elements in each column, and V is the measured variance about the mean of each column.

On this basis, we conclude that the small difference in means observed between the first and third columns is not statistically significant. We have no grounds for believing the figures in the first and third columns to be from a different population, and would be able to detect a difference in means of 0.1.

However, a null hypothesis that the column with circular obstacles also has the same mean can be rejected with 99.9% confidence (rejecting with 99.9% confidence for these data requires a difference between means of 0.257, as opposed to 0.27 between the first and second column, and 0.31 between the second and third). This means that we can be confident that, in this geometry, the flame-speed ratio between fuels is consistently different for round and rectangular obstacles.

Table 4 gives the ratio between flame speeds caused by obstacles of the same drag, for each fuel, and also for obstacles of the same blockage. For lower roof venting (60% or less), the two obstacle grids of similar drag give rise to similar flame speeds, whereas the two grids of similar blockage do not. Generally, grids with the same drag give similar flame speeds to within the uncertainties of estimating drag. It would be impossible to tell that the dependence at faster flame speeds is not simply on drag without the comparison between fuels described above.

For high roof venting (70 or 80%), the obstacle grids with similar drag differ in their effect. This may be because the mean flow inside the duct is significantly towards the roof vent, so that the flow encounters the grid elements at an angle. This would make the effective drag of the rectangular elements higher than that of the circular obstacles. Higher flame speeds are to be expected in this case.

Conclusions

The simplest conclusion from these experiments is that in this geometry at all but the highest levels of roof venting, obstacles with a similar drag give rise to *approximately* similar flame speeds. This is encouraging from the point of view of characterization of subgrid obstacles by drag in numerical simulators. At similar flame speeds and roof venting, the flowfield away from the combustion region is the same for the two different sets of obstacles, and we can infer that the burning rate behind the two sets of obstacles (which is caused by a similar flow rate) is similar. Conversely, the difference between flame speeds for obstacles of similar blockage implies that at the same flow rate these two sets of obstacles give rise to different burning rates (otherwise the flame speeds would end up the same). At a higher order of approximation, however, the flame-speed ratio between fuels depends on whether obstacles are circular or rectangular. This implies that the flame speed in this geometry cannot depend purely on obstacle drag for both fuels (since a circular grid of higher blockage can have the same drag as a rectangular grid). By following a similar argument to the above, we can conclude that the burning rate behind an obstacle does not depend *purely* on the obstacle drag, but has a fuel-dependent sensitivity to shape. This implies that the relative severity of obstacle configurations depends on fuel.

We conclude that flame speeds (and burning rates) are a more complicated function of turbulent statistics and fuel than can be accounted for by the dependence of burning rate on turbulence level (at a low resolution).

The most probable explanation for why coarse turbulent statistics are unable to characterize combustion behind these obstacles is the presence of strong shear layers around the corners of the rectangular obstacles. For example, with a 30 m/s flow, the shear layer coming off the rear corner of an obstacle is likely to be around 0.1 mm thick; this implies a local shear rate of 300,000 s^{-1}. This figure is far greater than that at which local extinction would occur for any of these fuels.

As the shear layer trails away from the corner, the stretch rate will decay, covering a whole range of values. In these circumstances, one could not accurately estimate the average burning rate using the burning rate at the average shear rate.

Summary

Drag is a far better parameter for characterizing obstacle effects than blockage ratio. Geometries with the same venting and blockage ratio can vary considerably in severity. Geometries with similar drag give rise to broadly similar deflagrations.

Drag characterization is, however, of limited accuracy. The inaccuracies in drag characterization could not be identified directly, but only by comparison between fuels.

Burning rates have a more complicated dependence on turbulence and fuel than can be accurately accounted for simply by considering a burning rate based on mean turbulence levels. In a numerical simulation, it is likely that the grid resolution would have to be fine enough to describe partly developed turbulent features, such as shear layers from corners, in order to reproduce such effects accurately.

The relative severity of an explosion in two different geometries can be fuel dependent.

References

[1]Hjertager B. H., Fuhre K. and Bjorkhaug M., *Journal of Loss Prevention in the Process Industries*, Vol. 1, 1988, pp. 197-205.

[2]Batchelor G. K. and Townsend A. A., "Decay of vorticity in isotropic turbulence," *Proceedings of the Royal Society*, 1947, pp. 534-555

[3]Teodorczyk, A. Lee J. H. S. and Krystantas R., "The Structure of Fast Flames in Very Rough Obstacle Filled Channels," 23rd Symposium on Combustion, *Combustion Institute*, 1990, pp. 735.

[4]Taylor P. H. and Hirst W. J. S., *Poster Presented at 22nd International Symposium on Combustion*, Seattle, WA, 1988.

[5]Kwok K. C. S., "Turbulence Effect on Flow round Circular Cylinders," *Journal of Engineering Mechanics*, Vol. 112, 1986, pp. 1171-1197.

[6]Taylor P. H., "Fast Flames in a Vented Duct," *Proceedings of the 21st International Symposium on Combustion*, 1986, pp. 1601-1608.

[7]Batchelor G. K., "An Introduction to Fluid Dynamics," *Cambridge University Press, Cambridge*, 1967, p. 341f.

[8]Hoerner S. F., "*Fluid-Dynamic Drag*," Published by the Author, 1958.

Corrections to Zel'dovich's "Spontaneous Flame" and the Onset of Explosion via Nonuniform Preheating

M. Short* and J. W. Dold†
University of Bristol, Bristol, United Kingdom

Abstract

Given an initial temperature disturbance in a thermally-sensitive exothermic reactant, Zeldovich[1] postulated that a supersonic "spontaneous flame" would arise. He estimated the movement of this flame as the continuum of local explosions that would emerge if each fluid particle were able to explode at its constant volume induction time calculated from its initial state. If interactions between adjacent particles were truly negligible, then this estimate would indeed provide an accurate and simple means of calculating the path of a spontaneous reaction wave.

However, to ignore all compressibility and expansion effects is, in general, unrealistic. Building on Zeldovich's ideas, Dold and Kapila[2] found the spontaneous emergence of a similar reaction wave (which they called an "induction flame") through the solution of a system of partial differential equations that model an induction process that includes chemical and compressible interactions. Thus, while being more accurate, their results are not nearly as straightforward to calculate.

Still using a thermally sensitive chemical model, this paper presents a corrected estimate for Zeldovich's spontaneous flame path that is very much closer to these more accurate calculations. Moreover, being based on an asymptotic compressible correction to Zeldovich's leading-order constant volume estimate, the formula we obtain is also simply expressed solely in terms of initial conditions. The full description of the reaction wave is given in terms of the initial data alone, including the possible first appearance of a sonic point at a known position, that heralds shock formation and the creation of a fully fledged Zeldovich–von Neuman–Döring detonation at the Chapman–Jouguet speed.

---

* Research Student, School of Mathematics.

† Research Fellow, School of Mathematics.

Introduction

In recent years there has been an interest, particularly among Russian workers,[3–5] in the transition to detonation due to temperature and concentration nonuniformities initially present in an explosive mixture. Nearly all practical systems will involve some nonuniformity in their distributions at an initial instant. If the initial distribution is very nearly homogeneous, then it is observed that constant volume explosion takes place in the system with almost uniform pressure rise throughout the volume.[3] On the other hand, if the gradient of the initial state is large enough a detonation wave is seen to form in the medium with substantially higher rises in pressure. This is of fundamental interest in the question of engine knock.

Most of the theory developed so far has evolved from the original work of Zeldovich[1] who postulated a spontaneous flame as the reaction wave which would be formed if every particle of reactant were to explode at the constant volume induction time appropriate to its initial state. If each particle is treated as a separate entity, independent of its neighbouring particles, the constant volume explosion time of each particle can be determined purely in terms of its initial temperature and concentration from classical theory. A locus of the constant volume explosion times can then be determined, giving rise to a path of a reaction wave which Zeldovich called a spontaneous flame.

The speed of this flame is then known simply in terms of the initial state. A purely supersonic wave can arise if the velocity of the flame is everywhere greater than that of the Chapman–Jouguet detonation velocity appropriate to the initial mixture at any point. Any spontaneous flame in this range is able to persist in travelling at uniformly supersonic speeds with no shock wave in its structure.[1] A limiting case in this range occurs when the reacting medium is initially uniform so that all particles explode at the same time in a genuinely constant volume evolution. This may be thought of as producing a wave of infinite speed, with lower (finite) spontaneous-flame speeds arising if the temperature or concentration gradient is increased.

Alternatively, a second case occurs where the initial disturbance gives rise to a spontaneous flame whose velocity falls below the Chapman–Jouguet velocity. The faster velocity is then able to dominate and take over in the movement of the reaction wave so that a transition from the spontaneous flame to a Chapman–Jouguet detonation can be anticipated. In numerical work, Zeldovich et al.[3,4] and Gel'fand et al.[5] observed exactly this type of transition to a "normal" detonation close to the position at which the spontaneous flame speed falls below the Chapman–Jouguet speed. More recent work by Dold and Kapila[6] has succeeded in describing in detail such a transition from a supersonic reaction wave through to a normal detonation using asymptotic techniques. For smaller initial gradients, the numerical work demonstrates the evolution of a near constant volume explosion, which can be thought of as a weak detonation.[1]

The most useful point about Zeldovich's approach is that the path and speed of the spontaneous flame are simply specified purely in terms of the initial disturbance. However, this does lead to inaccuracies because compressible interactions between adjacent particles and volumetric expansions are completely neglected—something that is not generally realistic. This leaves room for significant improvement. By including the full effects of compressibility in an asymptotic analysis of an induction process, Dold and Kapila[2] have recently been able to identify a more accurate analogue of Zeldovich's spontaneous flame, which they refer to as an "induction flame." However, in order to calculate the path and speed of this flame it becomes necessary to solve a system of partial differential (or characteristic[7]) equations under appropriate initial and boundary conditions. Normally this is a considerably more difficult and cumbersome task, although important features such as the position of transition to Chapman–Jouguet detonation are then much more precisely determined.

In reality, there will be at least some interaction between chemical and gas-dynamic effects. The degree of importance of these interactions in any system depends on the *local* initial distribution of the chemical and thermodymamic quantities about a particular point. In examining any chemical evolution about such a point there will be two independent factors of great importance, which can be determined at the initial instant. The first is the local acoustic length, which is representative of the length an acoustic disturbance can propagate during a typical local induction time. The second represents a local chemical length which gives a representation of the length scale over which comparable rates of chemical reaction are distributed. It turns out (see following section) that the degree of importance of compressibility in the system is measured by the ratio of the acoustic length to the chemical length. If this ratio is small, i.e. acoustic disturbances remain almost fixed, then chemical evolution is dominated by constant volume heating. Zeldovich's ideas (perhaps improved with suitable corrections for compressibility) then give a fairly accurate representation. On the other hand, if these two quantities are comparable in size there will be a significant chemical-acoustic interaction in the evolution. The full effects of compressibility must then be included and one must revert to the analysis of Dold and Kapila.[2,6]

In the following work we derive compressibility corrections to the notion of Zeldovich's spontaneous flame under the assumption that compressibility effects are locally small. Results are presented in the form of straightforward explicit formulae that retain the advantage of depending only on the initial data. These corrections are indeed small when predicted spontaneous speeds are much greater than the Chapman–Jouguet speed. In such cases, Zeldovich's constant volume formulae are themselves reasonably accurate. However, in many cases with an initial temperature and concentration disturbance where we predict significant compressible effects, we also find that the *corrected* spontaneous flame predictions duplicate the more precise induction flame calculations with surprising accuracy, even close to the point at which spontaneous speeds approach the Chapman–Jouguet speed. Because of this, these improved predictions can be used to estimate (say) the position of transition to strong detonation with considerably more accuracy than Zeldovich's original formula, while maintaining the valuable simplicity of straightforward dependence only on initial conditions.

Model

It is convenient to model the explosive as a polytropic fluid which undergoes chemical reaction by means of a single step isomolar exothermic Arrhenius reaction $F \to P$ with a large activation energy. Accordingly, we use a set of reactive Euler equations in an ideal gas to describe the flow. In terms of a mass-weighted Lagrangian coordinate ψ, these can be written as follows:

$$
\begin{gathered}
V_t - U_\psi = 0 \\
U_t + P_\psi = 0, \qquad VP_t + \gamma P U_\psi = -Q\lambda_t \\
\lambda_t = -\frac{\lambda}{t_0} e^{-T_A/T}, \qquad T = \frac{PV}{nR}, \qquad \psi = \int_{x_0(t)}^{x} \frac{dx}{V}
\end{gathered}
\tag{1}
$$

In this $x_0(t)$ is any conveniently chosen position that moves with the local fluid velocity $(dx_0/dt = U)$ in order to set the origin of the mass-weighted coordinate, ψ. Also, T, P, V, U, and λ are the temperature, pressure, specific volume, velocity, and a normalised reactant mass fraction, respectively; R represents the universal gas constant, n the number of moles per unit mass, Q the heat of reaction, T_A the activation temperature and γ the principal specific heats' ratio or adiabatic exponent. The frequency factor t_0^{-1} represents a pre-exponential rate constant in Arrhenius kinetics and reflects the rate of chemical change at very high temperatures.

In the same spirit as Zeldovich[1] we consider that these equations are to be solved subject to a set of initial conditions at $t = 0$,

$$
\begin{gathered}
T \equiv T_i(\psi), \quad P \equiv P_0, \quad U \equiv 0, \quad \lambda \equiv \lambda_i(\psi) \\
\text{and} \qquad V \equiv V_i(\psi) \equiv \frac{nRT_i(\psi)}{P_0}
\end{gathered}
\tag{2}
$$

in which pressure and velocity are taken to be initially uniform in space. It may be noted that nonuniform initial states for these quantities could also be considered in the analyses that follow. However, when the activation temperature is large $(T_A \gg T)$ the sensitivity of the system is most pronounced to temperature and concentration nonuniformities so that it is both simplest and most informative to consider only disturbances in temperature and concentration.

Induction Stage

To begin the examination of the induction stage of the reaction, we first of all start by considering an arbitrary location $\psi = \widetilde{\psi}$ in the explosive medium at which $T = \widetilde{T} = T_i(\widetilde{\psi})$ and $\lambda = \widetilde{\lambda} = \lambda_i(\widetilde{\psi})$ at the initial instant. If the activation energy is large, so that a small dimensionless parameter $\widetilde{\epsilon}(\widetilde{\psi}) = \widetilde{T}/T_A$ can be identified, then during the induction stage of the explosion all of the thermodynamic and chemical variables deviate by a relative amount of order $\widetilde{\epsilon}$ from their initial state. It is therefore useful to define a set of dimensionless perturbation variables such that

$$T/\widetilde{T} \sim 1 + \widetilde{\epsilon}\widehat{\phi}, \quad P/P_0 \sim 1 + \widetilde{\epsilon}\gamma p, \quad \lambda/\widetilde{\lambda} \sim \widehat{\Lambda}_i - \widetilde{\epsilon} q w$$

$$V/\widetilde{V} \sim 1 + \widetilde{\epsilon}\widehat{v}, \qquad U \Big/ \sqrt{\gamma P_0 \widetilde{V}} \sim \widetilde{\epsilon} u \tag{3}$$

$$\widehat{\Lambda}_i = \frac{\lambda_i}{\widetilde{\lambda}} \quad \text{and} \quad q = \frac{Q\widetilde{\lambda}}{\gamma n R \widetilde{T}}$$

Changes in velocity are anticipated as being of the order of $\widetilde{\epsilon}$ relative to the speed of sound. For initial concentration nonuniformities (represented by the function $\widehat{\Lambda}_i$,) to affect the analysis as much as initial nonuniformities in temperature, $\widehat{\Lambda}_i$ and $\widehat{\phi}$ would need to vary by comparable amounts. It is also convenient to scale time with respect to the constant pressure induction time that can be calculated from the initial state of the particle at $\psi = \widetilde{\psi}$, and to scale the coordinate ψ with respect to a suitable local scale-factor $\widetilde{\psi}_c$, such that

$$\chi = \frac{\psi - \widetilde{\psi}}{\widetilde{\psi}_c} \qquad \text{and} \qquad \tau = \frac{t}{\widetilde{t}_c} \qquad \text{where} \qquad \widetilde{t}_c = \frac{t_0 \widetilde{\epsilon}}{q} e^{T_A/\widetilde{T}} \tag{4}$$

Substituting the expansions (3) into Eqs. (1) now yields a set of induction perturbation equations

$$v_\tau - \mu u_\chi = 0, \qquad u_\tau + \mu p_\chi = 0$$
$$p_\tau + \mu u_\chi = e^\phi, \qquad w_\tau = e^\phi, \qquad \phi = \gamma p + v \tag{5}$$

where

$$\phi = \widehat{\phi} + \ln \widehat{\Lambda}_i, \qquad v = \widehat{v} + \ln \widehat{\Lambda}_i, \qquad \mu = \frac{\widetilde{t}_c \sqrt{\gamma P_0 \widetilde{V}}}{\widetilde{\psi}_c \widetilde{V}} \tag{6}$$

In this μ is a ratio of length scales which we discuss below.

Defining ϕ and v in this way leaves Eqs. (5) in the form of a well-known set of induction equations,[2] where $\widehat{\Lambda}_i$ does not appear explicitly. However, the role of ϕ now takes into account the initial presence of an 0(1) concentration variation (through $\widehat{\Lambda}_i$), and ϕ does not purely represent a temperature disturbance, as is usually the case.[2] From the initial conditions of (2) one obtains the set of initial conditions for the perturbation quantities:

$$\phi \equiv v \equiv \phi_i(\chi) = \widehat{\phi}_i(\chi) + \ln \widehat{\Lambda}_i(\chi) = \frac{T_i(\psi) - T_i(\widetilde{\psi})}{\widetilde{\epsilon} T_i(\widetilde{\psi})} + \ln\left(\frac{\lambda_i(\psi)}{\lambda_i(\widetilde{\psi})}\right) \tag{7}$$

$$\text{and} \quad p \equiv u \equiv w \equiv 0$$

at the initial time $\tau = 0$.

The scaling factor $\widetilde{\psi}_c$ is most sensibly taken to be chosen such that the initial perturbation $\phi_i(\chi)$ has first and second derivatives with respect to χ that are essentially of order one. By construction, a suitable definition would require

$$\widetilde{t}_c\left(\widetilde{\psi} + O(\widetilde{\psi}_c)\right) = O\left(\widetilde{t}_c(\widetilde{\psi})\right) \neq o\left(\widetilde{t}_c(\widetilde{\psi})\right) \tag{8}$$

In other words we ensure that, over the length scale $\widetilde{\psi}_c$, the relative induction times at each point remain of the same order of magnitude. Moreover, since we already have $\phi_i(0) = 0$ (by construction), this would mean that $\widetilde{\psi}_c$ is chosen to characterise a coordinate scale over which the initial perturbation of ϕ takes (positive or negative) values of order one. The dominating physical process in the chemical evolution over the coordinate scale $\widetilde{\psi}_c$ is then simply determined by the size of the factor μ in Eqs. (6).

With the definitions of $\widetilde{t}_c$ and $\widetilde{\psi}_c$ above, it can be seen that μ measures the ratio of an acoustic length scale $l_a = \widetilde{t}_c(\gamma P_0 \widetilde{V})^{1/2}$ (based on the induction time $\widetilde{t}_c$) to a chemical length scale $l_c = \widetilde{\psi}_c \widetilde{V}$ (over which the reaction-rates e^{ϕ} initially have a similar order of magnitude). If the temperature or concentration distribution around $\psi = \widetilde{\psi}$ causes μ to be much less than 1, then the chemical length l_c is much greater than the acoustic length l_a and acoustic disturbances caused by the chemical reaction are unable to propagate to any significant degree. This means that around $\widetilde{\psi}$ the reactant remains confined by its own inertia and that chemical effects dominate over acoustic effects. It is readily noted from Eqs. (5) that the specific volume perturbation v then remains practically fixed in time. This is the type of situation envisaged by Zeldovich[1] where the initial state is nearly homogeneous so that relatively small initial gradients in temperature are involved.

While this might be the case for some values of $\widetilde{\psi}$, the sensitivity of $\widetilde{t}_c$ to changes in initial temperature shows that it is still quite possible to find another region $\psi = \widetilde{\psi}_1$ (say) in the reacting mixture at which $\mu \approx 1$. Where this is the case, significant compressible and chemical interactions would become inevitable.

Constant Volume Induction—Zeldovich's Flame

To begin our analysis, we can firstly identify the solutions that result from taking $\mu = 0$ in Eqs. (5). In this limit, the effects of compressibility are completely neglected and Eqs. (5) are easily solved under the conditions of Eq. (7) to give

$$\begin{aligned} v &\sim v_0 = \phi_i \\ \phi &\sim \phi_0 = -\ln(e^{-\phi_i} - \gamma\tau) \\ p &\sim p_0 = (\phi_0 - \phi_i)/\gamma \\ u &\sim u_0 = 0 \\ w &\sim w_0 = p_0/q \end{aligned} \tag{9}$$

These results present the leading order behaviour of the perturbation quantities defined in Eqs. (3) in the limit where the chemical length-scale is much longer than the acoustic length-scale. It can be seen that chemical changes feed energy directly into pressure and temperature increases, and not into kinetic energy or volumetric expansion.

Significantly, the solution for ϕ_0 reveals a leading order estimate of a path in time and the mass coordinate space around $\psi = \widetilde{\psi}$ at which ϕ becomes infinite, namely where

$$\tau = \bar{\tau}(\chi) \sim \bar{\tau}_0(\chi) = e^{-\phi_i(\chi)}/\gamma \tag{10}$$

The gradient of $\bar{\tau}_0(\chi)$ or dimensionless inverse mass flux of this singularity path is then given by

$$\frac{d\bar{\tau}_0}{d\chi} = -\frac{\phi_i'(\chi)}{\gamma} e^{\phi_i(\chi)} \tag{11}$$

Of course the real, non perturbed temperature remains bounded at this path which rather heralds the appearance of much more rapid and substantial chemical change. Thus, in this limit, the time at which particles around $\widetilde{\psi}$ begin to react explosively depends purely on their initial state.

Rewriting the expressions for ϕ_0 and $\bar{\tau}$ in terms of dimensional variables gives

$$\begin{aligned} T &\sim T_i - \frac{T_i^2}{T_A} \ln\left(1 - t \times \frac{Q\lambda_i T_A/t_0}{nRT_i^2}\, e^{-T_A/T_i}\right) \\ \bar{t}\,(\psi) &\sim t_0\, \frac{nRT_i^2}{Q\lambda_i T_A} e^{T_A/T_i} \end{aligned} \tag{12}$$

At $t = \bar{t}(\psi)$ a path of spontaneous vigorous chemical reaction is predicted. This result is considerably more general than in Eq. (10) since it represents the path of a wave of strong chemical activity for the whole of the reactant, given the initial temperature distribution $T_i(\psi)$. It is not restricted only to the region about the selected point $\widetilde{\psi}$, but to all points at which $\mu(\widetilde{\psi}) \ll 1$. It should be noted that the expression for $\bar{t}(\psi)$ in Eq. (12) is precisely the constant volume induction time at each point which thus reproduces Zeldovich's definition.[1]

It is clear that Zeldovich's ideas are exact in the limit $\mu \to 0$. In itself, this extreme is wholly unrealistic except when the initial state of the medium is completely uniform. Otherwise, there would always be some compressible interaction between reacting layers.

Corrections for Compressibility

In order to find the degree of change that occurs when μ is small but non-zero, we need to consider higher orders in Eq. (9). We therefore expand the perturbation quantities in a regular series in powers of μ. This is most usefully done by firstly defining a new time-like variable

$$\zeta = -\ln\left(\bar{\tau}(\chi) - \tau\right) \tag{13}$$

that increases without bound as $\tau \to \bar{\tau}$. The distribution of induction times $\bar{\tau}(\chi)$ is then explicitly included in the formulation of the problem. Based on the leading order expansions found in Eq. (9) it is now sensible to consider higher order expansions in the form

$$\begin{aligned} v - \phi_i &\sim \mu v_1 + \mu^2 v_2 \\ \phi - \zeta + \ln\gamma &\sim \mu\phi_1 + \mu^2\phi_2 \\ p - \tfrac{\zeta - \phi_i - \ln\gamma}{\gamma} &\sim \mu p_1 + \mu^2 p_2 \\ u &\sim \mu u_1 + \mu^2 u_2 \\ w - \tfrac{\zeta - \phi_i - \ln\gamma}{\gamma q} &\sim \mu w_1 + \mu^2 w_2 \end{aligned} \tag{14}$$

to the order of μ^2. We should also expand the location of the singularity path $\bar{\tau}(\chi)$ in terms of μ, so that

$$\bar{\tau}(\chi) \sim \bar{\tau}_0(\chi) + \mu\bar{\tau}_1(\chi) + \mu^2\bar{\tau}_2(\chi) \tag{15}$$

to the same order of approximation.

Transforming Eqs. (5) to the $(\chi,\, \zeta)$ reference frame gives

$$\begin{aligned} &v_\zeta + \mu(\bar{\tau}' u_\zeta - e^{-\zeta}u_\chi) = 0 \\ &u_\zeta - \mu(\bar{\tau}' p_\zeta - e^{-\zeta}p_\chi) = 0 \\ &\phi_\zeta + (\gamma - 1)v_\zeta = \gamma e^{\phi - \zeta} \\ &q w_\zeta = e^{\phi - \zeta} \\ &\gamma p = \phi - v \end{aligned} \tag{16}$$

with the initial conditions that all of the left (or right) hand sides of Eq. (14) are zero at the initial instant when

$$\zeta = -\ln\bar{\tau}(\chi) \tag{17}$$

Using the asymptotic limit $\mu \to 0$, these equations can be solved asymptotically.

Without entering into the details of this procedure, the singularity path $\bar{\tau}(\chi)$ is found to take the asymptotic form

$$\bar{\tau}(\chi) \sim \frac{e^{-\phi_i}}{\gamma} - \mu^2 \frac{\gamma - 1}{4\gamma^4} \left(\frac{\phi_i''}{3} + \phi_i'^2 \right) e^{-3\phi_i} \tag{18}$$

to second order. Surprisingly, when μ is small, it turns out that the constant volume induction estimate in Eq. (10) is accurate to second order in powers of μ. The corresponding dimensionless inverse mass flux of the compressibility corrected flame path, $\bar{\tau}(\chi)$, becomes

$$\frac{d\bar{\tau}(\chi)}{d\chi} = -\frac{\phi_i'}{\gamma} e^{-\phi_i} - \mu^2 \frac{(\gamma - 1)}{4\gamma^4} \left(\frac{\phi_i'''}{3} + \phi_i'\phi_i'' - 3\phi_i'^3 \right) e^{-3\phi_i} \tag{19}$$

In its dimensional form the result for $\bar{\tau}(\chi)$ becomes

$$\begin{aligned} \bar{t}(\psi) \sim\ & t_0 \frac{nRT_i}{Q\lambda_i} \frac{T_i}{T_A} e^{T_A/T_i} \left[1 - \frac{\gamma - 1}{4\gamma} \frac{\gamma P_o}{V_i} \left(t_0 \frac{nRT_i}{Q\lambda_i} \frac{T_i}{T_A} e^{T_A/T_i} \right)^2 \right. \\ & \left. \times \left\{ \frac{T_A}{T_i^2} \left(\frac{T_i''}{3} + \frac{T_A}{T_i^2} T_i'^2 \right) + 2 \frac{\lambda_i'}{\lambda_i} \frac{T_A}{T_i^2} T_i' + \frac{\lambda_i''}{3\lambda_i} + \frac{2}{3} \left(\frac{\lambda_i'}{\lambda_i} \right)^2 \right\} \right] \end{aligned} \tag{20}$$

which provides a fairly straightforward extension of the constant specific-volume formula seen in Eq. (12). Note that the effect of initial concentration nonuniformities is automatically included in this expression. The inverse of the mass-flux $(\overline{m}(\psi))$ through the singularity path becomes

$$\frac{1}{\overline{m}(\psi)} = \frac{d\bar{t}(\psi)}{d\psi} \tag{21}$$

which can be evaluated by straightforward differentiation of Eq. (20).

The result in Eq. (18) for $\bar{\tau}(\chi)$ is locally asymptotically valid wherever

$$\mu^2 \left(\frac{\phi_i''}{3} + \phi_i'^2 \right) e^{-2\phi_i} \ll 1 \tag{22}$$

and the result in Eq. (20) for $\bar{t}(\psi)$ is more globally asymptotically valid throughout the region where

$$\begin{aligned} & \frac{\gamma - 1}{4\gamma} \frac{\gamma P_o}{V_i} \left(t_0 \frac{nRT_i}{Q\lambda_i} \frac{T_i}{T_A} e^{T_A/T_i} \right)^2 \\ & \times \left\{ \frac{T_A}{T_i^2} \left(\frac{T_i''}{3} + \frac{T_A}{T_i^2} T_i'^2 \right) + 2 \frac{\lambda_i'}{\lambda_i} \frac{T_A}{T_i^2} T_i' + \frac{\lambda_i''}{3\lambda_i} + \frac{2}{3} \left(\frac{\lambda_i'}{\lambda_i} \right)^2 \right\} \ll 1 \end{aligned} \tag{23}$$

or, more simply,

$$\overline{m}(\psi) \gg \sqrt{\gamma P_o / V_i} \tag{24}$$

Comparison with Numerical Solutions

In order to gauge the accuracy of the estimates in Eq.s (18) and (19) when these conditions are not satisfied, we select two generic cases for comparison with accurate numerical solutions of the path of the reaction wave (or perturbation singularity). Both have significant physical relevance for any realistic initial temperature variation, $T_i(\psi)$, and concentration variation, $\lambda_i(\psi)$.

The first arises in the kind of situation mainly envisaged by Zeldovich,[1] in which $\mu \ll 1$ around the position of some maximum temperature or concentration value. Since $\mu(\widetilde{\psi})$ varies with the position $\psi = \widetilde{\psi}$, we consider any point $\psi = \widetilde{\psi}_1$ at which we may take $\mu(\widetilde{\psi}) = 1$. Typically, if the initial temperature and concentration varies fairly smoothly as a function of ψ, then on the now relatively small scale of $\widetilde{\psi}_c$ it is only the first derivative of $T_i(\psi)$ that plays any significant role on the local initial temperature and concentration variation. Without loss of generality, we may take the choice of $\widetilde{\psi}_1$ to be made such that

$$\phi_i(\chi) = -\chi \qquad \text{and} \qquad \mu = 1 \tag{25}$$

with second derivatives of $\phi_i(\chi)$ being negligible on the local fine-scale measured by χ. Recalling that $\phi_i(\chi) = \widehat{\phi}_i(\psi) + \ln \widehat{\Lambda}_i(\psi)$, where $\widehat{\phi}_i$ represents the local initial temperature disturbance, such an initial distribution could arise with an initially uniform concentration ($\widehat{\Lambda}_i = 1$) leaving only a disturbance in temperature. Alternatively, temperature could be uniform initially, and a suitable form for the concentration distribution could be taken. In general circumstances there may be a contribution from both disturbances.

Zeldovich's expression [Eq. (10)] and the corrected expression [Eq. (18)] for the flame path are compared with a numerically calculated result for this initial temperature perturbation in Fig. 1. The method used for obtaining the numerical result is described elsewhere.[8] Zeldovich's estimate approaches the numerical solution only as ϕ_i increases (that is, as χ increases negatively). For order one values of χ, and particularly when χ is positive, it significantly overestimates the true time to self-ignition of any particle. By contrast, the estimate (18), which incorporates compressibility corrections, is much improved.

Equally important are the relative magnitudes of the flame-path gradients, $d\overline{\tau}/d\chi$. In the following section it will be shown that the location (if any) of a transition to detonation will be determined by a set value of $d\overline{\tau}/d\chi$. Fig. 2 shows the comparision of these gradients for the Zeldovich Flame, numerically calculated flame and corrected flame. Again the cor-

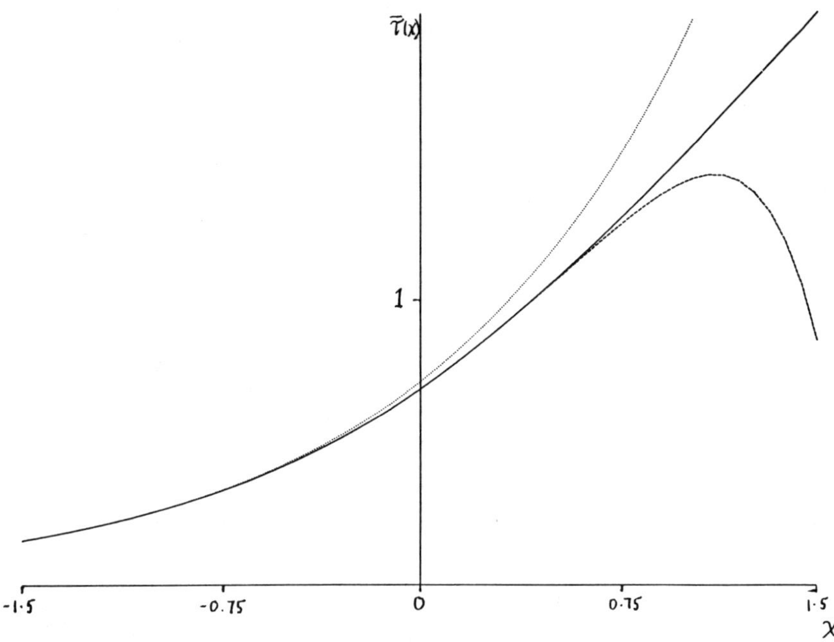

FIG. 1 Zeldovich's estimate (dotted line), the estimate with corrections for compressibility (dashed line), and the numerical calculation (solid line) of the spontaneous flame path for initial conditions with $\phi_i = -\chi$. The circle (∘) marks the position of minimum flame speed according to the estimate with corrections (at which $(d\bar{\tau}/d\chi)^{-1} \approx 1.203$).

rected estimate follows the true value of $d\bar{\tau}/d\chi$ over a significantly larger region than the Zeldovich flame path.

It may be noted[2] that the true singularity path must always remain supersonic, represented by a slope of $(d\bar{\tau}/d\chi)^{-1} > 1$. The point at which the corrected estimate (18) actually becomes poor enough to produce a minimum speed of the reaction wave is marked on Fig. 1 by a circle (∘). At this point the formula yields a reaction wave mass-flux of $(d\bar{\tau}/d\chi)^{-1} \approx 1.203$ which (in practical terms) is only barely supersonic. Even at this extreme point (beyond which the trend in the wave speed is wrong, and so should not be acceptable) the estimate for $\bar{\tau}$ is remarkably good.

In reality, because Chapman–Jouguet detonation speeds are significantly supersonic, a transition to a strong Chapman–Jouguet detonation happens well before this point is reached, where the estimate remains very good. Thus it can be seen that the expression (18) typically provides a very good representation of the path of the reaction wave at all points where the path has any physical significance. It is therefore able to predict the point at which a *normal* Zeldovich–von Neuman–Döring detonation

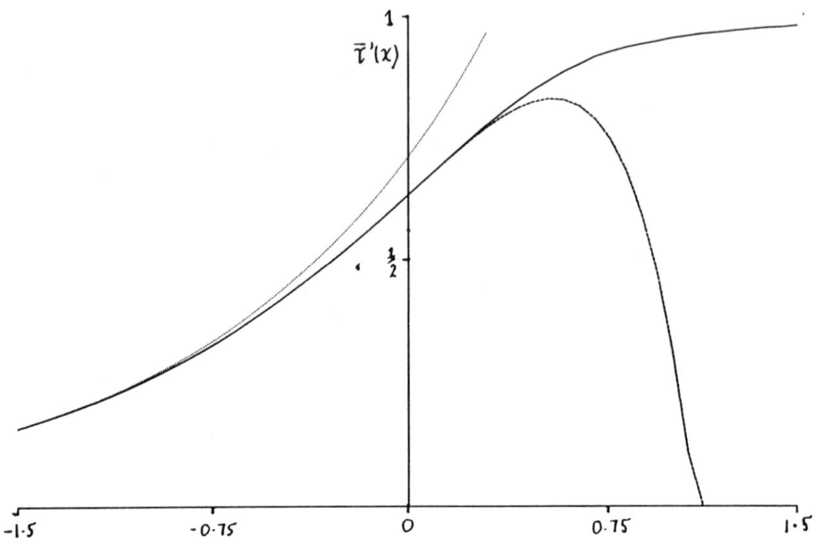

FIG. 2 Zeldovich's estimate (dotted line), the estimate with corrections for compressibility (dashed line), and the numerical calculation (solid line) of the gradient spontaneous flame path for initial conditions with $\phi_i = -\chi$. The inverse of this gradient represents the dimensionless mass-flux of the flame.

forms with significantly better accuracy than Zeldovich's original estimate, but still retains the advantage of being determined from the initial state.

The second generic case focuses on any position of maximum temperature or concentration, which we can take to be dominated by a locally quadratic variation of ϕ_i in ψ or χ. Knowing that the approximate estimates (10) and (18) are good when μ is small, it is sufficient once again to select cases in which μ is unity so that

$$\phi_i(\chi) = -a\chi^2 \qquad \text{and} \qquad \mu = 1 \tag{26}$$

for any order one positive constant a. Two comparisons for the values $a = 1$ and $\frac{1}{4}$ are shown in Fig. 3. Although the corrected estimate (18) is an improvement over the constant volume formula (10), the results in these cases can be seen to remain good over a reduced range as a is increased, where $d\bar{\tau}/d\chi$ approaches unity over a shorter distance. We can conclude that the compressibility-corrected version of Zeldovich's formula offers a significant improvement even when considerable compressibility effects are clearly present.

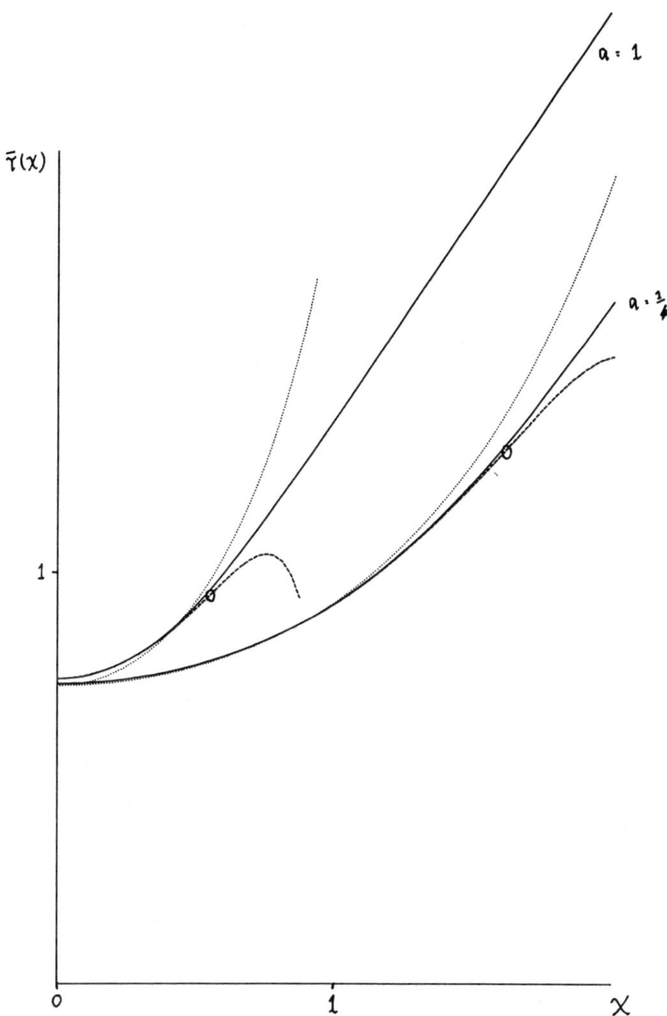

FIG. 3 Zeldovich's estimates (dotted lines), estimates with corrections for compressibility (dashed lines), and the numerical calculations (solid lines) of spontaneous flame paths for initial conditions with $\phi_i = -\chi^2$ and $\phi_i = -\frac{1}{4}\chi^2$. The circles ($\circ$) mark the positions of minimum flame mass-fluxes according to the estimates with corrections (at which $(d\bar{\tau}/d\chi)^{-1} \approx 1.505$ and $(d\bar{\tau}/d\chi)^{-1} \approx 1.345$, respectively).

Structure of the Reaction Wave

As well as the expression (20) for the path $t = \bar{t}(\psi)$ we can obtain a description for the structure of the reaction wave that emerges spontaneously and closely follows this path. In fact, in order to be able to determine any position of transition to a strong Chapman–Jouguet detonation, it is necessary to know something about this structure.

Perturbation Singularity Structure

As ζ becomes large, that is $\tau \to \bar{\tau}$ at any point, the terms proportional to $e^{-\zeta}$ in Eqs. (16) become negligible. In this limit, it is relatively simple to integrate Eqs. (16) which leads to the following linear relationships between chemical and thermodynamic quantities in the initial stages of the reaction waves, when pressure changes $P - P_0$ are not very large:

$$\begin{aligned} T - T_i &\sim T_i \frac{P - P_0}{P_0}\left(1 - \frac{P_0}{V_i}\bar{t}'^2\right) + \frac{T_i^2}{T_A}a(\psi) \\ \lambda_i - \lambda &\sim \frac{P - P_0}{Q/V_i}\left(1 - \gamma\frac{P_0}{V_i}\bar{t}'^2\right) + \frac{\gamma P_0 V_i}{Q}b(\psi) \\ U &\sim (P - P_0)\bar{t}' + \sqrt{\gamma P_0 V_i}\,\frac{T_i}{T_A}c(\psi) \end{aligned} \tag{27}$$

In this, the functions $a(\chi)$, $b(\chi)$ and $c(\chi)$ contribute only a small amount to the structure of the reaction wave and may be neglected. Nevertheless, they can be determined in a similar way to the formula (20) through solving Eqs. (16) and (17) under the limit $\mu \to 0$. Where this is not the case, the functions a, b and c may still be evaluated numerically.[8]

Reaction Wave Structure

In order to understand the fuller structure of the spontaneous reaction waves it is necessary only to recognise that relatively very fast reaction rates are involved (compared with those in the induction stage). This means that the waves become very thin and consequently quasi-steady in terms of any local wave-following coordinates. This being so, the governing Eqs. (1) can be used to show that the reaction waves follow Rayleigh line behaviour on Hugoniot curves.[2,7]

Accordingly, it can be shown that pressure, temperature, velocity and normalised reactant mass fraction satisfy the relations

$$\begin{aligned} nRT &\sim P\left(V_i - \frac{P - P_0}{m^2}\right) \\ (1 - \lambda)\,Qm^2 &\sim \left(P_0 + V_i m^2\right)(P - P_0) - \frac{\gamma + 1}{2}\left(P^2 - P_0^2\right) \end{aligned} \tag{28}$$

and

$$mU \sim P - P_0$$

where $m = m(\psi)$ is the mass-flux through the travelling full reaction wave (or induction flame[2]) and an obvious leading order connection with the initial conditions and the results (27) has been applied. It is straightforward to see that a matching of the gradients with respect to pressure between these formulae and the induction expressions (27) is achieved if

$$m(\psi) \sim 1 \Big/ \bar{t}'(\psi) \tag{29}$$

for $T_i \ll T_A$. That is, the movement of the path $t = \bar{t}(\psi)$ determines the leading-order mass flux (for small values of ϵ) through the full reaction wave.

Chapman–Jouguet Mass Flux

At the end of the chemical reaction one has that $\lambda = 0$. The middle of Eqs. (28) then provides a quadratic equation which may be solved for the final pressure behind the reaction wave. Both roots are the same at the Chapman–Jouguet mass flux, which can therefore be identified as

$$m_{CJ}^2(\psi) \sim \gamma \frac{P_0}{V_i} + (\gamma+1)\frac{Q\lambda_i}{V_i^2}\left(1 \pm \sqrt{1 + \frac{2\gamma}{\gamma+1}\frac{P_0 V_i}{Q\lambda_i}}\right) \tag{30}$$

with the upper supersonic root (positive sign) providing the relevant value. Because $(\gamma P_0/V_i)^{1/2}$ is the acoustic impedance, or sound-speed in terms of the mass-weighted coordinate ψ at the initial conditions, this result clearly yields a significantly supersonic mass flux. The critical value m_{CJ} is determined by the initial state of the chemical medium. As a guide, typical C-J Mach numbers, $m_{CJ}/(\gamma P_o/V_i)^{1/2}$, range between about 2 to 5.

Transition to Strong Detonation

It is precisely when (and if) the mass flux $m(\psi)$ decreases to the Chapman–Jouguet value $m_{CJ}(\psi)$ that a sonic point first appears in the flow through the reaction wave. Any lower speeds lead to the formation of a shock wave and a transition to strong detonation.[6,7] In order to determine any value of ψ at which this happens, it is necessary only to use Eqs. (29) and (21) to calculate $m(\psi)$ as a function of the initial conditions and to compare the result with the expression (30) for $m_{CJ}(\psi)$. Using the analysis presented in this paper, both of these expressions are given purely in terms of initial conditions. Except in mixtures with an unphysically low heat release, m_{CJ} is significantly supersonic. As a result (as seen in Figs. 1–3), the estimate for the flame path (20) normally remains good right up to the point when (and if) a strong C-J detonation is created. In cases where no strong detonation wave is predicted to form, the mass flux of the spontaneous flame again remains significantly supersonic, and thus Eq. (20) provides an excellent approximation to the path of the resulting flame (weak detonation).

Conclusions

In this paper we have built on Zeldovich's fundamental analysis[1] of the processes that lead to the formation of a strong detonation wave when a system is allowed to react from an initial temperature and concentration distribution. In doing this, we have attempted to place Zeldovich's ad-hoc use of a constant-volume explosion approximation in defining the path of a *spontaneous flame*, on a firmer footing.

In certain limits, the spontaneous appearance of a reaction wave certainly is very well described by a continuous distribution of constant specific volume self-ignitions. However, we have been able to produce significant improvements to this basic estimate by adding corrections for compressibility and thermal expansion. The resulting reaction wave is then much more accurately described in many cases. Where it fails is where Zeldovich's approximation is also very poor, and in fact one would then need to take recourse to numerical solutions of equations describing induction processes[9] in order to calculate flame paths with any accuracy. On the other hand, in the wide class of cases where the analysis holds good, flame paths are usefully given as simple explicit functions of the initial data.

We have taken the analysis a little further in order also to obtain an expression for the Chapman–Jouguet mass flux as a simple function of the initial state of the reacting mixture. Where the actual mass flux of the flame falls below this value the previously weak-detonation (i.e., shockless) structure of the spontaneous reaction wave becomes untenable and a transition to strong detonation (including the formation of a shock wave) becomes inevitable.[6]

References

[1] Zel'dovich, Ya. B., "Regime Classification of an Exothermic Reaction with Non-Uniform Initial Conditions," *Combustion and Flame*, Vol. 39, 1980, pp. 211–214.

[2] Dold, J. W. and Kapila, A. K., "Asymptotic Analysis of Detonation Initiation for One-Step Chemistry: I-Emergence of a Weak Detonation," submitted.

[3] Zeldovich, Ya. B., Librovich, V. B., Makhviladze, G. M. and Sivashinsky, G. I., "Development Of Detonation In a Non-Uniformly Preheated Gas," *Astronautica Acta*, Vol. 15, 1970, pp. 313–321.

[4] Zeldovich, Ya.B., Gelfand, B.E., Tsyganov, S.A., Frolov, S.M. and Polenov, A.N., " Concentration and Temperature Nonuniformities (CTN) of Combustible mixtures as a reason of Pressure Wave Generation," *11-th Colloquium (International) on Dynamics of Explosions and Reactive Systems,* Warsaw, 1988, pp. 89.

[5] Gelfand, B. E., Polenov, A. N., Frolov, S. M. and Tsyganov, S. A., " Occurence of Detonation In a Nonuniformly Heated Gas Mixture," *Fizika Goreniya i Vzryva,* Vol. 21, No. 4, 1985. pp. 118-123.

[6] Dold, J. W. and Kapila, A. K., "Asymptotic Analysis of Detonation Initiation for One-Step Chemistry: II-From a Weak Structure to ZND," submitted.

[7] Dold, J. W., "Emergence of a Detonation within a Reacting Medium", *Fluid Dynamical Aspects of Combustion Theory,* Longman, U.K., 1991, pp. 161–183.

[8] Dold, J. W., "Induction Period Generation of a Supersonic Flame Induction Flame", *Numerical Combustion,* Springer Verlag, 1989. pp. 245-256.

Numerical and Experimental Studies of Flame Propagation Through a Grid

G. O. Thomas* and R. J. Bambrey†
University of Wales, Aberystwyth, Dyfed, United Kingdom
and
B. H. Hjertager,‡ T. Solberg,§ and J.-E. Forrisdahl¶
Telemark Institute of Technology and Telemark Innovation Centre, TMIH Kjolnes, Porsgrunn, Norway

Abstract

A combined experimental and numerical program of studies of flame propagation through a grid is described. Unlike earlier flame propagation studies, where the objectives were to study the global acceleration of the flame, the present studies have concentrated on more fundamental studies of flame propagation through a single grid and its associated turbulent flow field. By establishing an initial flow over the grid by means of shock waves, well defined turbulence fields could be generated without any initial and complicating flame acceleration. This approach also allows a flow field to be generated independently of the mixture reactivity.

Results are presented of flame propagation in initially quiescent gas and also gas mixtures with initial flow velocities of up to 240 ms^{-1}. The

---

* Lecturer, Department of Physics.

† Research assistant, Department of Physics.

‡ Professor, Department of Process Engineering.

§ Senior Scientist, Group for Industrial Fluid Dynamics.

¶ Research Student, Department of Process Engineering.

gas mixtures tested were stoichiometric ethylene/oxygen with various nitrogen dilutions. Flame propagation and explosion development was monitored using schlieren and pressure gauges. The initial results have been compared to numerical simulations and are found to be in reasonable agreement.

Introduction

Although many studies exist on flame propagation by repeated obstacles, the majority have concentrated on the global nature of the phenomena. On a laboratory scale, studies such as those of Moen et al.[1,2] initially demonstrated that rapid flame acceleration could be induced in methane-air flames by placing repeated obstacles in the flow. This acceleration arises from an increase in the burning rate as the flame propagates into a turbulent flowfield. A positive feedback mechanism is then established as the flowfield induced ahead of the flame and hence the turbulence levels generated are directly related to the increase combustion rate of the reactants.

The degree of confinement is a critical factor in this process, as shown by Chan et al.[3], van Wingerden and Zeeuwen[4], and Taylor[5]. Increased confinement reduces the pressure relief of combustion products by venting, resulting in increased pressures and flow velocities which further increases combustion rates. Lee et al.[6] identified four propagation regimes of increasing velocity and showed that for sufficiently reactive mixture transition to detonation can eventually occur. On a larger scale, similar accelerations have been observed by Harrison and Eyre[7] and Hjertager et al.[8] for example, while transition to detonation has been observed by Moen et al. [9].

In parallel with these experimental studies, there have been significant advances in the modeling of explosions. The model described by Hjertager[10] is capable of simulating explosions in complex geometries, the results of which are in reasonable agreement with experimental observations. Such models are of great importance in assessing the hazards associated with potential industrial explosion scenarios. One important element in such numerical simulations is the turbulence model employed and the corresponding rate of turbulent combustion. Although all of the experimental tests undertaken to date provide data which is useful in assessing the correctness of such codes, the turbulence model is only one of several closely coupled mechanisms which give rise to the final experimental observables. It is desirable therefore to have some body of information which would allow the interaction between chemistry and gasdynamic effects to be initially decoupled so that their relative influence on explosion development from some defined initial condition could be assessed.

In the present paper some preliminary results of an experimental study of flame propagation over a single grid are presented. An initial steady flow could be established over the grid by means of shock-tube techniques and independently of mixture reactivity. The relative influence of induced turbulence levels and reactivity on flame propagation away from a line ignition source could therefore be studied separately. Also, three-dimensional numerical simulations of the experimental configurations have been undertaken and the results of some preliminary calculations are presented and compared to experimental observations.

Experimental Details

The experimental tests were performed in a 6.5 m long duralumin shock tube of internal cross-section 38 x 76 mm. The tube, shown in Fig. 1(a), was fitted with a window section which gave an observation section 230 mm long and 76 mm deep. The location of this section was such that there was minimum interference from the shock reflected from the closed end of the tube and from the reflected expansion wave from the driver. The 1 m long driver section, pressurized from 1-2 bar, was used to generate shocks with Mach numbers in the range of 1.36 to 1.89. In this way, steady state flows of 180 and 390 ms^{-1} could be established over the grid prior to ignition. The initial pressure of the flowing gas was always maintained at 200 Torr by suitable choice of the initial test section pressure.

Ignition was effected by means of a linear array of five sparks generated by the discharge of a 0.5 mF capacitor charged to 3.5 kV. The igniter arrangement is shown schematically in Fig. 1(b). It was constructed from a 2 x 3 mm section of double sided circuit board. Breaks could easily be machined into the conducting layer to give the desired spark gap locations. The external discharge circuit was such that at the operating pressures used breakdown occurred across the combined gap spacings. The gaps themselves were placed facing downstream with respect to the incident shock flow. This method was chosen in preference to a hot wire method as it provided a much more rapid, controllable, and repeatable source. Even if a wire could be heated rapidly, there were grave doubts as to the shot-to-shot repeatability as a result of random local hot spot generation along the wire under such rapid heating rates. Laser ignition was also suggested, as this would be truly nonintrusive, but the blast generated would be significant. Even with the present system, there is evidence of pressure waves emanating from the five sources.

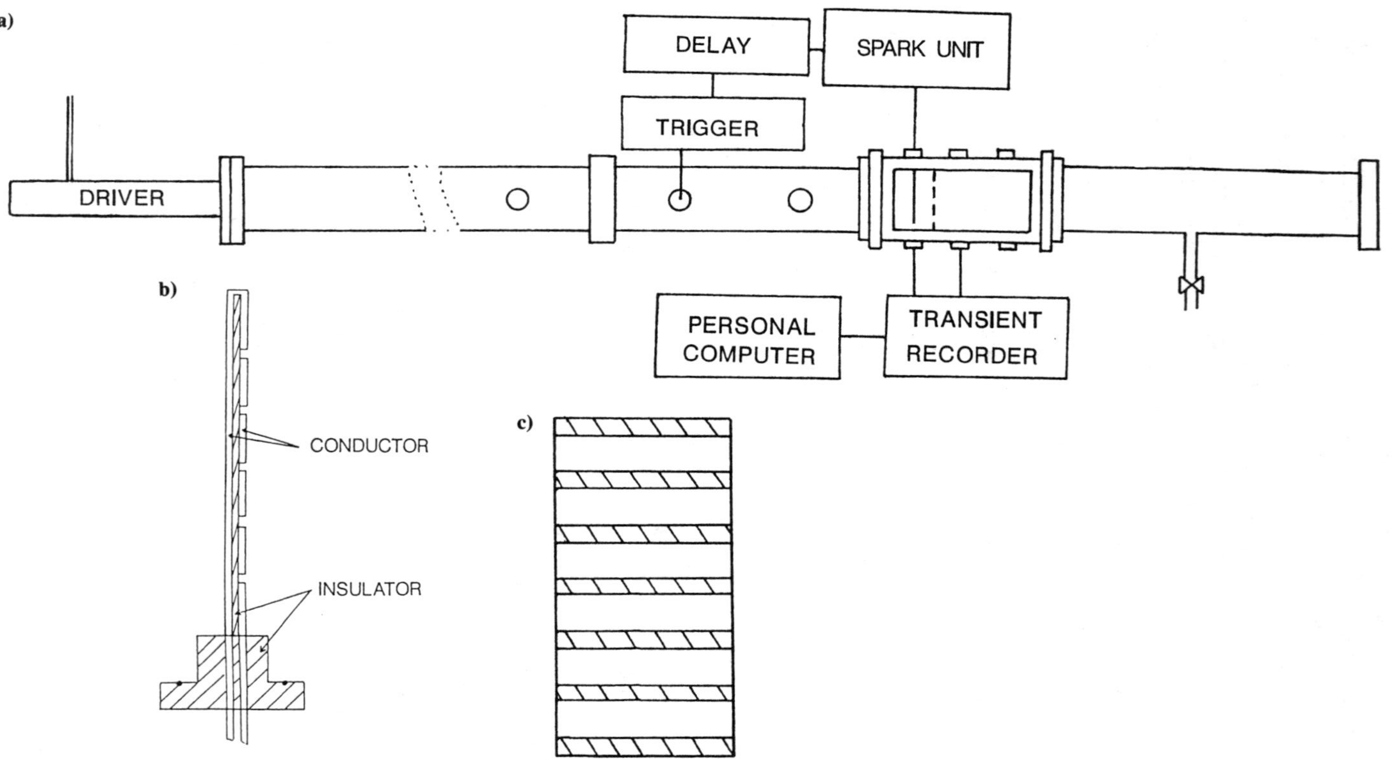

Fig. 1 Schematic diagram of a) shock tube layout, b) igniter rod and c) obstacle grid.

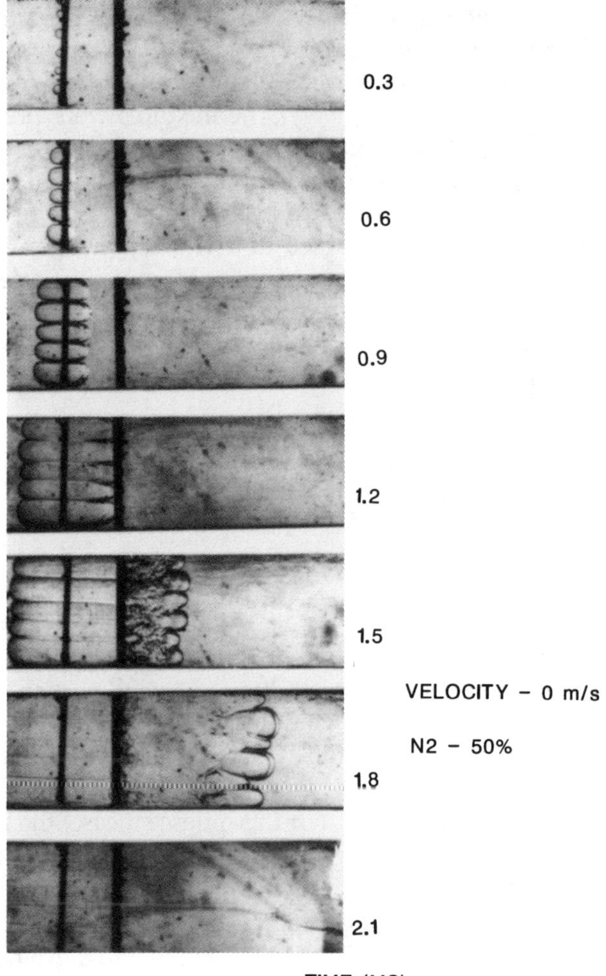

Fig. 2 Spark schlieren photographs of flame development in initially quiescent gas with grid. Mixture - $C_2H_4 + 3O_2 + 50\%\ N_2$, P_o - 200 Torr.

The grid employed for the present tests was fabricated from a 1 mm thick stainless steel sheet cut to the tube cross-section. A number of slots, 4 mm wide, were then machined as shown in Fig. 1(c) to give a grid with a blockage ratio of 37%. The plate was mounted in the window section supported by a frame fabricated from a similar sheet. For the present tests the location of the grid remained fixed, with ignition effected at a plane 38 mm upstream of the grid with respect to the incident shock. Further, the ignition line was placed at the center of the tube and aligned parallel to the longest wall of the tube cross-section. In preliminary tests with initially quiescent test mixtures, only the final 1.6 m section was used with a solid flange used to close the tube at a plane 80 mm away from the grid location.

The gas mixtures tested were stoichiometric ethylene/oxygen with 50, 60, or 73% nitrogen respectively. The mixtures were prepared in a separate vessel and admitted into the previously evacuated test section as required. Flame propagation was monitored using an eight spark schlieren camera system, while explosion pressures were monitored using PCB pressure gauges coupled to a Thurlby model 524 transient recorder controlled by a personal microcomputer. Three pairs of opposing ports were available for pressure measurements. This allowed the ignition source and one pressure gauge to be placed directly opposite, while a second pressure gauge was placed either 35 or 110 mm downstream of the grid.

Experimental Results

Initially Quiescent Gas

The initial studies undertaken were with an initially quiescent mixture of ethylene/oxygen diluted with 50% nitrogen at an initial pressure of 200 Torr. For these studies the major portion of the test section was not used. Only the window section was used, attached to a 1.6 m long section of the main tube. The grid was located 80 mm from the end flange with the ignition source located 38 mm from the grid and between the grid and the flange. For this quiescent case, the main spark unit was not required as the trigger spark energy was sufficient to give ignition. Pressure gauges were placed both at the ignition location and also 110 mm downstream of the grid.

Spark schlieren photographs of the flame development are shown in Fig. 2 at various times after ignition. The growth of the five individual flame kernels can be seen and the source is clearly not an ideal linear source in this case. Initially the flames fronts grow smoothly and the average flame velocity measured between the frames at 0.6 and 0.9 ms is 37 ms^{-1}. During this stage, between frames at 0.9 and 1.2 ms, flame contact with the glass windows can just be identified. As the flame passes through the grid, there is a rapid acceleration. The average flame velocities measured between the next three frames (1.2-1.8 ms) are of the order of 130-150 and 170-200 ms^{-1}, respectively. A surprising feature is that despite the development of quite intense turbulence immediately downstream of the grid, the leading edge of the flame front remains essentially laminar retaining the lobed characteristics of the initial flame kernels. Also of interest is that the lobes now correspond to the number of grid openings and not the spark kernels.

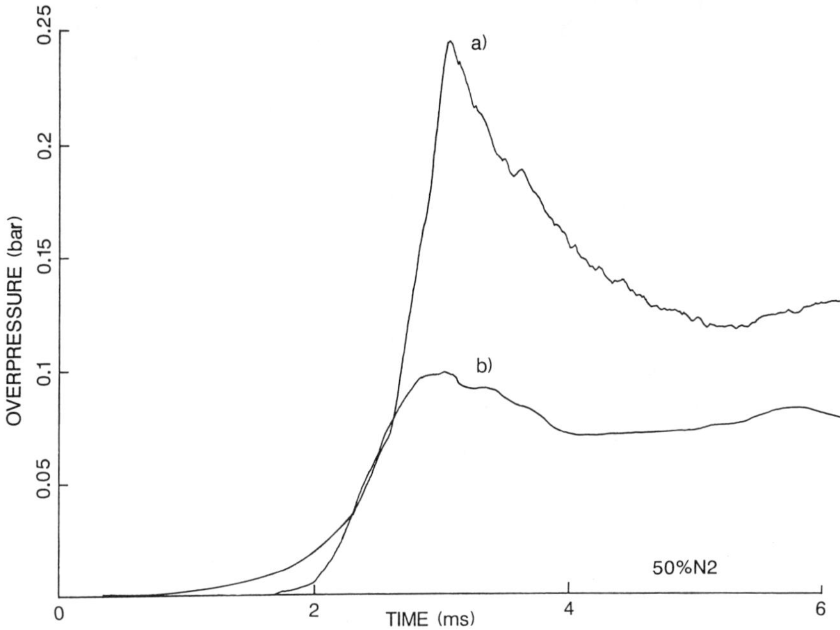

Fig. 3 Pressure records of flame development in initially quiescent gas (a) with grid and (b) without grid. Mixture - $C_2H_4 + 3O_2 + 50\% \ N_2$, P_o - 200 Torr.

The corresponding pressure records obtained during these tests are shown in Fig. 3 curve(a). The times corresponding to the individual frames shown in Fig. 2 are also indicated on this diagram by the arrows. These show clearly the rapid rise in pressure — between 1.2 and 1.8 ms — associated with the propagation of the flame through the turbulence field generated by the grid.

Initially Steady State Gas Flow

Spark photographs obtained with initially steady state gas flow velocity of 180 ms^{-1} over the grid is shown in Fig. 4 for 50% nitrogen dilution. For all of these tests, the initial pressure was chosen such that the shocked gas pressure was 200 Torr, irrespective of the shock and hence gas velocity. In this test, as with all other tests with flowing gases, but unlike the quiescent case, the multipoint nature of the ignition source is not significant and a linear flame front develops rapidly.

At the time of ignition, the incident shock can just be observed on the extreme right of the first frame. In the next two frames the flame front is observed to move downstream, but the mixture reactivity is such that a rod stabilized flame results. This stabilized flame was also observed in tests without the grid in place. At this point the flame is

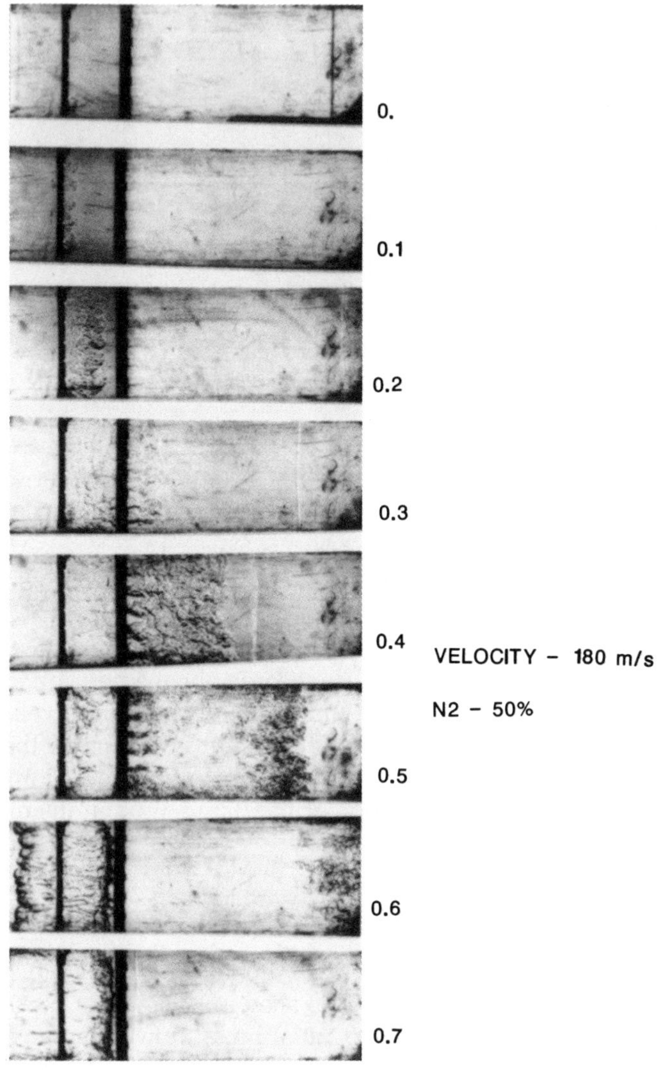

Fig. 4 Spark schlieren photographs of flame development in an initially flowing gas. Mixture - $C_2H_4 + 3O_2 + 50\% \ N_2$, P_o - 200 Torr, V_o - 180 ms^{-1}.

heavily wrinkled and the velocity of the leading edge is of the order of 195 ms^{-1}. In later frames the flame front has propagated through the grid and a distinct increase in the downstream turbulent combustion intensity is apparent. The average velocities over the four frames from 0.2 to 0.5 ms are 330, 450, and 460 ms^{-1} respectively. Also visible on certain frames is a weak pressure wave reflected from a partial obstruction downstream. There is a distinct loss of planarity as this approaches the flame front indicating the turbulent nature of the flowfield ahead of the flame. Repeat tests without this weak wave showed that in this case it did not significantly affect the flame propagation. In these tests, as with all others performed in this study, the test-to-test repeatability was found to be very good.

The downstream flame can be seen to have contacted the wall of the window by 0.5 ms, and from this point the is an increase in the combustion upstream which now rapidly develops, filling the available remaining observational volume. The corresponding pressures measured during this test at the ignition location and 38 mm downstream are shown in Fig. 5. Both gauges, records (a) and (b), show

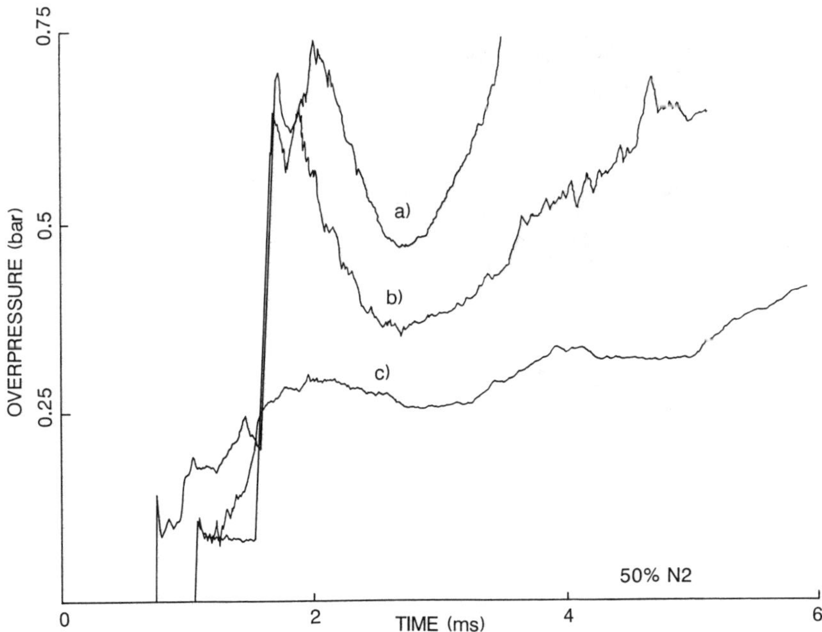

Fig. 5 Pressure records of flame development in an initially flowing gas with grid. a) at ignition location, 38 mm upstream; b) 35 mm downstream; and c) 35 mm downstream with no grid. Mixture - $C_2H_4 + 3O_2 + 50\% N_2$, P_o - 200 Torr, V_o - 180 ms^{-1}.

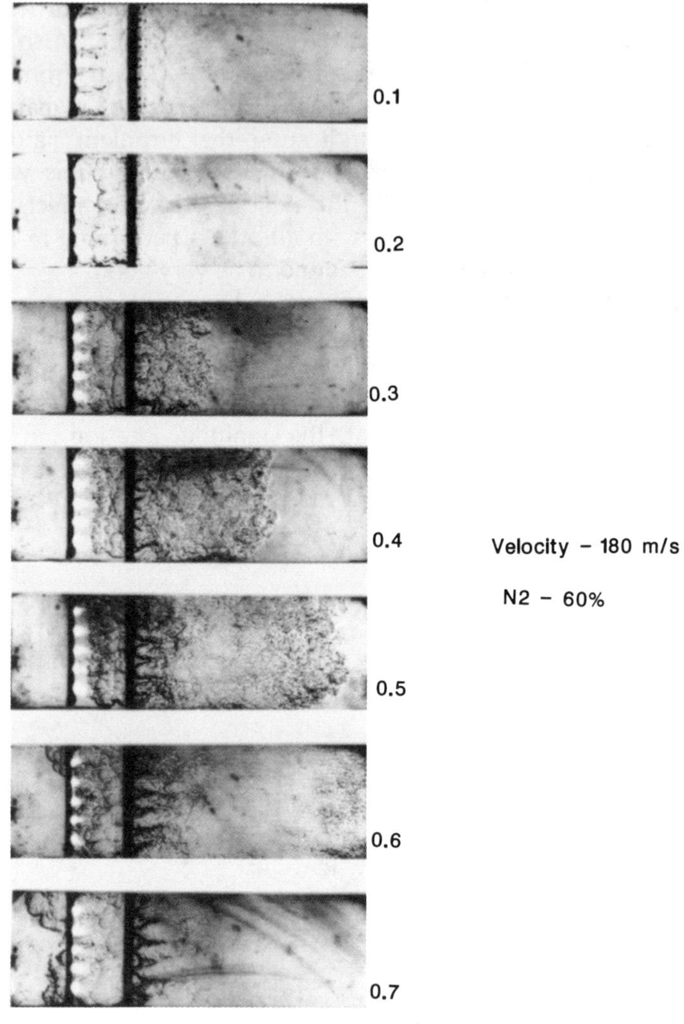

Fig. 6 Spark schlieren photographs of flame development in an initially flowing gas. Mixture - $C_2H_4 + 3O_2 + 60\% \ N_2$, P_o - 200 Torr, V_o - 180 ms^{-1}.

a very rapid pressure rise due to the combustion in the turbulent field. Also included for comparison is a pressure record obtained when there was no grid present, record (c).

Figure 6 shows a series of photographs of the flame propagation for a 60% nitrogen dilution, again for an initial steady state flow velocity of 180 ms^{-1}. In this case, after ignition, the flame is seen to detach from the igniter bar, increasing in intensity as it propagates through the grid. The average velocity of the leading edge of the downstream flame is again of the order of 450 ms^{-1}. Eventually, as the downstream combustion fills the tube cross-section, the upstream flame starts to propagate backwards away from the grid. At this point there still appears to be a net forward flow through the grid, as evidenced by the slow burning wave in the vicinity of the shear layer of the jet from the grid slits. The corresponding pressure records are shown in Fig. 7, where the spark ignition time can be clearly identified as a noise spike. The peak pressure and rate of pressure rise are significantly less in this case compared to the comparable 50% nitrogen test. For 73% nitrogen dilution, the flame was observed to completely detach from the igniter bar, and the entire combustion region was advected through the grid and downstream. There was little evidence of flame acceleration even

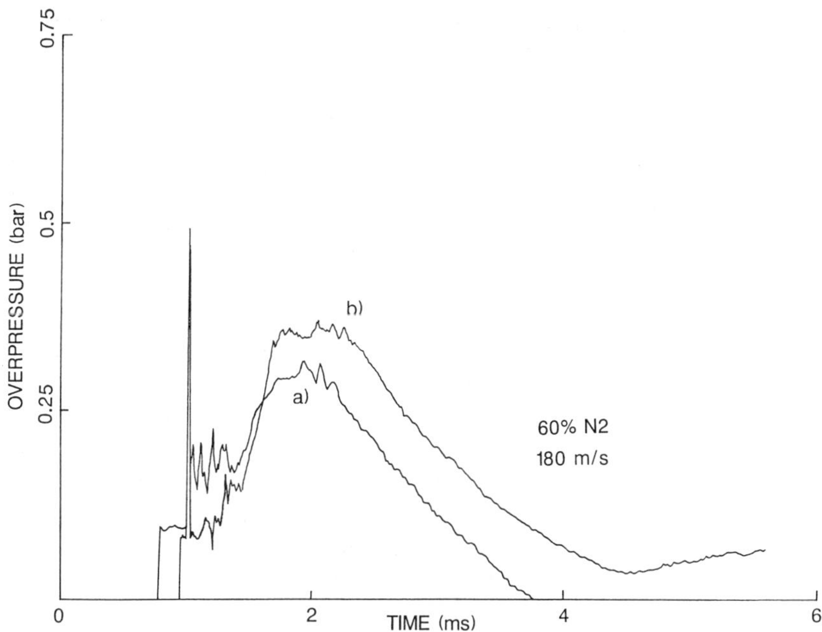

Fig. 7 Pressure records of flame development in an initially flowing gas. a) at ignition location 38 mm upstream; and b) 35 mm downstream. Mixture - $C_2H_4 + 3O_2 + 60\% N_2$, P_o - 200 Torr, V_o - 180 ms^{-1}.

when an initial gas flow velocity of 390 ms^{-1} was used, as shown in Fig. 8. In this case there was very little overpressure developed due to combustion; see Fig. 9. One marked feature on these records, observed to a lesser extent for lower flow velocities, is a reflected shock generated by the acoustic mismatch represented by the grid. In this case there is also a significant pressure drop across the grid.

Numerical Simulation

The numerical model used has been described in detail elsewhere[10], so only a brief description will be given here. The governing equations for mass, momentum, and energy, together with equations describing the combustion, are solved by specifying the relevant fluxes, combustion rates, and the appropriate boundary conditions. Combustion is treated as a single-step, irreversible reaction between fuel and oxidizer. The fluxes of momentum, energy, and fuel are expressed in terms of time-averaged gradients of the relevant variables and effective turbulent transport coefficients in the standard manner. These transport coefficients are determined by the k-e model of turbulence described in detail by Hjertager[10].

The combustion rate is modeled according to the "eddy-dissipation" concept for large, local, turbulent Reynolds numbers and a quasilaminar combustion model for low Reynolds numbers. The turbulent combustion rate is proportional to the density, the inverse of the turbulence turnover time, and the limiting mass-fraction of fuel, product, and oxygen, with a reaction rate constant as proportionality factor. A detailed description of this model has been reported by Hjertager[11] who has used it to simulate several large-scale explosions.

The model is incorporated into a 3D computer code named EXSIM. For the present calculations, two computational grids were used for each of the quasistatic and flowing cases. The first calculation domain was composed of 190 x 8 x 21 nodes corresponding to physical lengths of 0.75 m and cross-section 19 and 38 mm. The second modeled a tube of length 3.04 m using an expanded grid (expansion ratio 1.1) of 122 nodes in the x direction and again 8 x 21 for the other two coordinates. In both cases there were planes of symmetry parallel to the tube axis and both parallel and normal to the line of the igniter.

The results of a simulation of ignition with an initially quiescent mixture are shown in Fig. 10 and were obtained with a 50% nitrogen mixture for three reaction rate constants (20, 10, and 5). These are also compared with the corresponding experimental pressure history. As would be expected, the higher reaction rates give the fastest pressure rise and the highest pressures. In all cases, however, the peak pressure

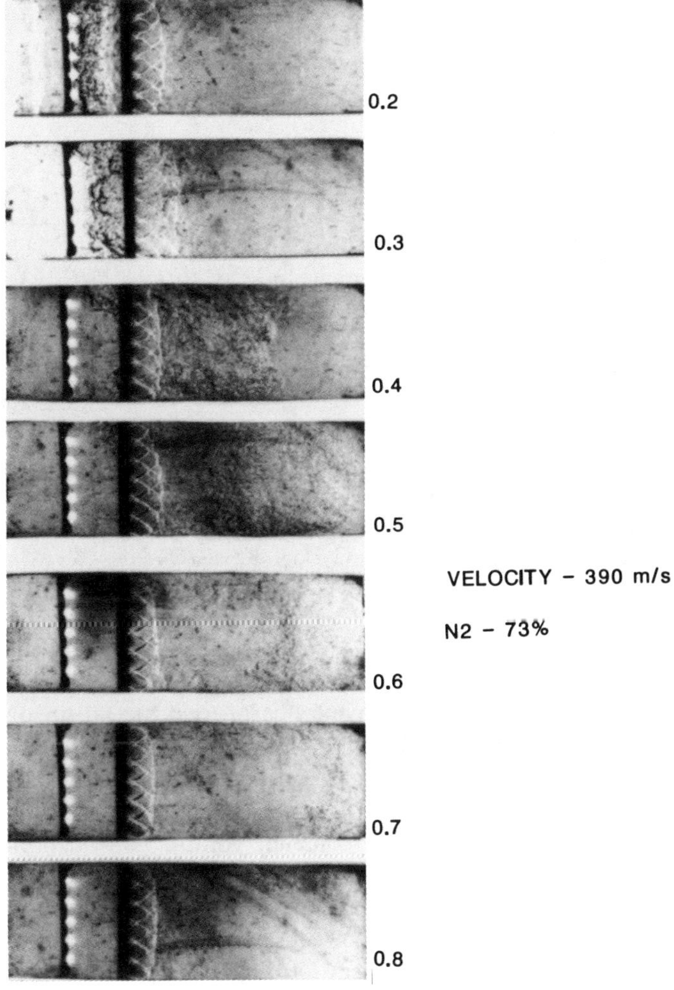

Fig. 8 Spark schlieren photographs of flame development in an initially flowing gas. Mixture - $C_2H_4 + 3O_2 + 73\%\ N_2$, P_o - 200 Torr, V_o - 390 ms^{-1}.

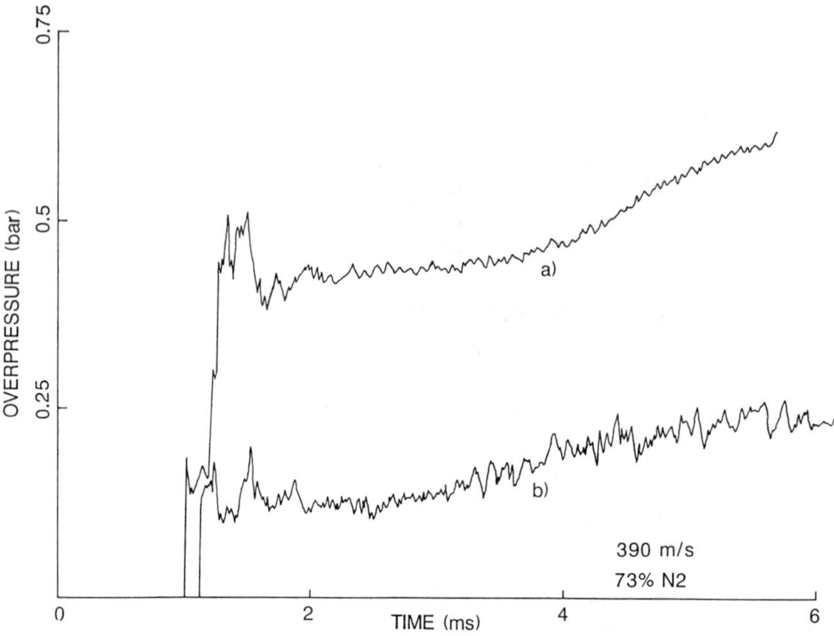

Fig. 9 Pressure records of flame development in an initially flowing gas. a) at ignition location, 38 mm upstream; b) 38 mm downstream; and c) 35 mm downstream with no grid. Mixture - $C_2H_4 + 3O_2 + 73\% \ N_2$, P_o - 200 Torr, V_o - 390 ms^{-1}.

occurs later than in the experimental case. Also the fall in pressure as the combustion propagates into the weakening turbulence field beyond the grid is slightly prolonged in the numerical case. Later, the pressure increases again as the combustion wave fills the test volume, although heat losses limit the maximum constant volume combustion pressure attained.

The numerical results for the flowing case, initial shock velocity 450 ms^{-1} and gas velocity 180 ms^{-1}, are compared with the corresponding experimental pressure gauge output in Fig. 11, again for the same three reaction rate constants. A rate factor of the order five gives good agreement with the experimental pressure history, although once again there is a discrepancy in the absolute timings. Initially this case was modeled with a line source to effect numerical ignition. Great difficulty was experienced, however, in obtaining flame propagation away from the source. The problem was only overcome when a blockage, corresponding to that used in the actual experimental case, was placed before the ignition nodes. Only with this configuration could reasonable ignitions be obtained with initially flowing cases.

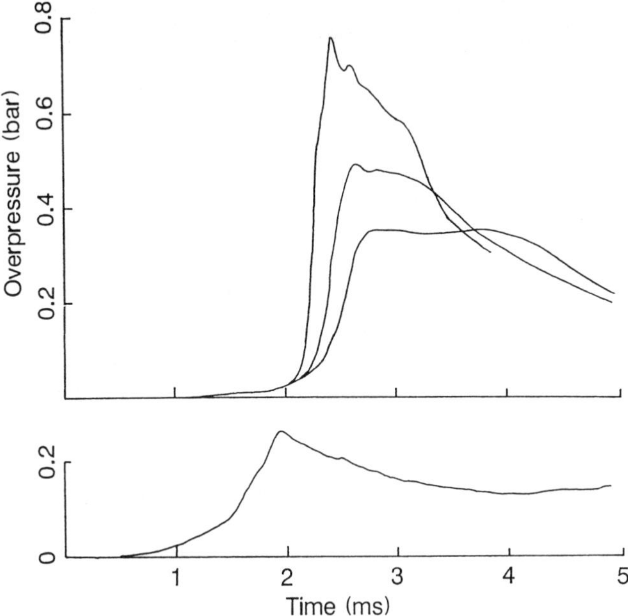

Fig. 10 Comparison of numerical simulation using three reaction rates (20, 10, and 5) with an experimental record. Initially quiescent gas. Mixture - C_2H_4 + $3O_2$ + 50% N_2, P_0 - 200 Torr.

Discussion

In the present studies, a major objective was to develop a parallel experimental and numerical program of studies of small-scale explosion development. In this way it was hoped that as many as possible of the experimental and modeling parameters required could be determined in advance. Thus, for example, the obstacles chosen were baffles as opposed to cylinders. The latter require a significantly finer mesh to model the surface accurately and are more computationally expensive.

The simplest initial case that can be attempted is that of ignition of initially quiescent gas. Ignition near a closed end was chosen as the levels of induced flow did not generate significant flow velocities, and hence overpressure, due to turbulence when there was no restraining end closure. A comparison of the measured and calculated overpressures determined for this configuration has been given in Fig. 10. Here, the measured experimental profile exhibits a shorter time delay to the peak pressure than any of the numerical predictions, although the rate of pressure rise is greater in the numerical cases for all of the reaction rates used. One possible cause for this observed difference could arise from the fact that in the numerical simulation the

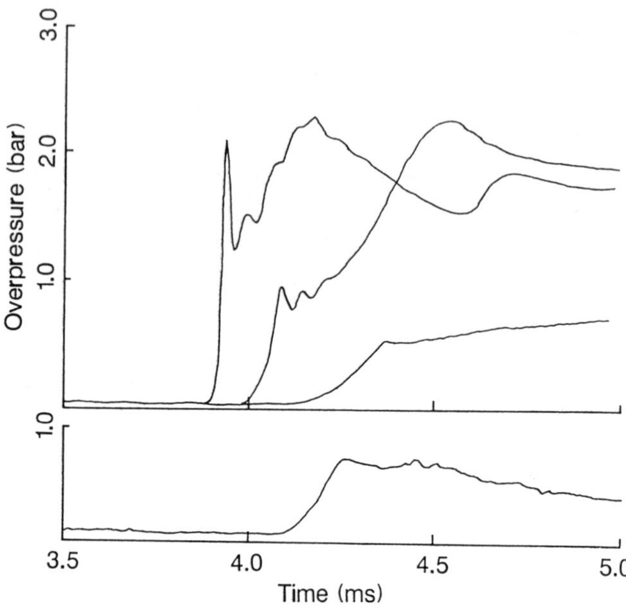

Fig. 11 Comparison of numerical simulation using three reaction rates (20, 10, and 5) with experimental record. Initially flowing gas. Mixture - $C_2H_4 + 3O_2$ + 50% N_2, P_0 - 200 Torr, V_o - 180ms^{-1}.

source is truly linear whereas in the present experimental tests the source was multipoint. Such a difference in geometry will obviously have an effect on the dynamics of the initial flame growth. The importance of such multidimensional source effects on the early stages of propagation was recently noted by Barr[12] in simulations using the vortex dynamics code.

Compared with a truly cylindrical flame, rather than a given flame radius, the surface area of the present five kernels will initially be less than the linear case. However, this situation reverses when the ideally spherical flame kernels touch. Beyond this point, the surface area increases relative to the ideal linear case. This, coupled with the perturbing effect of the flowfield, causes an increase in the rate of flame growth, faster rates of pressure rise, and given the venting restrictions through the obstructing grid, somewhat higher peak pressures in the experimental case.

An inspection of the numerical simulations for the varying reaction rates also shows that the inital presure rise is the same in all cases. Divergence occurs when the combustion encounters the grid and the turbulent burning model comes into play.

In general, however, the simulation shows good qualitative agreement with the observed flame propagation characteristics, with rapid acceleration in the vicinity of the grid, which results in flow reversal upstream of the grid followed by a further reversal as the remaining upstream mixture burns. Given the somewhat diffuse nature of the reaction zone in the numerical solution and the difficulty in assigning a comparable zero time origin, there is reasonable agreement in the flame front location between experiment and simulation.

The turbulence field in the vicinity of the grid predicted by the numerical solution would also appear to be qualitatively in agreement with the experimental results obtained by Christill and Leuckel[13] for steady state flows over a grid with 36% blockage ratio, but with cylindrical objects. The rapid establishment of a relatively homogeneous turbulence field indicated by the numerical solution (of the order of 2-2.5 grid widths i.e. 8-10 mm downstream) is confirmed by the experimental schlieren photographs obtained with shock flow. The influence of this induced turbulent flowfield in the initially quiescent case can be seen in Fig. 3 curve (b), which is the experimental profile obtained without the grid in place. The initial pressure growth is very similar in both cases up to the point where the flame reaches the grid, which gives rise to rapid change increase in the pressure as it enters the turbulence field.

For the tests with initially flowing gases, the experimental results show the expected increased rates of combustion with increasing mixture reactivity. The influence of the turbulence field can be seen in Fig 12. Here the pressure records from initially quiescent and initially flowing gases are compared. In both cases there is a distinct increase in rate of pressure rise as the flame enters the turbulent region downstream of the grid. The numerical simulation indicates a mean flow velocity of some 40 ms^{-1} at the grid for the initially quiescent case, compared to 180 ms^{-1} in the flowing case. Due to difficulties in obtaining steady shocks at a lower Mach number and the lack of turbulence data for the flowing case, it is not possible at present to make generalized correlations of peak pressure and rates of pressure rise with reactivity and mean flow and turbulence levels.

The pressure-time records for the flowing case are qualitatively in agreement with the experiments, although it is necessary to use a lower value of reaction rate constant of proportionality than that normally used (of the order of 15). Otherwise, the calculations led to an overprediction of the peak pressures by up to three times. This immediately indicates that the combustion model predicts too large a reaction rate for the 50% dilution mixture. In additon, the calculations show at least two contradictory aspects that may influence the peak pressures. The first is that turbulent combustion is predicted in the wake of the igniter bar, where laminar combustion is observed experimentally. As a result, a pressure buildup is observed in the

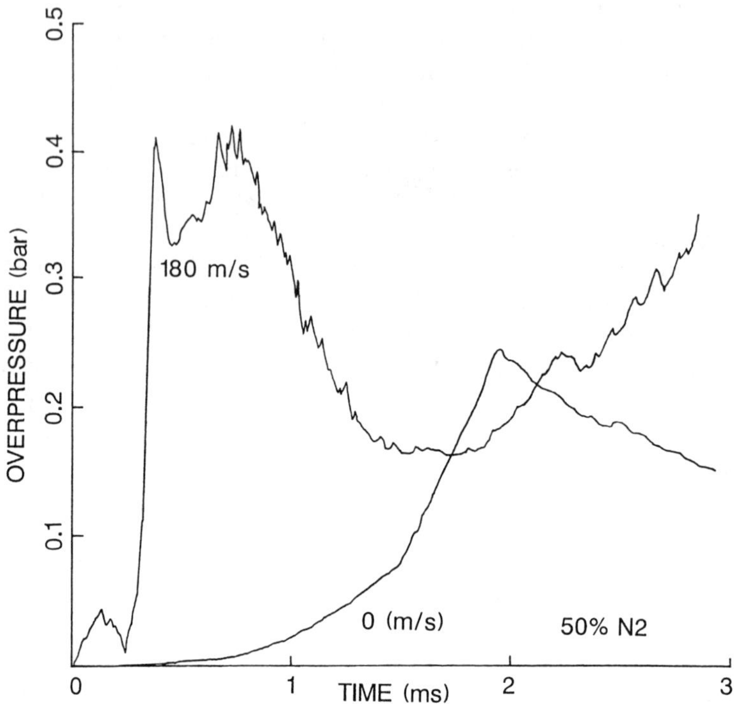

Fig. 12 Comparison of pressure histories for flame propagation through grid for initially quiescent gas and with initial gas velocity of 180 ms^{-1}. Mixture - $C_2H_4 + 3O_2 + 50\%\ N_2$, P_o - 200 Torr.

calculations before the flame encounters the grid. The second aspect is associated with a second flame burning upstream. As the pressure rises because of combustion behind the grid, the flow in front of the grid is reversed. A thin zone of large flow gradients is thus produced, where fuel is transported downstream and products transported upstream. The second peak of the pressure-time records are due to the upstream combustion. By reducing the reaction rate constant, the calculated pressure peaks show better agreement with the experimental records. As the reaction rate is reduced, however, the pressure rise is smoother. Moreover, the pressure peaks produced by the combustion behind the grid are overshadowed by the combustion upstream.

Conclusions

A technique which allows an initial decoupling of the gasdynamics and chemistry in explosion development studies has been demonstrated to be feasible. Such a study, together with detailed three dimensional numerical studies, can provide a greater insight into the mechanisms of flame acceleration in the rapidly varying flow field in the wake of obstacles. Such studies also provide data which are a severe test for numerical models used for predicting such reactive flows.

Acknowledgments

RJB was funded under an SERC/DTI Teaching Company Scheme. Work on gas explosions at Telemark was financially supported by Shell Research Ltd. The authors gratefully acknowledge receipt of a NATO travel award in support of the present work.

References

[1]Moen, I. O., Donato, M., Knystautus, R., and Lee, J. H., "Flame Acceleration Due to Turbulence Produced by Obstacles", *Combustion and Flame* Vol. 39, 1980, pp 21 .

[2]Moen, I. O., Donato, M., Knystautus, R., Lee, J. H. and Wagner, H. Gg., "Turbulent Flame Propagation and Acceleration in the Presence of Obstacles", *Gasdynamics of Detonations and Explosions,* edited by . R. Bowen, N. Manson, A. K. Oppenheim, and R. I. Soloukhin, Vol. 75, Progress in Astronautics and Aeronautics, AIAA, New York, 1981.

[3]Chan. C., Moen, I. O. and Lee, J. H., "Influence of Confinement on Flame Acceleration Due to Repeated Obstacles", *Combustion and Flame* Vol. 49, 1983, pp 27.

[4]Van Wingerden C. M. J. and Zeeuwen, J. P., "Investigation of the Explosion Enhancing Properties of a Pipe Rack Like Obstacle Array", Dynamics of Explosions, edited by. J. R. Bowen, J. C. Leyer, and R. I. Soloukhin, Vol. 106, Progress in Astronautics and Aeronautics, AIAA, New York, 1985.

[5]Taylor, P. "On the Role of Partial Confinement in the Generation of Fast Flames", *Combustion Science Technology,* Vol. 44, 1985, pp 161.

[6]Lee, J. H., Knystautus, R. and Chan, C., "Turbulent Flame Propagation in Obstacle-Filled Tubes", Twentieth Symposium (International) on Combustion, The Combustion Institute, 1984, 1984, pp. 1663.

[7]Harrison, A. J. and Eyre, J. A., "The Effect of Obstacle Arrays on the Combustion of Large Premixed Gas/Air Clouds', *Combustion Science Technology,* Vol. 52, 1987, pp 121.

[8]Hjertager, B. H., Fuhre, K., Parker, S. J. and Bakke, J. R., "Flame Acceleration of Propane-Air in Large-Scale Obstructed Tube", *Dynamics of Shock Waves, Explosions, and Detonations,* edited by R. Bowen, N. Manson, A. K. Oppenheim, and R. I. Soloukhin, Vol. 75, Progress in Astronautics and Aeronautics, AIAA, New York, 1985.

[9]Moen, I. O., Sulmistras, A., Hjertager, B. H., and Bakke, J. R., "Turbulent Flame Propagations and Transition to Detonation in large Fuel-Air Clouds", Twenty-first Symposium (International) on Combustion, The Combustion Institute, 1986, pp. 1617.

[10]Hjertager, B. H., "Simulation of Gas Explosions", *Modeling, Identification and Control*, Vol. 10, 1989, pp. 227.

[11]Hjertager, B. H., "Explosions in Off-shore Modules", paper presented at Intitution of Chemical Engineers Hazards XI Symposium, UMIST, Manchester, England, April 16-18, 1991.

[12]Barr, P. K., "Acceleration of a Flame by Flame-Vortex Interactions", *Combustion and Flame*, Vol. 82, 1990, pp. 111.

[13]Christill, M. and Leuckel, W, 'Experimental Investigations Concerning the Influence of Turbulence on the Flame Front Velocity of Fuel Gas/Air Mixture Deflagrations' Paper presented at 12th Int. Colloquium on Dynamics of Explosions and Reactive Systems, Ann Arbor, Michigan, July 23-28, (1989)

Experimental Study of Large-Scale Unconfined Fuel Spray Detonations

V. I. Alekseev,* S. B. Dorofeev,† V. P. Sidorov,‡ and B. B. Chaivanov§
I. V. Kurchatov Institute of Atomic Energy, Moscow, Russia

Abstract

Detonation evolution in gasoline and kerosene spray clouds in air is investigated. Unconfined clouds are produced by dispersing the liquid through a nozzle. The clouds acquire a semicylindrical shape 1100–1500 m^3 in volume, 3.0–7.5 m in radius, and 15–20 m long. The important dynamic detonation parameters (minimum initiation energy, detonation velocity, detonation pressure, critical radius of direct initiation, and minimum height of the detonable cloud) were determined. Detonation properties are found to be strongly depend on the fuel vapor content. The minimum height of a semicylindrical detonable gasoline-air mixture is found to be about 3 m. In spite of the low detonability of purely heterogeneous mixtures, detonation of a kerosene spray was achieved because of the large scale of these tests. The minimum mass of charge initiating self-sustaining detonation in this mixture proves to be ten times as large as that for a gasoline-air mixture, and the critical cloud height is found to be about 6.5–7.0 m.

Introduction

It is of great interest to study detonation processes in heterogeneous fuel-air mixtures in connection with recurrent accident explosions in industry and fundamental interest in understanding explosion hazards in fuel-air mixtures. Former investigations of liquid fuel spray detonations discussed

---

*Staff Scientist, Induced Chemical Reactions Laboratory, Department of Chemical Physics.

†Head of Laboratory, Induced Chemical Reactions Laboratory, Department of Chemical Physics.

‡Senior Research Scientist, Induced Chemical Reactions Laboratory, Department of Chemical Physics.

§Head of Department, Department of Chemical Physics.

in detail by Dabora, Gelfand et al. and Sichel in their reviews[1–4] cannot give a comprehensive picture of the complex physical processes in such a system. For heterogeneous mixtures as well as for gaseous ones both detonability and detonation propagation conditions are of primary interest. Minimum energy of direct detonation initiation by a powerful source is usually considered to be a measure of mixture detonability. There is a lack of systematical information about these features of heterogeneous mixtures of liquid fuels in literature.

Unconfined detonation is known as most demonstrative in studying detonability of combustible mixtures. As for insensitive fuels such as hexane, decane, kerosene, and others, unconfined detonation has been achieved only in sprays of hexane and only when vaporization of the fuel in a cloud has been allowed to occur.[5,6] To judge from published data, fuels with a low vapor pressure (kerosene, decane, and others) could not be detonated at all. Detonability of heterogeneous mixtures under confined conditions has also been found to depend on the vapor content.[5,7] However, there is information on successful detonation of decane aerosols in oxygen[8] and decane-air mixtures with additions of some nitrocompounds.[6] Experiments on systematical investigation of two-phase fuel-air mixtures detonability, especially in case of low vapor fuels, can be expected to be scaled up in comparison with the tests undertaken earlier. The present paper reports recent results of large scale experiments of unconfined two-phase detonations in mixtures involving examples of fuel systems with both high and low vapor pressure – sprays of gasoline and kerosene, accordingly. Submitted data amplify and generalize former experimental investigations implemented by the authors in this direction.[9,10]

Experimental Details

Dynamics of processes initiated in atmospheric dispersions of motor gasoline A-76 and kerosene TS-1 (analog of Jp-1 and Jet A-1) by high explosive charges has been investigated. Aerosol fuel-air clouds of reproducible properties were generated by means of a disperser (see Fig. 1) consisting of a hermetic cistern with 0.4-m^3 capacity with a siphon and horizontal tubes having adjustable slot nozzles. Burning of a smokeless powder charge in the cistern caused impulsive fuel injection at high pressure through a number of nozzles, jet atomization, and formation of an aerosol cloud. The cloud could take a nearly semicylindrical shape of 8–20 m in length and up to 8 m in radius. Mean fuel concentration in the cloud depends upon both fuel quantity and mass of powder charge. The mean droplet size was estimated by a cloud settling rate and was found to be approximately 50 and 100 μm for gasoline and kerosene clouds, respectively. TNT charges located 0.2–3.3 m above ground were used to initiate blast processes in clouds. Initiation delays (from the very beginning of dispersion) varied from about 0.8 to 4.0 s in the tests.

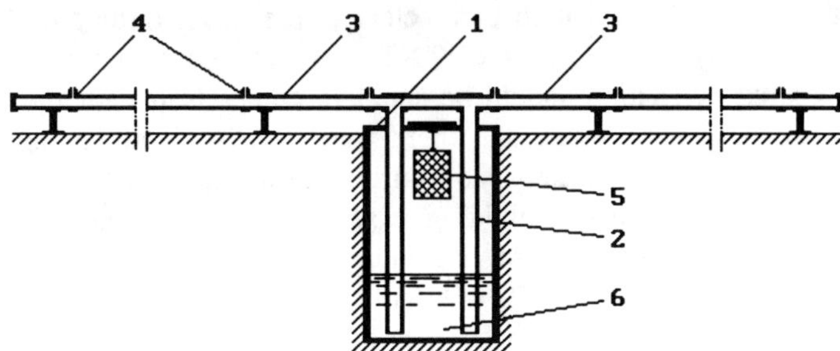

Fig. 1 Scheme of the disperser: 1) hermetic cistern; 2) siphon; 3) horizontal tubes; 4) slot nozzles; 5) powder charge; 6) fuel.

During the experiments, the function of overpressure over time was recorded by piezoelectric transducers located on the ground along the cloud (see Fig. 2). Generation and development of the blast processes were visualized by means of both high-speed shooting (up to 4000 frames/s) and video taping (50 frames/s).

Results and Discussion

Minimum Initiation Energy

The character of blast processes in clouds of fuel-air mixtures depends on initiator mass. In some cases these processes were decayed, the shock came off the reaction zone, and the rest of the cloud was burned down. In other cases when self-sustained detonation was initiated, it had a constant

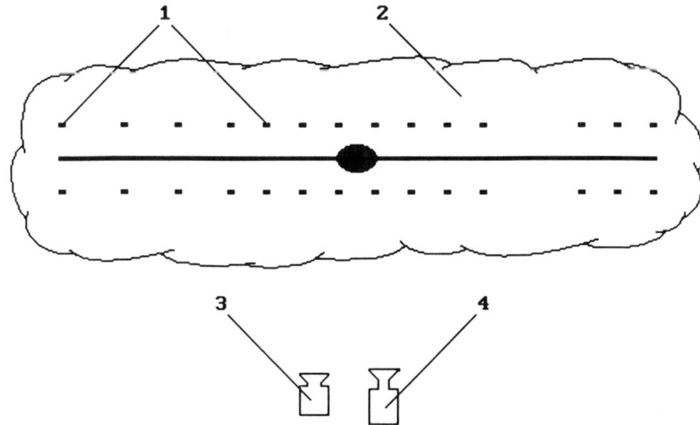

Fig. 2 Top view of the experimental area: 1) piezoelectric transducers; 2) fuel-air cloud; 3) high speed camera; 4) video camera.

propagation velocity and high parameters of the wave. Initiation charges were up to only a 7-kg mass of TNT in the tests so that an area of overpressured detonation should be small in comparison with the whole volume of the cloud. As a result there were exactly identified the failed blast processes characterized by slow burning of the most part of the cloud. Marginal initiation energy dividing cases of sustained detonation and decaying blast wave propagation (minimum energy of detonation initiation E_m) was determined in the series of tests.

The detonability of two-phase heterogeneous mixtures is known to depend strongly upon the fuel vapor content. Vapor phase accumulation may be expected to take place with time in the gasoline-air mixture in contrast with the kerosene-air one. To determine the effect of gasoline vapor quantity in the mixture upon its detonability, the initiation delay was varied from a minimum of 0.8 s to a maximum of 4 s. Proceeding from both the quantity of fuel dispersed and the volume of the cloud formed, one could estimate that mean fuel concentration in clouds ranged from about 0.08 to 0.12 kg/m^3. Figure 3 shows that minimum energy of detonation initiation depends sharply enough on the time of vapor phase accumulation. It could be seen that threshold saturation of the cloud by the gasoline vapor occurred within about 1.7 s. When increasing the initiation delay from this value up to 4 s, the minimum initiation energy is as good as permanent. Greater delays could not be set in the tests, as gravity settling caused considerable change of mixture composition. Because it is found

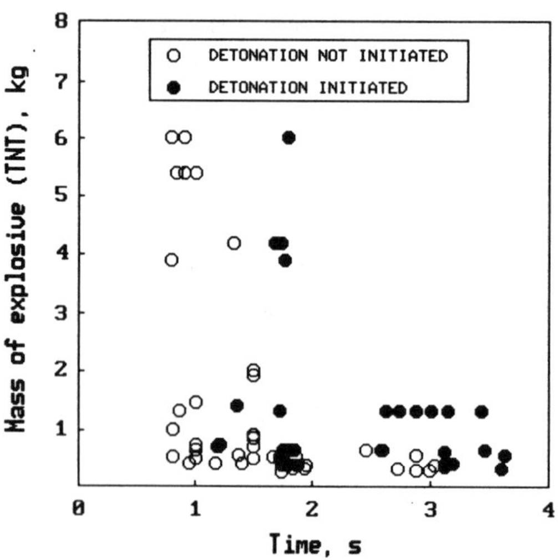

Fig. 3 Experimental values of mass of explosive against the time of vapor accumulation (• - detonation initiated; ○ - detonation not initiated); mean fuel concentration is within the range of 0.08–0.12 kg/m^3.

steady, detonation of the gasoline-air mixture can be readily initiated by the charge of not less than a 0.3-kg mass of TNT. The minimum initiation energy dependence on the average fuel concentration was obtained for the mixtures high in vapor content (see Fig. 4). The vapor accumulation time ranged from 2.4 to 3.4 s in these experiments. The dependence had a typical U-shaped form with the minimum displaced to the mixtures rich in fuel that is known to be characteristic for heterogeneous mixtures. The mixture of about 0.1-kg/m^3 fuel content turned out to be the most sensitive to detonation initiation.

As for the kerosene-air mixture, minimum energy of detonation initiation was found not to depend on the initiation delay. This fact confirms the content of fuel vapor in the mixture to be low enough. The minimum mass of TNT required for detonation initiation in such a mixture turned out to be 4.5 kg. Average fuel concentration in the clouds was about 0.1 kg/m^3. An increase in fuel concentration up to 0.14 kg/m^3 as well as its decrease down to 0.08 kg/m^3 caused the minimum initiator mass to increase to 7 kg.

Explosion Processes Propagation Character

Detonation processes initiated in clouds had nondecaying character. Some essential peculiarities of detonation propagation in the mixtures under discussion were found out by means of pressure records and high-speed shooting. The detonation wave underwent up to four large

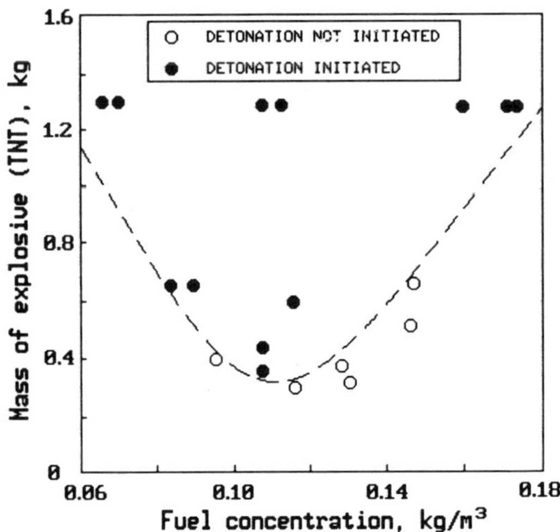

Fig. 4 Experimental values of mass of explosive against mean fuel concentration in the cloud (• - detonation initiated; ○ - detonation not initiated); the vapor accumulation time is within the range of 2.4–3.4 s.

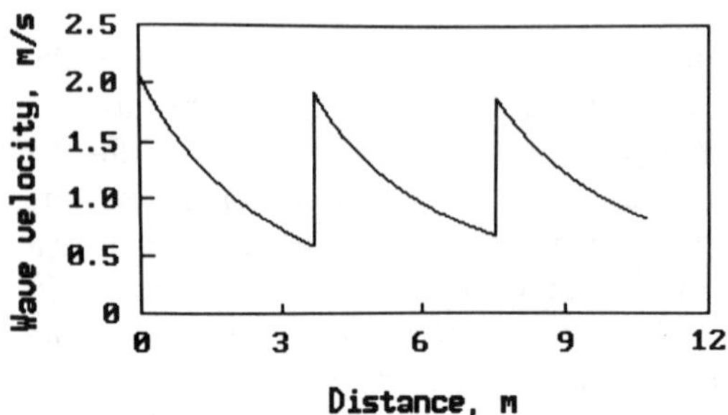

Fig. 5 Oscillations of leading shock wave velocity with distance.

oscillations while propagating along the cloud of kerosene-air mixture (see Fig. 5) with wave velocity drops down to 700–900 m/s (see Fig. 6). The period of the oscillations was about 2–3 s. The average velocity of self-sustained detonation propagation was 1000–1300 m/s. Maximum overpressure in the shock front was changed over a range of (12–35) 10^5 Pa in accordance with wave velocity oscillations. High maximum pressure values (up to 60 10^5 Pa) registered by some transducers could be explained by an essential distortion of the detonation front while reinitiated under wave propagation in a pulsed manner. Thus individual transducers recorded parameters of the wave falling down to ground surface under various angles.

Analogous pulsing detonations were observed in the gasoline-air mixture, when initiation delays were less than 1.8 s. Steady detonations were initiated without any noticeable velocity oscillations; if delays were more than this value, wave parameters turned out to be similar to

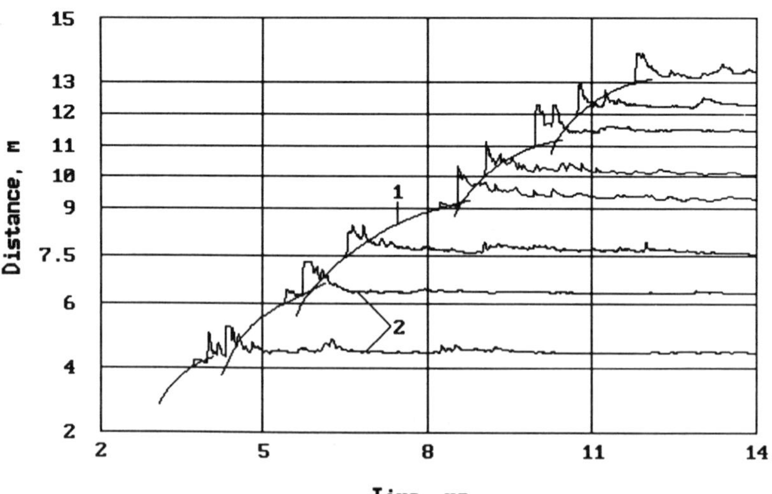

Fig. 6 (*x-t*) diagram of explosion process: 1) shock-wave trajectory; 2) pressure-time history.

thermodynamic ones – the propagation velocity was of 1500–1800 m/s and the overpressure was of $(18–30)\cdot10^5$ Pa.

To identify modes and completeness of explosive transformation of the mixture air blast wave parameters were recorded near a cloud boundary, where the blast wave pressure was about $(7–15)\cdot10^5$ Pa, and in the distant zone (up to 40 m). In the tests with sprays of gasoline (the fuel mass – 90 kg), pressure and impulse TNT equivalents (at the distance of 40 m) of the explosion turned out to be 250–300 kg and 450–550 kg, correspondingly. As for sprays of kerosene (the fuel mass – 80 kg), these values were about 230–270 and 400–450 kg. A wide range of the pointed values is explained mainly by a cloud displacement caused by wind just before initiation. Detonation character of explosions as well as high transformation completeness (they are, however, somewhat less in the case of kerosene sprays in comparison with gasoline ones) are born out by these values of pressure and impulse equivalents, accordingly.[11,12]

Thus, blast processes initiated in sprays of both kerosene and gasoline can be identified as detonations due to such factors as nondecaying propagation character, constant and high average wave velocity, pressure of the shock, and pressure near the cloud border, as well as pressure and impulse in the distant zone. Relatively low average wave velocity in the kerosene-air mixture over the Chapman-Jouguet one and large oscillations may be connected with a near-limit transversal dimension of the cloud.

Critical Dimensions Under Detonation Initiation and Propagation

One of the important features of mixture detonability is the critical radius in the case of direct initiation R_c. The initiating wave velocity has a minimum value at this distance. Stable detonation parameters are known in gaseous detonations to be settled at the distance of two critical radii. The estimation of critical radius value for gasoline-air mixtures could be expected obtainable, if an initiator is located more than 3 m above ground. This value proved to be approximately of about 2m. In the case of kerosene-air mixtures the critical radius of spherical detonation could not be estimated because of the reflected wave influence on the initiation process.

The minimum height (radius) of the detonable semicylindrical cloud r_m was determined thanks to tests with clouds differing in height. This value proved to be of 2.5–3.0 m for gasoline sprays and of 6.5–7.0 m for kerosene sprays. Thus, all of the detonations of kerosene sprays were near-limit, as transversal dimensions of clouds were close to critical (the cloud height was not more than 8 m). The low velocity of detonation propagation in kerosene sprays could only be explained by this factor. In this case the detonation velocity fell by 25% as for gaseous mixtures under critical conditions.[13]

Link Between Critical Parameters

The minimum detonation initiation energy is known to be connected with such representative linear scales as a critical tube diameter (d_c) for transformation of planar detonation in a circular tube to spherically diverging detonation, a critical initiation radius, and a multiheaded detonation cell size. Therefore, comparison of the critical parameters obtained by the empirical way allows the evaluation of their trustworthiness based on conventional presentations. The following semiempirical relations connecting values of E_m and R_c with the detonation cell length b could be used for this purpose[14–16]:

$$R_c = (8-12)b \tag{1}$$

$$E_m = 50\rho_0 D_0^2 b^3 \tag{2}$$

where ρ_0 is the mixture density and D_0 is the steady detonation wave velocity. The values of b for near-stoichiometric gasoline-air mixtures calculated according to Eqs. (1) and (2) turned out to be similar enough, $b = (0.17–0.25)$ m. The value of b for kerosene sprays was found by Eq. (2) to be about 0.5 m, which makes it difficult to define R_c experimentally. The value of E_m was assumed equal to the explosion heat of the critical initiation charge.

The above-mentioned values of b involve effective cell sizes calculated by Eqs. (1) and (2). These values were not determined experimentally; furthermore, they could not be obtained empirically in the case of near-limit detonation. As for unconfined detonations of gaseous mixtures, an increased cell size is observed in comparison with one for detonations in tubes. The cell size value is further increased while approached detonation propagation limit, cellular structure is degenerated and pulsing regime of detonation propagation takes place. By analogy with gaseous detonations, the pulsing regime observed in kerosene sprays can be considered as resulting from the cellular structure degeneration when detonation propagation is near-limit.

The minimum diameter of a detonable gaseous unconfined charge d_f is known to be connected with the value of d_c.[13,18] One can obtain the correlation $d_f = (1.6–3.5)d_c$.[13,17–19] It should be noted that the value of d_f depends upon such factors as mixture composition and diffusion on mixture borders, which results in ambiguity of d_f estimation. Moreover, d_f is not less than $14b$ (or 20λ, where λ is a cell width) and can exceed the pointed value twice as much if mixture composition is changed due to mixture border diffusion. Thus, expected values of r_m ($r_m = d_f/2$) are 1.4–2.8 m and 3.5–7.0 m for gasoline and kerosene sprays, respectively. These values do not run counter to the ones obtained experimentally.

The preceding comparison of the experimental values of E_m, R_c, and r_m points out their conformity to each other within conventional presentations about critical conditions of detonation initiation and propagation. No peculiarities of the link were found between the critical parameters in connection with heterogeneous character of mixtures (within the accuracy of the experimental data).

Summary

The present tests clearly show that insensitive liquid fuels such as gasoline and kerosene can be detonated in aerosol form. Fundamental parameters characterizing detonation initiation conditions and detonation propagation in these mixtures were found (minimum initiation energy, the minimum transversal dimension of detonable unconfined clouds).

The present series of tests confirmed that failure to detonate sprays of kerosene and decane in earlier experiments[6,9,10] has accounted for deficient transversal dimensions of clouds.

Detonation processes initiated in sprays of kerosene were pronounced pulsing. According to the above analysis, such a near-limit regime is expected to be connected with the near-critical transversal dimension of the cloud (6–8 m). The minimum mass of the initiation charge was found to be of 4.5-kg TNT.

Detonation initiation sensitivity of heterogeneous gasoline-air mixtures depends strongly upon the time of fuel vapor phase accumulation; threshold saturation of the cloud by fuel vapor occurs within about 1.7 s. Detonation properties of the gasoline-air mixture turned out to be similar to the properties of the kerosene mixture in the case of short initiation delays. A critical energy of direct initiation for gasoline sprays saturated by the fuel vapor (the minimum initiator mass – 0.3-kg TNT) proved to be close to the corresponding value for gaseous propane-air mixtures. Detonation initiation was experimentally found to be possible over a wide range of gasoline concentration: 0.06–0.17 kg/m^3.

References

[1]Dabora, E. K., and Weinberger, L. P., "Present Status of Detonations in Two Phase Systems," *Acta Astronautica*, Vol. 1, 1974, pp. 361–372.

[2]Borisov, A. A., and Gelfand, B. E., "Review of Papers on Detonation in Two Phase Systems," *Archive of Thermodynamics and Combustion*, Vol. 7, Feb. 1976, pp. 273–287.

[3]Dabora, E. K., "Fundamental Mechanisms of Liquid Spray Detonations," *Proceedings of the International Conference on Fuel-Air Explosions*, Montreal, Canada, Univ. of Waterloo Press, 1982, pp. 245–264.

[4]Sichel, M., "The Detonations of Sprays: Recent Results," *Proceedings of the International Conference on Fuel-Air Explosions*, Univ. of Waterloo Press, Montreal, Canada, 1982, pp. 265–304.

[5]Bull, D. C., McLeod, M. A., and Mizner, G. A., "Detonations of Unconfind Fuel Aerosols," *Gasdynamics of Detonations and Explosions*, edited by J. R. Bowen, N. Manson, A. K. Oppenheim, and R. I. Soloukhin, Vol. 75, Progress in Astronautics and Aeronautics, AIAA, New York, 1981, pp. 48–60.

[6]Benedick, W. B., Knystautas, R., Lee, J. H. S., and Tieszen, S. R., "Detonation of Unconfined Large Scale Fuel Spray-Air Clouds," *Proceedings of the 12th International Colloquium on the Dynamics of Explosions and Reactive Systems*, Univ. of Michigan, Ann Arbor, MI, 1989.

[7]Lu, P. L., Slagg, N., Fishburn, B. D., and Ostrowski, P., "Relation of Chemical and Physical Processes in Two-Phase Detonations," *Acta Astronautica*, Vol. 6, 1979, p. 815.

[8]Bowen, J. R., Ragland, K. W., Steffes, F. J., and Loflin, T. G., "Heterogeneous Detonation Supported by Fuel Fogs or Films," *Proceedings of the 13th International Symposium on Combustion*, The Combustion Institute, Pittsburgh, PA, 1971, p. 1131.

[9]Alekseev, V. I., Dorofeev, S. B., Sidorov, V. P., and Chaivanov, B. B., "Study of Detonation Initiation in Motor Fuels Sprayed in Air," Collected Scientific Works 1988, I. V. Kurchatov Institute of Atomic Energy, Moscow, USSR, 1989, pp. 90–91.

[10]Alekseev, V. I., Dorofeev, S. B., Sidorov, V. P., and Chaivanov, B. B., "Experimental Study of Detonation Initiation in Motor Fuels Sprayed in Air," Preprint IAE-4872/13, Atominform, Moscow, USSR, 1989.

[11]Borisov, A. A., Gelfand, B. E., Gubin, S. A., Odintsov, V. V., and Shargatov, V. A., "Parameters of Air Blast Wave Under Different Regims of Explosive Transformation of Combustible Gaseous Mixtures," *The Chemical Physics*, USSR, Vol. 5, 1986, pp. 670–679.

[12]Fishburn, B. D., "Some Aspects of Blast from Fuel-Air Explosives," *Acta Astronautica*, Vol. 3, 1976, pp. 1049–1065.

[13]Borisov, A. A., Mikhalkin, V. N., and Homik, S. V., "Detonation of Gaseous Mixtures in Unconfined Cylindrical Charge," *Reports of USSR Academy of Sciences*, USSR, Vol. 296, Jan. 1987, pp. 88–91.

[14]Edwards, D. H., Hooper, G., and Morgan, J. M., "An Experimental Investigation of Spherical Detonations," *Acta Astronautica*, Vol. 3, 1976, p. 117.

[15]Lee, J. H. S., Knystautas, R., and Guirao, C. M., "The Link Between Cell Size, Critical Tube Diameter, Initiation Energy and Detonability Limits," *Proceedings of the International Conference on Fuel-Air Explosions*, Univ. of Waterloo Press, Montreal, Canada, 1982, pp. 157–188.

[16]Lee, J. H. S., "Dynamic Parameters of Gaseous Detonations," *Annual Review of Fluid Mechanics*, Vol. 16, 1984, pp. 311–336.

[17]Vasilyev, A. A., Mitrofanov, V. V., and Topchiyan, M. E., "Detonation Waves in Gases," *The Physics of Combustion and Explosion*, USSR, Vol. 23, May 1987, pp. 109–131.

[18]Vasilyev, A. A., and Subbotin, V. A., *Dynamics of Continuous Medium*, USSR, Vol. 62, 1983, pp. 32–38.

[19]Vasilyev, A. A., and Zuck, D. V., *The Physics of Combustion and Explosion*, USSR, Vol. 22, 1986, pp. 82–88.

Investigation on Blast Waves Transformation to Detonation in Two-Phase Unconfined Clouds

V. I. Alekseev,* S. B. Dorofeev,† V. P. Sidorov,‡ and B. B. Chaivanov§
I. V. Kurchatov Institute of Atomic Energy, Moscow, Russia

Abstract

The evolution of shock waves in a nonuniform combustible mixture was experimentally studied. Unconfined heterogeneous detonations were produced in kerosene-air mixture clouds 1100 m^3 in volume. The galloping mode of detonation propagation was studied in the first series of tests. Detonation reinitiation was found to be due to fast pressure wave acceleration in mixtures with an induction time gradient. In the second series of tests, mixture volumes with a nonuniform induction time distribution were created by the blast from a high explosive charge, with the shock wave decoupling from the reaction zone in the course of its propagation. Evolution of weak shock waves spreading through this volume was studied. The tests showed that the Zeldovich mechanism is essential for the processes of detonation onset and propagation under these conditions.

---

*Staff Scientist, Induced Chemical Reactions Laboratory, Department of Chemical Physics.

†Head of Laboratory, Induced Chemical Reactions Laboratory, Department of Chemical Physics.

‡Senior Research Scientist, Induced Chemical Reactions Laboratory, Department of Chemical Physics.

§Head of Department, Department of Chemical Physics.

Introduction

Detonation initiation processes in fuel-air mixtures are traditionally divided into direct initiation by an impulsive energy source, for example a high explosive (HE) charge, and deflagration-to-detonation transition (DDT). Zeldovich et al. were the first to point out existence of a new detonation initiation mechanism[1] by numerical modeling of the flow in the reactive media with temperature gradient. They showed the opportunity for quick formation of the detonation wave under spontaneous ignition of the nonuniform mixture. Such a mechanism is directly connected to the existence of an induction time gradient in the reactive mixture. Induction time nonuniformity results in the appearance of the chemical reaction spontaneous front and pressure waves formation. Velocity of the spontaneous front depends upon the induction time distribution. Quick amplification of the pressure wave occurs in the case of synchronization between chemical energy release and gas dynamics processes. Unlike direct initiation, detonation is initiated due to amplification of a weak pressure wave without a stage of overpressure detonation.

Theoretical investigations of the new initiation mechanism[2–8] show that explosion processes can be caused by different kinds of nonuniformities in the mixture. Criteria of "coupling" between gasdynamics processes and chemical energy release (criteria of quick amplification of a weak pressure wave) are formulated by Gelfand et al.[8] In essence, they determined the range of mixture parameters distributions, under which onset of the spontaneous explosion process could be expected. Another aspect of the problem was explored by Dorofeev et al.,[9,10] who estimated conditions under which the spontaneous explosion process in the nonuniform mixture leads to self-sustained detonation onset in the undisturbed part of the mixture. These conditions are connected to minimum energy changes of the mixture state in the nonuniform zone, and the authors offered a way to estimate conditions of detonation initiation by the gradient mechanism on the basis of minimum blast initiation energy data.

Lee et al.[11] were the first to reveal the action of the gradient mechanism experimentally. Anomalously quick development of an explosion in extremely sensitive mixtures ($H_2 + Cl_2$, $H_2 + O_2$ with addition of Cl_2) was observed. The process was the result of ultraviolet irradiation creating nonuniform distribution of free radicals concentration in the mixture. It was shown that a shock wave of maximum amplitude can be obtained under some optimum intensity of irradiation. However, development of the explosion process was accompanied by the continuing action of irradiation complicating results interpretation. The widespread name of the pressure waves amplification mechanism, shock wave amplification by coherent energy release (SWACER), was offered by the authors of Ref. 22. In the tests of Knystautas et al.,[12] mixture parameter nonuniformity was created by a jet of combustion products. It was found that spontaneous ignition in such a system leads to quick formation of the detonation wave.

Thus there is conclusive experimental evidence of the existence of "new" detonation initiation mechanism by means of pressure wave formation and its quick amplification in nonuniform mixtures. However, these experimental results are obtained in the extremely sensitive mixtures under conditions of confined volume, and their interpretation is not clear for all the cases.

Analyzing experimental results of both DDT and direct initiation, Lee and Moen[13] assumed that necessary conditions for the detonation initiation by gradient mechanism are created during the last stage of DDT (local explosion) or between the reaction zone and the leading shock in case of the direct initiation. Confirmation of this assumption would create some basis for universal description of different processes of detonation development.

Spatial distribution of the induction time can be created, for example, by a blast wave propagating through the mixture. Spontaneous reaction front is formed behind the leading shock if wave intensity is high enough. Evolution of such a system depends upon blast wave width and intensity, mixture composition, and some other factors. The self-sustained detonation wave in the mixture is known as a complex consisting of the shock and the spontaneous front of the chemical reaction initiated by the shock, with average velocity of the complex remaining constant. Another characteristic regime in such a system is the decaying explosion process. In this case the shock comes from the reaction zone, and its intensity decreases and becomes insufficient to sustain formation of the spontaneous reaction front.

The variety of induction time distributions in the mixture as a result of blast wave propagation allows us to expect other modes of explosion processes to take place. The possibility of pressure wave formation and quick amplification in the nonuniform mixture within a certain interval of spontaneous reaction front velocity (or ignition delay gradient) excites particular interest. The present paper reports recent results of experimental investigations on nonstationary explosion processes. Experiments have been carried out in mixtures with induction time distributions formed by decaying blast waves propagating through the combustible mixture.

Experimental Details

The main part of the tests were carried out in unconfined clouds of a heterogeneous kerosene-air mixture characterized by low detonability. Kerosene TS-1 (analog of Jp-1 and Jet A-1) was used. Some tests (with two charges inside the cloud) were carried out using gasoline A-76 as fuel. Clouds of the mixture were generated by a special facility[14,15] and had a semicylindrical shape of 15–17 m in length and 6–8 m in radius. Their volume was about 1100 m^3. The mean fuel concentration in clouds was close to stoichiometric.

During the experimental processes, overpressure was recorded by piezoelectric transducers located on the ground along the cloud. Generation and development of blast processes were visualized by means of both high-speed shooting (up to 4000 frames/s) and video recording (50 frames/s). Because of low sensitivity of the test mixture and, consequently, large-scale detonation processes, it became possible to register blast wave evolution in detail by pressure transducers spaced 1--2 m apart.

Results and Discussion

In the first series of tests, detonation propagation in the cloud was studied under critical conditions; initiation charge mass and transversal dimensions of the cloud were close to critical. To initiate explosion processes in clouds, HE charges were used. A charge was usually located near the border of the cloud. The minimum TNT charge mass required for initiation of a self-sustained detonation process in the mixture was found to be about 4.5 kg. It was shown that clouds of 6–8 m in height were detonable. The resulting explosion processes had strongly pronounced pulsing characters.

Figure 1 shows a characteristic fragment of process history. Oscillations of the detonation wave recurred in 2–3 ms, whereas the detonation complex disintegrated during every oscillation with decreasing leading shock velocity [down to 700–900 m/s (shock trajectory is shown by the solid curve)]. The trajectory of the chemical reaction spontaneous front approximately shown by the dashed curve is specified by both consecutive

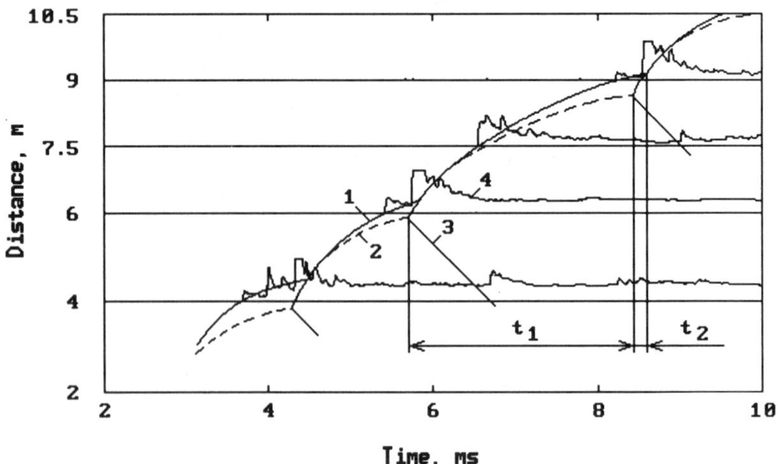

Fig. 1 Typical part of (*x*-*t*) diagram of pulsing detonation. 1) trajectory of leading shock wave; 2) approximate trajectory of chemical reaction front; 3) trajectory of retonation wave; 4) piezoelectric transducer recordings.

ignition of shock-pressed mixture volumes and flow behind the wave front. After delay time t_1, sharp acceleration of the chemical reaction front is observed and a pressure wave is formed and quickly amplified. That causes the explosion process to be reinitiated, and wave velocity increases up to a value considerably exceeding C-J one. Also, a retonation wave propagates in the opposite direction. Duration of every oscillation consists of an interval t_1 of mixture preparation by the leading shock front and explosion wave development time t_2. The second explosion wave is formed in the nonuniform mixture and overtakes the leading shock.

Decrease of the initiator mass down to the critical value results in a more expressive process (see Fig. 2). The decaying blast wave comes from the reaction zone at a distance of 3–4 m. When the mixture preparation time (about 10 ms) is up, sharp acceleration of the chemical reaction front is observed and a pressure wave is formed and amplified in a zone of nonuniformity. An essential peculiarity of this process consists of the fact that the leading shock intensity is low enough (0.1–0.2 MPa) by the moment of reaction front acceleration; that is, an initial stage of wave acceleration occurs practically in the nonpressed mixture. Thus, pressure wave acceleration is defined mainly by properties of the mixture with nonuniform distribution of induction time.

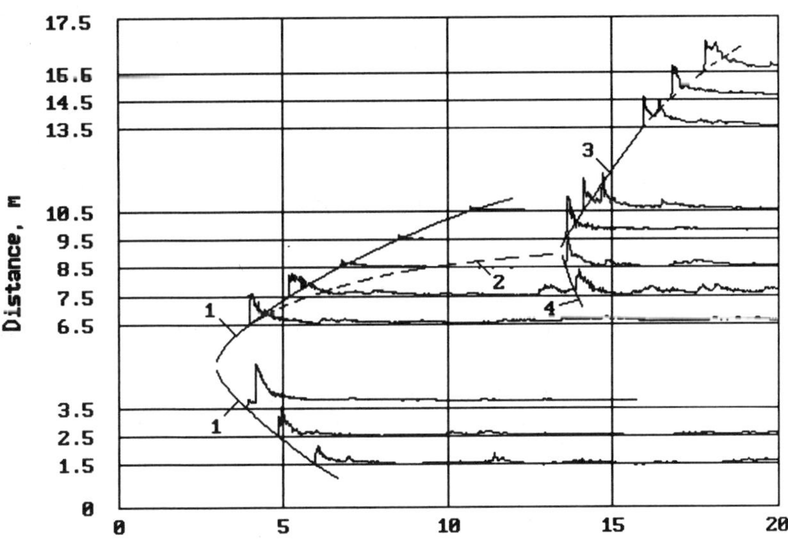

Fig. 2 (x-t) diagram of critical initiation mode. 1) trajectory of initiating shock wave; 2) approximate trajectory of chemical reaction front; 3) detonation wave trajectory (solid line) and air blast wave outside the cloud (dashed line); 4) retonation wave trajectory.

What is the real reason of reaction front acceleration after the expiration of a certain delay time (2–10 ms depending on the leading wave intensity)? Submitted tests show that sharp change of flow character can be explained only by altering the mixture condition between the shock and the reaction zone. In essence, distribution of ignition delay time behind the blast wave front is constantly varied while the wave propagates. Pressure wave formation and quick amplification become possible due to this distribution, and they occur by Zeldovich's mechanism, when the mentioned delay time is up. Perhaps saturation of fuel vapor in the cloud plays an important role in formation of the necessary delay time distribution.

It is necessary to note that the words "sharp acceleration of the chemical reaction front" must not be taken literally. Perhaps ignition delay time distribution between the shock and the reaction zone is not monotonous over mixture nonuniformity and interaction between weak pressure waves. In this case local spontaneous ignition of the mixture volume before the chemical transformation front could occur. Furthermore, such a situation is conceivable when the leading wave cannot sustain spontaneous reaction, and reaction propagation could be expected to result from turbulent combustion. In this case, turbulent mixing of reaction products and reagents might be the cause of spontaneous ignition (turbulent combustion regime is also depended on the blast wave history). A probable strong curvature of the reaction front and the fact that the process is essentially three-dimensional [(x-t) diagrams involve one-dimensional "projections" of the process] must be also taken into account. All of these factors could explain the appearance of retonation waves (Figs. 1 and 2). Results of high-speed shooting show that the starting "point" of second explosion processes development always lies near the front of reaction. The possibility of local spontaneous ignition also cannot be excluded. This notice does not alter the above interpretation of the experimental results. An important question remains to be seen – a role of turbulent flame possibly formed before reinitiation.

The next series of experiments has been carried out to study the evolution of weak blast waves from an outer source in nonuniform mixtures. In these tests, mixture volume with nonuniform induction time distribution was formed as a result of explosion process disintegration. The mass of HE charges using for blast wave generation was less than 40% of critical one. The blast wave of low intensity (0.2–0.3 MPa) was propagated through this zone (the secondary charge was located out of the cloud). The time of blast wave coming into the zone of non-uniform induction time distribution was varied in the tests.

It was found that explosion process development strongly depends on the time the wave enters this zone. If the wave comes into the zone of nonuniformity 2–4 ms after the first HE charge blasting, outer blast wave quick amplification followed by self-sustained detonation takes place. Corresponding (x-t) diagrams plotted in conformity with data obtained

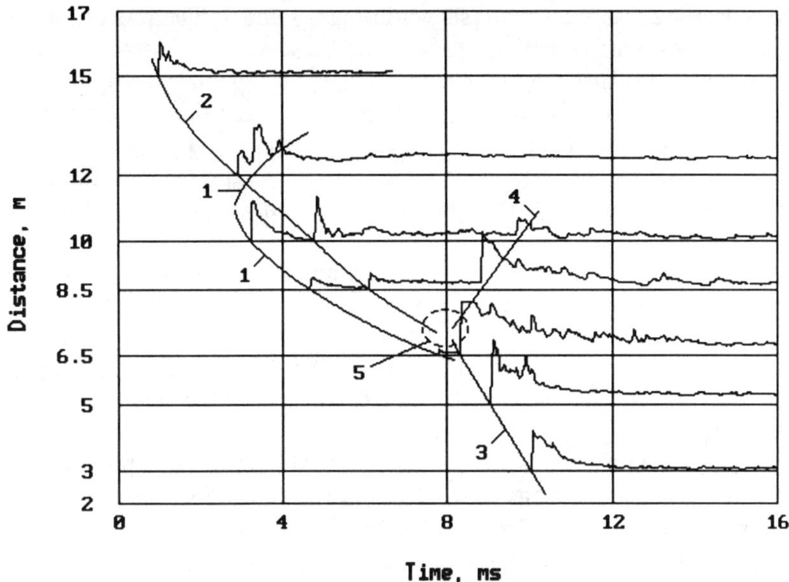

Fig. 3 (*x-t*) diagram of detonation initiation under the action of secondary wave. 1) trajectory of primary shock wave: 2) trajectory of outside shock wave passed through the cloud; 3) detonation wave trajectory; 4) retonation wave trajectory; 5) detonation formation field.

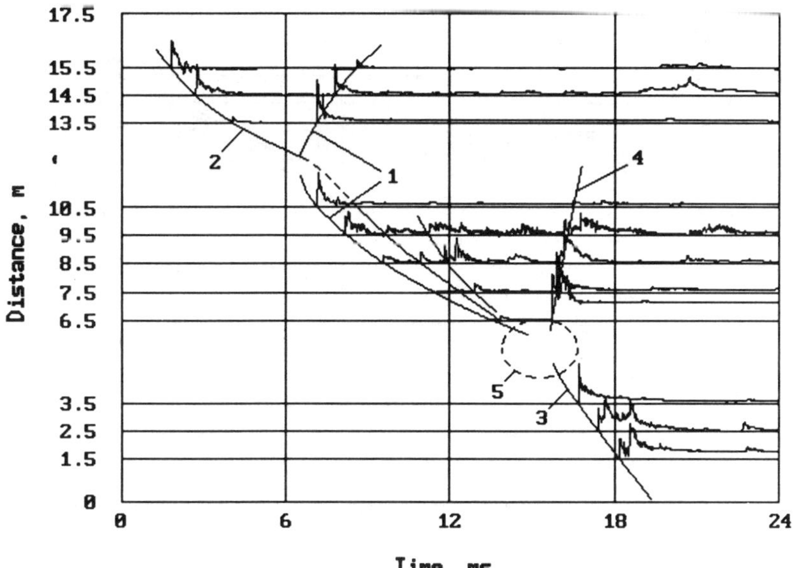

Fig. 4 (*x-t*) diagram of detonation initiation under the action of secondary wave: legend is the same as in Fig. 3.

from pressure transducers are shown in Figs. 3 and 4. The process shown in Fig. 4 has more complicated character. Blast wave 2 caused a series of local explosions that may be connected with the spontaneous ignition of mixture volumes behind the shock. However, because the mixture volume was located between 6.5 and 9.5 m, pressure waves could not be amplified. Formation of the detonation complex (caused by pressure disturbances quick amplification) and retonation wave appearance are observed in field 5.

If the outer wave does not come within the pointed time interval, decaying wave propagation through the mixture is observed (see Fig. 5). This interval of 2–4 ms was determined by high-speed shooting because large enough distances between pressure transducers and their location on the ground did not permit definition of an exact time and location of detonation onset.

Experimental results indicate development of the explosion process to be essentially non-one-dimensional. Blast wave amplification was started near the intersection of the reflected wave 1 (Fig. 6a) from the first charge and wave 2 from the second charge. Accelerating part 3 of wave 2 (Fig. 6b) is notable for bright luminescence peculiar to a detonation wave. Quick amplification of the pressure wave occurs within a comparatively small volume (field 5 in Figs. 3 and 4) of 1-m characteristic dimension and detonation complex 3 (Fig. 6c) is formed and propagates over the rest of the cloud volume.

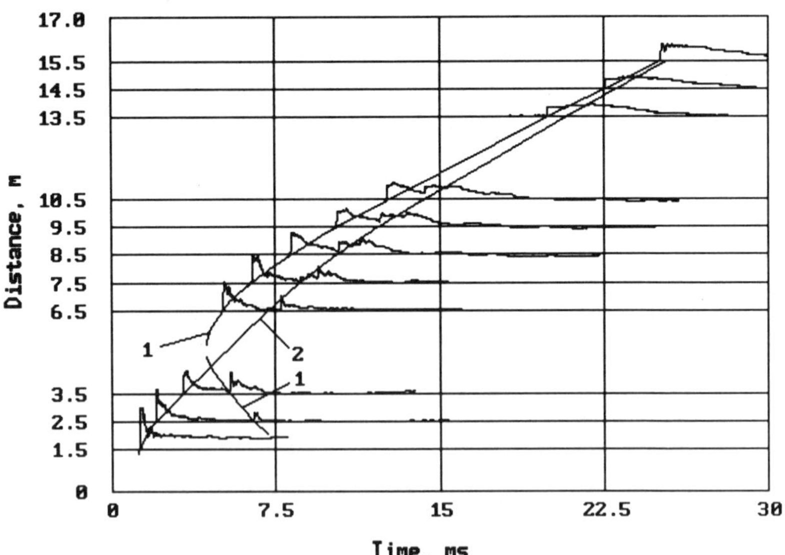

Fig. 5 (*x*-*t*) diagram of decaying explosion process: 1) trajectory of primary shock wave; 2) trajectory of outside shock wave passed through the cloud.

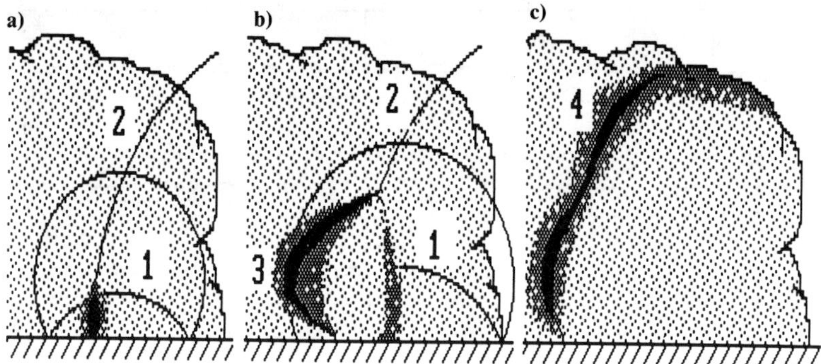

Fig. 6 Evolution of the shock wave passed through the nonuniform mixture (obtained by high speed shooting): 1) shock wave front resulting in nonuniformity formation; 2) front of the low intensive shock wave (ΔP = 0.2–0.3 MPa) passed through nonuniform mixture field; 3) accelerating part of the shock wave; 4) detonation wave.

An inherently similar effect has been observed in the third series of tests when critical conditions of detonation initiation by two HE charges are studied. The tests of this series were run with the gasoline-air mixture. Total critical initiation mass M of two charges depending on the distance Δ between them was investigated. The value of M is equal to the critical initiator mass of 0.3 kg of TNT if $\Delta = 0$ and must be twice as much in cases of large enough distances Δ , when the mutual influence of the charges is not essential. Results of experiments presented in Fig. 7 demonstrate nonmonotonous dependence of total critical mass on the distance between charges. There exists an optimum distance, providing the minimum total

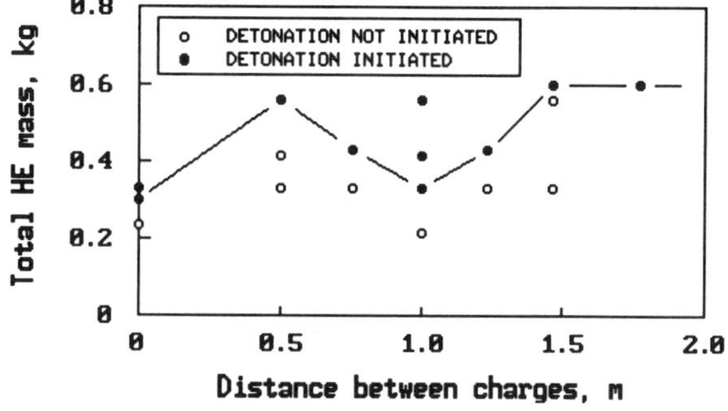

Fig. 7 Experimental values of total mass of explosive against the distance between charges.

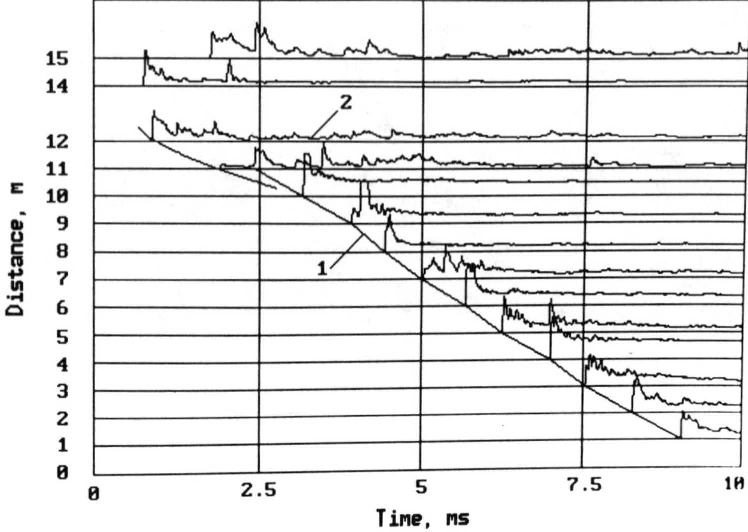

Fig. 8 (x-t) diagram of detonation initiation under the two charges blasting: 1) shock trajectory; 2) pressure-time history.

critical mass (which is also equal to the critical initiator mass for the gasoline-air mixture).

Detonation onset under initiation by two HE charges of mass close to critical one occurs due to interaction of decaying blast waves generated by these charges. The blast wave from one of the initiators propagates through the mixture preconditioned by the wave from another initiator. Quick formation of the detonation complex followed by self-sustained detonation in the rest of the cloud volume is observed just as in the previous series of tests. Detonation onset takes place at a distance of 2–3 m from the place of initiators blasting. The typical (x-t) diagram of the process is shown in Fig. 8. Results of one-dimensional numerical modeling of the similar problem[9,10] can serve as an additional confirmation of the above experiments interpretation.

Conclusion

The process of secondary wave amplification followed by formation of the detonation wave does not essentially differ from the processes of reinitiation under pulsing detonation propagation, direct initiation by the critical charge, and detonation onset due to two-charge initiation. In all of the described processes, disintegration of the shock-reaction zone complex results in formation of the zone of nonuniform induction time distribution. In this zone quick amplification of pressure waves can be possible under certain conditions. In spite of the remaining question about a role of possible turbulent flame formation, this fact allows for a conclusion that the above-considered phenomena are based on the general physical mechanism first mentioned by Zeldovich and experimentally revealed by Lee (the SWACER mechanism). The key significance of the mechanism was shown in simple systems without any action of additional factors (ultraviolet irradiation, active chemical additions, and others), while nonuniformity was caused by shock-wave propagation in unconfined clouds of the combustible mixture low in sensitivity.

Acknowledgment

The authors would like to thank J. H. S. Lee for the interesting and useful discussion.

References

[1]Zeldovich, J. B., Librovich, V. B., Makhviladze, G. M., and Sivashinskiy, G. I., "On Detonation Onset in the Nonuniformly Heated Gas," *Journal of the Applied Mechanics and Technical Physics*, USSR, Feb. 1970, pp. 76–84.

[2]Zeldovich, J. B., Gelfand, B. E., Tsyganov, S. A., Frolov, S. M., and Polenov, A. N., "Concentration and Temperature Nonuniformities (CTN) of Combustible Mixtures as a Reason of Pressure Waves Generation," *Proceedings of the 11th International Colloquium on Dynamics of Explosions and Reactive Systems*, Univ. of Technology, Warsaw, Poland, 1988, p. 89.

[3]Gelfand, B. E., Frolov, S. M., and Tsyganov, S. A., "Spontaneous Generation of Shock and Detonation Waves Under Reactive Medium Expansion," *Fundamental Problems of the Shock Wave Physics*, Vol. 1, Part 1, Chernogolovka, USSR, 1987, p. 144.

[4]Gelfand, B. T., Polenov, A. N., Frolov, S. M., and Tsyganov, S. A., "On Detonation Onset in the Nonuniformly Heated Gaseous Mixture," *The Physics of Combustion and Explosion*, USSR, Vol. 21, July-Aug. 1985, pp. 118–123.

[5]Gelfand, B. E., Frolov, S. M., Polenov, A. N., and Tsyganov, S. A., "Detonation Onset in Systems with Nonuniform Temperature and Concentration Distribution," *The Chemical Physics*, USSR, Vol. 5, Sept. 1986, pp. 1277–1284.

[6]Gelfand, B. E., Frolov, S. M., and Tsyganov, S. A., "On Mechanism of Explosions in Gas-Pumping Units of Main Pipelines," *The Physics of Combustion and Explosion*, USSR, Vol. 24, May-June 1989, pp. 101–103.

[7]Makhviladze, G. M., and Rogatykh, D. I., "Initial Temperature and Concentration Nonuniformities as a Reason of Explosion Character of Chemical Reaction in a Combustible Gas," Preprint No. 321, Institute of Mechanics Problems of USSR Academy of Sciences, Moscow, USSR, 1988.

[8]Gelfand, B. E., Makhviladze, G. M., Rogatykh, D. I., and Frolov, S. M., "Criteria of Spontaneous Onset of Explosion Regimes of a Reaction on Nonuniformities of Self-Ignition Induction Time Distribution," Preprint No. 424,

Institute of Mechanics Problems of USSR Academy of Sciences, Moscow, USSR, 1989.

[9]Dorofeev, S. B., Kochurko, A. S., and Chaivanov, B. B., "Detonation Onset Conditions in Spatially Nonuniform Combustible Mixtures," Preprint IAE-4871/13, Atominform, Moscow, USSR, 1989.

[10]Dorofeev, S. B., Kochurko, A. S., and Chaivanov, B. B., "Detonation Onset Conditions in Spatially Nonuniform Combustible Mixtures," *Proceedings of the 6th International Symposium on Loss Prevention*, Vol. 4, Oslo, Norway, 1989, pp. 22-1 – 22-19.

[11]Lee, J. H. S., Knystautas, R., and Yoshikawa, N., "Photochemical Initiation of Gaseous Detonation," *Acta Astronautica*, Vol. 5, Nov.–Dec. 1978, pp. 971–982.

[12]Knystautas, R., Lee, J. H. S., Moen, I. O., and Wagner, H. G., "Direct Initiation of Spherical Detonation by a Hot Turbulent Gas Jet," *Proceedings of the 17th International Symposium on Combustion*, The Combustion Institute, Pittsburgh, PA, 1979, pp. 1235–1245.

[13]Lee, J. H. S., and Moen, I. O., "The Mechanism of Transition from Deflagration to Detonation in Vapor Cloud Explosions," *Progress in Energy and Combustion Science*, Vol. 16, Pergamon Press, Great Britain, 1980, pp. 359–389.

[14]Alekseev, V. I., Dorofeev, S. B., Sidorov, V. P., and Chaivanov, B. B., "Study of Detonation Initiation in Motor Fuels Sprayed in Air," Collected Scientific Works 1988, I. V. Kurchatov Institute of Atomic Energy, Moscow, USSR, 1989, pp. 90–91.

[15]Alekseev, V. I., Dorofeev, S. B., Sidorov, V. P., and Chaivanov, B. B., "Experimental Study of Detonation Initiation in Motor Fuels Sprayed in Air," Preprint IAE-4872/13, Atominform, Moscow, USSR, 1989.

Dynamics of Gas Explosions in Vented Vessels: Review and Progress

Vladimir Molkov,* Anatoly Baratov,† and Alexander Korolchenko‡
All-Russia Scientific Research Institute for Fire Protection, Balashikha-6, Moscow, Russia

Abstract

The model and the system of dimensionless differential equations of the dynamics of a turbulent gas deflagration in a venting vessel are described. A comparison with models and a study of the results of other authors is made. The two most important parameters of the model are turbulization factor χ and venting parameter W (similarity group)

$$W \propto \frac{\mu F}{V^{2/3}} \frac{C_{ul}}{S_{ui}}$$

The dependencies of turbulization factor χ on such parameters as 1) a vessel volume and its shape, 2) a relative size of a relief vent $(F/V^{2/3})$ and an initial cover availability, 3) a maximum permissible pressure in a vessel and on the conditions of efflux and cover bursting, and 4) availability and type of obstacles in the vessel, are defined by the inverse problem method (results of 30 experiments in different volume vessels, including experiments with gas efflux through venting duct to receiving vessel are given). Turbulization factor χ design formula is suggested. It is emphasized that the "cube root law" doesn't work and must be modified at least to form $(dp/dt)\cdot V^{1/3}/\chi$=const. Approximate algebraic equations for safe vent area design (or maximum explosion pressure in an unclosed vessel design for the inverse problem) are given. The Le Chatelier Brown principle is discovered

---

*Department Head, Fire and Explosion Hazard of Substances.
†Professor, Occupational Safety Subfaculty of Moscow Civil Engineering Institute.
‡Deputy Chief, Russian Scientific Testing Centre of Fire Safety.

and described, according to which gas combustion dynamics in an unclosed vessel responds to external changes in process conditions in such a way as to weaken the effect of external influence.

Nomenclature

a = vessel radius, m
A = venting area part occupied by combustion products at a given moment
C = speed of sound, m·s^{-1}
d = relief vent diameter, m
E_i = combustion products expansion coefficient at initial conditions
f = flame front area, m^2
F = venting area, m^2
K = coefficient taking into account mixture burnout in a receiving vessel
L_d = relief duct length, m
m = mass, kg
m_o = temperature factor
M = molecular mass, kg·kmol^{-1}
n = relative mass, $n=m/m_i$
n_o = baric factor
p = pressure, Pa
r_b = flame radius, m
r = dimensionless flame radius, $r=r_b/a$
$R^{\#}$ = parameter, characterizing efflux
R^Y = parameter, characterizing efflux in the Yao model
S_u = burning velocity, m·s^{-1}
t = time, s
T = temperature, K
V = vessel volume, m^3
W = venting parameter, $W = \dfrac{1}{(36\pi_o)^{1/3}\sqrt{\gamma_u}} \dfrac{\mu F}{V^{2/3}} \dfrac{C_{ui}}{S_{ui}}$
$W^Y = 3W(8/E_i^7)^{1/6}$ = venting parameter in the Yao model
$W^C = 3W[2\gamma/(\gamma-1)]^{1/2}$ = venting parameter in the Crescitelli model
γ = adiabatic factor
ε = thermokinetic factor, $\varepsilon = m_o + n_o - m_o/\gamma_u$
μ = discharge coefficient
ξ = coefficient taking into account heat mass exchange in a receiving vessel
π_o = pi number
π = dimensionless pressure, $\pi = p/p_i$
$\Delta\pi$ = dimensionless overpressure, $\Delta\pi = \pi - 1$
ρ = density, kg·m^{-3}
σ = dimensionless density, $\sigma = \rho/\rho_i$
τ = dimensionless time, $\tau = t \cdot S_{ui}/a$

$\tau^Y =$ dimensionless time in the Yao model, $\tau^Y = E_i^{2/3}\tau$
$v =$ gas velocity
$\chi =$ turbulization factor (ratio of areas)
"χ"= ratio of velocities

Subscripts
$a =$ space into which the mass of gas vented
$b =$ burnt gas
$c =$ critical state
$e =$ maximum closed vessel value
$i =$ initial state
$m =$ maximum value
$r =$ receiving vessel
$s =$ spherical flame front
$t =$ turbulent flame front or burning velocity
$u =$ unburnt gas
$v =$ venting
$2 =$ second peak pressure rise at the calculated pressure-time relation

Introduction

The problem of venting of gas explosions has been a matter of close study since the beginning of last century and was originally bound up with accidents in coal mines[1]. According to statistical data analysis found in *Industrial Risk Insurers,*[2] the share of explosions in chemical and oil process industries is 67% of the total accidents, and their damage constitutes 85% of the total. About 40% of the total explosions are the result of deflagration burning inside process equipment and buildings.

During the last decades the studies performed by Mandey,[3] Streltchuck and Ivaschenko,[4] Yao,[5] Pasman,[6] Sapko,[7] Anthony,[8] Bradley and Mitcheson,[1] Korotkich and Baratov,[9] Crescitelli et al.,[10] Molkov and Nekrasov,[11] Molkov et al.,[12] Baratov et al.,[13] and others involved the problem of gas combustion dynamics modeling in unclosed vessels and that of safe venting area design of process equipment and industrial rooms.

One of the most in-depth investigations in this field was conducted by Bradley and Mitcheson,[1] where results of 54 tests were generalized. To determine an explosion overpressure the investigators proposed graphic relations $\Delta p_m = f(\overline{A}/\overline{S}_0)$. However, many aspects of the problem (in particular, the burning in adjacent vessels) have not been thoroughly studied until now, though there were interesting works carried out by Heinrich[14] more than 20 years ago.

Flame propagation along a homogeneous gaseous mixture in a spherical vessel away from the central ignition point is considered according to the mathematical model adopted here. From the ignition moment, $t = 0$, until the moment of venting, t_v, gas burning occurs in a closed vessel. The model and relations proposed by Babkin et al.[15] are valid for this period. At the moment t_v an instantaneous venting of the vessel takes place accompanied by the flow through a relief vent into the space with constant pressure p_a or along an exhaust duct into a reservoir with initial pressure p_{ir}. We neglect only a minor unburnt gas efflux period just after the vent rupture, because with the same pressure differential, the higher the gas temperature the higher the gas flow velocity ($v \propto T^{1/2}$). Either burnt gas or their unburnt gas mixtures discharge depending on both the ignition source location relative to the vent and the venting pressure. As a result of vent rupture and gas dynamic processes the flame front can be deformed. The "surface" turbulent combustion model is used by modeling. For its convenience it is assumed that after venting the flame front retains a spherical shape but its burning velocity becomes $S_t = (f_t/f_s) \cdot S_u = \chi \cdot S_u$.

One of the main goals of the present investigation is to determine multiparametric dependence of turbulization factor χ on burning conditions in a vented vessel. As long ago as the 1950s B. Lewis emphasized that finding a method for determining instantaneous flame front surface is very important.

Modeling

The authors of all known papers, devoted to modeling of gas explosion in an unclosed vessel, propose various incomepatible quations. Let us dwell upon those papers in which, to determine turbulization factor χ, the results of calculations according to proposed models are compared with experimental ones. On the basis of mass conservation law and the assumption of adiabatic gas compression (in the approximation $\gamma_u = \gamma_b = \gamma$), Yao[5,8] derived a set of equations for dimensionless rates of change in pressure, burnt gas mass, and unburnt gas mass, respectively:

$$\frac{d\pi}{d\tau^Y} = 3\chi\gamma(E_i - 1)\pi^{1-2/3\gamma} n_b^{2/3} \pi^{2-2/\gamma+no} - \gamma W^Y [(1-A)E_i^{1/2} + AE_i] \pi^{1-1/\gamma} R^Y$$

$$\frac{dn_b}{d\tau^Y} = 3\chi\pi^{1/3\gamma}\, n_u^{2/3}\, \pi^{2-2/\Gamma+no} - AR^Y W^Y$$

$$\frac{dn_u}{d\tau^Y} = -3\chi\pi^{1/3\gamma}\, n_b^{2/3}\, \pi^{2-2/\gamma+no} - (1-A)R^Y W^Y E_i^{1/2}$$

where, for subsonic flow, when $\pi < [(\gamma+1)/2]^{\gamma/(\gamma-1)}$,

$$R^Y = \left[\frac{\gamma}{\gamma-1}(\pi^{1-1/\gamma} - 1)\right]^{1/2}$$

and for sonic flow,

$$R^Y = \left[\frac{\gamma}{2}\left(\frac{2}{\gamma+1}\right)^{(\gamma+1)/(\gamma-1)} \pi^{(\gamma+1)/\gamma}\right]^{1/2}$$

Calculations were carried out assuming $A = n_b/(n_b + n_u)$ and $\mu = 0.6$ (with the exception of the case when one of the sides in the cubic vessel served as a relief cross section taking μ=1). Burning velocity change during explosion development was estimated from the equation $S_u=S_{ui}(T/T_i)^2(p/p_i)^{no}$.

Experiments[5] with 5% propane-air mixtures were performed in a cubic chamber with a 0.76-m^3 volume and in two cylindrical chambers 0.91 m in diameter and 0.91 and 1.82 m in length. The lower explosion pressures with smoothly opening relief cross sections were explained as caused by a lower level of turbulization effect. Experimental relationships $p(t)$ with two characteristic pressure peaks were compared with design ones. When χ=2-4 design maximum explosion pressures become equal to experimental values. The chief disadvantage of the work and the model used in it lies in time discrepancy of experimental and calculated second pressure peaks equalling in value.

To determine turbulization factor χ S. Crescitelli[10] compared maximum explosion pressures obtained experimentally in a 1.7-m^3 vessel by Harris and Briscoe[16] and on a design basis. To calculate a portion of combustion products in the vessel in the mathematical model, one applied the "simplified heat balance equation" $n_b=(\pi-1-n_v+\eta)/(\pi_e-1)$, where $n_v=m_v/m_i$ was a relative venting mixture mass, and a dimensionless variable η took into account, unlike previous investigations, not only an internal venting gas energy, but also gas expansion work inside the vessel. In the case of unburnt gas flow S. Crescitelli[10] propose the following set of equations:

$$\frac{d\pi}{d\tau} = 3\chi(\pi_e-1)\ [1-\frac{\pi_e(1-n_v)-\pi-\eta}{\pi^{1/\gamma}(\pi_e-1)}]^{2/3}\pi^{\varepsilon+1/\gamma} - \frac{dn_v}{d\tau} - \frac{d\eta}{d\tau}$$

$$\frac{dn_v}{d\tau} = W^C\ \pi^{(\gamma+1)/2\gamma}(\frac{\pi_a}{\pi})^{1/2\gamma}\ [1-(\frac{\pi_a}{\pi})^{1-1/\gamma}]^{1/2}$$

$$\frac{d\eta}{d\tau} = (\gamma\pi^{1-1/\gamma}-1)\ \frac{dn_v}{d\tau}$$

If there is sonic flow, i.e. $\pi_a < \pi_c = \pi\,[2/(\gamma+1)\,]^{\gamma/(\gamma-1)}$, then substitution $\pi_a=\pi_c$ is made into the second equation. With flowing burnt gas the second and the third equations are somewhat modified. The comultiplier $(\rho_b/\rho_u)^{1/2}$ appears in the right-hand side of the equation for $\frac{dn_v}{d\tau}$. The third equation takes the form

$$\frac{d\eta}{d\tau} = [\gamma\ \frac{\pi-(1-n_v-n_b)\pi^{\gamma-1)/\gamma}}{n_b} - \pi_e\,]\ \frac{dn_v}{d\tau}$$

Laminar burning velocity change, as in the case with our paper, was described by the the equation $S_u=S_{ui}\,\pi^{\varepsilon}$. Similarly to the investigations of Yao[5] calculations were carried out with turbulization factor χ unchanging during pressure development. Owing to the uncertainty in the discharge coefficient μ Crescitelli et al.[10] determined the modified turbulization factor $\chi'=\chi/\mu$ (this is a feasible substitution in equations of the model with simultaneous introduction of the modified dimensionless time $\tau\cdot\mu$). The authors note correctly that when pressure in the second peak is less than the venting pressure π_v the determination of χ' is impossible in the absence of the experimental dependence $p(t)$. When μ=0.6, determined turbulization factors lie in $\chi=\chi'\cdot\mu$=3-27 range. Though obtained values χ in general agree with real ratio S_t/S_u change range[17], they exceed results of similar investigations of Yao[5] and Pasman et al.[6] and are too high for a relatively small 1.7-m^3 vessel and efflux into the atmosphere. Differences become even greater when μ>0.6. This can be explained, in particular, because the model of Crescitelli et al.[10] is only an "approximation." Though the comparison of the experiment with the design only in a maximum explosion pressure is a matter of interest, it is approximate and does not allow final decisions as to the model perfection to be made.

A set of equations for gas explosion in an unclosed vessel adopted in our work was derived from energy and mass conservation equations, constant vessel volume and ideal gas equations, and also the equation of adiabatic gas compression and extension in a vessel[18]. Hence, the derivation of the set of equations quoted in the paper is the most valid. The set of equations used in the present investigation involves dimensionless differential equations of energy conservation, burnt gas and unburnt gas mass change, respectively, and also the equation of pressure change in a receiving vessel

$$\frac{d\pi}{d\tau} = 3\pi\frac{\chi Z\pi^{\varepsilon+1/\gamma_u}(1-n_u\pi^{-1/\gamma_u})^{2/3} - \gamma_b W[(1-A)R_u^{\#}+AR_b^{\#}(\frac{\pi^{1/\gamma_u}-n_u}{n_b})]}{\pi^{1/\gamma_u} - \frac{\gamma_u-\gamma_b}{\gamma_u}n_u} \quad (1)$$

$$\frac{dn_b}{d\tau} = 3\ [\chi\pi^{\varepsilon+1/\gamma_u}(1-n_u\pi^{-1/\gamma_u})^{2/3} - AR_b^{\#}W] \quad (2)$$

$$\frac{dn_u}{d\tau} = -3\ [\chi\pi^{\varepsilon+1/\gamma_u}(1-n_u\pi-1/\gamma_u)^{2/3}+(1-A)R_u^{\#}W] \quad (3)$$

$$\frac{d\pi_r}{d\tau} = 3\xi\,[K\gamma_u(1-A)R_u^{\#} + \gamma_b AR_b^{\#}(\pi^{1/\gamma_u}-n_u)/n_b]\,\pi^{1-1/\gamma_u}\frac{p}{p_{ri}}\frac{V}{V_r}W \quad (4)$$

where

$$Z = \gamma_b\ [E_i - \frac{\gamma_u}{\gamma_b}\frac{\gamma_b-1}{\gamma_u-1}]\,\pi^{(1-\gamma_u)/\gamma_u} + \frac{\gamma-\gamma_u}{\gamma_u-1}$$

and characterizing efflux parameter for subsonic flow

$$R^{\#} = [\frac{2\gamma}{\gamma-1}\pi\sigma\ [(\frac{p_a}{p_i\pi})^{2/\gamma}-(\frac{p_a}{p_i\pi})^{(\gamma+1)/\gamma}]]^{1/2}$$

and for sonic flow, when $\pi \geq \frac{p_a}{p_i}(\frac{1+\gamma}{2})^{\gamma/(\gamma-1)}$,

$$R^{\#} = [\gamma(\frac{2}{\gamma+1})^{(\gamma+1)/(\gamma-1)}\pi\sigma]^{1/2}$$

Before venting, the flame propagation occurs in a closed vessel: μ=0 and $\chi \approx 1$ (if the original mixture is not turbulized). Because of the lack of data in burning velocity the values S_{ui} and ε (necessary for the relationship $S_u = S_{ui}\,\pi^{\varepsilon}$) can be estimated in terms of the relation $p(t)$ up to the venting moment by the method proposed in the work[11].

To determine a pressure impulse (pressure integral in time) it is necessary to take account of only burnt gas efflux period after the completion of burning. Here one should substitute χ=0, n_u=0, A=1 (as the initial values n_b, π, and τ reached up to the moment of complete gaseous mixture burnout inside the vessel are used) into the explosion dynamics equations.

A simple analysis of explosion dynamics equations for a closed vessel (W=0) shows that the "cube root law" does not work and should be modified into the form $(dp/dt)\cdot V^{1/3}/\chi$=const by turbulization factor χ introduction. At the present knowledge level it is considered inexpedient to use the "law" for computing venting systems of process equipment and industrial rooms.

As follows from dimensionless differential explosion dynamics equations cited above, the design relation $p(t)$ and, hence, maximum explosion pressure, do not depend on dimensional values F, V, S_{ui}, etc. at the constancy of similarity groups W, χ, ε, γ_u, γ_b, E_i, $\pi_v = p_v/p_i$, $\pi_a = p_a/p_i$ or when venting into a receiving vessel ξ, K, V/V_r, p_i/p_{ri} (instead of π_a).

Gas combustion dynamics in an unclosed vessel with known gas properties is defined only by two unknown parameters, namely turbulization factor χ and discharge coefficient μ. The last coefficient is the only unknown in the venting parameter W.

Turbulization Factor Determination

The turbulization factor χ and discharge coefficient μ were defined by the inverse problem method. Optimization of design relations $p(t)$ with experimental ones employed some mechanisms following from analysis of the set of differential equations (1-4) in the mathematical model. Thus, when χ/μ=const maximum explosion pressure does not change, and time for maximum pressure attainment varies in inverse proportion to the turbulization factor χ. At μ=const the rise in χ causes the rise in maximum explosion pressure and the decrease in time of its attainment, as illustrated in Fig. 1. In accordance with the above the uniqueness (within the error of the method) of turbulization factor χ and discharge coefficients μ, determination by the inverse problem method can be deduced.

One investigated the dependence of turbulization factor χ on various parameters and burning conditions, including vessel volume and its shape, venting degree $(F/V^{2/3})$, the presence of obstacles turbulizing the burning (Fig. 2), and vent rupture and efflux conditions. The running trials are given in Table 1. Acetone was basically used as test fuel. Some experiments were conducted using near-stoichiometric air mixtures of propane and hexane, 0.0215-m^3 and 0.027-m^3 vessels with a 1:1 ratio of length to diameter, 2 m^3 (2.6:1 ratio), 10 m^3 (3.4:1 ratio) 11 m^3 (1.45:1 ratio, vent 1.8 m × 0.7 m)

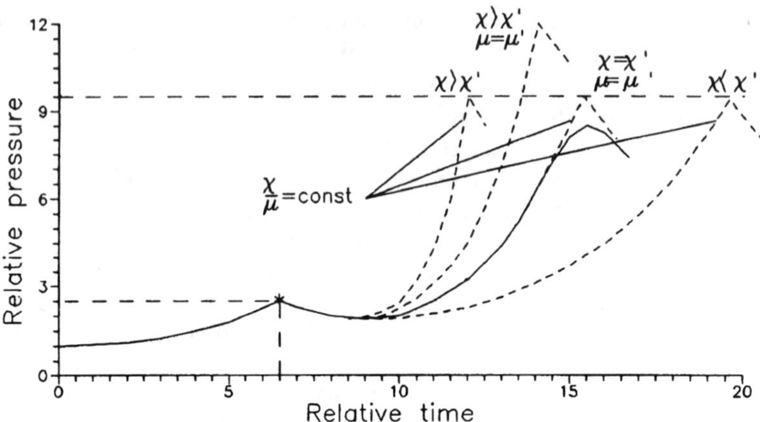

Fig. 1 Influence of turbulization factor χ and discharge coefficient μ on explosion dynamics: * - vent opening; χ′ and μ′ - optimal values; - - - - design; ——— experiment.

were used. Mixture ignition took place in the center of the vessel, and in a number of cases, near the vent or opposite closed end of the vessel. In special test series unburnt mixture and combustion products efflux occurred not into the atmosphere, but through an exhaust pipeline into the vacuumed receiving vessel. In the number of tests water was discharged into the flux to suppress an intensive burning in the pipeline. The initial pressure was p_i = 0.1 MPa in the explosion vessel and p_{ri} = 0.02 MPa in the receiving vessel. Cinegram processing showed that joint unburnt mixture and combustion products flow ($A=r^2$) occurred in the case of central ignition and low-venting pressures. At high-venting pressures ($\pi_v>2$) or ignition source location near the vent, only combustion products (A=1) left the vessel.

Approximate Safe Vent Area Design

Approximate formulas for minimal safe vent area design can be obtained on the base of model equations (1-3) respectively for subsonic and sonic ($2 \leq \pi_m \leq \pi_e$) efflux:

$$W = \frac{\chi(E_i - 1)}{E_i^{1/2}\sqrt{\pi_m - 1}} \tag{5}$$

$$W = 0.9\,\frac{\chi(\pi_e - \pi_m)}{E_i^{1/2}} \tag{6}$$

where π_m = maximum pressure, that a sheath of a vessel can withstand (maximum explosion pressure in case of inverse problem) and π_e = adiabatic explosion pressure in a closed vessel. Relief vent diameter determination

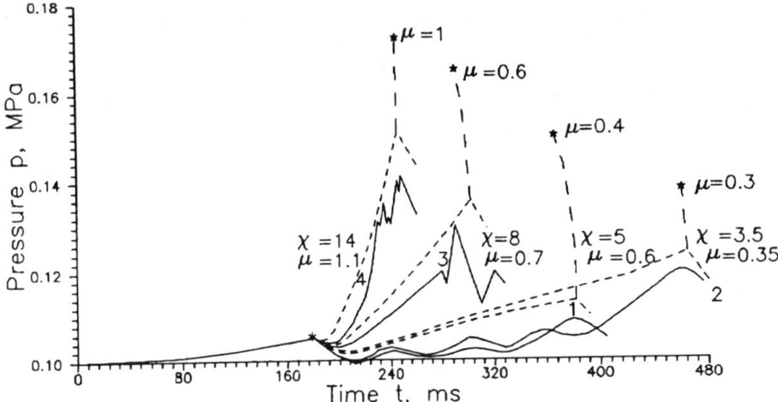

Fig. 2 Influence of obstacles turbulizing the burning on explosion dynamics in 11-m^3 volume vessel: 1 - N27, 2 - N28, 3 - N29 (mesh size 500 mm), 4 - N30 (mesh size 125 mm) in Table 1; * - vent opening; ⋆ - minimal value of μ at design series with χ=const; – – – – second peak position; - - - - - design; ——— experiment.

error when using Eqs. (5) and (6) is near 10% in comparison with exact computer solution of differential explosion dynamics equations.

Similar formulas have been proposed earlier by other authors, but only for small overpressures. Lack of turbulization factor χ in design formulas is a main drawback of previous investigations. If turbulization factor χ is introduced into the formulas, suggested by other authors, and these formulas are transformed to Eq. (5) form, then the discrepancy in comparison with Eq. (5) will be the occurrence of a comultiplier on the right side: E_i (Sapko et al.[7]), $0.27\mu E_i^{1/2}$ (Korotkich and Baratov[9]), 0.5 (Vodyanik[19]). It can be seen that in some cases differences may be significant.

Results of 54 experimental investigations in spherical, cubic, and cylindrical shaped vessels of 10^{-3}-200 m^3 volume (from 1:1 to 4.4:1 ratio of length to diameter) were generalized by Bradley and Mitcheson.[1] Graphic dependencies recommended for determination of explosion overpressure on venting parameter value are given at Fig. 3 for uncovered (1) and covered (2) vent areas.

The transition from Bradley's venting parameter $\overline{A}/\overline{S}_o$ to the venting parameter W in this work can be made according to the formula $W=(\overline{A}/\overline{S}_o)\cdot(E_i-1)/\gamma_u^{1/2}$. Since graphs 1 and 2 (see Fig. 3) were obtained on the basis of large experimental data generalization, they can be regarded today as the most reliable in those cases, when there is no certain concrete recommendations, in particular relating to degree of combustion turbulization. For turbulization factor estimation the design data, obtained using Eqs. (5) and (6) with χ variation, were compared with Bradley's graphic dependencies. Coincidence was reached for small explosion overpressures (Δp_m<0.1 MPa or $\Delta\pi$<1) with χ=2 for uncovered vent areas and with χ=8 for covered ones. For relatively large explosion overpressures $\Delta\pi_m \geq 1$ coinci-

Table 1 Values of χ, μ, and ξ determined by the inverse problem method for various burning and venting conditions

N	V, m^3	d, m	p_v, MPa	V_r, m^3	L_d, m	H_2O g	S_{ui}, sm/s	χ	μ	ξ	p_2, MPa	t_2, ms
1	0.0215	0.05	0.100	-	-	-	$S_{ui}\cdot\chi$=32		0.67	-	0.17	149
2	-"-	-"-	0.120	-	-	-	30	1	0.67	-	0.16	159
3	-"-	-"-	0.123	-	-	-	29.5	1	0.67	-	0.152	167
4	-"-	-"-	0.220	-	-	-	30.5	1.5	0.87	-	0.225	141
5	-"-	-"-	0.240	-	-	-	29.5	1.5	0.92	-	0.20	146
6	-"-	-"-	0.265	-	-	-	29.1	1.45	0.92	-	0.205	143
7	-"-	-"-	0.290	-	-	-	29.5	1.4	0.95	-	0.212	136
8	-"-	-"-	0.300	-	-	-	29.5	1.4	0.96	-	0.224	134
9_o	2.0	0.2	0.115	-	-	-	31.5	1.75	0.92	-	0.40	570
10	-"-	-"-	0.175	-	-	-	31.5	1.6	0.92	-	0.48	374
11	-"-	-"-	0.290	-	-	-	31.5	1.2	0.96	-	0.49	363
12	0.027	0.05	0.123	-	-	-	29	1	0.67	-	0.17	166
13^m	-"-	-"-	0.120	0.05	1.83	-	26	3.7	0.63	0.57	0.60	86
14	-"-	-"-	0.125	-"-	2.35	-	28	3.7	0.77	0.50	0.54	83
15	-"-	-"-	0.125	-"-	2.35	-	28	3.7	0.78	0.40	0.45	123
16	-"-	-"-	0.265	-"-	2.35	-	29	1.7	0.93	0.40	0.29	127
17^m	-"-	-"-	0.242	-"-	1.83	-	29	1.65	0.52	0.54	0.54	113
18	-"-	-"-	0.140	-"-	1.83	25	26	1.15	0.43	0.30	0.29	153
19	-"-	-"-	0.155	-"-	1.83	8	24	1.2	0.48	0.70	0.24	164
20	2.0	0.2	0.115	3.2	4	-	32.5	2.1	0.65	0.30	0.53	350
21	-"-	0.2	-"-	-"-	10	-	33.5	2.2	0.50	0.50	0.62	330
22	-"-	0.38	-"-	-"-	10	-	29.5	3.1	0.42	0.60	0.315	325
23	10.0	0.50	0.110	12.0	25	-	32	4	0.99	0.35	0.51	410
24	-"-	-"-	0.105	-"-	-"-	+	27	2.5	0.71	0.30	0.38	635
25^h	-"-	-"-	-"-	-"-	-"-	+	38	2.5	0.99	0.35	0.407	420
26^h	-"-	-"-	-"-	-"-	-"-	+	30	2.5	0.99	0.35	0.305	585
27^p	11.0	1.8x0.7	0.105	-	-	-	34	5	0.60	-	0.113	382
28^p	-"-	-"-	-"-	-	-	-	34	3.5	0.35	-	0.122	464
$\overline{29}^p_c$	-"-	-"-	-"-	-	-	-	34	8	0.70	-	0.135	300
$\overline{30}^p_c$	-"-	-"-	-"-	-	-	-	34	14	1.1	-	0.151	245

p = propane; h = hexane; o = ignition near the relief vent; c = ignition at the closed end of the vessel; $^{-}$ = grid is installed in the vessel (mesh size N29 - 500 mm, that N30 - 125 mm); m = incomplete membrane opening (effective venting area μF decrease is taken into account by way of changing μ at F=const).

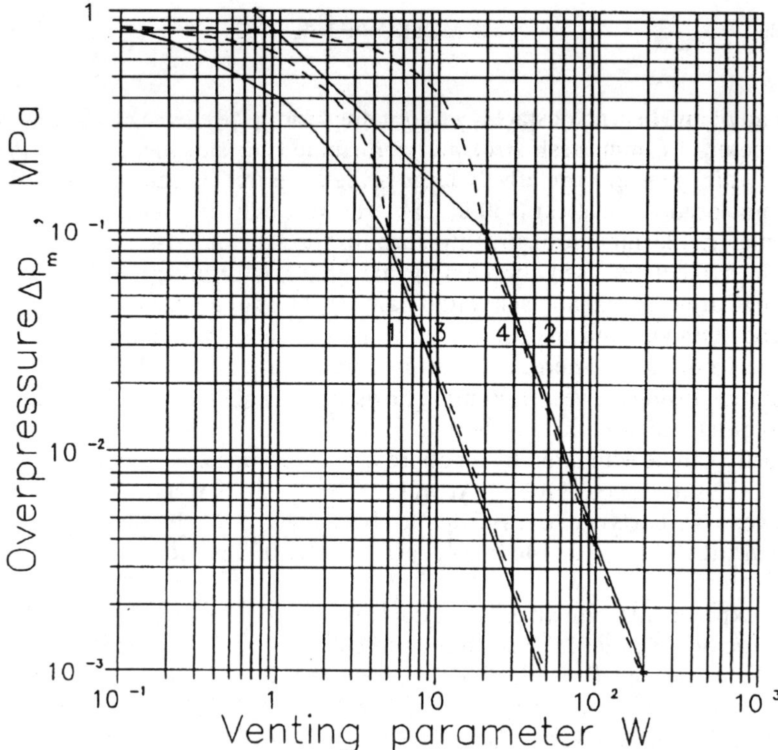

Fig. 3 Dependence of explosion overpressure on venting parameter: 1,2 - safe recommendations (D. Bradley and A. Mitcheson, 1978) respectively for uncovered and covered vent areas; 3,4 - design graphs, obtained by usage of formulas (5), (6) with χ=2 and χ=8.

dence is reached with χ=0.8-2 for uncovered vent areas and with χ=2-8 for covered ones.

The explosion pressure difference between cases of uncovered and covered vent areas can be explained by means of a unified approach, described in this work. That consists, in particular, of invariability of Eqs. (5) and (6) for all conditions and of taking into account the influence of these conditions and different parameters on maximum explosion pressure in an unclosed vessel through value of turbulization factor χ. Actually, as it follows from the process cinegrams, in the case of uncovered vent area flame can make only smooth stretching in the direction of a vent. Value χ=2, obtained for that case, seems reasonable if the occurrence and development of flame cellular structure is taken into consideration. In the case of covered vent area (for example, bursting diaphragm) the amplitude of disturbing affect on a flame surface is significantly higher, and Bradley's recommendations are described by design formula (5) with χ=8.

Discussion

A scaling factor has a strong influence upon the turbulization factor χ. Under laboratory conditions[20] a maximum ratio of visible burning velocity ahead of and behind obstacles was obtained for three meshes (0.8-mm wire diameter, 1.6-mm mesh size) and contributed for stoichiometric mixture C_2H_2-air "χ"=12. In relatively large-scale experiments conducted in a 11-m^3 unclosed vessel (test N30 in Table 1) with only one grid (18-mm wire diameter, 125-mm mesh size) the turbulization factor χ=14.

The symbol "χ" used above underlines that the turbulization factor χ is a ratio of areas of real and simulated flame fronts, but not a ratio of velocities ahead of and behind an obstacle as "χ" is. The value "χ" characterizes local flame front zone, whereas χ is an integral characteristic defining energy evolution inside the vessel. Venting flows through obstacles and out the vent can contribute to a considerable excess of "χ" above χ. Subsonic venting flow velocity can be estimated by an approximate formula $v \approx \sqrt{(2\Delta p)/(\gamma p_o)}\,C$, where C is the speed of sound. At pressure differential Δp 1 or 10 or 100 Torr between two parts of one vessel divided by a screen obstacle, the flow velocity constitutes 14 or 45 or 140 $m \cdot s^{-1}$, respectively. It has a strong influence on "χ" value.

In experiments with an exhaust duct and a receiving vessel the turbulization factor χ is practically constant and equal to χ=3.7-4 off dependence on the vessel volume. Cinegrams of the process showed that a few milliseconds short-term inverse discharge of reacting gas from the duct into combustion products within the vessel turbulizes the flame front having a shape which is near to the vessel shape. Given results affirm theoretical statements of Schchelkin[21] and Karlovits et al.[22] According to the Schchelkin model,[21] at a fairly high ratio of pulsating to burning velocity (v'/S_u) the turbulent burning velocity approximately equals the pulsating one $(S_t \approx v')$ and explicitly does not depend on the burning velocity. In the case of turbulence generated by the turbulent flame, the maximum pulsating velocity amounts to[2] $v'=S_u(E_i-1)/\sqrt{3}$ for a near-stoichiometric carbonhydrogen-air mixture χ=3.5-4, which matches experimental data (experiments N13-15, 25 in Table 1).

A somewhat higher maximal turbulization factor χ=4-6 can be obtained by the point ignition under isotropic turbulence produced, for instance, by fans[23]. Further rise in the isotropic turbulence intensity causes flame extinction at the Karlovits-Kovazhny-Klimov criteria $(v'/S_u)\cdot(\delta_l/l) \approx 10\text{-}20$ (l = turbulence scale; δ_l = a laminar flame front depth).

Maximum burning acceleration is realized by combining large-scale and small-scale turbulences. Thus, large-scale turbulence increases the total flame surface by large-scale eddies formation, and small-scale turbulence accelerates the flame at every relatively local front region of a large-scale eddy. The investigation of Moen et al.[24] can be presented as a laboratory example of simultaneous influence of "large-scale" eddies and small-scale turbulence on the increase in flame speed. Visible velocity of a flame front spread reached 130 $m \cdot s^{-1}$ for a stoichiometric methane-air mixture ("χ"=24 times increase in the flame front rate as compared to the case without obstacles).

For the first time the approach adopted has allowed the analog of the Le Chatelier-Brown principle to be discovered, according to which gas combustion dynamics in an unclosed vessel responds to external changes in process conditions in such a way as to weaken the effect of external influence. In accord with the theory proposed, an explosion overpressure is $\Delta\pi_m \propto (\chi/\mu F)^2$. Tenfold increase of a vent area F in order to reduce an explosion overpressure in an 11-m^3 vessel is "compensated" by twofold increase in the turbulization factor χ. It is equivalent to only 5 times an increase of the area (when other parameters are constant). Physical phenomena explanation is simple enough according to which disturbing influence on the flame front increases with the venting area. Another example (see Fig. 2) may be "compensation" of turbulization factor χ rise by discharge coefficient μ rise when introducing obstacles into a vessel. As mentioned above explosion overpressure is proportional to (χ/μ) , but not simply χ. Decrease in mesh size of a turbulizing grid contributes to 1.75 times rise of χ. However, due to simultaneous rise in the discharge coefficient μ the ratio χ/μ increases only 1.11 times. "Compensative" influence of μ on χ - tells on mixtures in hand when χ>5. At a high gas rate inside the vessel the discharge coefficient μ may be somewhat greater than 1. The gas discharge equations, used for the derivation of explosion dynamics equations, were explained as obtained with the gas rate equal to 0 ahead of the vent (this supposition work well for vents relatively small towards the vessel surface).

Turbulization factor dependence on the vessel volume, degree of its venting $(F/V^{2/3})$, maximum pressure π_m (which a sheath of a vessel can withstand), and a number of other conditions can be expressed by the

Table 2 Empirical coefficients for turbulization factor χ calculations

Burning and venting conditions	Empirical coefficients			
	a_1	a_2	a_3	a_4
Vessel volume $V\leq 10$ m^3; degree of venting $F/V^{2/3}\leq 0.25$	0.15	4	1	0
Vessel volume $V\leq 200$ m^3; $1<\pi_m<2$:				
- uncovered vent area	0	0	2	0
- covered vent area	0	0	8	0
Vessel volume $V\leq 200$ m^3; $2\leq\pi_m<\pi_e$:				
- uncovered vent area	0	0	0.8	1.2
- covered vent area	0	0	2	6
Vessel volume $V\leq 10$ m^3; degree of venting $F/V^{2/3}\leq 0.04$; availability of relief duct; $1<\pi_m<2$:				
- without sprinkling of venting gases	0	0	4	0
- with sprinkling of venting gases	0.15	4	1	0

formula

$$\chi = (1+a_1 V)(1+a_2 \frac{F}{V^{2/3}})(a_3+a_4 \frac{\pi_e-\pi_m}{\pi_e-2}) \tag{7}$$

in which the coefficients a_1, a_2, a_3, a_4 are defined in Table 2.

References

[1]Bradley, D., and Mitcheson, A., "The Venting of Gaseous Explosions in Spherical Vessels," *Combustion and Flame*, Vol. 32, March 1978, pp. 221-236 and pp. 237-255.

[2]Davenport, J. A., "Explosion Losses in Industry," *Fire Journal*, Vol. 75, Jan. 1981, pp. 52-53, 55-56, 71-72.

[3]Mandey, G., "The Calculation of Venting Areas for Pressure Relief of Explosions in Vessels," *Second Symposium on Chemical Process Hazards (Instn Chem. Engrs)*, Manchester, England, 1963, pp. 46-54, .

[4]Streltchuck, N. A., and Ivaschenko, P. F., "Design of Loads on Building Constructions from Explosions of Gas-Air Mixtures," *Fire Prevention and Extinguishing*, Pub. 3, VNIIPO, Moscow, 1966, pp. 3-19.

[5]Yao, C., "Explosion Venting of Low-Strength Equipment and Structures," *Loss Prevention*, Vol. 8, 1974, pp. 1-9.

[6]Pasman, H. J., Groothuisen, Th. M., and Gooijer, P. H., "Design of Pressure Relief Vents," *Loss Prevention and Safety Promotion in the Process Industries*, edited by C.H. Buschman, New-York, 1974, pp. 185-189.

[7]Sapko, M. J., Furno, A. L., and Kuchta, J. M., "Flame and Pressure Development of Large-Scale CH_4-Air-N_2 Explosions (Buoyancy Effects and Venting Requirements)," *US Bureau of Mines*, Rept. 8176, Washington, DC, 1976.

[8]Anthony, E. J., "The Use of Venting Formulae in the Design and Protection of Building and Industrial Plant from Damage by Gas or Vapor Explosions," *Journal of Hazardous Materials*, Vol. 2, 1977/78, pp. 23-49.

[9]Korotkich, N. I., and Baratov, A. N., "Discharge Coefficient Calculations with Due Account of Rooms Unfailure by Explosion Combustion of Vapor and Gas-Air Mixtures," *Combustibility of materials and chemical extinguishing means*, Pub. 5, VNIIPO, Moscow, 1978, pp. 3-15.

[10]Crescitelli, S., Russo, G., and Tufano, V., "Mathematical Modelling of Relief Venting of Gas Explosions: Theory and Experiments," *3rd International Symposium on Loss Prevention and Safety Promotion in Process Industries*, Basel, Vol. 3, 1980, pp. 16/1187-16/1197.

[11]Molkov, V. V., and Nekrasov, V. P. "Burning Velocity of Acetone-Air Mixture," *Physics of Combustion and Explosion*, Vol. 17, March 1981, pp. 45-49.

[12]Molkov, V. V., Nekrasov, V. P., Baratov, A. N., and Lesnyak S. A., "Turbulent Gas Burning in an Unclosed Vessel," *Physics of Combustion and Explosion*, Vol. 20, Feb. 1984, pp. 28-33.

[13]Baratov, A. N., Molkov, V. V., and Agafonov, V. V., "Mechanisms of Homogeneous Gaseous Mixtures Combustion in Unclosed Vessels," *Archivum Combustionis*, Vol. 8, Febr. 1984, pp. 179-195.

[14]Heinrich, H. J., "Design of Pressure Discharge Valves Used for Explosive Plant Protection in Chemical Industry," *Chemie-Ing.-Techn.* , Vol. 38, Heft 11, 1966, pp. 1125-1133.

[15]Babkin, V. S., Babushok, V. I., and Suyushev, V. A., "Dynamics of Turbulent Gas Combustion in a Confined Vessel," *Physics of Combustion and Explosion*, Vol. 13, March 1977, pp. 354-358.

[16]Harris, G. F. P., and Briscoe, P. G., "The Venting of Pentane Vapor-Air Explosions in a Large Vessel," *Combustion and Flame*, Vol. 11, March 1967, pp. 329-338.

[17]Andrews, G. E., Bradley, D., and Lwakabamba, S. B., "Turbulence and Turbulent Flame Propagation - A Critical Appraisal," *Combustion and Flame*, Vol. 24, March 1975, pp. 285-304.

[18]Molkov, V. V., and Nekrasov, V. P. "Gas Combustion Dynamics in a Vented Vessel with a Constant Volume," *Physics of Combustion and Explosion*, Vol. 17, April 1981, pp. 17-24.

[19]Vodyanik, V. I., *Explosion Protection of Equipment*, Tekhnika, Kiev, 1979.

[20]Dorge, K. J., Pangritz, D., and Wagne,r H. G. "Experiments on Velocity Augmentation of Spherical Flames by Grids," *Acta Astronautica*, Vol. 3, 1976, pp. 1067-1076.

[21]Schchelkin, K. I., "About Burning at Turbulent Flow," *Journal of Technical Physics*, Vol. 13, No. 9-10, 1943, pp. 520-530.

[22]Karlovits, B., Denniston, D. W. Ir., and Wells, F. E., "Investigation of Turbulent Flames," *Journal of Chemical Physics*, Vol. 19, May 1951, pp. 541-547.

[23]Karpov, V. P., and Severin, E.,S., "Turbulent Velocities of Gaseous Mixtures Burnout for Describing Burnout in Engines," *Combustion of Geterogeneous and gaseous systems*, Department of Chemical Physics Institute, the USSR Academy of Science, Chernogolovka, 1977, pp. 74-76.

[24]Moen, J. O., Donato, M., Knystautas, R., and Lee, J. H., "Flame Acceleration Due to Turbulence Produced by Obstacles," *Combustion and Flame*, Vol. 39, Jan. 1980, pp. 21-32.

Chapter II. Dust Explosions

Detonation Processes in Dusty Mixtures of Different Oxygen Contents

Marek Wolinski,* Marek Kapuscinski,† and Piotr Wolanski‡
Warsaw University of Technology, Warsaw, Poland

Abstract

This paper presents the results of an experimental investigation of oxygen influence on initiation of detonation, detonation wave parameters, structure, and limits. Tests were carried out with the use of vertical detonation tube (4.5 m long, L/D=57) in gaseous oxidizer containing 95%, 70%, 50%, and 30% of oxygen. Combustion was initiated by a shock wave generated in a small auxiliary shock tube mounted on the top of the detonation tube. Construction of the test stand allowed taking streak pictures of the flame self-acceleration, as well as to measuring pressure variations and dust particles temperature. Experiments were done for two fractions of dust of Egyptian brown coal, and transition to detonation was observed for not less than 50% of O_2 in the mixture.

Introduction

The increase of interest in the field of dusty mixture detonation resulted in a series of experimental

---

*Research Scientist, Institute of Heat Engineering.
†Also Ph.D. Student, Mercantile Marine University, Szczecin, Poland.
‡Professor, Institute of Heat Engineering.

works conducted at the Institute of Heat Engineering.[1-3] Those modeling studies were carried out in atmospheres of nearly pure oxygen (95% of volume) and allowed determination of influence of dust properties on the transition to detonation as well as on the detonation wave parameters and structure. However, there are evidences of detonation waves propagating in mixtures of dust and air,[4-9] and it has become necessary to compare and extend the laboratory scale results obtained in oxygen atmospheres to real (air) conditions.

The aim of this work was to study processes of detonation in mixtures of organic dust containing different amounts of oxygen and to determine the influence of the oxidizer composition on the detonation wave parameters.

Experiments

Experimental works were conducted in 4.5 m long vertical detonation tube with an internal diameter of 80 mm. Construction and operation of the test stand have been described in previous papers.[1-3] However, it is necessary to restate more important details.

Construction of the tube allowed recording the flame propagation process with the use of a drum camera an also measuring pressure and light emission profiles in the flame front. Pressure measurements were done by means of the piezoquartz pressure transducers KISTLER and the three-wavelength pyrometer allowed measuring light emission and calculation of the temperature in the flame front. Time profiles of pressure and light emission were recorded with use of Tektronix digital storage oscilloscopes. Construction of the tube allowed creation atmospheres of different oxygen content inside the tube and described tests were carried out at levels of 95%, 70%, 50%, and 30% volume of oxygen, under ambient conditions. Improvements in the dust dispersion system influenced more intense and efficient creation of a dust-oxidizer mixture that resulted in better accuracy of obtained results. Combustible mixture was ignited by a shock wave generated in a small auxiliary detonation tube (1 m long, with a 20 mm internal diameter), in consequence of H_2 + 0.5 O_2

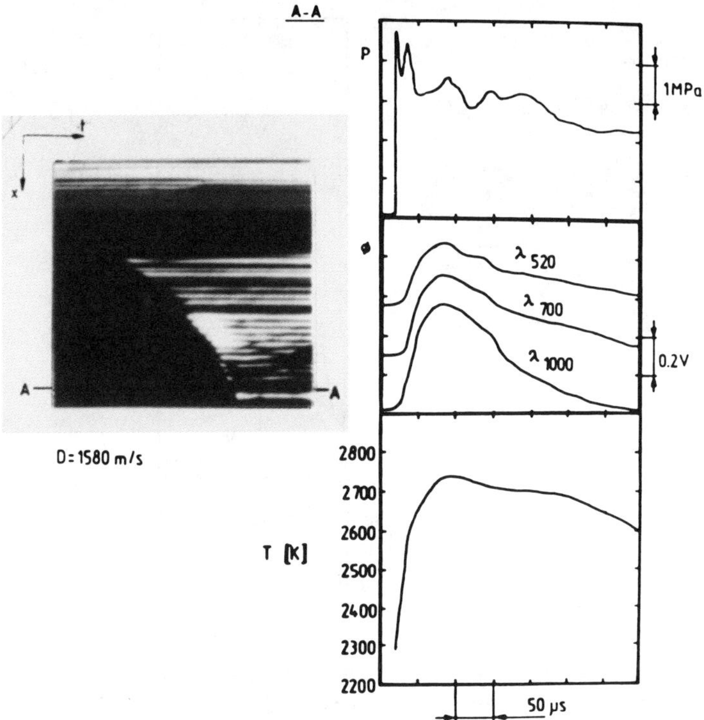

Fig. 1 Streak picture of the flame acceleration to detonation as well as pressure and luminosity (λ=1000, 700, 520 nm) profiles with calculated profile of the dust particles temperature in the detonation front. 88<d_p≤102 μm, c=0.6 kg/m^3, 95% volume of oxygen.

mixture explosion. Initial pressure in that tube was 0.4 MPa for the majority of experiments; 0.6 MPa was applied in only a few cases of less reactive mixtures.

In a presented part of the study two fractions of dust of brown coal were used: 88<d_p≤102 μm and 150<d_p≤200 μm. Undertaken analysis showed the following composition of dust: fixed carbon comprised 39.1% of weight; volatile matter, 54.5%; ash, 3.7%; moisture, 2.7%; while C comprised 77.76% of weight; H, 6.42%; N, 1.13%; S, 2.96%; and O - 11.73%. Heat of combustion for tested dust was found equal to 30.2 MJ/kg.

Results and Discussion

Experiments were conducted for decreasing oxygen content in the mixture and started at 95%. Exemplary

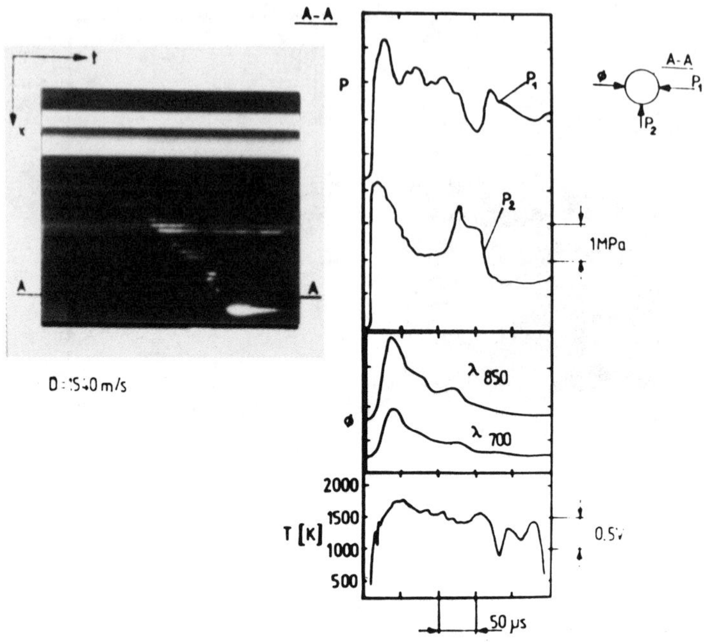

Fig. 2 Streak picture of the flame acceleration to detonation as well as pressure and luminosity (λ=700, 850 nm) profiles with calculated profile of the dust particles temperature in the detonation front. $150<d_p\leq200$ μm, c=1.05 kg/m^3, 50% volume of oxygen.

results obtained during the tests are presented in the following figures.

Figure 1 shows streak picture of the flame acceleration to detonation in a mixture containing c=0.6 kg/m^3 of dust (particles 88-102 μm) and 95% volume of O_2. That dust concentration was found to be limiting: below that value detonation was not observed. It can be seen that the flame acceleration process was nonsteady at the tube length. Ignition spots ahead the flame front were visible as well as related to them changes of the flame velocity. Recorded pressure profile shows existence of a rather strong pressure wave which propagated in the detonation front.

Further decrease of the oxygen content in the mixture resulted in an increase of the transition distance (distance between igniter and point at which flame achieves detonation regime) and, as suggested particles temperature profiles, resulted in decrease of

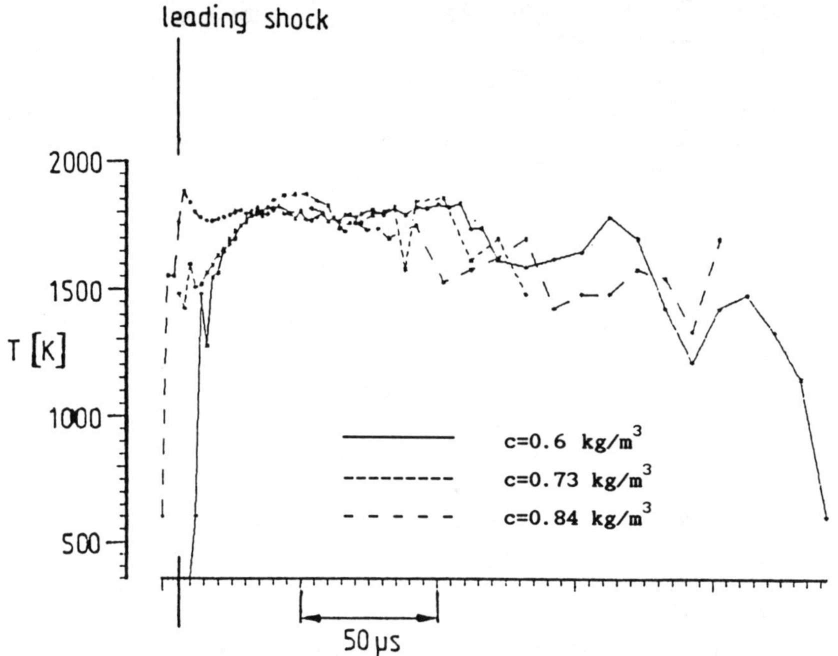

Fig. 3 Profiles of the dust particles temperature in the detonation front. $88<d_p\leq102\mu m$, 50% volume of oxygen.

the reaction zone length. It was also found that for lower oxygen contents flame acceleration process had rather smooth character (without intense ignitions ahead flame front), but structure of the detonation wave was more complicated. However, oxygen content equal to 30% was found insufficient to support flame acceleration to detonation and increase of the initiator pressure up to 0.6 MPa caused only significant intensification of the flame acceleration process.

Processes recorded in a mixtures of 150-200 μm particles with 95% volume of oxygen showed that transition to detonation was smoother than for smaller particles and took place with rather small velocity changes and weak ignition spots ahead of the flame front. Also, recorded pressure profiles did not reveal significant nonuniformities in the reaction zone. However, substantial variations of the particles temperature were observed. In the case of 70% of oxygen

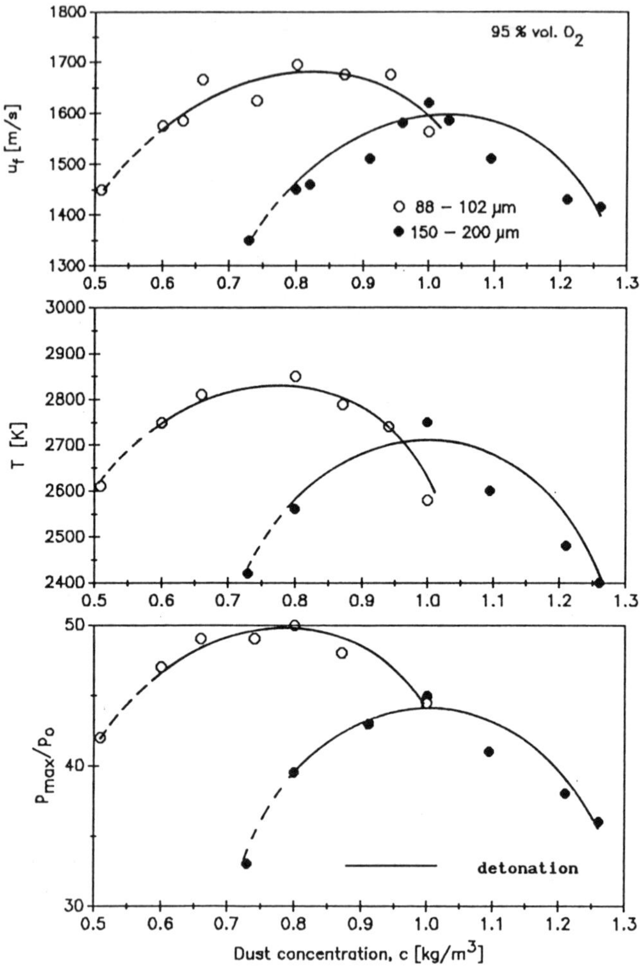

Fig. 4 Influence of the dust concentration on flame velocity, maximum temperature of dust particles, and maximum pressure in the flame front at 95% volume of oxygen in the mixture.

in the mixture, the flame acceleration process occurred with repeated ignitions ahead of the flame front and accompanying velocity changes. Simultaneously recorded pressure and luminosity profiles revealed intense nonuniformities and pressure waves propagating in the reaction zone. However, for 50% of oxygen (Fig.2) the flame acceleration process had more gradual character, with only one ignition spot ahead flame front. Also, temperature profiles showed significant decrease of value of particles temperature.

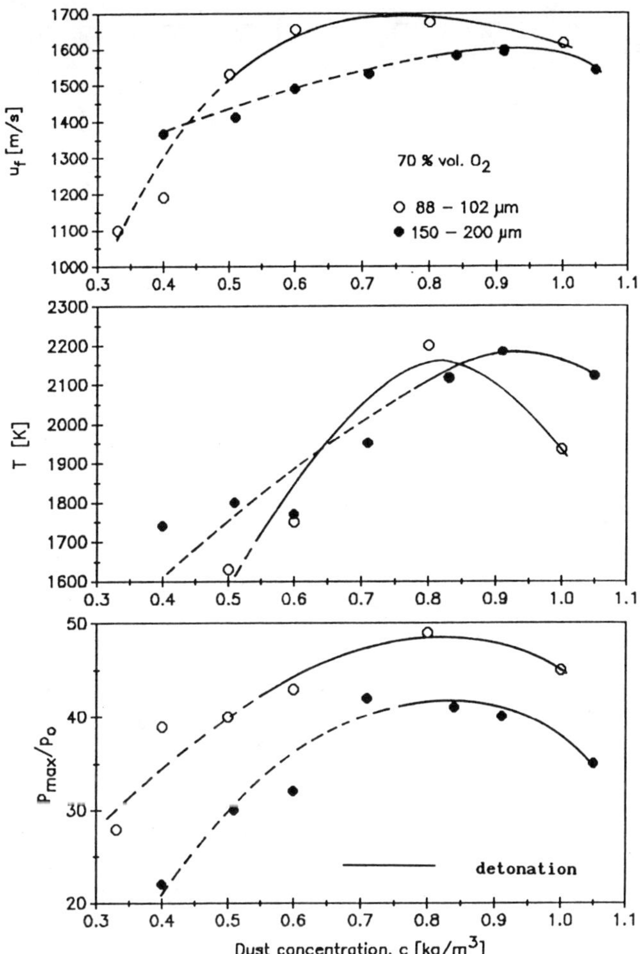

Fig. 5 Influence of the dust concentration on flame velocity, maximum temperature of dust particles, and maximum pressure in the flame front at 70% volume of oxygen in the mixture.

Similar to case of smaller particles, oxygen content equal to 30% of volume was found insufficient to support flame acceleration to detonation, and increase of initiator pressure up to 0.6 MPa did not force the transition to detonation process.

Figure 3 presents experimentally measured profiles of particle temperature in the reaction zone (88-102 μm dust particles, 50% of oxygen). The temperature rise can be seen even slightly ahead of the leading shock

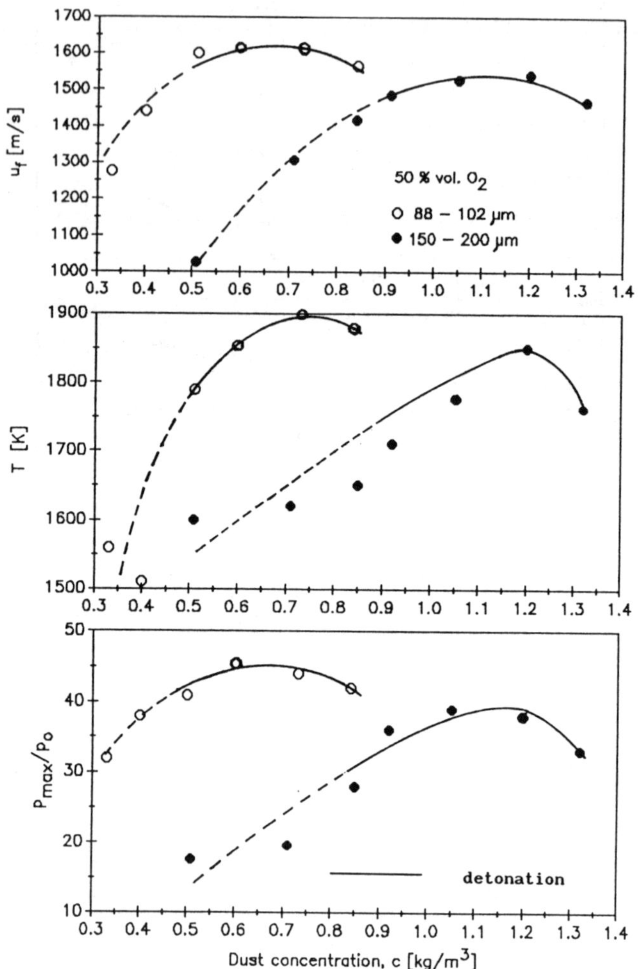

Fig. 6 Influence of the dust concentration on flame velocity, maximum temperature of dust particles, and maximum pressure in the flame front at 50% volume of oxygen in the mixture.

wave position. It shows some existence of spinning-like mode of the detonation wave propagation for this case.

Figures 4-7 present a summary of measurements done for 95%, 70%, 50%, and 30% of oxygen, respectively, for both fractions of dust particles. At every oxygen content, similar influence of dust concentration on the flame parameters (i.e., flame front velocity, maximum temperature of particles and maximum pressure in the flame front) was observed: increase of the parameters values up to a certain maximum (at optimum dust

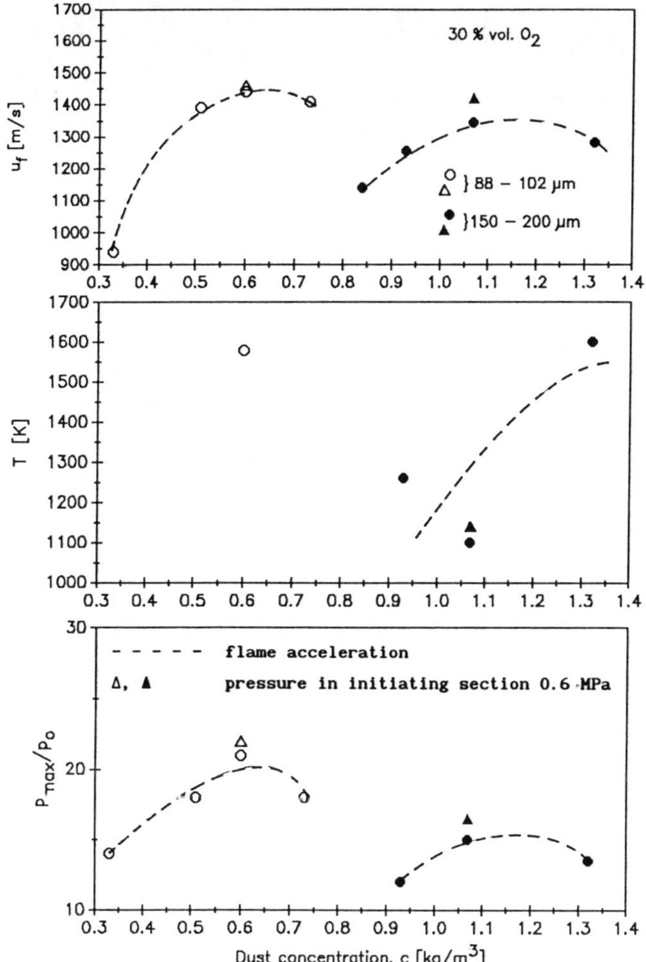

Fig. 7 Influence of the dust concentration on flame velocity, maximum temperature of dust particles, and maximum pressure in the flame front at 30% volume of oxygen in the mixture.

concentration range) and then decrease. Also, significant influence of dust particle size on the flame parameters could be seen: values of the parameters are lower for greater dust particles.

Figure 8 presents influence of oxygen content in the mixture on maximum flame velocity, maximum temperature of dust particles, and maximum pressure in the flame front. It can be seen that values of maximum flame

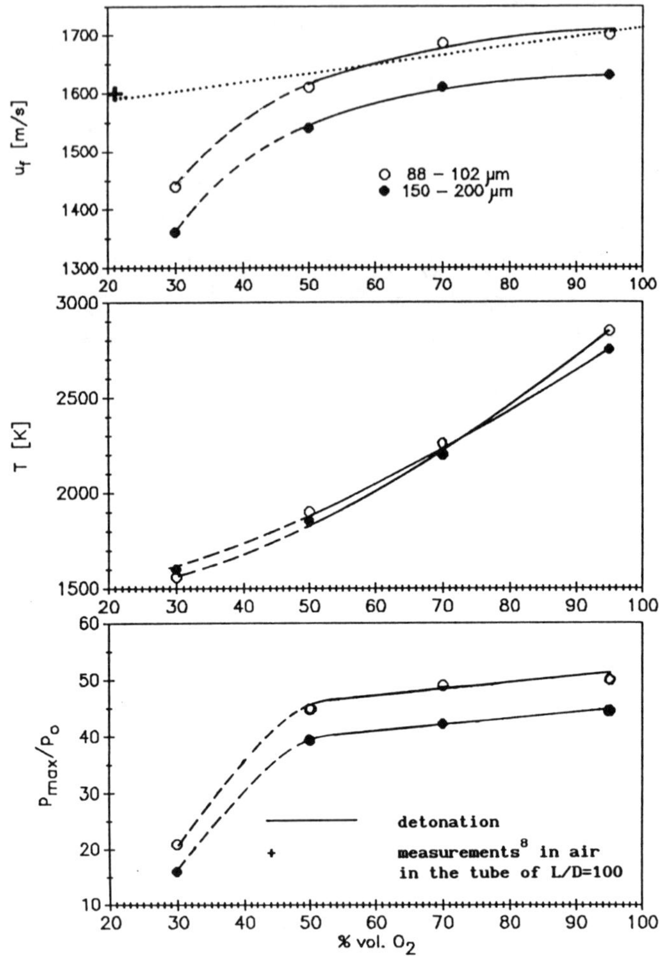

Fig. 8 Influence of oxygen content in the mixture on maximum flame velocity, maximum temperature of dust particles, and maximum pressure in the flame front.

velocities decreased slightly with decrease of oxygen content in the range of 95-50% of volume. Only further decrease of oxygen content in the mixture caused significant change of maximum flame velocity. One can also observe agreement of obtained results with measurements[8] of detonation wave velocity for the same dust in air, in the tube of L/D=100. The effect of oxygen content was similar for both particle sizes; however, for greater particles lower values of detonation velocity were obtained. Values of maximum

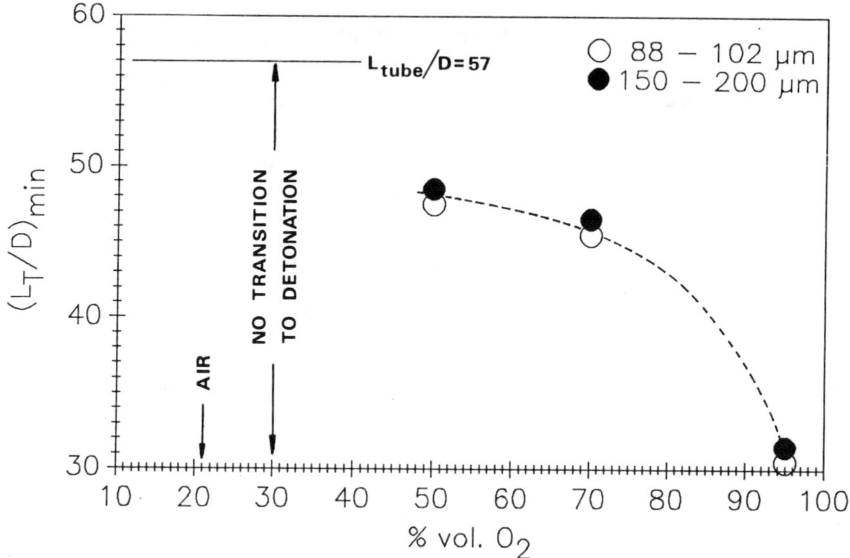

Fig. 9 Influence of oxygen content in the mixture on minimum distance for transition to detonation for investigated dust.

temperatures of particles differed only for 95% of oxygen and decrease of oxygen content caused significant decrease of the temperatures -- to 1600 K at 30% of oxygen. Maximum pressures in the flame front decreased slightly with a decrease of oxygen content in the range of 95-50% of volume (similarly to the decrease of detonation velocity). Only a decrease of oxygen content to 30% (and total change of the character of investigated processes) caused significant decrease of maximum pressure values. Lower values of maximum pressures were recorded for greater particles, but also in that case the effect of oxygen concentration was similar for both particle sizes.

Figure 9 shows the influence of oxygen content in the mixture on minimum distance from ignition source, which is necessary for transition to detonation. In all cases, value of that parameter was greater for 150-200 μm particles, and for both fractions the distance increased while oxygen content decreased. Transition to detonation was found impossible in mixtures of 30% of oxygen (and less) -- for existing laboratory conditions. However, obtained results can suggest that flame self-acceleration with transition to detonation may be possible in the longer tube.

Conclusions

Investigations of oxygen content influence on dusty detonation processes were carried out using the vertical detonation tube. Two different fractions of dust of Egyptian brown coal were tested. Rather weak shock wave was used as an ignition source for the dust-gas mixture, what allowed observation of the flame self-acceleration. During the test streak pictures of the flame self-acceleration were taken and pressure as well as temperature measurements in the dusty flame front were made.

Obtained results are summarized as follows:

1) Detonation was found for oxygen content in the mixture greater than 50%. However, obtained results suggest that for investigated dust, in the case of the flame self-acceleration, could be possible to achieve detonation regime in air in the longer tube.

2) In the range of detonation, changes of oxygen content in the mixture influenced mostly the dust particles temperature; the influence on the wave velocity and maximum pressure in the wave was less significant (only slight decrease of values of those parameters was observed with decrease of oxygen content in the mixture).

3) For all investigated oxygen concentrations the particle size had similar effect on the measured flame parameters: greater values of the flame velocity, maximum particles temperature, and maximum pressure in the flame front were found for smaller particles.

4) For all investigated oxygen concentrations it was found that increase of the dust particle size also caused rise of the optimum dust concentration (regarding the flame front parameters) and minimum dust concentration (regarding the transition to detonation).

Acknowledgments

The financial support for this work was made by the Department of Technical Development and Applications of Poland (under Program No 02.18).

References

[1]Fangrat, J., Glinka, W., Wolański, P., and Woliński, M., "Detonation Structure in Organic Dust-Oxygen Mixtures", *Archivum Combustionis*, Vol. 7, No 3/4, 1987, pp. 321-332.

[2]Dahab, O. M., Kapuściński, M., Wolański, P., and Woliński, M., "Detonation Processes in Egyptian Brown Coal Dust-Oxygen Mixtures", *Archivum Combustionis*, Vol. 9, No 1/4, 1989, pp. 197-204.

[3]Dahab, O. M., Kapuściński, M., and Wolański, P., "Influence of Dust Parameters on the Detonation Velocity, Structure, and Limits", Proceedings of the 12th ICDERS, AIAA, New York, 1990 (in press).

[4]Cybulski, W., "Detonation of Coal Dust", *Bulletin de l'Academie des Sciences*, Vol. 19, 1971, pp. 37-41.

[5]Kauffman, C. W., Wolański, P., Ural, E. A., Nicholls, J. A., and Van Dyk, R., "Shock Wave Initiated Combustion of Grain Dust", Proceedings of the International Symposium on Grain Dust, Manhattan, Kansas, Oct. 2-4, 1979, pp. 164-190.

[6]Bartknecht, W., *Explosions: Course, Prevention, Protection*, Springer Verlag, Berlin, Heidelberg, New York, 1980.

[7]Gardner, B. R., Winter, R. J., and Moore, M. J., "Explosion Development and Deflagration-to-Detonation Transition in Coal Dust/Air Suspensions", The Combustion Institute, pp. 335-343.

[8]Wolański, P., Sacha, W., and Zalesiński, M., "Effect of Dust Concentration on Detonation Parameters in Grain Dust-Air Mixtures", *Proceedings of the 4th International Colloquium on Dust Explosions*, Warsaw, Poland, 1991, pp. 355-370.

[9]Zhang, F., and Groenig, H., "Detonation Structure of Corn Starch Particles-Oxygen Mixture", *Dynamics of Explosions*, Progress in Astronautics and Aeronautics series, AIAA, New York, 1990 (in press).

Measurements of Cellular Structure in Spray Detonation

J. Papavassiliou,* A. Makris,* R. Knystautas,† and J. H. S. Lee†
McGill University, Montreal, Quebec, Canada
and
C. K. Westbrook‡ and W. J. Pitz‡
Lawrence Livermore National Laboratory, Livermore, California 94550

Abstract

The cellular structure of heterogeneous detonations in a low vapor pressure fuel (decane) droplet mixture with oxygen and nitrogen was studied in the present investigation. The aerosol was generated by an ultrasonic nebulizer and the fuel concentration of the mixture was regulated by monitoring the volume flow rate of oxygen and nitrogen through the nebulizer. The vertical detonation tube is 64 mm in diameter and 3 m long and ignition was by a powerful spark (120 joules stored energy) or by a high explosive detonator. Velocity was measured with ionization probes, pressure by a PCB piezoelectric transducer and cell size by a smoked metallic foil inserted either into the top end or at the center of the detonation tube. The initial pressure of all the experiments was 1 atmosphere. In order to compare the time scales associated with the physical processes of droplet breakup, heat transfer, evaporation and mixing, experiments were also

---

*Graduate Student, Department of Mechanical Engineering.

†Professor, Department of Mechanical Engineering.

‡Physicist, Computational Physics Division.

carried out in the tube heated to 100°C and 185°C, using electrical heating tape to ensure a homogeneous gas-phase mixture of decane-oxygen-nitrogen. Comparison of the cell size for the same mixture in the cold and the heated tube permits one to separate the time scales associated with the physical processes from the chemical kinetic rate processes. The results from the heated tube for the homogeneous vapor phase decane detonations are similar to those for the common gaseous fuels in the alkane group (i.e., ethane, propane, butane). Corresponding results for the heterogeneous case (cold tube) of aerosol decane detonation indicate that the cell size is larger by a factor of about two for the present case of 5 μm particle size. The measurements of cellular structure obtained experimentally have been compared to the computed results determined using the ZND chemical kinetic detonation model. The detonation cell size was chosen as being 60 times the calculated induction length.

Introduction

Although detonations in liquid fuel sprays in air have been studied for more than two decades,[1-6] their precise mechanisms of propagation have not been resolved. It is generally recognized that the true test of the detonability of a fuel-air mixture is in the unconfined mode. Most of the unconfined detonations in liquid spray-air mixtures have been carried out with high vapor pressure fuels (e.g., propylene-oxide, hexane) and unconfined detonations in low vapor pressure fuels such as decane had not been observed until recently. It was assumed that only low vapor pressure nitrated hydrocarbon liquid spray-air mixtures could be detonated in the unconfined mode.[7] It has been found, however, with a cloud size of 8-20 m in length and 8 m in radius, that both gasoline and kerosene of droplet sizes of 50 μm and 100 μm respectively, are detonable in air.[8] Decane in the form of a fine mist and in pure oxygen has also been detonated in tubes by Bowen et al.[9] Using a weak ignition spark instead of a powerful igniter, they observed transition from deflagration to detonation, indicating the relative ease to detonate the mixture. They also reported a spinning wave structure suggesting a cellular nature of the detonation front similar to homogeneous gas phase mixtures. In view of the usefulness of the cell size as a fundamental length scale, it would be of great value to determine the cell size for heterogeneous detonations. By increasing the initial temperature, the concentration of the fuel in the vapor phase can be controlled. In this manner, the characteristic time scales involved in the physical processes (i.e., droplet breakup time, mixing time) can be separated from the chemical time scales (i.e., chemical kinetics). Thus an estimate of the characteristic length scale of a spray detonation and the influence of vapor pressure and droplet size can be assessed. In this paper, cell size measurements of decane-oxygen-nitrogen mixtures at an initial pressure of 1 atmosphere are determined. The tube can be heated to 100°C and 185°C to permit the detonation cell size to be measured in the homogeneous vapor phase. The corresponding experimental results are compared with a chemical kinetic detonation model.

Experimental Details

The experiments for the present investigation were performed in a vertical stainless steel shock tube which was 3 m long with an internal diameter of 6.4 cm. Ignition of the homogeneous phase mixtures was at the bottom of the tube via the discharge of a 0.6 μf capacitor (20 kV). The mixtures in the heterogeneous phase were ignited by a high explosive detonator in conjunction with a 50 cm long Shchelkin spiral. A schematic diagram of the experimental setup is given in Fig. 1. Ionization probes positioned along the tube length were used to determine the average flame propagation velocity and a piezoelectric pressure transducer was mounted for pressure measurements. In addition, a metallic smoked foil was placed at the top end or at the center of the tube for each trial so as to record the cell size λ in the case of detonation. Settling of the mist can be observed 10 cm from the top end of the tube. Thus the smoked foil was placed at the center of the tube for all the trials with the mixture in the heterogeneous phase so as to obtain the representative cell distribution along the foil.

Liquid decane was used as the fuel because of its low vapor pressure at ambient conditions (1 Torr at 16.5°C). The decane fuel was introduced from the bottom of the tube as an aerosol. The decane mist was produced by a nebulizer (Ultra-Neb 99 Ultrasonic Nebulizer). A schematic diagram of the nebulizer is given in Fig. 2. The free surface of the decane was excited by a piezoelectric transducer (1.63 MHz) causing the formation of ripples on the surface which eventually

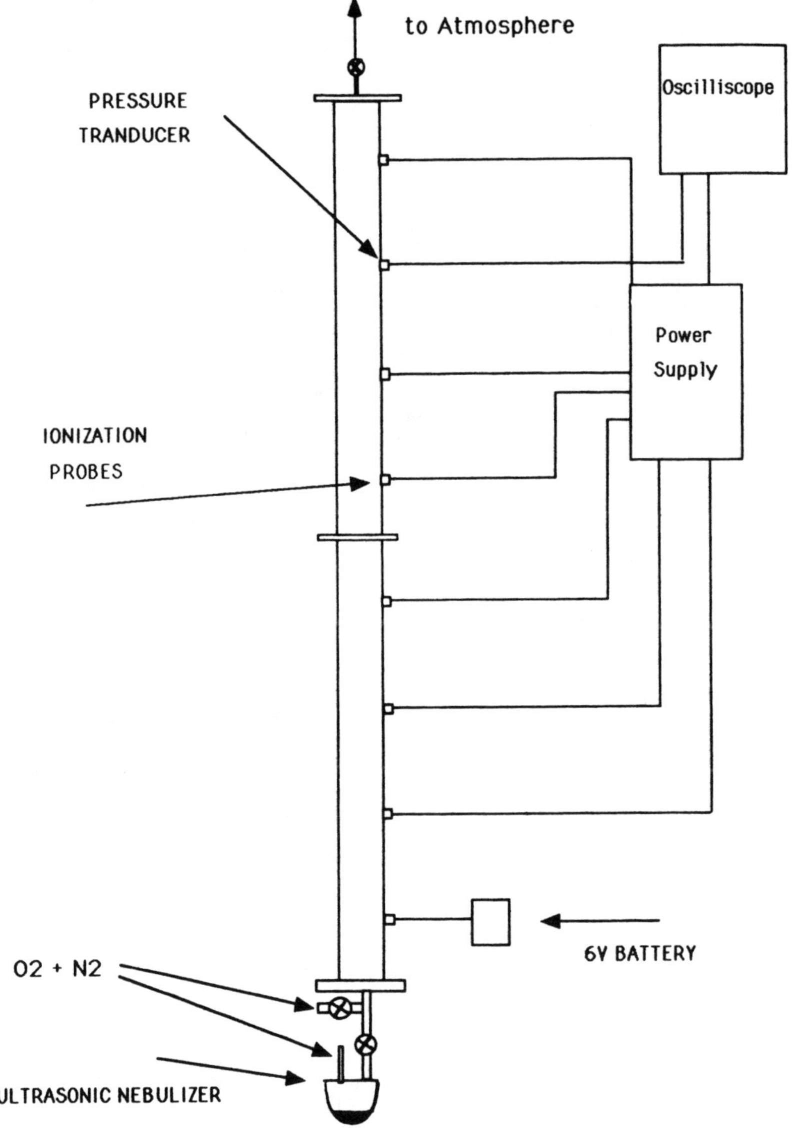

Fig. 1 Schematic diagram of vertical detonation tube apparatus.

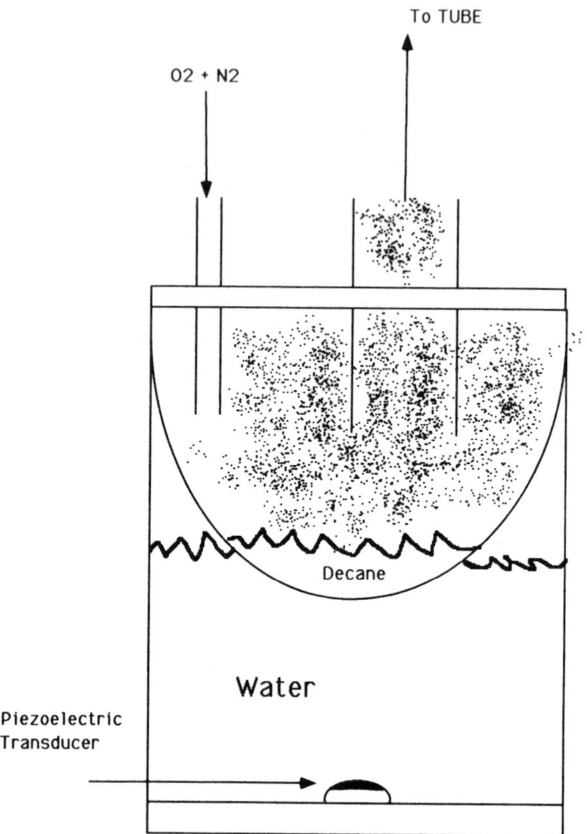

Fig. 2 Schematic diagram of Nebulizer System.

grow enough to form droplets. The resulting decane mist was then entrained by the flowing oxygen and appropriate nitrogen dilution and carried into the tube. For each trial, the nebulizer cup was filled with a fixed mass of decane fuel (10 g). The volume flowrates of oxygen and nitrogen were controlled by rotary type flowmeters to produce the desired equivalence ratio, ϕ and nitrogen dilution ratio, β (where $\beta = N_2 / O_2$). A rough calibration of the mass of decane leaving the nebulizer, with volume flowrate of O_2 and N_2 passing through it, was done initially. This allowed an estimate of what gas flowrate to use to obtain a given ϕ. However, due to limitations of the nebulizer system, the actual ϕ was determined after each trial by measuring the actual mass that left the nebulizer cup and entered the tube. Furthermore, it was not possible to obtain equivalence ratios approaching unity for low dilution ratios.

Experiments were performed both in a cold and a hot tube to assess the effect of the vapor phase on the detonability of decane. In all cases, an equivalent of four tube volumes of the appropriate oxygen-nitrogen mixture to be used for the trial were flushed through the tube from a port downstream of the nebulizer. The gas flow was then diverted to the nebulizer for a predetermined time period, corresponding to an equivalent flushing of two tube volumes at the required ϕ. The gas flow and nebulizer were then immediately turned off, all valves to the tube were closed, and ignition was effected at an initial pressure of 1 atm. Due to settling of the fine decane mist with time, ignition had to occur immediately so as to minimize large decane concentration gradients along the tube. Thus, for the decane-aerosol-oxidizer mixtures, solid explosive detonators were used for ignition.

The detonability of decane as an aerosol was compared to that for decane in the vapor phase from the hot tube experiments. The tube was heated to 100°C and insulated with fiberglass wool. Several heating tapes were wound along the tube length to ensure a uniform temperature distribution. Also, the gaseous fuel was mixed with the gaseous oxidizer by the use of a high temperature recirculating pump. The pump was insulated and preheated to 100°C. In other experiments, the tube was preheated by the heating tapes to an average temperature of 185°C which is above the boiling point for decane of 174.1°C, prior to flushing the decane-oxygen-nitrogen mixture through the tube. To ensure that all the decane in the tube is in the vapor phase at ignition, the mixture at both 100°C and 185°C was maintained for one-half hour within the heated vessel after flushing, while periodically venting to relieve any pressure increase above atmospheric resulting from heating.

A typical oscillogram of the pressure signal and triggering of the ionization probes on the tube from a decane-oxygen detonation in the cold tube, is shown in Fig 3. As can be observed, the pressure record (lower trace) is typical of a self-sustained detonation.

Results and Discussion

Successful initiation and stable self-sustained propagation of detonation along the vertical tube was achieved in decane spray-oxygen mixtures for a range of stoichiometries investigated. This is illustrated in Fig. 4, which plots the experimentally measured

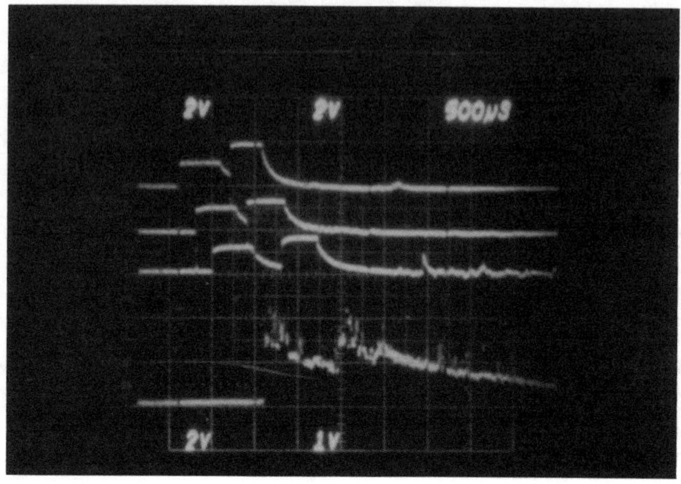

Fig. 3 Typical pressure trace and velocity record of a detonation in a decane aerosol oxygen mixture.

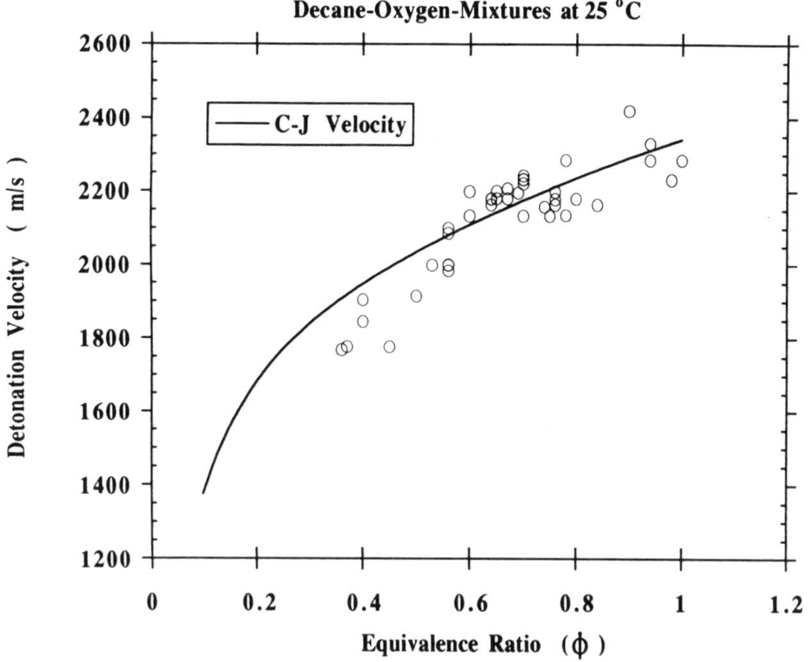

Fig. 4 Results for detonation velocity in decane aerosol-oxygen mixtures as a function of equivalence ratio, ϕ.

detonation velocity and compares the results with the Chapman-Jouguet equilibrium theory. The results are for fuel-oxygen equivalence ratios ranging from $0.36 < \phi < 1.0$ and corresponding to initial conditions of 25°C and 760 Torr pressure for the mixture. The agreement between theory and experiment is very good (to within less than 10%). As expected, the detonation velocity increases with increasing equivalence ratio consistent with the energetics of the medium.

Figure 5 displays results for detonation velocity in decane spray-oxygen-nitrogen mixtures as characterized by $\beta = N_2 / O_2$, the dilution parameter. Increasing β means a progressively more diluted mixture with inert N_2. Theoretical C-J results are presented for two stoichiometries, namely $\phi = 0.5$ and $\phi = 1.0$, which covers the range of experimental conditions. The theoretical curves again show the expected trend of decreasing detonation velocity with increasing dilution (i.e., increasing β). The experimental results are for a cold tube corresponding to an aerosol-oxidizer mixture at 25°C and 760

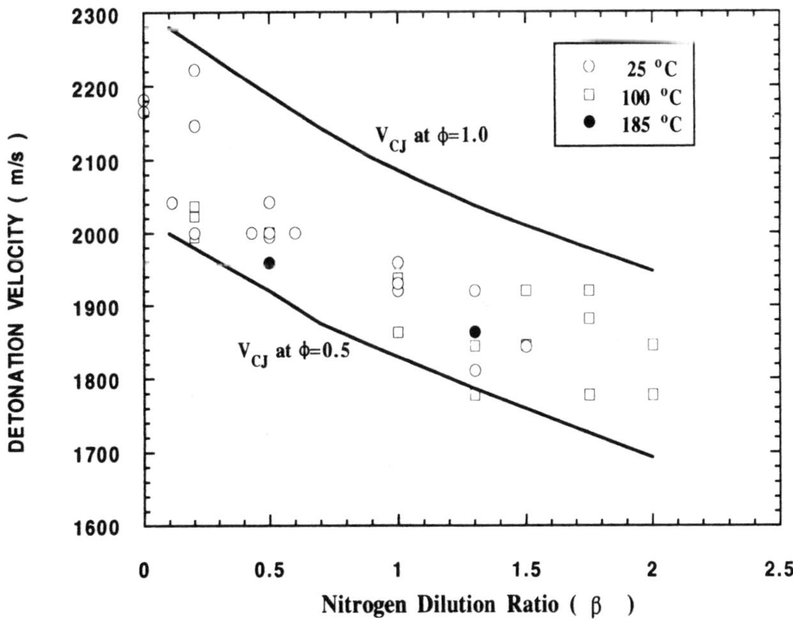

Fig. 5 Results for detonation velocity in decane aerosol-oxygen-nitrogen mixtures as a function of nitrogen dilution ratio, β.

Torr initial conditions and for the hot tube corresponding to gaseous fuel-oxidizer mixture at 100°C and approximately 185°C both at 760 Torr initial pressure. The experimental results cover a stoichiometry range $0.6 < \phi < 1.54$. However, once the atomized fuel has been introduced, the actual ϕ of the mixture can be determined a posteriori. The theoretical curves clearly bracket the experimental results for the cold and hot tube conditions. It must be pointed out that for the hot tube case, the effective density of the mixture is reduced since the mixture must be vented to 760 Torr during the heating process. However, the energy per unit mass remains invariant. Since the detonation velocity is proportional to the square root of the specific energy, it is not surprising that the experimental results are insensitive to tube heating as far as detonation velocity is concerned. The important conclusion that can be drawn from the results in Figs. 4 and 5 is that in aerosol detonations, the equilibrium parameters behave similarly to gas-phase detonations.

Figure 6 shows the results of detonation pressure for decane aerosol-oxygen mixtures at an initial temperature of 25°C and an initial

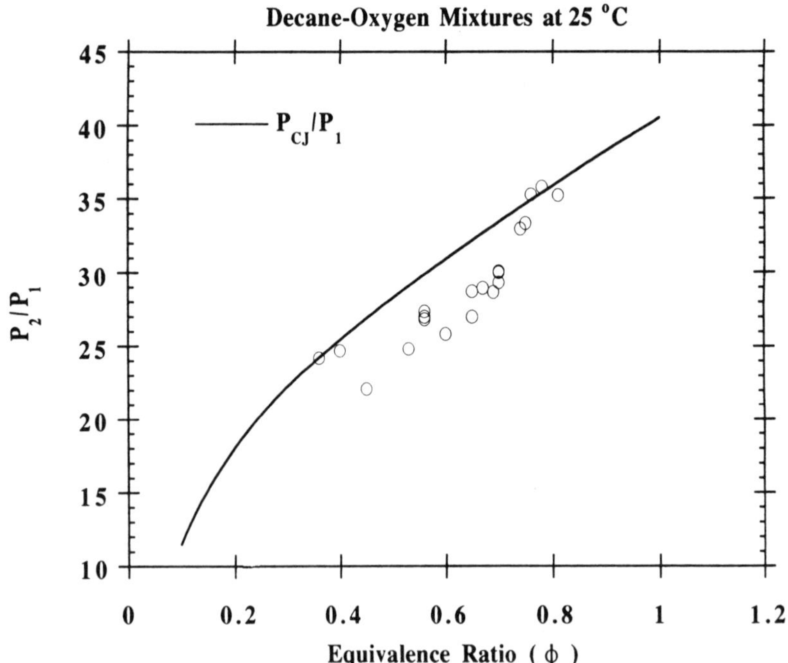

Fig. 6 Results for detonation pressure in decane aerosol-oxygen mixtures as a function of equivalence ratio, ϕ.

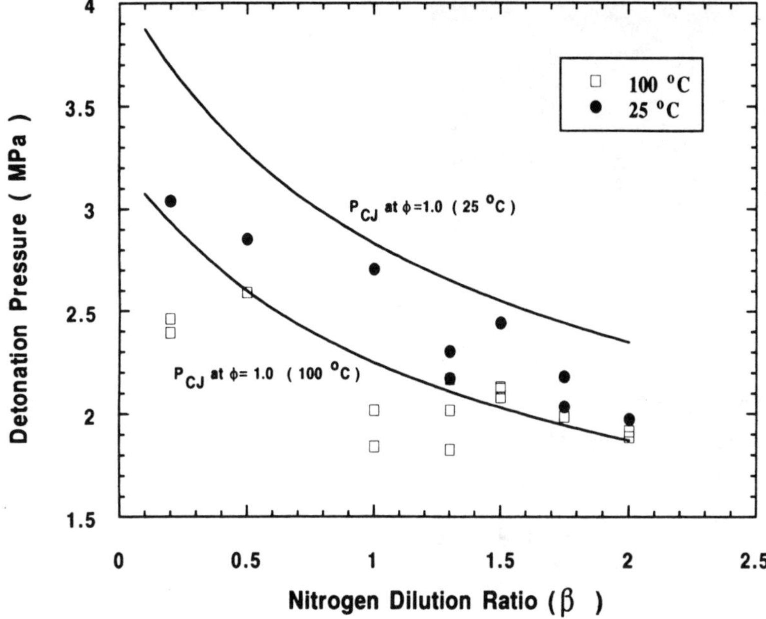

Fig. 7 Results for detonation pressure in decane-oxygen-nitrogen mixtures as a function of nitrogen dilution ratio, **β**.

pressure of 750 Torr. The results cover a stoichiometry range of $0.36 < \phi < 0.94$. As the equivalence ratio increases the detonation pressure increases. This is consistent with the energetics of the medium. Figure 7 shows the results of detonation pressure for decane-oxygen-nitrogen mixtures both in the heterogeneous phase (25°C initial temperature) and in the homogeneous phase (100°C initial temperature). The theoretical curves show the expected trend of decreasing detonation pressure with increased dilution ratio at stoichiometric composition. The results are for a stoichiometry range of $0.73 < \phi < 1.54$. For all the dilution ratios investigated the detonation pressure is lower in the hot tube case than in the cold tube. This comes about due to a decrease in the initial density of the mixture since the tube was vented so as to maintain an initial pressure of 1 atm. in the heated vessel. Nevertheless, for both the cold tube and the hot tube experiments the detonation pressure decreases with increased nitrogen dilution ratio consistent with the energetics of the medium.

Figure 8 shows a typical smoked foil record of a decane spray-oxygen detonation. The remarkable regularity of the cellular

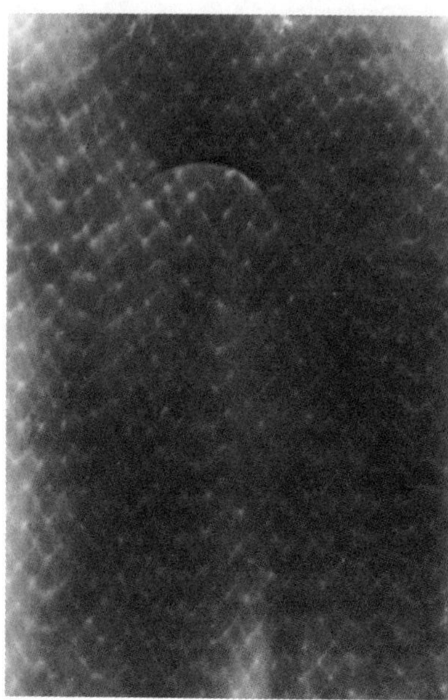

Fig. 8 Typical smoked foil record of a decane aerosol-oxygen detonation.

structure for such aerosol detonations is to be particularly noted. Using such foils, detonation size, λ, was measured and the results are plotted in Figs. 9, 11, and 12. In Fig. 9 the variation of cell size λ with equivalence ratio φ for decane aerosol-oxygen detonations is shown. The cell size decreases with increasing φ and the trend is identical to the typical U-shaped curves of cell size vs. equivalence ratio for hydrocarbon fuel-air detonations.[10] The minimum value which is obtained as one approaches $\phi = 1$ for decane aerosol-oxygen is about $\lambda = 4$ mm. This is to be compared to the data for the most insensitive gaseous hydrocarbon butane-oxygen detonations previously reported.[11] Thus, aerosol detonations in decane are about four times more insensitive than gas phase detonation in butane. The most insensitive hydrocarbon gaseous fuel is methane for which $\phi = 1$ in oxygen, has a cell size λ of approx. 4mm.

Figure 10 shows a typical smoked foil record in a decane-oxygen-nitrogen homogeneous mixture at an initial temperature of 100°C. Figure 11 shows cell size measurements for decane-oxygen-nitrogen mixtures as a function of nitrogen dilution ratio for the cold and hot mixtures. The cold tube results are at 25° and 760 Torr initial pressure, while the hot tube results are at 100°C and again at 760

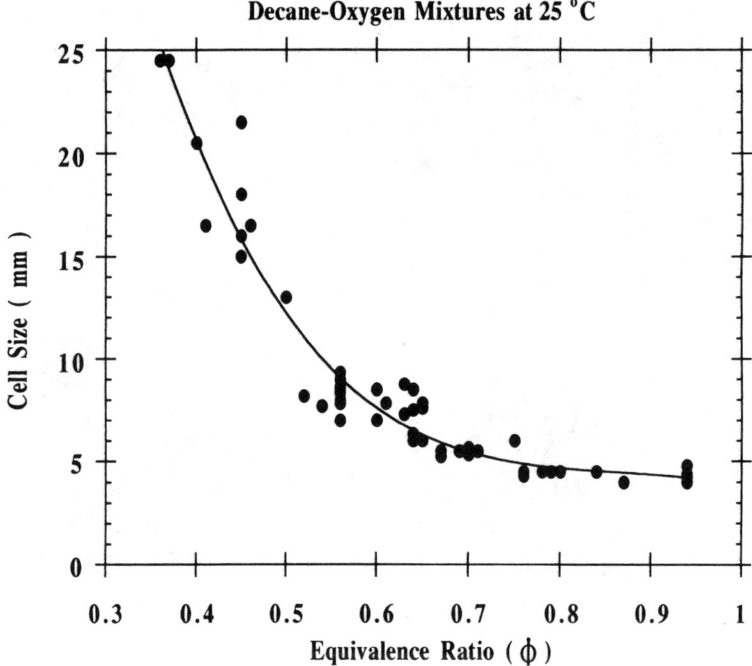

Fig. 9 Cell size measurements, λ for decane aerosol-oxygen mixtures plotted as a function of equivalence ratio.

Torr initial pressure. Also included are the results from Sandia National Laboratories hot detonation tube (HDT)[12] corresponding to decane-air mixtures (β = 3.76) at 100°C and 760 Torr initial pressure. To be noted is the trend for cell size which is similar to that reported for the critical tube diameter[11] namely that $d_c = 13\lambda$. If one extrapolates the data for the cold tube to the fuel-air composition in the same manner, then one obtains a cell size typically λ = 78 mm which is almost twice the value of λ = 42 mm from the 0.43 m dia. hot tube.[12]

Moreover, the extrapolated cold decane aerosol-air result falls in the same range of cell size as that for the most insensitive alkane-air detonation cell size measurements reported earlier.[10] For example, for decane aerosol-air we estimate by extrapolation λ = 78 mm, whereas for propane-air λ = 54 mm, ethane-air λ = 67 mm, and butane-air λ = 60 mm, all at ϕ = 1.

Figure 12 shows the cell size for the hot tube at 185°C initial temperature in comparison with the cold tube for decane-oxygen-

Fig. 10 Typical smoked foil of decane vapor (100°C)-oxygen-nitrogen detonation.

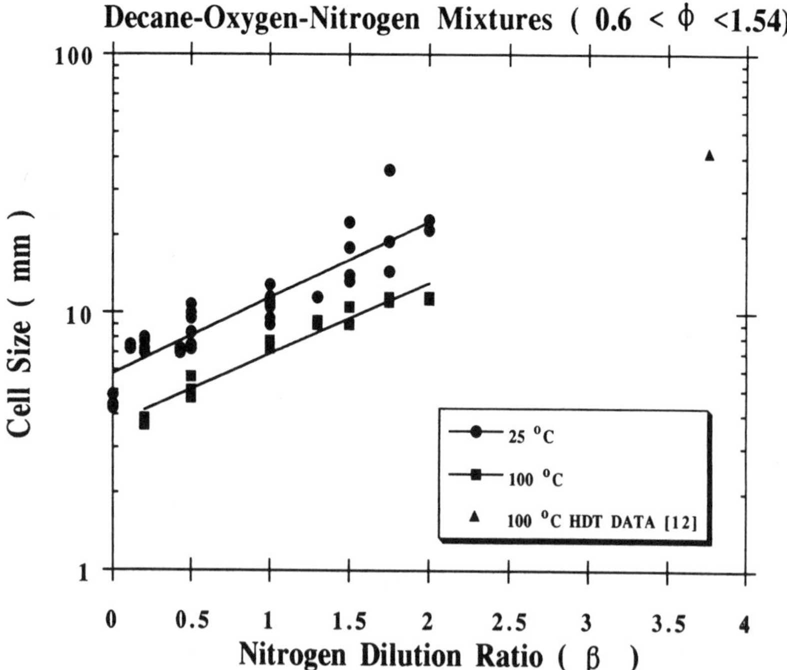

Fig. 11 Cell size measurements λ for decane-oxygen mixtures at various nitrogen dilution ratios, β, for both the cold (25°C) and hot tube (100°C).

nitrogen mixtures at various dilution ratios. Again the cell size for the homogeneous gas phase mixture is smaller than that for the heterogeneous mixtures for all the dilution ratios investigated.

Detonation Model

The general approach used relies on the Zeldovich-von Neumann-Doring (ZND) model. The initial conditions used for the chemical kinetics calculation are those at the von Neumann point immediately behind the leading shock wave. These conditions, including the temperature T_1, pressure P_1, and particle velocity v_1 are calculated from the shock conservation relations as described by Westbrook[13] and Westbrook and Urtiew.[14]

For the kinetics calculations, the reactive mixture is assumed to remain at a constant volume over its reaction time. The ignition delay time is defined in terms of its temperature time history. Most of the mixtures examined undergo a large temperature increase of abut

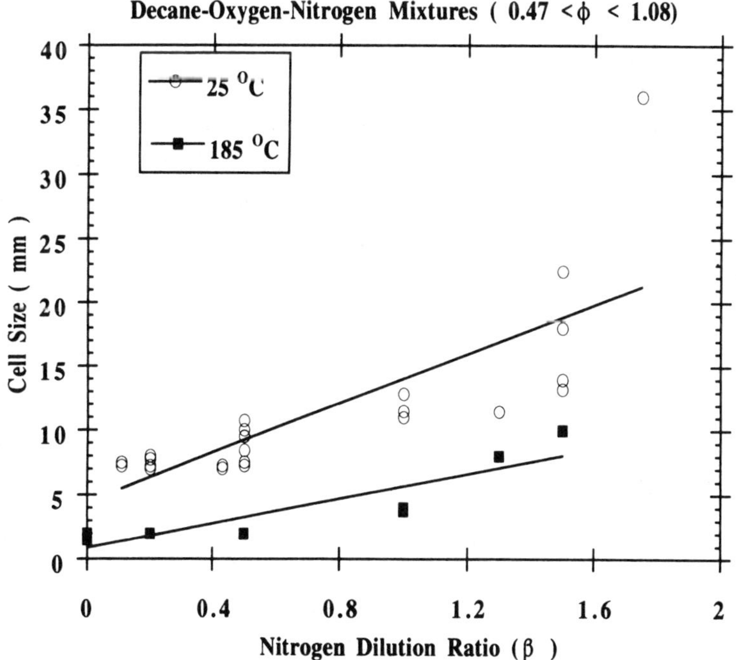

Fig. 12 Cell size measurements, λ, for decane-oxygen mixtures at various nitrogen dilution ratios, β, for both the cold (25°C) and hot tube (185°C).

2000°K, and the ignition delay time is defined as the time of maximum rate of temperature increase. This coincides roughly with the time at which the temperature has completed about half of its total increase. In addition to the induction time τ, it is useful to define the induction length Δ, where $\Delta = (D_{CJ} - v_1)\tau$.

As a result of these simplifications, the computed induction times and induction lengths should be considered as characteristic time and length scales, not as a precise description of the state of a gas element through the detonation front. Shepherd[15] examined H_2-air mixtures and compared the differences between following the Rayleigh line during the induction period and the use of the constant volume approximation. For the type of induction length defined here (that approximately measures the thermally neutral portion of the reaction zone), he found the differences to be very small.

The relationship between the computed induction length and the experimental scales of interest like the cell width is not well understood. Initial studies,[13-14] assumed that these quantities were directly proportional. However, the quantity A had one range of values for hydrocarbon-air mixtures and another range for hydrocarbon-O_2 mixtures.[13] Shepherd[15] showed that A also varies with equivalence ratio for H_2-air mixtures. In spite of these difficulties, the computed induction length can be a useful analytical tool for comparison of measured cell widths. This has been demonstrated extensively for hydrogen and hydrocarbon fuel detonations.[12,13,14,16] For the results reported here, a value of 60 was used for A. This gives the best agreement between the experimental and the computed results for the homogeneous mixtures.

The central element in the present model is the detailed chemical kinetic mechanism which includes the elementary reactions and reaction rate expressions to be integrated by the numerical model. The chemical kinetic mechanism contains 101 species and 785 reversible reactions to treat the high temperature oxidation of n-decane. The mechanism is too lengthy to report here, but is briefly described below.

The chemical kinetic mechanism was assembled and reaction rates were estimated in the same manner as described by Axelsson et al.[17] where the oxidation of n-octane was considered. The current mechanism was assembled by adding the reactions necessary to treat n-decane to an existing reaction mechanism for n-octane. The reactions added included all the possible decomposition reactions for n-decane. Next, we added the site specific abstraction of H-atoms from n-decane by H, O, OH, O_2, HO_2, CH_3, CH_3O, CH_3O_2, C_2H_5, C_2H_3, and $C_{10}H_{21}$. Also included were the decyl ($C_{10}H_{21}$) radical decomposition reactions and decyl reacting with O_2. Each of the five types of decyl radicals were considered. Reactions consuming all the various decanes formed from decyl radicals were included, but their rates were approximate and their products were speculative in some cases. The thermochemistry for the C_{10} species was calculated using THERM[18] which uses bond additivity methods. The thermochemistry was used to calculate reverse reaction rates. The calculated thermochemistry compared very well with the measured values for stable C^{10} species whose thermochemistry is well known.

Using the detonation model described above the cell size was computed for decane-oxygen and decane-oxygen-nitrogen mixtures. Figure 13 plots the cell size obtained experimentally for the decane aerosol-oxygen mixtures and the theoretical curve obtained from the

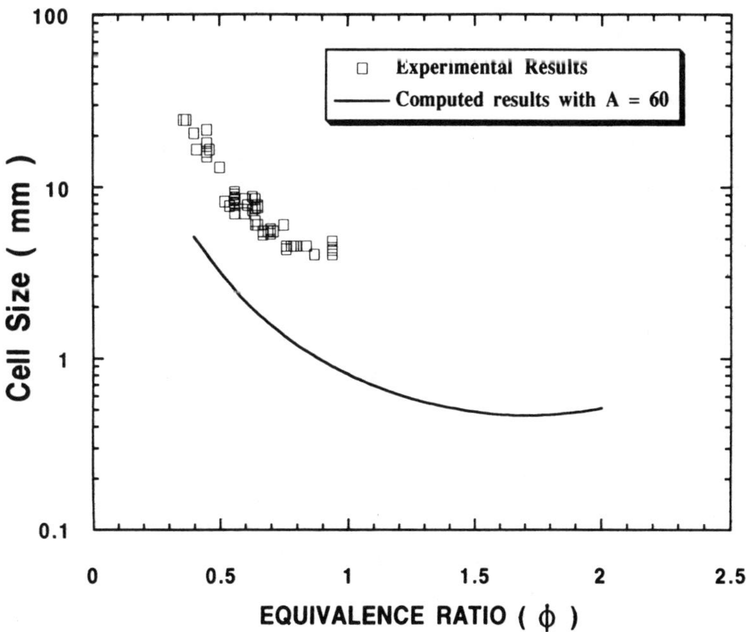

Fig. 13 Experimental and computational cell size, λ, for decane-oxygen-mixtures 25°C.

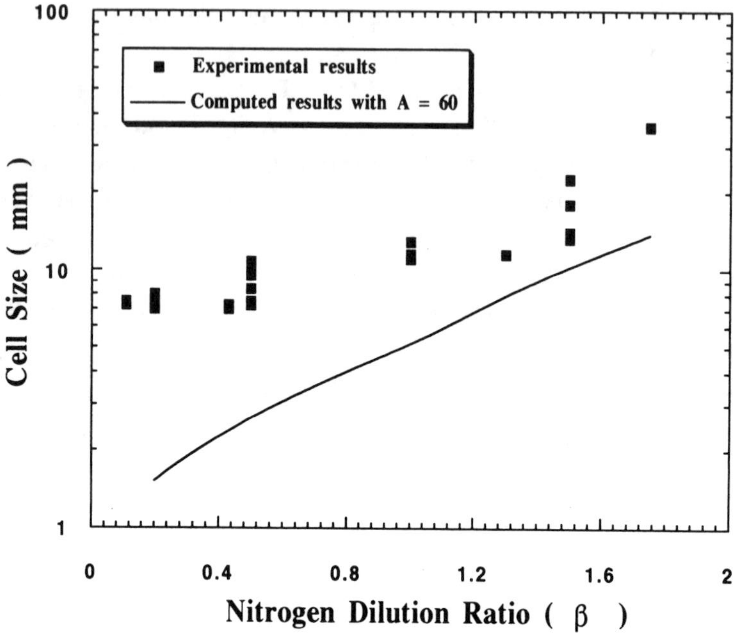

Fig. 14 Experimental and computational cell size, λ, for decane-oxygen-nitrogen mixtures 25°C.

kinetic detonation model. The computed cell size for stoichiometric decane-oxygen at 25°C initial temperature does not agree with the experimentally determined cell size because the detonation model fails to take into account the physical processes present that can potentially increase the length scale of the detonation. The theoretically computed cell size only considers kinetic rate processes thus indicating that the length scale is increased by 5 times, (i.e. $\lambda_{experimental}$=4.0 mm and $\lambda_{computed}$=0.8 mm at approx. ϕ=1.0 with A=60), in decane aerosol-oxygen mixtures for the present 5 μm droplet size. Figure 14 shows both the experimental and theoretical cell size for decane-oxygen-nitrogen mixtures at 25°C initial temperature for various nitrogen dilution ratios. The theoretical results correspond to an equivalence ratio of 0.9. Again the experimental cell size is greater than the computed cell size when a constant of proportionality of 60 is used between the induction length and cell size. Figure 15 shows the experimental and computational results for the mixture in the

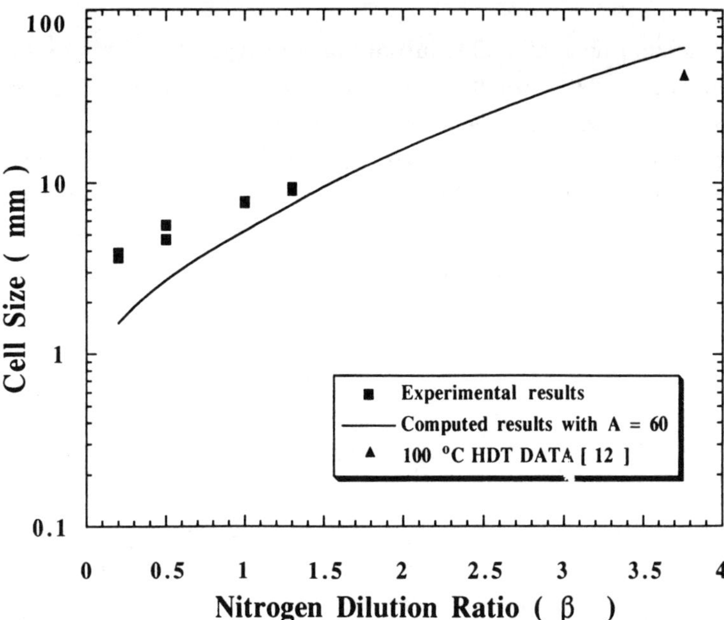

Fig. 15 Experimental and computational cell size, λ, for decane-oxygen-nitrogen mixtures 100°C.

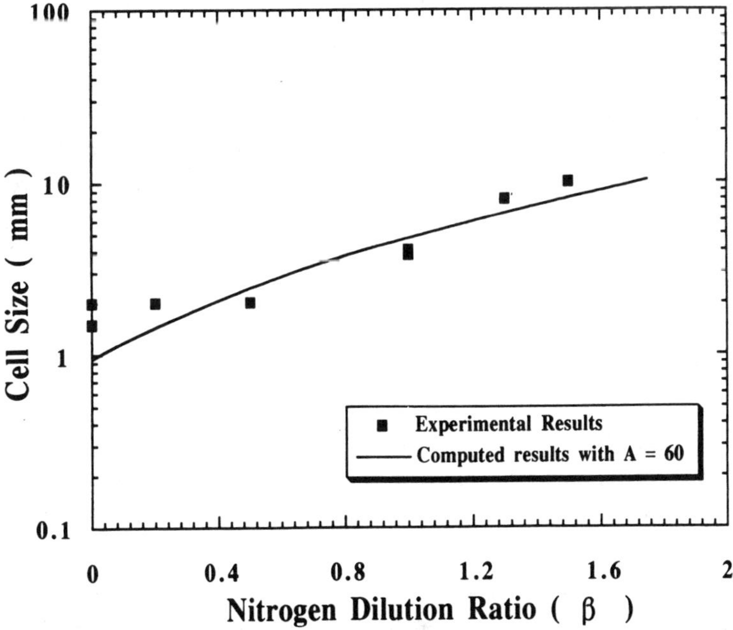

Fig. 16 Experimental and computational cell size, λ, for decane-oxygen-nitrogen mixtures 185°C.

homogeneous phase at 100°C initial temperature. Also included is the experimentally determined cell size for decane-air obtained at Sandia National Laboratories.[12] With a constant of proportionality of $A = 60$ the agreement between the computed results and the experimental results is best. Figure 16 shows both the computed and experimental results for cell size at 185°C initial temperature at various nitrogen dilution ratios. The experimentally measured cell size approaches the value computed by the detonation model. Thus, the chemical kinetic model is in good agreement with the experimental findings when the cell size is considered to be 60 times the induction length.

Conclusions

The present paper demonstrates that in a purely heterogeneous liquid spray-air detonation, the cellular structure is similar to that observed for homogeneous gas mixtures. This suggests that for heterogeneous detonations the mechanism of propagation is the same; i.e., transverse waves are necessary for the coupling between the shock and the reaction zone. It is to be noted, however, that the particle size of the decane droplets in the present experiment is small (i.e., of the order of 5 μm mean diameter). For larger diameter droplets, shock reflections and diffraction from the droplets as well as the turbulence generated in the wake flow behind the droplet may be sufficient to assist the self-sustenance of the detonation. This mechanism has been demonstrated in the work of Wolanski et al.[19] where detonations in CH_4-air mixtures can be obtained readily in the presence of large sand grains distributed in the mixture. This is also similar to the quasi-detonation process in rough tubes[20] where shock reflections assist the propagation of the detonation. Thus it is not clear that larger diameter particles must necessarily render the mixture less sensitive. The decane vapor-air detonation result of Tieszen et al.[12] (i.e., $\lambda = 42$ mm) are in accord with the cell size for the alkane group (e.g., butane $\lambda = 60$ mm, ethane $\lambda = 67$ mm, propane $\lambda = 54$ mm. However, for the heterogeneous phase decane detonation, the cell size is almost a factor of two larger (i.e., $\lambda = 78$ mm). Thus the physical processes for droplet breakup, heat transfer, evaporation, and mixing require a length scale of almost the same order of magnitude as the chemical kinetic processes for the case of the present 5 μm size droplet. It should be noted that the time (or length) scales for the physical processes cannot be scaled linearly with droplet size nor can one even postulate a lesser sensitivity where the droplet size is larger. It is interesting, however, to note that even for very small size aerosols, the characteristic length scale for detonation is increased by almost a factor of two. Since initiation energy is proportional to the cube of the cell size, we note that a 5 μm liquid spray requires an increase of the initiation energy by almost an order of magnitude. This accounts for the difficulty in detonating unconfined decane-air mixtures. The measurements of cellular structure of the heated detonation tube (185°C and 100°C) are in good agreement with the computed results when a constant of proportionality of 60 is used to define the relationship between cell size and induction zone length. Therefore, when the mixture is in the homogeneous phase the kinetic rate processes determine the induction zone length. However, when the fuel is detonated in the aerosol form the chemical kinetic model cannot predict the induction zone length since physical processes also take place. The cell size of heterogeneous detonations can only be determined experimentally. The experimentally determined cell size for stoichiometric decane aerosol-oxygen for the present 5 μm droplet size is 5 times greater than the computed results again indicating that the physical processes present in heterogeneous detonations substantially increase the induction zone length.

Acknowledgments

We are truly grateful to Professor J. R. Bowen for providing us with literature relevant to this study. We are also grateful to Dr. Marshall Berman of Sandia National Laboratories, New Mexico, for his interest in and support of the present work. We would also like to acknowledge D. Tam for his assistance in the initial phase of the experimental work. At McGill University this work was sponsored by the Natural Sciences and Engineering Research Council of Canada under Grants A-3347 and A-7091 and by the Department of National Defence of Canada under Contract 01SG.W7702-6-2521-A.

References

[1]Dabora, E. K., Ragland, K. W., and Nicholls, J. A., "A Study of Heterogeneous Detonations," *Astronautica Acta*, Vol. 12, 1966, pp. 9-16.

[2]Dabora, E. K., Ragland, K. W., Ranger, A., and Nicholls, J. A., "Two Phase Detonations and Drop Shattering Studies," *National Aeronautics and Space Administration*, CR 72225, 1967.

[3]Dabora, E. K., Ragland, K. W., and Nicholls, J. A., "Drop-Size Effects in Spray Detonation," *Proceedings of the 12th Symposium (International) on Combustion*, The Combustion Institute, Pittsburgh, Pa., 1969, pp. 19-26.

[4]Ragland, K. W., Dabora, E. K., and Nicholls, J. A., "Observed Structure of Spray Detonations," *Physics Fluids 11*, 1968, pp. 2377-2388.

[5]Bull, D. C., McLeod, M. A., and Mizner, G. A., "Detonation of Unconfined Fuel Aerosols," *Gasdynamics of Detonations and Explosions*, Vol. 75, 1981, pp. 48-60, AIAA Inc., New York.

[6]Smeets, G., "Spray Detonation Studies Using Laser Doppler Techniques," Institut Franco-Allemand de Recherches de Saint-Louis, Report ISL CO215/85, 1985.

[7]Benedick, W. B., Tieszen, S. R., Knystautas, R., and Lee, J. H., "Detonation of Unconfined Large Scale Fuel Spray-Air Clouds," *Dynamics of Detonations and Explosions: Detonations*, Progress in Astronautics and Aeronautics, Vol. 133, 1989, pp. 297-310, AIAA Inc., Washington, D.C.

[8]Alekseev, V.I., Dorofeev, S. B., Sidorov, V. P., and Chaivanov, B. B., Experimental Study of Detonation Initiation in Motor Fuels Sprayed in Air, Preprint IAE-4872/13, Moscow, USSR, Atominform, 1989.

[9]Bowen, J. R., Ragland, K. W., Steffes, F. J. and Loflin, T. G., "Heterogeneous Detonation Supported by Fuel Fogs or Films," *Proceedings of the 13th Symposium (International) on Combustion*, The Combustion Institute, Pittsburgh, Pa., 1971, pp. 1131-1139.

[10]Knystautas, R., Guirao, C., Lee, J. H., and Sulmistras, A., "Measurements of Cell Size in Hydrocarbon-Air Mixtures and Predictions of Critical Tube Diameter, Critical Initiation Energy, and Detonability Limits," *Dynamics of Shock Waves, Explosions, and Detonations*, Progress in Astronautics and Aeronautics, Vol. 94, 1985, pp. 23-37, AIAA, New York.

[11]Knystautas, R., Lee, J. H., and Guirao, C. M., "The Critical Tube Diameter for Detonation Failure in Hydrocarbon-Air Mixtures," *Combustion and Flame*, Vol. 48, 1982, pp. 63-83.

[12]Tieszen, S. R., Stamps, D. W., Westbrook, C. K., and Pitz, W. J., "Gaseous Hydrocarbon-Air Detonations," Sandia National Laboratories Report SAND 88-1925J, 1988.

[13]Westbrook, C. K., "Chemical Kinetics of Hydrocarbon Oxidation in Gaseous Detonations," *Combustion and Flame*, Vol. 46, 1982, pp. 191-210.

[14]Westbrook, C. K., and Urtiew, P.A., "Chemical Kinetic Prediction of Critical Parameters in Gaseous Detonation," *Proceedings of the 19th Symposium (International) on Combustion*, The Combustion Institute, Pittsburgh, Pa., 1983, pp. 615-623.

[15]Shepherd, J. E., "Chemical Kinetics of Hydrogen-Air-Diluent Detonations," *Dynamics of Explosions*, Progress in Astronautics and Aeronautics, Vol. 106, 1986, pp. 263-293, AIAA Inc., New York.

[16]Westbrook, C. K., Pitz, W. J. and Urtiew, P. A., "Chemical Kinetics of Propane Oxidation in Gaseous Detonations," *Dynamics of Shock Waves, Explosions and Detonations*, Progress in Astronautics and Aeronautics, Vol. 94, 1985, pp. 151-174, AIAA Inc., New York.

[17]Axelsson, E. I., Brezinksy, K., Dryer, F. L., Pitz, W. J., and Westbrook, C. K., "Chemical Kinetic Modeling of the Oxidation of Large Alkane Fuels: n-Octane, *Proceedings of the 21st Symposium (International) on Combustion*, The Combustion Institute, Pittsburgh, Pa., 1986, pp. 783-793.

[18]Ritter, E. R. and Bozzelli, J. W., "THERM: Thermodynamic Property Estimation for Radicals and Molecules," Department of Chemical Engineering, Chemistry and Environmental Science, New Jersey Institute of Technology, Newark, N.J., International Journal of Chemical Kinetics, in press, 1990.

[19]Wolinski, M., and Wolanski, P., "Gaseous Detonation Processes in the Presence of Inert Particles," *Archivum Combustionis*, Vol. 7, 1987, pp. 353-370.

[20]Teodorczyk, A., Lee, J. H. S., and Knystautas, R., "Photographic Studies of the Structure and Propagation Mechanisms of Quasi-Detonations in a Rough Tube," *Dynamics of Detonations and Explosions: Detonations*, Progress in Astronautics and Aeronautics, Vol. 133, 1991, pp. 223-240, AIAA Inc., Washington, D.C.

Experimental Investigations of Accelerating Flames and Transition to Detonation in Layered Grain Dust

Y.-C. Li,* C. G. Alexander,† P. Wolanski,‡ C. W. Kauffman,§
and M. Sichel§
University of Michigan, Ann Arbor, Michigan 48109

Abstract

The processes of flame acceleration and transition to detonation in layered grain dust/air mixtures are studied in a 70-m-long horizontal tube with a 30-cm inside diameter. A 500 g/m^3 cornstarch cloud, ignited by a detonation tube filled with the 100-kPa stoichiometric H_2/O_2 mixture at room temperature, serves as the ignition source throughout the study. Tests were conducted with three types of grain dusts: corn dust, cornstarch, and Mira Gel. The effects of moisture content, dust layer width, dust path length, and dust concentration are considered. Detonation velocities as high as 1650 m/s and overpressures as high as 3.5 MPa have been obtained in a Mira Gel/air mixture. Results are presented which demonstrate clear evidence that, under certain favorable conditions, the transition to quasidetonation or detonation can be achieved in layered grain dust/air mixtures.

Introduction

Dust explosion has resulted in numerous deaths and injuries and in significant property damage since the first reported dust explosion happened in a flour mill of Turin, Italy in 1785. This problem, however, did not draw enough attention in the scientific forefront until the late 1970s. Since then tremendous efforts have been made to understand the explosion hazards of various kinds of combustible dusts. The investigations are primarily focused on the subjects of ignition, turbulent effects, deflagration, and detonation. Generally, grain dust explosions are initiated by relatively small primary explosions. These primary explosions act as ignition sources and transit to secondary explosions which are detonation waves (it should be noticed that the phrase "secondary explosion" used here does not imply any meaning of "minor" or "less important.") In secondary explosions, combustion waves can travel at supersonic speeds

---

* Graduate Student, Department of Aerospace Engineering.
† Undergraduate Student, Department of Aerospace Engineering.
‡ Visiting Professor, Warsaw University of Technology, 00-665 Warsaw, Poland.
§ Professor, Department of Aerospace Engineering.

accompanied by sharp pressure rises that devastate workplaces. Therefore, research concerning the phenomenon of secondary explosions, especially those of deflagration-to-detonation transition (DDT), remain of prime interest to the scientists dedicated to understanding combustible dust explosions.

The first evidence that detonation might occur in a dust/air mixture was obtained in 1971 when Cybulski[1] first observed DDT in coal dust. The investigation showed that the detonation velocity was between 1600 and 2000 m/s and that the pressure in the detonation front was higher than 4.1 MPa. Between 1971 and 1975, Bartknecht[2] reported the detonative combustion in organic dust/air mixtures with detonation velocities in the range of 1300-2500 m/s and maximum pressures 0.7-2.8 MPa.

Kauffman et al.[3,4] first demonstrated the possibility of a grain dust/air mixture detonation in laboratory conditions. They conducted the experimental research on a vertical detonation tube in which the dust was suspended in an oxidizer and showed flame waves traveling at a velocity of 1530 m/s while generating overpressures of 2.5 MPa. Many more experimental studies carried out by Wolanski et al.[5,6] and by Zhang and Groenig[7] demonstrated the detonation or quasidetonation of grain dust/air mixtures with velocities ranging from 1100 to 1550 m/s. Gardner et al.[8] observed the detonation velocity and overpressure for coal dust transition to detonation to be 2850 m/s and 8.15 MPa, respectively.

In laboratory research dust is dispersed, suspended in a gas oxidizer, or layered on the walls of a tube. Most research results[2-8] were carried out with the first two methods because of their favorable conditions for supporting the flame accelerating process. However, dust particles are usually layered on the surfaces of structures. The last method then appears to simulate more closely the real situation in industry and mines. The flame acceleration tube (FAT), the first experimental apparatus devoted to the research of *layered* combustible dusts in laboratory conditions, was originally established at The University of Michigan in the early 1980s. Extensive research was conducted at this facility with many fruitful results.[9-11] However, because during that period the dust path length could not be long enough due to the limitation of the tube length (L = 36 m or L/D = 120), detonation phenomena were not observed.

This paper aims to reveal the processes of deflagration wave propagation and deflagration-to-detonation transition in layered grain dust. Three types of grain dusts were employed: corn dust, cornstarch, and Mira Gel. Special attention is paid to the effects of moisture content, the dust layer width, and concentration on detonation parameters such as quasidetonation or detonation velocity and maximum pressure.

Experimental Apparatus

The FAT is actually a "slim" and long horizontal pipe with one end closed and the other opened to the atmosphere. A schematic diagram of the FAT is shown in Fig. 1. It is a steel tubing with a length of 70 m and an i.d. of 30 cm. It has a working pressure up to 20 MPa, is about 3 m above the laboratory ground, and is firmly sustained by clamps and steel cables. The FAT is instrumented with static pressure transducers, dynamic pressure transducers, photodiodes, and chromel-alumel thermocouples with a 76-μm junction diameter at eight stations along the tube. The location for each station is also shown in Fig. 1. A 4-wavelength optical pyrometer is mounted near the end of the tube. Six pressure switches are distributed along the last 20 m of the FAT. The data acquisition system is composed of a 64-channel, 2-kHz analog-to-digital converter: a LeCroy 8212A system and an IBM System 9000. Four Tektronix 7D20 digital oscilloscopes are employed to record the signals from PCB piezoelectric pressure transducers and photodiodes at the last two stations. The initiator assembly, as shown in Fig. 2, consists of a 2.4-m long and 5.4-cm i.d. detonation tube and a 3-m-long V-channel within the FAT. The detonation tube, separated from the FAT by a Mylar diaphragm, is filled with H_2/O_2 mixture. A specially designed cart, equipped with a 6-liter dust pan, an auger, two motors (one drives the auger and the other the cart wheels), and a fan, is used to travel inside the FAT and load the dust with a desired concentration which can be adjusted by varying the auger speed.

Operationally, the dust was loaded into the FAT with the desired concentration by the cart. The dust mass loading was not necessarily the same along the whole tube. In the initiator, cornstarch was laid along the V-channel. The amount of the dust was weighted such that a dust cloud with a 500-g/m^3 concentration could be generated above the V-channel. The detonation tube was filled with the stoichiometric H_2/O_2 mixture in the pressure of 100 kPa abs. High pressure air (2.1 MPa) was discharged through the V-channel suspending that dust. After a 0.25-s delay, a glow plug ignited the gas mixture, which then ignited the suspended dust cloud, which then led to the combustion of the layered dust in the rest of the FAT.

Three kinds of dusts were tested in the current study. All experiments were conducted with air as the oxidizer and at atmospheric pressure and room temperature. The sizes of the dust and the air moisture contents are given in Table 1. The moisture content of the air qualitatively represents that of the dust.[11] Corn dust and cornstarch were tested in both wet and dry conditions and in two different layered patterns: wide and narrow. The dry condition refers to

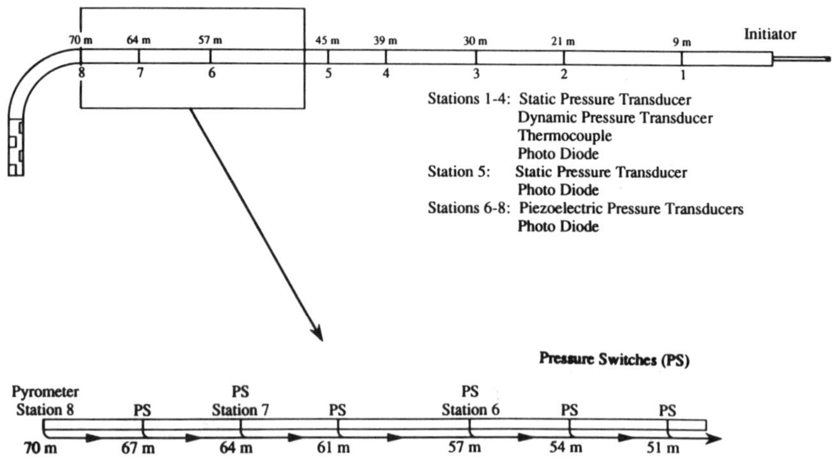

Fig. 1 Schematic diagram of the flame acceleration tube (FAT).

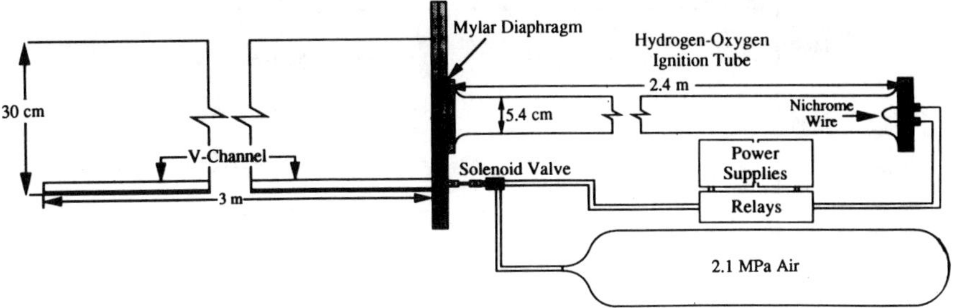

Fig. 2 Schematic diagram of the initiator assembly.

the dust in the FAT being dried by dry laboratory air (dew point as low as -32 °C) for 15-20 hours. The wet condition simply refers to the lack of a drying process. The wide or narrow patterns depend on whether or not the fan of the cart is in operation. The path length refers to the length of the layered dust, excluding that of the V-channel. Cabosil, a very fine silicon-based and volatile-like dust, was added into cornstarch at 1% by weight of the total amount of the dust. Mira Gel has the same chemical composition as cornstarch. Nevertheless, the different manufacturing processes lead to the different dust characteristics in the explosions.

Results and Discussion

A comparison of maximum pressure P_{max} among four different corn dust runs at the eight stations is given in Fig. 3. Dry dust and wet dust were layered in both wide and narrow patterns for a path length of 27 m. As shown in Fig. 3, P_{max} for the narrow and dry dust is 1.3-2 times higher than that of narrow and wet dust, whereas P_{max} for the wide and wet dust is 2-4 times higher than that of narrow and wet dust. The dust moisture content and dust layer width are of importance for this result. In an ordinary situation, the overwhelming majority of dust particles do not exist individually but agglomerate together. The moisture content strongly affects the cohesiveness of dust agglomerations, and so the induced convective flow is easier to break the dust agglomerations with

Table 1 Dusts tested in the current study

		Air Moisture Content (%)	
Dust	Sizes (μm)	Dry	Wet
Corn	All Sizes	0.020-0.115	0.240-0.790
Cornstarch[a]	< 75	0.028-0.141	0.196-1.113
Mira Gel	< 150	0.008-0.141	0.917-1.394

[a] With 1% cabosil additive by weight.

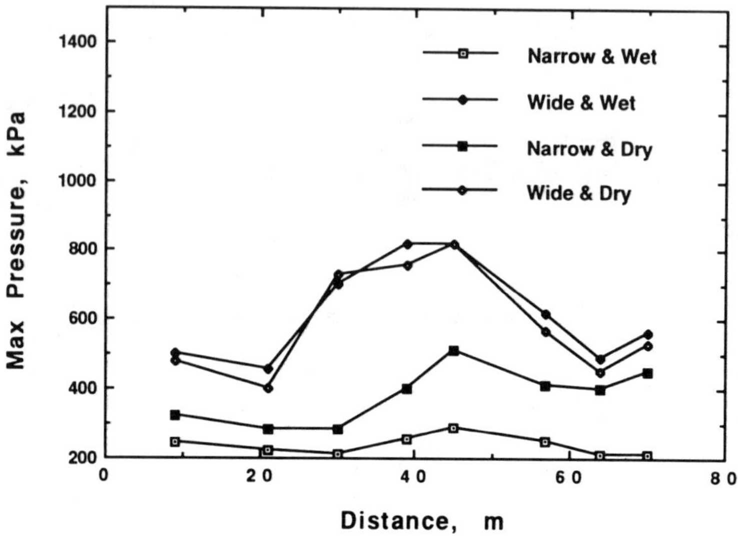

Fig. 3 Maximum pressure along with the tube for corn dust, concentration 500 g/m^3, path length 27 m.

low moisture content than those with high moisture. Moreover, the moistures in the dust also absorb the heat energy to vaporizing themselves, which in turn reduce the strength of the dust combustion process. Increasing the width of the dust layered, incidentally, increases the particle numbers exposed to the convective flow and therefore improves the combustion conditions. Note that the highest maximum pressure occurs at x = 39-45 m rather than at x = 9-30 m in which the dust was layered. This is because the initial shock wave and primary explosion waves generate a strong convective flow, carrying out the layered dust towards the end of the FAT. Therefore, it is reasonable to expect the highest maximum pressure in the downstream of the layered path.

For cornstarch static overpressure is illustrated in Fig. 4 as a function of the time at the eight stations. The conditions for this run were as follows: the concentration was 500 g/m^3, the path length was 64 m, and the dust was wet and layered in a wide pattern. An x-t diagram for the flame front and initial shock path is shown in the same figure. Pressure is normalized with the atmospheric pressure at the time the run is made, distance with the inner diameter of the FAT, and time with the time taken by an acoustic wave to travel the length of the tube. The first disturbance arriving at any station is the shock wave resulting from the rupture of the initiator diaphragm, as can be seen in Fig. 4. The slope at any point of the flame trajectory represents the flame velocity relative to the laboratory frame at that point. The flame velocity continuously accelerates along the tube. From x = 9-45 m the combustion of the dust is still in its deflagration stage where pressure slopes gently, but the pictures of pressure are completely changed from x = 57-70 m. Pressure waves coalesce into shocks, and maximum pressure sharply jumps to about 2.3 MPa at x = 64 m and about 2.6 MPa at x = 70 m, i.e., the tube exit. The absolute velocity of the flame fronts

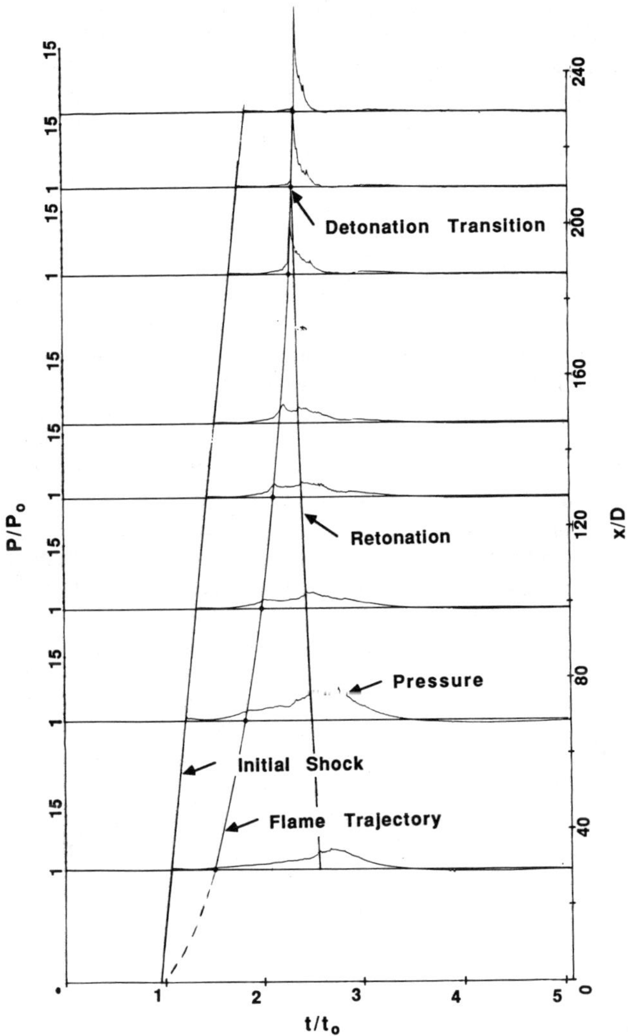

Fig. 4 Pressure histories, initial shock path, flame trajectory, and retonation wave for cornstarch, concentration 500 g/m^3, path length 64 m, wet dust, wide layered pattern.

reaches about 1330 m/s. The flame front becomes closely coupled to a peak on the pressure trace near the end of the FAT; undoubtedly, a quasidetonation, whose velocity and pressure are considerably lower than those of classical Chapman-Jouguet detonation, takes place during this run. The quasidetonation is caused by the roughness of the tube which causes substantial losses of momentum and energy.[12] The point of DDT can be roughly spotted at around $x = 64$ m. A retonation wave emerges at the DDT point and moves into the burnt mixture, generating a disturbance at each station, as shown in Fig. 4. Similar to the homogeneous reactive systems, this is a typical case of "explosion in the explosion," a phenomenon first observed by Oppenheim et al.[13,14] in gaseous mixtures, in the heterogeneous reactive system.

In Fig. 5 flow velocity histories of the same run described in the above paragraph are shown at different stations. Velocity is normalized with an acoustic wave speed at the time the run was made. The flame trajectory is also shown for reference. Like the pressure histories in the last figure, the first disturbance arriving at any station is due to the initial shock wave. This results in an induced convective flow, which in turn entrains the layered dust along after the flow. The kinetic energy of the flow is quickly transferred to the dust particles, which results in a velocity decrease immediately behind the initial shock at each station as can been seen in Fig. 5. Then pressure waves generated by the accelerating flame reinforce the flow and cause the velocity to increase again. The maximum velocity is reached once the flame front arrives, since the latter is always coupled to the peak of pressure waves. This is more apparent in the last two stations due to the coalescence of the pressure waves and formation of the leading shock of detonation. From the analysis of the experimental data, dust entrainment is the key factor that dominates the process of flame acceleration in layered dust. Dust particles are entrained by this induced convective flow and presuspended in the tube, which provides an environment supporting the following flame acceleration. More detailed descriptions on dynamics of dispersion of layered dust by a shock wave can be found in Refs. 10 and 15. It should be noted that the data processing of velocity measurement involves the temperature measurement by means of thermocouples. Because of the defects in the latter, as will be explained later, the velocity can only be illustrated qualitatively. It should also be noted that once flame fronts pass the flow scenario behind the flames become very complicated due to the three-dimensional and unsteady nature of the compression waves and flames; therefore, the velocity measurement once the flame front passes should be disregarded.

Figures 6-9 present the results of Mira Gel. The result of a typical run for Mira Gel is given in Fig. 6. The concentration in the FAT was 400 g/m^3, the path length was 64 m, and the dust was dry and layered in a wide pattern. The whole picture is quite similar to that of Fig. 4. The maximum pressure and flame velocity at the FAT exit are about 2.8 MPa and 1370 m/s, respectively. The quasidetonation once again occurs in this run.

A fine plot of pressure/time variation and photodiode response is shown in Fig. 7. This is the PCB pressure transducer output at $x = 64$ m recorded by the oscilloscope. To enhance the dispersion of the dust by the initial shock waves and primary explosion waves, two voids, i.e., sections without layering any dust, were adopted. One was 9 m long after the initiator; the other was 3 m long

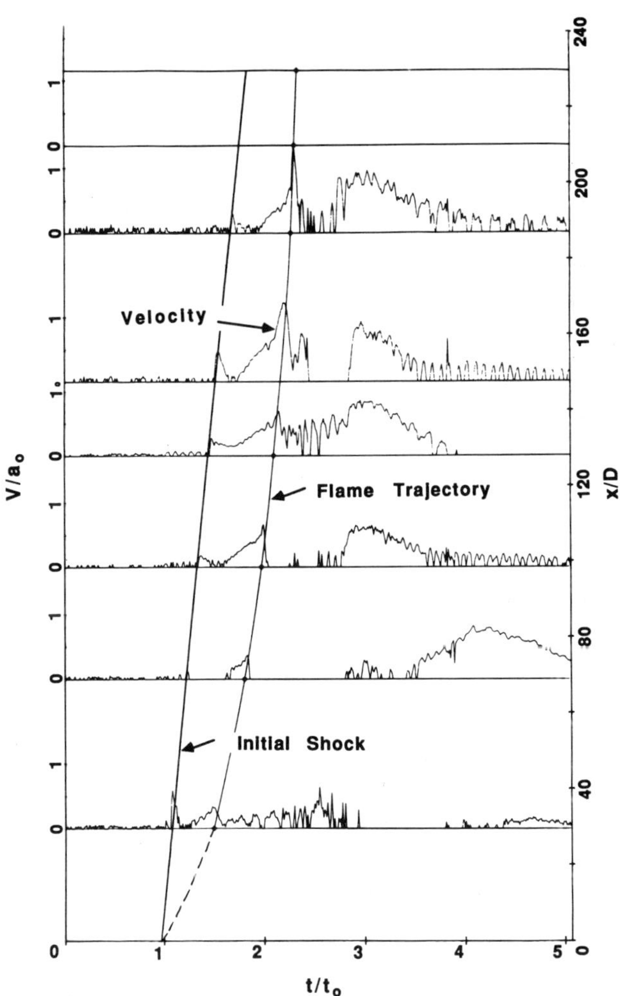

Fig. 5 Velocity histories, initial shock path, and flame trajectory for cornstarch, concentration 500 g/m³, path length 64 m, wet dust, wide layered pattern.

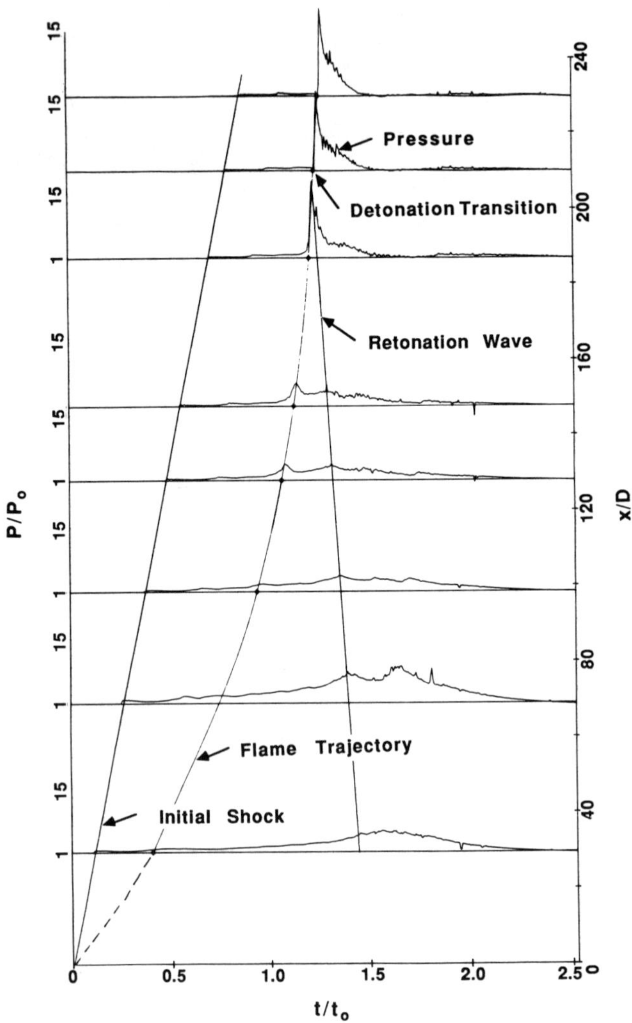

Fig. 6 Pressure histories, initial shock path, and flame trajectory for Mira Gel, concentration 400 g/m^3, path length 64 m, dry dust, wide layered pattern.

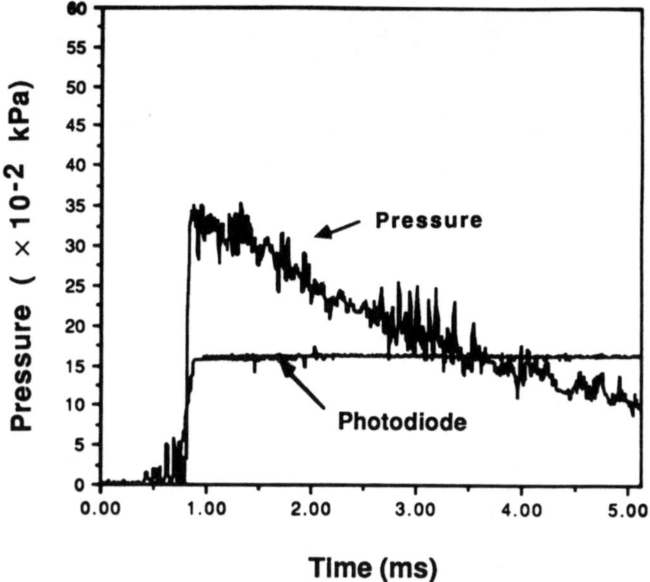

Fig. 7 Pressure history and photodiode respond at x = 64 m for Mira Gel, concentration 500 g/m^3, path length 64 m, wet dust, wide layered pattern, with two voids: 0-9 m and 30-33 m. Corresponding shock velocity is 1630 m/s.

after 30 m of path length. The concentration for this run was 500 g/m^3, the path length was 64 m, and the dust was dry and layered in a wide pattern. As can be seen from Fig. 7, the pressure jumps to about 3.5 MPa immediately after the photodiode signal passes. Again the fact that the flame front closely accompanies the pressure peak indicates that a detonation wave has emerged during this run. Despite the instrumental noises, strong evidence of the transverse pressure waves can be seen in Fig. 7. This indicates that the Mach stem structure which commonly exists in self-sustaining detonation waves in gaseous mixtures may also exist in dust /air mixtures.

Figures 8 and 9 contain temperature data from a single run for Mira Gel. Figure 8 shows temperature profiles measured by thermocouples at x = 9 and 21 m. Flame trajectory and initial shock are also shown for reference. The temperature is normalized with the room temperature. The concentration was 350 g/m^3, the path length was 67 m, and the dust was dry and layered in a wide pattern. Similar to the pressure profiles shown in Figs. 4 and 6, the initial shock generates the first disturbance on those of temperature. However, temperature rises due to initial shock are negligible compared with those due to the flame fronts. Unlike pressure in Figs. 4 and 6, the temperature always answers the flame front and rises immediately after it passes. It should be noted that due to the limitation of the sizes of the thermocouples, slow time response is expected, and so it is not quantitatively reliable for temperature measurements. Furthermore, since the FAT is not heat-isolated, the heat losses through the wall of the FAT may not be negligible. A thermal boundary layer must exist next to the wall. The thermocouple probes, unfortunately, can only measure the

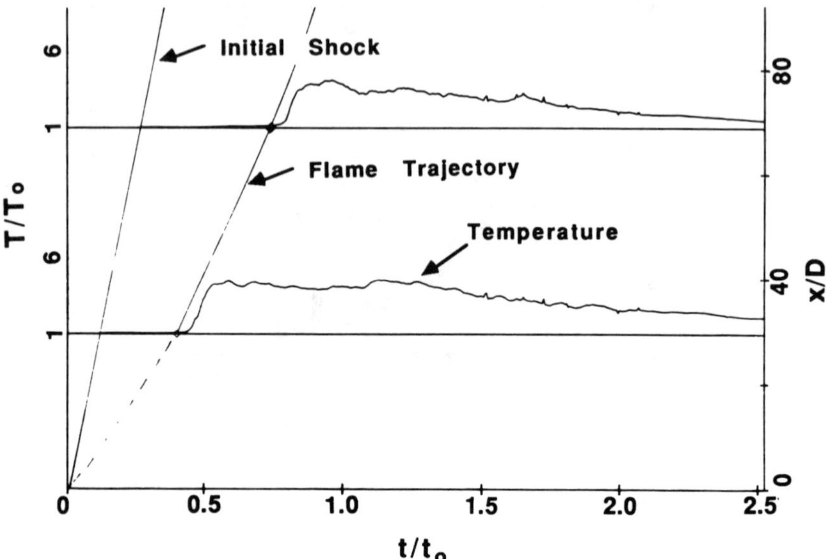

Fig. 8 Temperature profiles measured by thermocouples at $x = 9$ and 21 m for Mira Gel, concentration 350 g/m^3, path length 67 m, wet dust, wide layered pattern.

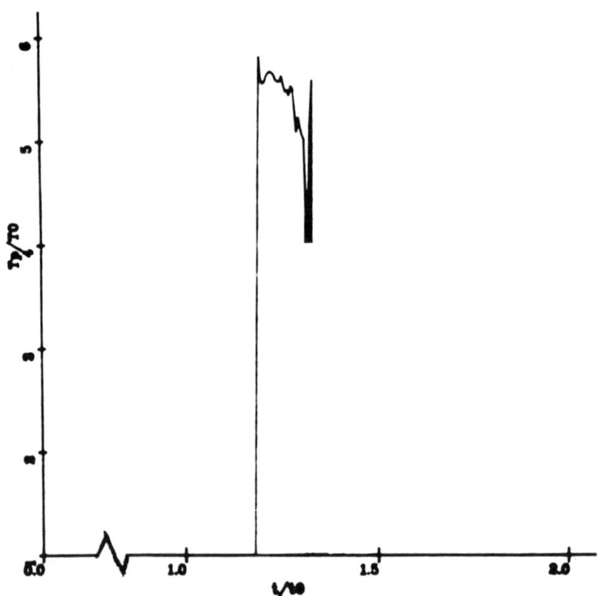

Fig. 9 Particle temperature measured by an optical pyrometer at $x = 70$ m for Mira Gel, concentration 350 g/m^3, path length 67 m, wet dust, wide layered pattern. Temperature below 1200 K is truncated.

temperature within this boundary layer, which implies that the temperature reading from the thermocouples is not the real temperature in the core region (namely, the main stream of the combustion waves). Nevertheless, it is still qualitatively meaningful for having the general trends for the temperature. Figure 9 contains temperature data obtained from the 4-wavelength pyrometer located at x = 70 m. The data processing involves the ratio of signals, and the temperature is set arbitrarily to zero once one of signals is lower than a threshold value. Hence, only the temperature above a certain value (e.g., 1200 K) can be recorded in Fig. 9. The temperature sharply jumps to about 1860 K, and temperatures below 1200 K are truncated from the picture.

The main experimental results for three different dusts are summarized in Table 2. Because of the extreme complexity and unpredictable feature of the dust explosion processes,[16] the experiments are not always repeatable and the data scatter in a relatively wide range. The data for corn dust were collected under the condition of a path length of 39 m, and, hence, it is inappropriate to compare them with the data of cornstarch and Mira Gel. Detonations were not observed in the corn dust runs. This is because the path lengths were not long enough to cause DDT. The experiments conducted among all three types of dusts show that there exists a minimum path length to provide necessary conditions for DDT. This minimum path length is in the range of 45-55 m or, in the nondimensionalized scale, x/D = 150-167. This is probably the reason why DDT was not achieved in the previous work in the FAT,[9-11] when it was 36-m-long or x/D was equal to 120. Among the favorable conditions supporting the flame acceleration, the moisture content and width of the dust (as discussed earlier) play very important roles. Besides these conditions the origin and nature of dust cannot be overlooked. Unlike premixed gas mixtures, dust mixtures vary in explosibility with their origins and natures. As can be seen in Table 2, cornstarch exhibits weaker explosive parameters compared to Mira Gel, although both dusts possess the same chemical compositions. Furthermore, by observing the processes of the flame acceleration, one can see DDT is easy to achieve if the flame front is far behind the initial shock in the front part of the FAT. In other words, a well-dispersed dust cloud provides a good opportunity for reaching detonation. This has been supported by the runs with the voids inserted in the dust layer. These voids slow the flame speed in the front part of the FAT and hence leave sufficient time for the initial shocks and pressure waves to entrain the layered dust. The following flames greatly benefit from this well-dispersed dust cloud and accelerate more rapidly in the rear part. As can be seen in Table 2, Mira Gel reaches its strongest explosion at a detonation velocity of 1650 m/s and a maximum pressure of 3.5 MPa with the concentration of 500 g/m^3. The differences among the different concentrations, however, does not seem significant.

Conclusions

The current study has provided a valuable insight into the development of layered grain dust/air explosions in laboratory conditions and has shown strong evidence that, under certain conditions, the combustion of layered grain dust produces rapidly accelerating flames and is capable of resulting in quasidetonations or detonations. Both wide-layered pattern and dry dust contribute favorable conditions to the combustion process of grain dust. The

Table 2 The data of major parameters for different dusts

Dust	Concentration (g/m3)	Flame velocity[a] (m/s)	Shock velocity[a] (m/s)	Quasi-detonation (detonation) velocity (m/s)	Maximum Temperature (K)	P (MPa)	(Pmax) (MPa)
Corn[b]	500				~1700	~0.5	(10)
Cornstarch	500	1078-1330	1085-1300	1050-1200	1640-1910	~1.9	(2.6)
Mira Gel	250	925	~1200	~1150	1670-1850	~0.9	
	300	1200	1250	1150-1200	~1700		(2.8)
	350	1300	1370	1250-1300	~1800	~2.0	(2.9)
	400	1000-1368	1170-1368	1100-1300	1620-1760	1.4-1.9	(2.8)
	500	1368-1714	1520-1714	1350-1650	1700-1730	~2.1	(3.5)

[a] Relative to laboratory frame.
[b] Path length < 39 m. Detonations not observed.

dust entrainment and dispersion apparently play a prominent role in the dust combustion. Of the dusts tested in the current study, Mira Gel seems the most explosive dust, followed by cornstarch and corn dust. It has been found that a least path length is necessary to afford DDT happening in the FAT. A layered dust explosion is intrinsically a time-dependent process, due to the competition between dust dispersion by the shock waves and the accelerating combustion waves. Evidence of transverse waves, which is common for the detonation structure in gaseous mixtures, has been found in layer dust explosions. More evidence for the transverse waves will be provided in another publication. High velocity of the detonation wave and the associated high rise in pressure obtained in this study show these dusts indeed pose a great potential hazard for the industry manufacturing such products.

Acknowledgments

This research is co-sponsored by the National Institute of Occupational Safety and Health of the U. S. Department of Health and Human Services under Grant OH01122-08. The authors would like to express their appreciation to A. S. Harbaugh for his technical assistance.

References

[1] Cybulski, W. B., "Detonation of Coal Dust," *Bulletin of the Polish Academy of Sciences,* Vol. 19, 1971, pp. 37-41.

[2] Bartknecht, W., *Explosions*, translated by H. Burg and T. Almond, Springer-Verlag, New York, 1981, Pt. I, Chap. 3.

[3] Kauffman, C. W., Wolanski, P., Ural, E., Nicholls, J. A., and Van Dyk, R., "Shock Wave Initiated Combustion of Grain Dust," *Proceeding of the First International Symposium on Grain Dust*, Manhattan, KS, 1979, pp. 164-190.

[4] Kauffman, C. W., Wolanski, P., Arisoy, A., Adams, P. R., Maker, B. N., and Nicholls, J. A., "Dust, Hybrid, and Dusty Detonation," *Dynamics of Shock Waves, Explosions, and Detonations,* edited by J. R. Bowen, N. Manson, A. K. Oppenheim, and R. I. Soloukhin, Vol. 94, Progress in Astronautics and Aeronautics, AIAA, New York, 1984, pp. 221-240.

[5] Wolanski, P., "Detonation in Dust Mixtures," *Proceeding of Shenyang International Symposium on Grain Dust Explosions*, Shenyang, China, 1987, pp. 568-598.

[6] Fangrat, J., Glinka, W., Wolanski, P., and Wolinski, M., "Detonation Structure in Organic Dust-Oxygen Mixtures," *Archivum Combustionis,* Warsaw, Vol. 7, 1987, pp. 321-332.

[7] Zhang, F., and Groenig, H., "Detonation Structure of Corn Starch Particles-Oxygen Mixtures," *Dynamics of Detonations and Explosions: Detonations,* edited by A. L. Kuhl, J.-C. Leyer, A. A. Borisov, and W. A. Sirignano, Vol. 133, Progress in Astronautics and Aeronautics, AIAA, New York, 1990, pp. 342-355.

[8] Gardner, B. R., Winter, R. J., and Moore, M. J., "Tests of Explosion Development in Coal Dust/Air Suspensions," *Twenty-First Symposium (International) on Combustion,* The Combustion Institute, Pittsburgh, PA, 1986, pp. 335-343.

[9] Kauffman, C. W., Srinath, S. R., Tezok, F. I., Nicholls, J. A., and Sichel, M., "Turbulent and Accelerating Dust Flames," *Twentieth Symposium (International) on Combustion,* The Combustion Institute, Pittsburgh, PA, 1984, pp. 1701-1708.

[10] Srinath, S. R., Kauffman, C. W., Nicholls, J. A., and Sichel, M., "Secondary Dust Explosions," *Industrial Dust Explosions,* edited by K. L. Cashdollar and M. Hertzberg, American Society of Testing and Materials, ASTM STP 958, Philadelphia, PA, 1987, pp. 90-106.

[11] Srinath, S. R., "Flame Propagation Due to Layered Combustible Dust," Ph.D. Thesis, University of Michigan, Ann Arbor, MI, 1985.

[12] Zel'dovich, Y. B., Borisov, A. A., Gelfand, B. E., Frolov, S. M., and Mailkov, A. E., "Nonideal Detonation Waves in Rough Tubes," *Dynamics of Deflagrations and Reactive Systems: Heterogeneous Combustion,* edited by A. L. Kuhl, J. R. Bowen, J.-C. Leyer, and A. A. Borisov, Vol. 114, Progress in Astronautics and Aeronautics, AIAA, New York, 1988, pp. 211-231.

[13] Oppenheim, A. K., Laderman, A. J., and Urtiew, P. A., "The Onset of Retonation," *Combustion and Flame,* Vol. 6, 1962, pp. 193-197.

[14] Urtiew, P. A., and Oppenheim, A. K., "Experimental Observations of the Transition to Detonation in an Explosive Gas," *Proceedings of the Royal Society of London,* Vol. A295, 1966, pp. 13-28.

[15] Boiko, V. M., Papyrin, A. N., Wolinski, M., and Wolanski, P., "Dynamics of Dispersion and Ignition of Dust Layers by a Shock Wave," *Dynamics of Shock Waves, Explosions, and Detonations,* edited by J. R. Bowen, N. Manson, A. K. Oppenheim, and R. I. Soloukhin, Vol. 94, Progress in Astronautics and Aeronautics, AIAA, New York, 1984, pp. 293-301.

[16] Wolanski, P., "Deflagration and Detonation Combustion of Dust Mixtures," *Dynamics of Deflagrations and Reactive Systems: Heterogeneous Combustion,* edited by A. L. Kuhl, J.-C. Leyer, A. A. Borisov, and W. A. Sirignano, Vol. 132, Progress in Astronautics and Aeronautics, AIAA, New York, 1990, pp. 3-31.

Enhancement and Generation of Detonations Using Dust Layers

J. Sheng,* C. W. Kauffman,† M. Sichel,‡ P. Wolanski§ and N. A. Tonello¶

University of Michigan, Ann Arbor, Michigan 48109

Abstract

The results of an experimental study of the generation of a detonation in a layer of dust adjacent to gaseous oxygen, and the enhancement of a gaseous detonation by an adjacent layer of dust are presented. In both cases a layer of corn starch dust of controllable thickness is deposited at the bottom of the low pressure section of a shock tube and the initiation energy is supplied by igniting a stoichiometric H_2-O_2 mixture in the high pressure section of the tube. The pressure variation behind the detonations was measured with pressure transducers, and pressure switches were used to measure the detonation velocity. To measure detonation enhancement the shock tube, which is 3.81cm by 6.35 cm in cross section area and 6.1 m long, was filled with H_2-O_2 mixtures of different equivalence ratios, and a dust layer was deposited on the bottom of the tube. Experiments were conducted for different oxygen concentrations, mixture pressures and dust loading ratios. With stoichiometric H_2-O_2 no dust reaction was observed due to the lack of oxygen required for the reaction of the dust, but with lean mixtures a detonation-like regime was observe. The measured results were also compared with those computed using the NASA Gordon-McBride Code (CET86), assuming that the dust-gas mixture is homogeneous. The objective of the dust layer detonation experiments was to achieve detonation using an oxidizer mixture as close as possible to air in oxygen content. Dust was deposited along the entire length of the low pressure section of the shock tube and an O_2-N_2 mixture was introduced at atmospheric pressure. Parameters that can be varied are dust concentration, oxygen concentration, and initiation energy. So far combustion has only been observed for very rich mixtures (high dust loading) in the presence of pure oxygen using high initiation energies.

---

*Postdoctoral Researcher, Department of Aerospace Engineering.

†Associate Professor, Department of Aerospace Engineering.

‡Professor, Department of Aerospace Engineering.

§Visiting Professor, Department of Aerospace Engineering.

¶Graduate Student, Department of Aerospace Engineering

Introduction

Dust explosions were first recognized as such in coal mines in 1844 by Faraday.[1] However, this interpretation was not accepted until the beginning of this century. [2] With the rapid growth of powder and mining technology, there has been a significant increase in serious explosions associated with combustible dust. In most cases, these explosions can be considered as deflagrations, but detonative combustion involving dust-oxidizer mixtures has also been observed since early 1970,[3] and extensive research studies have concentrated on this area. Kauffman et al.[4] studied the detonability of suspended oat dust-air mixtures, and Tulis [5,6] and Veyssiere [7,8] investigated the detonation of aluminum particles dispersed in air. The results from the large-scale experiment of Gardner et al[9] indicate transition from deflagration to detonation in premixed coal dust-air mixtures, with a maximum pressure and velocity of 81.5 bar and 2850 m/s; however, in these experiments a stable detonation has not been reached. However, Zhang and Gronig [10] observed a stable self-sustained detonation with a spin structure in premixed corn starch dust-oxygen mixtures. They concluded that transverse waves have a dominant effect in stable detonations in dust-gas mixtures, just as in the case of gaseous detonations. In many practical cases the dust is in the form of a layer at the bottom of a tube or tunnel so that mixing of this layer with the ambient gaseous oxidizer becomes a crucial step in the initiation and propagation of detonations. Work dealing with this type of dust detonation, sometimes referred to as layered detonations, is sparse, but will be discussed in detail in this paper. Since dust explosions involving detonations cause the most serious property damage and personal injury, and are more difficult to control, it is important to understand the mechanisms pertaining to dust detonations of both the premixed and layered types.

In many cases, such as in underground coal mines, explosions occur in a mixture of combustible dust and gaseous explosives. Such "hybrid" detonations have been investigated both experimentally and theoretically by Kauffman et al.,[4], Veyssiere,[7,8], Afanasieva et al,[11] and Khasainov and Veyssiere.[12] In the hybrid mixture, both gases and dust are reactive. The structure of such hybrid detonations depends mainly on the reactivities of the gaseous explosive and the combustible particles in the mixture. The majority of hybrid detonation studies considered only premixed dust-gas mixtures; nevertheless, there is very little work dealing with hybrid detonations in which the dust exists as a layer below the gaseous explosive, and this class of hybrid detonation forms the subject of the present paper.

Experimental Setup

A shock tube, which is shown schematically in Fig. 1, has been used to investigate the detonation enhancement by layers of corn starch dust. In this tube, which is 6.1 m. long and has a rectangular cross section of 6.35 cm by 3.81 cm, the gas composition and pressure as well as the dust loading ratio can be controlled. A 1.52 m-long tube with a circular cross section 7.37 cm in diameter, which can be used as an initiator, is attached to the driver end of the shock tube and is connected to it by a 0.31m long transition section. The other end of the shock tube is open to the atmosphere. The initiator is separated from the shock tube by a mylar diaphragm and is charged with mixtures of stoichiometric hydrogen and oxygen at different pressures. The initiator also can be used as an extension of the main shock tube by omitting the diaphragm

Pressure transducers (Kistler 603 B1) located upstream, in the middle, and downstream of the dust layer were employed to record the pressure history behind the detonation wave, and home made pressure switches were used to measure the wave velocity. The signals from the pressure switches and pressure transducers provided the input for a timing circuit labeled WIDGET in Fig. 1, which generates a step signal as the wave passes each of these devices. The resultant time of passage data was used to determine wave propagation velocities. The data were sampled using a CAMAC digital data acquisition system.

A thin layer of corn starch 2 m long was deposited on the bottom of the tube starting 3.3 m. from the diaphragm separating the initiator and main shock tube. The tube above the layer was filled with different gaseous mixtures. The gas mixtures used in the experiments were: 1) air, 2) pure O_2, 3) $2H_2$+O_2, and 4) H_2+O_2. For the pure oxygen experiments the initiator was charged with $2H_2$+O_2 + He at 5 and 7.67 atm.. The initiator was not separated from the test section by a diaphragm in cases (2) and (4) when the gas is detonable, except otherwise stated.

Preliminary Analysis

Before starting the experiments, the NASA Gordon-McBride Code[13] was run to estimate the detonation properties of cornstarch-air, cornstarch-O_2, and cornstarch-H_2-O_2 mixtures. Since the Gordon MacBride code does not have provisions for dealing with dust combustion per se, it was assumed in the

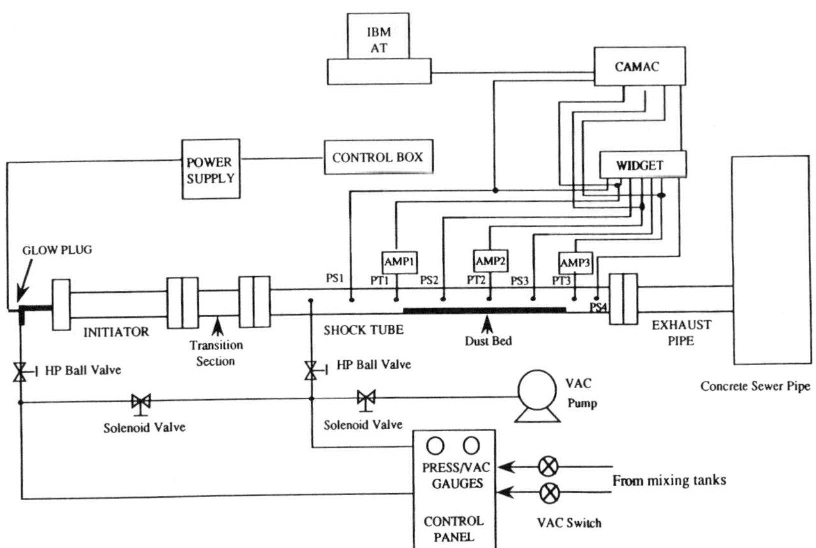

Fig. 1 Experimental setup.

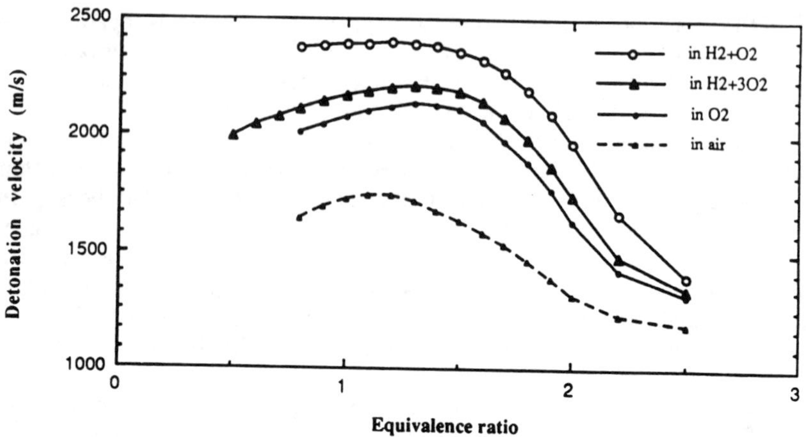

Fig. 2 Detonation velocity of cornstarch dust in different gas mixtures, calculated using The NASA Gordon-MacBride program.

calculations that the dust-gas mixtures are homogeneous. The computed detonation velocity and pressure ratio are plotted in Figs. 2 and 3, respectively.

Figure 3 shows that the cornstarch-O_2 mixture results in the maximum detonation pressure ratio with a value of 52 at an equivalence ratio of 1.5. The detonation pressure ratio of the cornstarch-H_2+3O_2 mixture ranks second, followed by the mixture of cornstarch-H_2+O_2 and cornstarch-air. It should also be noted that the maximum detonation pressure ratios occur at different values of the equivalence ratio for the different dust-gas mixtures. Figure 2 indicates that the maximum detonation velocity occurs for the cornstarch-H_2+O_2 mixture, while the maximum detonation velocity of cornstarch-H_2+3O_2 is the second highest, followed by the cornstarch-O_2 and cornstarch air mixtures. Again, the

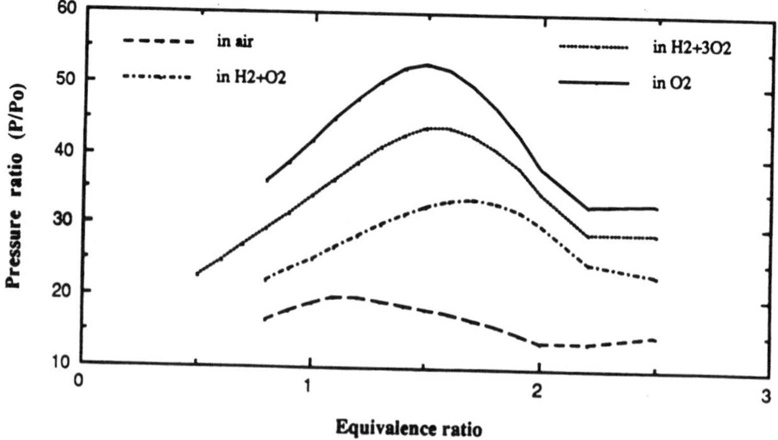

Fig. 3 Detonation pressure ratio of cornstarch dust in different gas mixtures calculated using NASA Gordon-MacBride program.

maximum detonation velocity occurs at different equivalence ratios for the different mixtures.

Experimental Results

Experiments were first conducted to establish whether layers of cornstarch dust in oxygen could be made to detonate in the shock tube used in these experiments. Transition from deflagration to detonation in the case of corn starch layers in air has been observed in this laboratory in a flame acceleration tube (FAT) 70 m long and 30 cm in diameter. With the same dust-loading ratio (500 gm/m3) as in the FAT, no chemical reaction was observed with mixtures of air at 1 atm, nor with dust-loading ratios of up to 5 kg/m3 in pure oxygen at 1 atm and 2 atm. A reasonable explanation is that with such a small and short tube, the dust layer does not have enough time to disperse before the initiating shock wave reaches the end of the tube. In the FAT experiments it was found that dust ignition did not occur until 0.1 seconds after the shock first came into contact with the dust layer, while in the much shorter shock tube, the shock wave propagates to the end of the shock tube within 5 ms in case of the air or oxygen charge. Another reason for the failure to observe reaction is that in the relatively short and small shock tube only a small fraction of the dust layer is actually dispersed and mixed with the gaseous oxidizer above it so that the resultant mixture is below the lean detonation limit. It is necessary to mention that the dust-loading ratio as used in the experiment is the dust concentration that would result if the entire dust layer were uniformly mixed with the gas above it. The effective, or actual, loading ratio will generally be much smaller than this value.

Based on these considerations the dust loading was increased until detonation was observed at a nominal dust loading of 10 kg/m^3, a value which is 20 times that required for detonation in the FAT. From the pressure trace at the center of the dust layer, shown in Fig. 4b, it can be seen that the initiating shock wave ignites the cornstarch-oxygen mixture after a delay of about 1 ms. The sharp pressure rise behind the initiating shock evident in the pressure trace at the end of the dust layer, shown in Fig. 4c, indicates that transition to detonation has occurred. The pressure trace from the transducer at the leading edge of the dust layer, shown in Fig.4a, indicates the existence of retonation wave about 4 ms after the passage of the initiating shock wave, which may be associated with the transition to detonation in the initiator. It is necessary to mention that in all these figures pressure profiles are normalized with the initial pressure, i.e. P/P_0.

The detonation enhancement by a layer of cornstarch was investigated by putting a layer of corn starch on the bottom of the tube and charging the tube with H_2-O_2 mixtures of different equivalence ratios at various pressures as the primary detonating medium. When a stoichiometric H_2-O_2 mixture was used, no dust reaction was observed due to a lack of oxygen required for reaction with the dust. Figure 5 shows the pressure profile behind a detonation in a mixture of H_2+O_2 at an initial pressure of 1 atm with a dust loading ratio of 5 kg/m3. The pressure trace does not indicate any corn starch reaction. Even with an increased initial pressure of 2 atm and the same dust-loading ratio, a comparison of the pressure profile with a loading ratio of 5 kg/m3 and that without dust indicated

a)

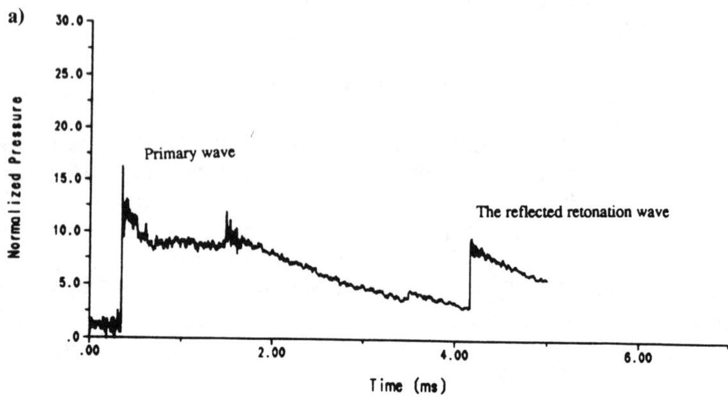

b)

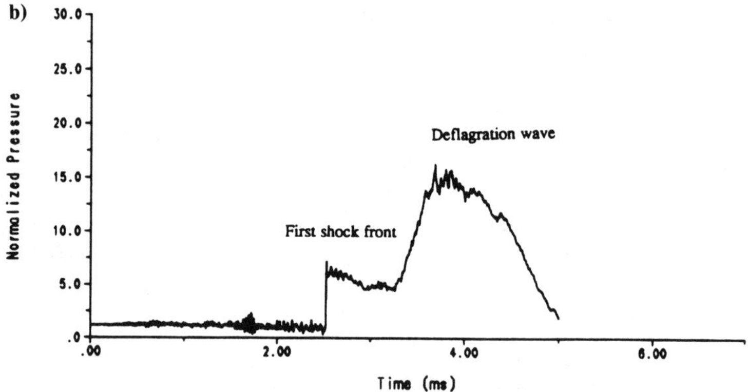

c)

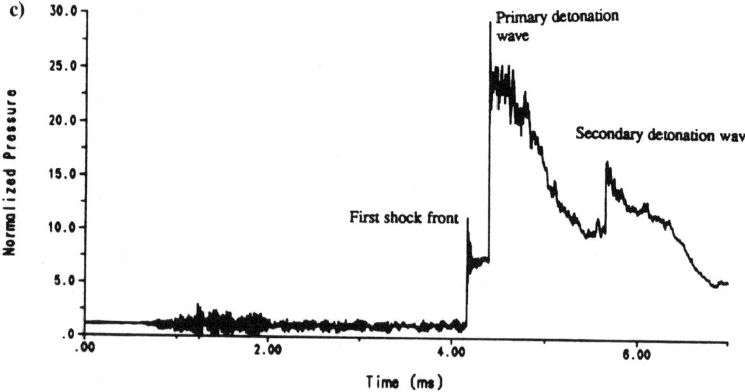

Fig. 4 Detonation development in O_2 and cornstarch layered on the bottom of the shock tube. a) pressure profile at the beginning of dust bed; b) pressure profile in the middle of dust bed; c) pressure profile at the end of dust bed.

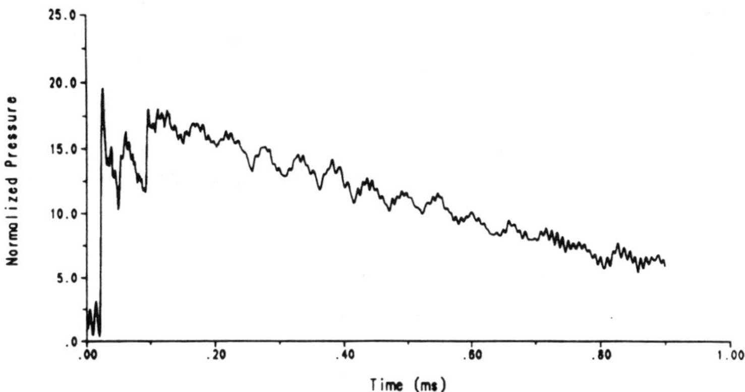

Fig. 5 Profile of detonation pressure in the middle of the dust bed in a cornstarch-H2+O2 mixture. The initial pressure is 1 atm., and the dust-loading ratio is 5 kg/m^3.

that the layered dust did not react. The profile of detonation pressure at the end of the dust bed in the cornstarch-H_2+O_2 mixture with an initial pressure of 2 atm and a dust loading ratio of 5 kg/m3 is shown in Fig. 6.

The conditions in the initiator section were the same as in the main tube in the above experiments. In light of the above negative results, a sequence of experiments was conducted in which stronger initiation was provided by charging the initiator with $2H_2$+O_2+He to 5 atm. A mylar diaphragm was used to separate the initiator and the test section, in which the conditions were the same as those which produced the pressure trace shown in Fig. 6. The pressure variation at the downstream edge of the dust layer near the end of the shock tube, measured under these conditions, is illustrated in Fig. 7, which indicates that the initial detonation wave is enhanced by the presence of the layer of cornstarch dust. The first front has an overpressure ratio of 26.6 (with an initial pressure of 2.0 atm, the pressure is then 51.2 atm). About 62 ms later, a second front can be observed with a maximum pressure of 41.3 atm and an overpressure ratio of 1.6. It is conjectured that this second front is also a detonation, which is a consequence of the fact that the strong initial detonation wave creates strong turbulence at the very beginning of the dust bed. As a result the additional dust which is entrained produces a detonable dust- O_2 + H_2O mixture.

The dust bed thickness was then increased to promote additional dust entertainment. With a dust loading ratio of 10 kg/m3 and an H_2+O_2 mixture at 2 atm, the resultant pressure profile at the end of the dust bed is illustrated in Fig. 8. It is obvious that the cornstarch reacts with the extra oxygen. There is a delay of about 174 ms before the appearance of the second front compared to 62 ms for the loading of 5 Kg/m^3, as shown in Fig.7. The overpressure ratio of the first front is 24.1 with a maximum pressure of 48.3 atm, while that of the second one is 1.9 with a maximum pressure of approximately 34 atm indicating a stronger front than in the last case. The overpressure ratio of the second front

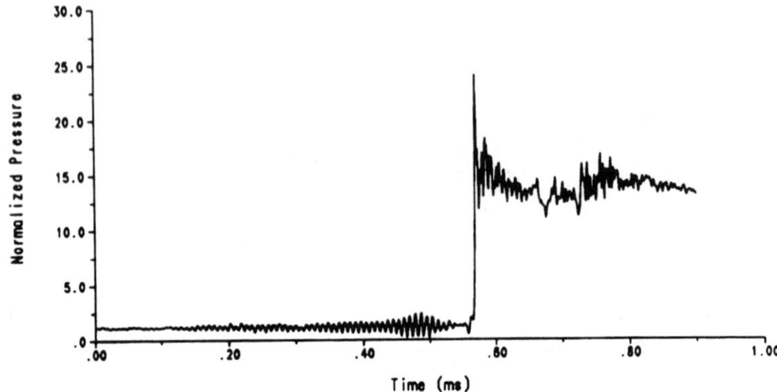

Fig. 6 Profile of detonation pressure at the end of dust bed in cornstarch-H_2+O_2 mixture. The initial pressure is 2 atm., and the dust-loading ratio is 5 kg/m^3.

is based on the pressure behind the initial detonation wave. The absolute pressure increase is close to 16 atm. The velocity of the leading front was found to be 2335 m/s. These results indicate that the increase in the thickness of the dust bed will increase the dust entrainment, which in turn enhances the initial detonation. Experiments have also been carried out for leaner mixtures of hydrogen and oxygen above the dust layer and enhancement of the detonation by the presence of the dust was also observed in these cases.

Although the pressure profiles indicated that the layered dust participates in the detonation of premixed H_2+O_2, the experiments did not show any significant influence of the dust on the initial detonation velocity; however, the

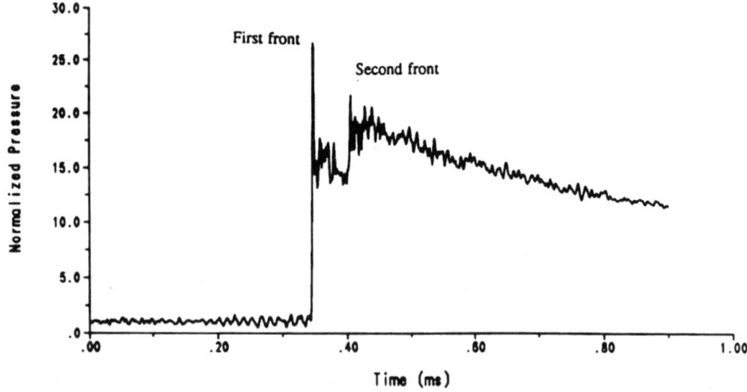

Fig. 7 Profile of detonation pressure at the end of dust bed in cornstarch-H_2+O_2 mixture. The initial pressure in the test section is 2 atm., the initial pressure of the initiator is 5 atm, and the dust-loading ratio is 5 kg/m^3.

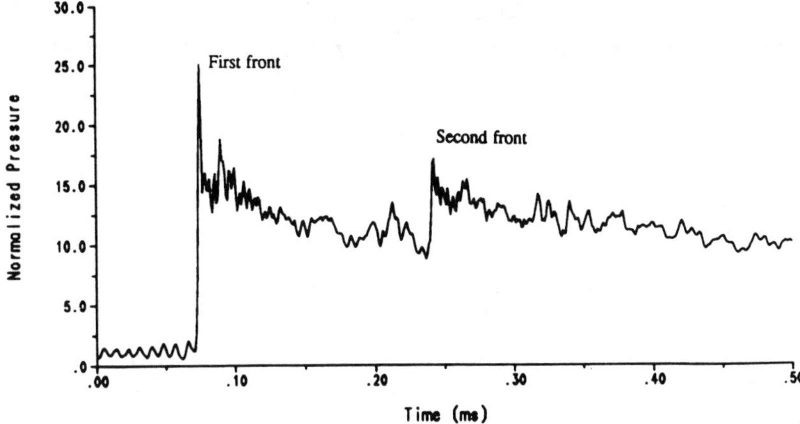

Fig. 8 Profile of detonation pressure at the end of the dust bed in cornstarch-H_2+O_2 mixture. The initial pressure in test section is 2 atm. The initial pressure of the initiator is 5 atm, and the dust-loading ratio 10 kg/m^3.

presence of dust was found to accelerate the detonation for the leaner mixtures H_2+3O_2 , and H_2+4O_2 . A likely explanation is that in the case of H_2+O_2 the amount of oxygen available for dust combustion is limited resulting in a large dust ignition delay. Thus, while the combustion of the dust increases the maximum pressure behind the wave, it occurs too late to influence the propagation of the leading shock front. A similar phenomenon was observed by Veyssiere[7,8]. For the leaner mixtures the reduction if the Chapman Jouguet velocity of the gaseous mixture and the increased oxygen available for dust combustion decreases the ignition delay so that the combustion of the dust can directly influence the wave propagation velocity.

Conclusions

Experiments have been conducted for different oxygen concentrations, H2-O2 mixtures, pressures, and dust-loading ratios in the test section, . When either air or a stoichiometric H_2-O_2 mixture was used, no dust reaction was observed. When an initially strong shock wave generated by the detonation initiator propagates down the tube charged with pure oxygen and with a sufficiently thick layer of cornstarch dust, an accelerating deflagration which undergoes transition to a detonation-like wave was observed. With lean H_2-O_2 mixtures, a relatively high dust loading, and a sufficiently strong initiator, the gaseous detonation was enhanced by the dust layer and a secondary detonation-like wave was observed to follow the primary wave. For the equimolar mixture of $H_2 + O_2$ the change in the velocity of the leading detonation front was not significant compared with the velocity measured in the absence of dust. An increase in the detonation velocity was observed for leaner H_2-O_2 mixtures. The above observations need to be confirmed using a longer layer of corn starch dust.

Acknowledgments

This work was conducted under subcontract to the Illinois Institute of Technology Research Institute. A. J. Tulis was the project monitor.

References

[1] Cybulski, W., *Coal Dust Explosions and Their Suppression*,Warsaw,Translated by Z. Zienkiewicz,451, Warsaw, Poland, 1975. p.451.

[2] Rice, G. S., and H. P. Greenwald, *Coal-Dust Explosibility Factors Indicated by Experimental Mine Investigations 1921 to 1929*, Bureau of Mines TP 464, 1929.

[3] Cybulski, W., "Detonation of Coal Dust," *Bulletin De L Academie Polonaise des Sciences*, Vol. 19, 1971, pp.37-41.

[4].Kauffman, C. W., et al.," Dust, Hybrid, and Dusty Detonations," *Dynamics of Shock Waves, Explosions and Detonations*, edited by J.A. Bowen, N. Manson, A.K. Oppenheim, and R.I. Soloukhin, Vol. 94, *Progress in Aeronautics and Astronautics*, AIAA, New York, 1984, pp 221-240.

[5]Tulis, A.J., "On the Detonation of Unconfined Aluminum Particles Dispersed in Air," *Shock Tubes and Waves*, Magnus Press, The Hebrew University, Jerusalem, Israel, 1980, p.539.

[6]Tulis, A.J., and Selman, J.R., "Unconfined Aluminum Particle Two-Phase Detonation in Air," *Dynamics of Shock Waves, Explosions and Detonations*, edited by J.R. Bowen, N. Manson, A.K. Oppenheim, and R.I. Soloukhin, Vol. 94, *Progress in Aeronautics and Astronautics*, AIAA, New York, 1984, pp. 277-292.

[7]Veyssiere, B., " 'Double-Front' Detonations in Gas-Solid Particle Mixtures," *Dynamics of Shock Waves, Explosions and Detonations*, edited by J.R. Bowen, N. Manson, A.K. Oppenheim, and R.I. Soloukhin, Vol. 94, *Progress in Aeronautics and Astronautics*, AIAA, New York, 1984, pp. 264-276.

[8] Veyssiere, B., "Structure of the Detonation in Gaseous Mixtures Containing Aluminium Particles in Suspension," *Dynamics of Shock Waves, Explosions and Detonations*, edited by J.R. Bowen, J.-C. Leyer, and R.I. Soloukhin, Vol. 106, *Progress in Aeronautics and Astronautics*, AIAA, New York, 1986, pp. 522-544.

[9]Gardner, B. R., Winter, R. J. and Moore, M. J., "Explosion Development and Deflagration-to-Detonation Transition in Coal Dust/Air Suspensions," *Twenty-First Symposium (International) on Combustion*, The Combustion Institute, Pittsburgh, PA 1986, pp. 335-343.

[10] Zhang, F. and Gronig, H., "Transition and Structure of Dust Detonations," *Lavrentyev Readings on Mathematics, Mechanics and Physics*, Novosibirsk, USSR, 1990, in press.

[11]Afanasieva, L .A., Levin, V. A., and Tunik,Y. V., " Multifront Combustion in Two Phase Media," *Dynamics of shock Waves, Explosions and Detonations*, edited by J.R. Bowen, N. Manson, A.K. Oppenheim, and R.I. Soloukhin, Vol. 87, *Progress in Aeronautics and Astronautics*, AIAA, New York, 1983, pp. 394-413.

[12] Khasainov, V.A. and Veyssiere, B., "Steady, Plane, Double-Front Detonations in Gaseous Detonable Mixtures Containing a Suspension of Aluminium Particles," *Dynamics of shock Waves, Explosions and Detonations*, edited by ?????, *Vol. 114, Progress in Aeronautics and Astronautics*, AIAA, New York, 1988, pp. 284-299.

[13] Gordon, S, McBride, B.J. and Zeleznik, F. J., "Computer Program for Calculation or Complex Chemical Equilibrium Compositions, Rocket Performance, Incident and Reflected Shocks, and Chapman Jouguet Detonations", *NASA ReportSP-273*, National Aeronautics and Space Administration, 1973.

Detonability of Organic Dust-Air Mixtures

F. Zhang* and H. Grönig†
Stosswellenlabor, RWTH Aachen, Germany

Abstract

Detonability of corn starch and anthraquinone dust suspended in air is investigated in a horizontal tube having a diameter of 141 mm and a test section length of 17.4 m. The results show that stable detonations can exist in dust-air mixtures and that it exhibits low-mode spin structure. For corn starch the transition to single-head spin detonation is observed only if the initial pressure p_0 is increased to more than 2 bar. Below this value, although spin structure can be enforced to appear by means of very strong initiation, it will degenerate into a smooth structure during the propagation. Thus, assuming that for a stable detonation the minimum tube diameter d_{min} is proportional to p_0^{-m} with $m \sim 0(1)$, for corn starch-air mixtures at atmospheric pressure $d_{min} \sim$ 0.28 m is estimated. In anthraquinone-air mixtures the transition is observed at p_0 = 1 bar, and therefore d_{min} = 0.14 m. For these kinds of dusts both the lean and rich limit exist.

---

* Assistant Professor.

† Professor, Department of Mechanical Engineering.

Introduction

When a detonation propagates into an oxidizing gas containing organic dust with a high percentage of volatiles, the following sequence mainly appears: (1) after penetrating through the shock front, the particles are heated, resulting in their pyrolysis and devolatilization; (2) mixing of the liberated volatiles with the surrounding oxygen and their subsequent combustion occur reciprocally; and (3) for lean mixtures any remaining solid is burnt, or for rich mixtures the oxygen is exhausted. Both signify the termination of the combustion zone, whereas in the latter case, the volatiles can still be liberated from remaining solid particles up to a certain extension behind the combustion zone.

However, for fine aluminum dust without volatiles suspended in a gas, a detonation can consist of another sequence: (1) after crossing the shock front, the particles are heated with subsequent melting and oxidizing; (2) surface reactions of the condensed materials heat the particles themselves, further leading to their gasification; (3) the gasificated intermediate aluminum oxides are mixed with the surrounding oxygen and undergo combustion alternately; and(4) for lean mixtures the fuel is used up, or for rich mixtures the oxygen is exhausted, meaning again the termination of the combustion zone. In the latter case, remaining condensed aluminum oxides can be further gasificated if the local temperature exceeds their boiling temperature. For very lean mixtures only, the sequence can be terminated by the surface reactions of the condensed intermediate aluminum oxides and no gasification occurs.

Spin detonation probably is the general mode of stable propagation in reactive dust-oxidizing gas mixtures in round tubes.[1] The reasons for this insensitivity are two-fold. First, immediately behind the shock front, the gas temperature jumps within a few molecular mean free paths to a new state corresponding to the shock strength. On the other hand the dust temperature increases continuously from the frozen state, i.e., the initial state, thus causing an increase of the induction time. Second, pyrolysis, surface oxidation, and diffusion mixing have their limited rates which also cause the reaction time to increase.

In tubes transition to stable detonation can be achieved in dust-air mixtures, if the diameter and ratio of length to diameter are sufficiently large. Bartknecht[2] reported in 1972 some detonation-like waves in organic and aluminum dust-air mixtures. Using a 0.4 m diameter by 40 m-long tube pressures of 21 to 25 bar were observed. Tulis and Selman[3] investigated detonability of aluminum dust-air mixtures in a tube 0.152 m in diameter and 5.5 m in length. For flake aluminum dust with specific surface areas of 3 to 4 m^2/g, the observed velocities ranged from 1350 to 1640 m/s for dust concentrations of 250 to 900 g/m^3. However, they reported no correlation between dust concentration and propagation velocity. An improved work on aluminum-air detonations was recently carried out by Borisov et al.[4] Using a tube 0.122 m in diameter and 4.2 m in length, the observed pressure profiles showed a spin structure. In the case of flake dust (specific surface area: 0.9 m^2/g), they reported detonation velocities of 1760 to 1810 m/s and pulsating peak pressures behind the spin head of 33 to 35 bar for dust concentrations between 210 to 400 g/m^3. For 1 μm spherical aluminum particles (specific surface area: 2.2 m^2/g), they found detonation velocities of 1420 to 1810 m/s and pressures of 19 to 27 bar for concentrations of 140 to 750 g/m^3. The observed maximum velocity of 1810 m/s is close to the maximum CJ detonation value of 1970 m/s. Whether the detonation propagates stably is unknown because of the strong initiation energy and the small length-to-diameter ratio (L/d = 34.4). In both works of Tulis et al. and Borisov et al., the detonation velocities are insensitive to the concentrations ranging from near stoichiometric (~ 310 g/m^3) to rich mixtures. This can be attributed to the fact that the constant amount of oxygen can only cause a constant amount of gasificated intermediate aluminum oxides to be burned up in rich mixtures.[5]

For the case of organic dust, Gardner et al.[6] reported an excellent work on the transition process in bituminous coal dust-air mixtures with particle sizes of less than 100 μm. In a tube 0.6 m in diameter and 42 m in length the velocity reached 2850 m/s, and a peak pressure of 81.5 bar was observed at the tube end. There the wave represents an overdriven detonation. However, this study might indicate that in dust-air mixtures, a transition to stable detonation is possible if the tube length were sufficiently extended. Sacha, Wolanski, and Zalesinski[7] recently reported a detonation study on corn starch, wheat, oats, and brown coal dust-air mixtures in a 0.13 m diameter by 13 m-length tube. By means of a H_2-O_2 detonation driver for initiation with 4.5 to 5 bar initiator pressure, they observed velocities ranging from 1530 to 1600 m/s and peak pressures of 25 to 32 bar. However, the displayed pressure profiles in a distance of about 4 m show a degenerating wave structure in the corn starch-air mixture. Thus the waves seem to be unstable. Zhang and Grönig[8] reported some transition results on corn starch-air mixtures in a tube 0.141 m in diameter and 17.4 m in length. The detonation onset occurred only at an initial pressure of more than 2 bar, and it exhibited a spin structure.

The present study is a continuing work of Ref. 8. Detonability of corn starch and anthraquinone dust suspended in air is investigated in detail. The minimum tube diameter d_{min} required for transition to a stable detonation is obtained, partly based on the assumption, that the rule $d_{min} \propto p_0^{-m}$ with $m \sim O(1)$ is valid for organic dust detonations in a certain range of initial mixture pressures p_0.

Experimental Setup

The equipment is the same as described previously[9] and composed of a horizontal tube 0.141 m in diameter with a test section length of 17.4 m. A dust-air suspension is prepared in the test section. The initiation pulse is produced by a $2H_2 + O_2$ detonation driver and injected into the test section

Table 1 Size distribution and heat of combustion of the studied dusts

Corn starch

Formula	Average size μm	Heat of combustion $p_0 = 1$ atm, $T_0 = 2$ °C kcal/mol
$[C_6H_{10}O_5]_n$	10	− 677

Size distribution (particle percentage)				
< 5 μm	5 - 10 μm	10 - 15 μm	15 - 20 μm	> 20 μm
18	50	29	2	1

Anthraquinone

Formula	Average thickness x length μm x μm	Heat of combustion $p_0 = 1$ atm, $T_0 = 20$ °C kcal/mol
$C_{14}H_8O_2$	6 x 22	− 1544

Size distribution (particle percentage)				
Thickness/length	<10 μm	10 - 20 μm	20 - 30 μm	> 30 μm
< 2 μm	22			
2 - 4 μm	15	15	4	
4 - 6 μm	3	9	3	1
6 - 8 μm	1	5	4	3
8 - 10 μm		1	3	4
> 10 μm			2	5

either as a planar wave over the whole cross-section or as a free jet through a circular orifice 20 mm in diameter. Kistler 603B transducers are used to measure pressures and wavefront velocities.

Two kinds of organic dusts were studied, i.e., corn starch and anthraquinone. The particle size distributions and some properties are presented in Table 1. Photographs of the particles obtained by electron microscope are shown in Fig. 1.

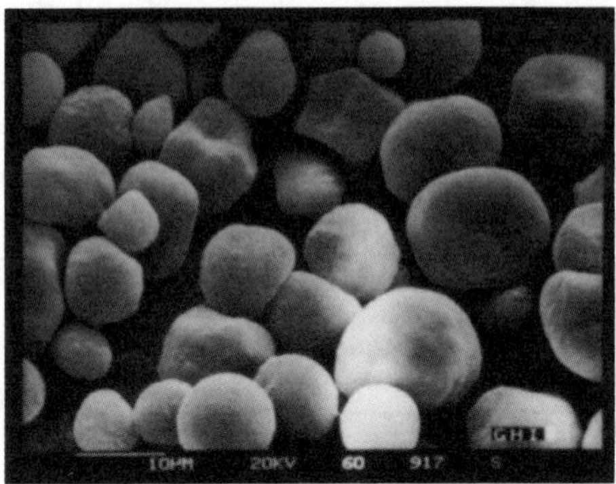

a) Corn starch

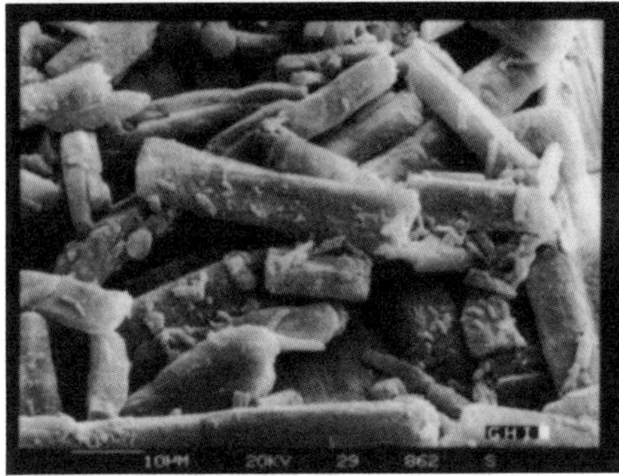

b) Anthraquinone

Fig. 1 Photographs of dust obtained by electron microscope.

Detonability of Corn Starch-Air Mixtures

First, experiments at an initial pressure p_0 = 1 bar were carried out in corn starch-air mixtures with equivalence ratio of $\varphi = 2$ under different initiation energy. The results are summerized in Fig. 2. By means of jet initiation with an initiator pressure of p_i = 6 bar or source initiation energy of E_0 = 31.8 kJ, deflagration propagation along the tube is quite mild. From $x/d \approx 80$ a rapid transition causes the propagation velocity to jump to a new level of $D \approx 1100$ m/s, which is maintained up to the end of the test section. The pressure in this propagation regime, illustrated in the pressure profile of number 5, displays a smooth structure. This phenomenon is similar to that in rough tube tests,[10] called 'choked' regime due to $d_1/\lambda < 1$ (d_1: orifice diameter of

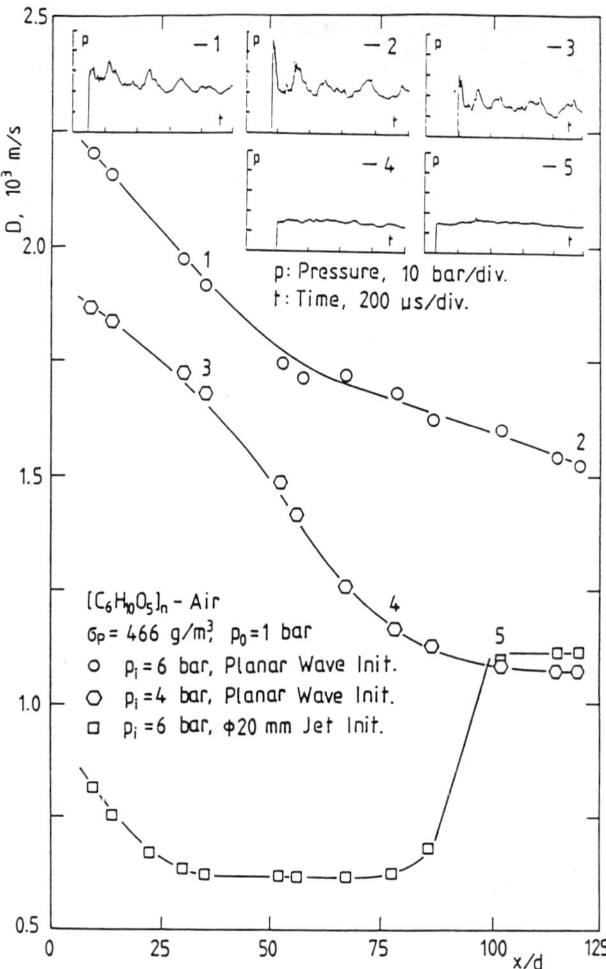

Fig. 2 Combustion propagation in a corn starch-air mixture of φ = 2 and p_0 = 1 bar with different initiation energy.

orifice rings spaced in the tube, λ: cell width), where the deflagration velocity is found to be close to the sound speed of the hot products. By using the planar wave initiation with p_i = 4 bar or E_0 = 624 kJ, the combustion shows a forced spin structure at the first stage as indicated in the pressure profile of number 3, then decays quickly to the »choked« regime with the smooth structure of the wave front (pressure profile of number 4). Although the spin structure can be enforced to exist in the whole test section by using very strong initiation, for instance, planar wave initiation with p_i = 6 bar or E_0 = 964 kJ (pressure profiles of number 1 and 2), the propagation velocity decays continuously. One can expect that up to a certain extended distance where the strong influence of initiation disappears, the spin structure will degenerate into the smooth structure, just as shown in the case of p_i = 4 bar. Thus, the forced spin induced by strong initial conditions is unstable due to $d < \lambda/\pi$. According to the above results, one comes to the conclusion that in a tube of $d \leq 141$ mm, a stable detonation is not able to be achieved for the corn starch suspended in air at atmospheric pressure.

However, since spin detonation is expected in dust mixtures, the validity of the rule

$$d_{min} \propto p_0^{-m} \text{ with } m \sim 0(1) \tag{1a}$$

for fine dust mixtures can be assumed. Here d_{min} denotes the minimum tube diameter required for transition to a stable detonation.

Figure 3 shows the result of transition to detonation in a corn starch-air mixture of p_0 = 2.5 bar and φ = 2 with planar wave initiation of p_i/p_0 = 1. The transition process is the same as discussed in corn starch-oxygen mixtures[1]. There the sequence of a complete detonation evolution is distinguished in four stages: (1) initial particle ignition; (2) compression wave amplification by coherent energy release; (3) unsteady reaction shock; (4) spin detonation. Although in Fig. 3 the slow transition stage 2 is amplified by the initiation energy, its physical property of compression wave amplification by reaction is obvious. In the fast transition stage 3, a leading shock wave is closely coupled to the exothermic reaction zone behind it. A characteristic feature of this reaction shock is that within this stage, the one-dimensional smooth pressure wave structure turns into a three-dimensional one. A detailed observation of the reaction shock can be found in Ref. 11. In Fig. 3, the prop-

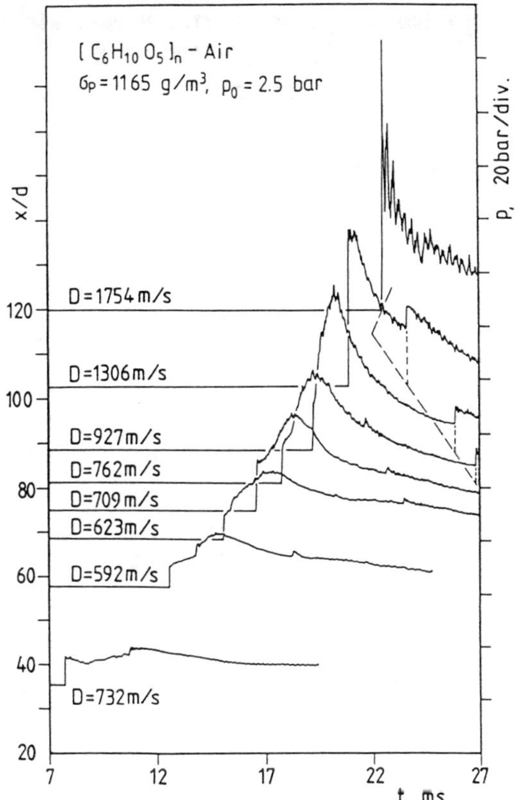

Fig. 3 Transition to detonation in a corn starch-air mixture of φ = 2 and p_0 = 2.5 bar.

agation distance of the reaction shock has a value of $x_{rs}/d \approx 20-25$. Within the stage of the reaction shock the onset of an overdriven detonation accompanied by a retonation wave occurs at $x/d \approx 115$.

The detonation formed, with planar wave initiation of $p_i/p_0 = 1.2$, exhibits a single-head spin structure, shown in Fig. 4 which is observed around the tube wall at $x/d = 118.4$. Detonation data for different dust concentrations are summarized in Table 2 with $p_i/p_0 = 1.2$. In Table 2 σ_p denotes the dust concentration, D the detonation velocity, p_s the pulsating peak pressure behind the spin head, p_A the spin head pressure, and n the number of spin heads. Both the lean and rich limit exist. The maximum velocity lies at $\varphi = 2.5$.

Table 2 Detonation data of corn starch dust-air mixtures (p_0 = 2.5 bar, T_0 = 295 ± 3 K)

φ	σ_p g/m^3	D m/s	p_s bar	p_s/p_0	p_A bar	p_A/p_0	n
1.72	1000	1456	66	26.4			lean limit
2.0	1165	1773	76	30.4	125	50	1
2.50	1450	1972	95	38.0	220	88	1
3.0	1750	1854	83	33.2	150	60	1
3.43	2000	1440	63	25.2			rich limit
2.0*	1165	1542	62	24.8	120	48	
2.5*	1450	1730	80	32.0	200	80	

* Direct initation with p_i/p_0 = 2.4

Whereas in the corn starch-oxygen mixtures it lies at φ = 1.8. For homogeneous gaseous mixtures the maximum velocity occurs near the stoichiometric composition. The shift of maximum velocity towards rich mixtures for corn starch suspended in air is more severe than in oxygen.

The detonation onset with planar wave initiation of p_i/p_0 = 1 - 1.2 occurs near the tube end. Thus, the above summerized detonations are still overdriven. This fact can also be seen in Fig. 4 in which the spin structure is immature. Table 2 also contains detonation results of direct initiation at p_0 = 2.5 bar with a critical initiator pressure of p_i/p_0 = 2.4 or an effective critical initiation energy of E_{ec} = 7.9 MJ/m^2. In contrast to the instability of the forced spin in Fig. 2, the direct initiated spin detonation here remains stable in further propagation. The observed maximum velocity of 1730 m/s has a deviation of 2% to the maximum CJ detonation velocity (1758 m/s). The maximum CJ detonation pressure of 54 bar is close to the pulsating mean pressure behind the spin head at φ = 2.5.

By decreasing the initial pressure to p_0 = 2 bar, the wave evolution is similar to Fig. 3. However, for φ = 2 only

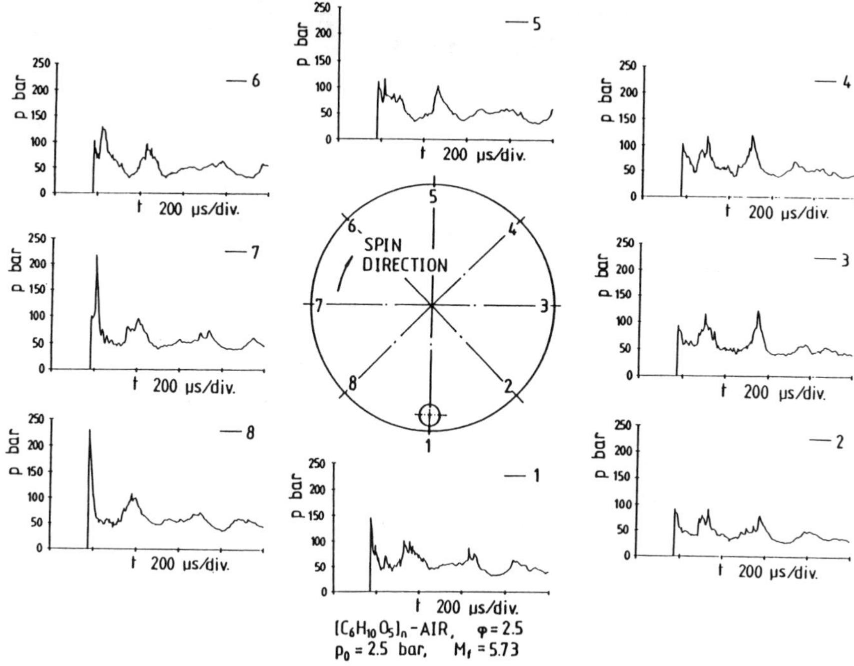

Fig. 4 Pressure profiles around the tube wall in a corn starch-air mixture of φ = 2.5 and p_0 = 2.5 bar.

a transition up to the reaction shock with D = 1410 m/s and p = 35 bar which is still accelerating is observed within the tube length. Thus, it is possible to reach the single spin detonation at p_0 = 2 bar if the tube length is extended. Further decreasing the initial pressure, one cannot obtain the transition stage of the accelerating reaction shock. Thus, from the above results and the rule (1), an estimation of $d_{min} \sim 0.28$ m for the stable detonation of corn starch in air at atmospheric pressure is derived.

Detonability of Anthraquinone Dust-Air Mixtures

In order to obtain detonations of dust-air mixtures at atmospheric pressure in the present tube of 141 mm in diameter, anthraquinone dust is chosen due to its K_{St}-value of about twice as much as that of corn starch. This point will

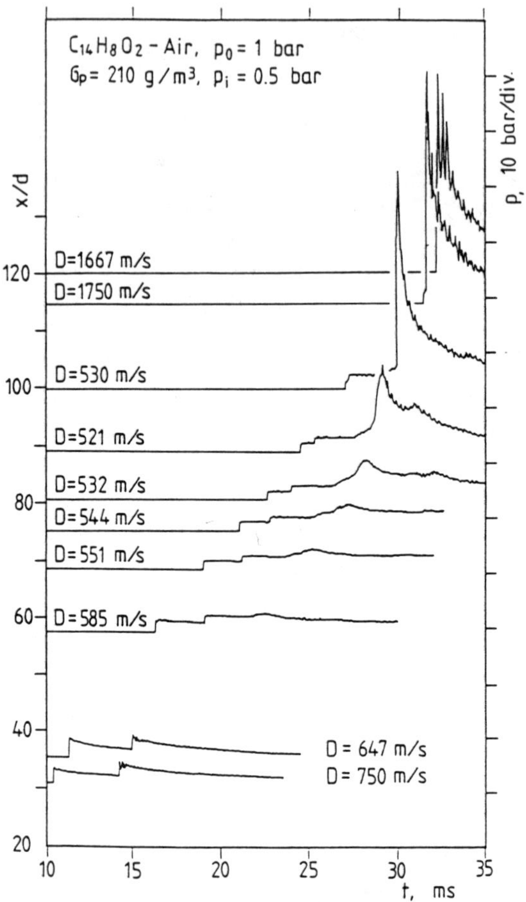

Fig. 5 Transition to detonation in an anthraquinone dust-air mixture of φ = 1.75 and p_0 = 1 bar.

be discussed further. Transition to detonation at p_0 = 1 bar is achieved with planar wave initiation of p_i/p_0 = 0.5, shown in Fig. 5. (The shock wave following the leading wave front is due to the re-reflected initiation wave from the end wall of the initiation section.) The transition process is the same as that in corn starch mixtures. The slow transition stage in Fig. 5 is clearer than in Fig. 3 due to the weaker initiation energy. At x/d = 90-100 the reaction shock forms, and the transition turns into the fast stage. Within the stage of the reaction shock the onset of an overdriven

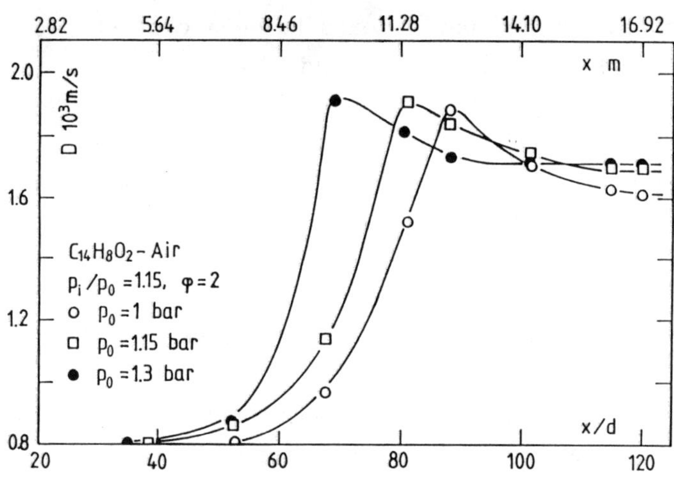

Fig. 6 Wavefront velocities versus propagation distance in anthraquinone dust-air mixtures at different initial pressures.

detonation occurs, resulting in a three-dimensional wave front structure. Thus, spin detonation is formed. The propagation distance of the reaction shock is of $x_{rs}/d \approx 20 - 25$. It is shown in Fig. 6 that the greater the inital pressure p_O the earlier the onset of detonation occurs. Increasing the inital pressure causes an increase of the detonation velocity (Fig. 7). However, the velocity tends to reach a limit as the initial pressure is further increased. Some detail of the single-head spin structure reached is shown in the pressure profiles around the tube wall at x/d = 118.4 in Fig. 8. It is obvious that there is a transverse wave around the tube circumference. The tail wave here is not as regular as that in the case of corn starch mixtures. This may be due to kinetic properties of the mixture and the geometry of the flake dust, which results in anisotropy. The detonation data for different dust concentrations are summarized in Fig. 9 and Table 3. Again, the detonation possesses both the lean and rich limit. The observed maximum velocity of 1650 m/s at p_O = 1 bar has a deviation of 7% from the maximum CJ detonation velocity (1775 m/s), whereas the observed maximum velocity of 1722 m/s at p_O = 1.15 bar has a deviation of 3% from the maximum CJ detonation velocity (1778 m/s). The larger deviation at p_0 = 1 bar appears to be a consequence of weak mixtures near the limiting condition of detonability. The maximum CJ detonation pressure of 23 bars at p_0 = 1.15 bar is close to the observed pulsating mean pressure behind the wavefront at φ = 1.45 (see Fig. 8). Based on the experiments of p_0 = 1 bar, d_{min} = 0.14 m is obtained for the stable detonation of anthraquinone in air at atmospheric pressure.

Estimate of Dynamic Parameters

For homogeneous gaseous mixtures, Dove and Wagner[14] suggested that the single-head spin should be the lowest stable detonation mode in round tubes. From this point of view, the existence of the single-head spin in a tube is a criterion for the minimum tube diameter d_{min}, regarded as a limiting necessary boundary condition for the transition to a stable detonation. The minimum tube diameter is associated with the normal cell width in the multicellular detonation by the empirical criterion

$$\lambda = \pi d_{min}$$

proposed first by Kogarko and Zeldovich[15] in 1948.

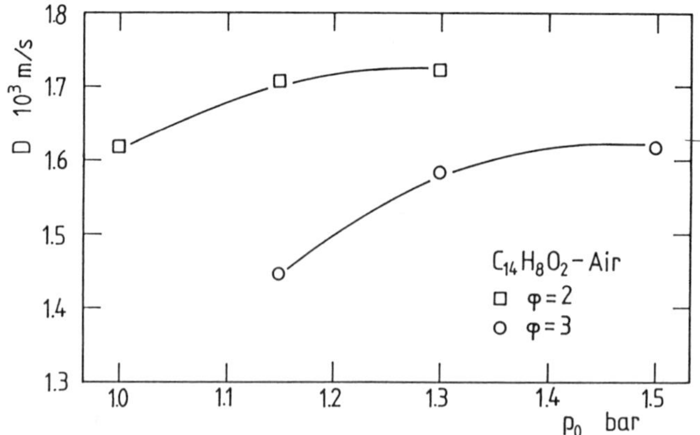

Fig. 7 Detonation velocities versus initial pressures in anthraquinone dust-air mixtures.

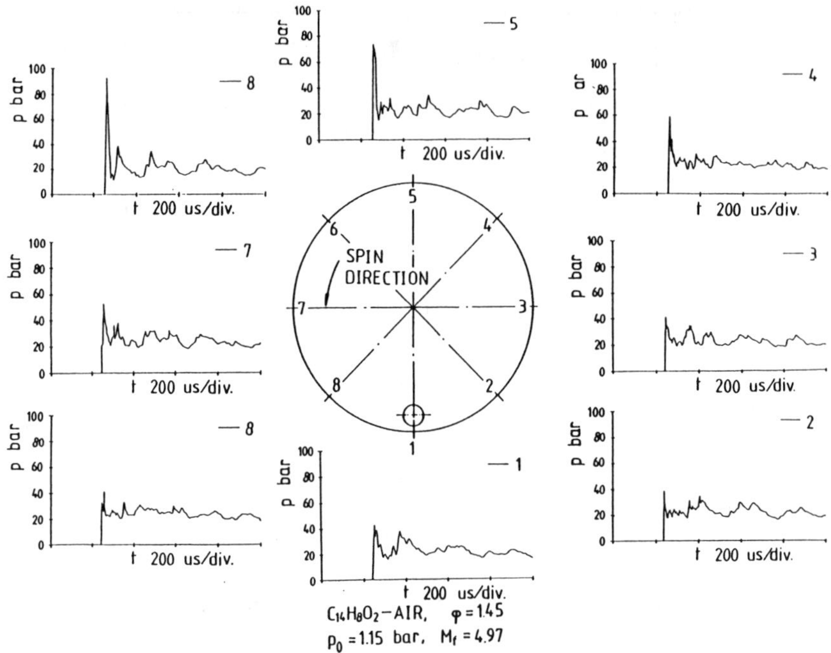

Fig. 8 Pressure profiles around the tube wall in an anthraquinone dust-air mixture of φ = 1.45 and p_0 = 1.15 bar.

Dust-air mixtures are insensitive mixtures. Only low-mode spin detonations have been observed so far as the stable detonation mode. Thus, $\lambda = \pi d_{min}$ gives an estimate of the cell width for equivalent conditions in a much larger tube. Based on this estimate of the cell width one can calculate the critical diameter d_c and the critical initiation energy E_c, assuming that $d_c \approx 13\,\lambda$ and a suitable relation for E_c in gaseous fuel-air mixtures are approximately valid in organic dust-air mixtures. Here the surface energy model of Lee et al.,[16] i.e.,

$$E_c = \frac{2197}{16}\,\pi \rho_0 U_{CJ}^2 I \lambda^3$$

will be used. ρ_0 denotes the initial density, U_{CJ} the CJ detonation velocity, and I the blast wave energy integral. Although those assumptions are needed to be proved in the case of organic dust fuels, the calculations are meaningful

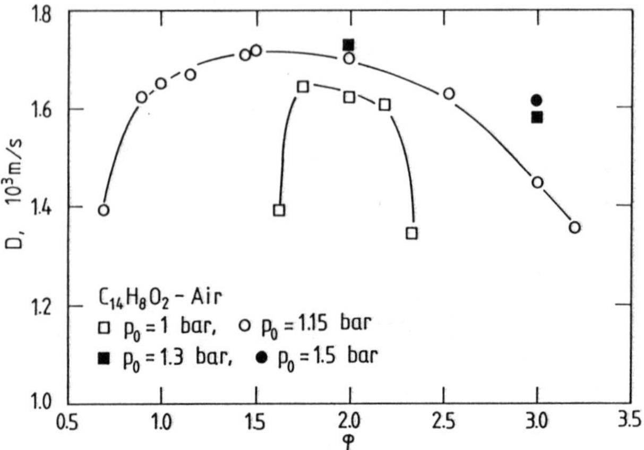

Fig. 9 Detonation velocities versus equivalence ratio in anthraquinone dust-air mixtures.

in at least an estimative sense. Thus one derives $\lambda \approx 0.88$ m, $d_c \approx 11.4$ m, and $E_c \approx 1570$ MJ for corn starch-air detonation of $\varphi = 2.5$ and $p_0 = 1$ bar; and $\lambda \approx 0.44$ m, $d_c \approx 5.7$ m, and $E_c \approx 160$ MJ for anthraquinone dust-air detonation of $\varphi = 1.75$ and $p_0 = 1$ bar. In these assessments the maximum CJ velocity are used. These values of E_c correspond to 367 kg and 37 kg of tetryl for corn starch and anthraquinone dust, respectively. This leads to the conclusion that direct initiation of unconfined detonations in organic dusts suspended in air is extremely difficult.

On the other hand, in the case of closed vessel dust explosions in air, a cubic law

$$K_{St} = V^{\frac{1}{3}} \left(\frac{dp}{dt}\right)_{\max} \tag{1b}$$

has been used in Europe to classify the explosion severity of different dusts. According to the combustion theory for spherical vessels by Lewis and von Elbe,[17] the K_{St}-value can be derived as a function of the maximum pressure rise and the burning velocity. Thus, the K_{St}-value provides a measure of the reaction rate. However, gravity settling of the particles results in a strong apparatus-dependence on the

Table 3 Detonation data of anthraquinone dust-air mixtures (T_0 = 295 ±3 K).

φ	p_0 bar	σ_p g/m^3	D m/s	p_s bar	p_s/p_0	p_A bar	p_A/p_0	n
1.63	1	195	1390	20	20			lean limit
1.75	1	210	1650	27	27	56	56	1
2.0	1	240	1619	26	26	52	52	1
2.21	1	265	1607	25	25	52	52	1
2.33	1	280	1337	18	18			rich limit
0.70	1.15	97	1391	24	20.9			lean limit
0.90	1.15	125	1620	32	27.8	60	52.2	1
1.0	1.15	138	1656	34	29.6	62	53.9	1
1.16	1.15	160	1669	34	29.6	63	54.8	1
1.45	1.15	200	1711	37	32.2	98	85.2	1
1.50	1.15	207	1722	37	32.2	98	85.2	
2.0	1.15	276	1704	33	28.7	74	64.3	2
2.54	1.15	350	1630	29	25.2	57	49.6	
3.0	1.15	414	1446	25	21.7	52	45.2	1
3.20	1.15	440	1350	20	17.4			rich limit
2.0	1.3	312	1716	37	28.5	80	61.5	2
3.0	1.3	468	1584	32	24.6	98	75.4	
3.0	1.5	540	1614	35	23.3	100	66.7	

K_{St}-value. For instance, turbulence produced by dispersion, ignition delay time, vessel geometry, and volume can significantly influence the K_{St} value. It was concluded from numerous experiments in spherical vessels of different sizes that a 20 l sphere is the minimum volume for determining explosion parameters.[13] The K_{St}-values of about 60 kinds of dust, observed in both the 20 l sphere and a 1 m^3 cylindrical vessel ($L/d \approx 1$), are in good agreement (Fig. 10). Thus, both vessels have been used as reference vessels to determine the explosion parameters.

Experimental data of d_{min} and K_{St} are summarized in Table 4. There K_{St}-values of corn starch, anthraquinone and aluminum are observed with the 20 l sphere and the 1 m^3 cylindrical vessel ($L/d \approx 1$). Due to the lack of the K_{St}-

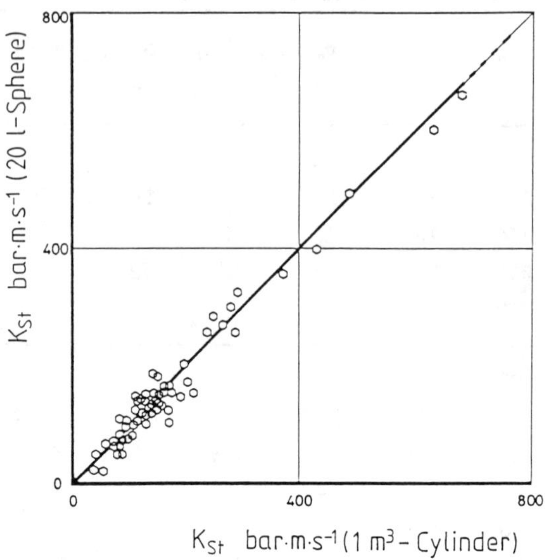

Fig. 10 Comparison of K_{St}-values in a 20 l sphere and in a 1 m cylinder ($L/d \approx 1$). Courtesy of W. Bartknecht.[13]

value for U.S.W. sub-bituminous coal, the K_{St}-value of brown coal obtained in a 5.6 l cylindrical vessel is cited,[12] since both coal dusts have approximately the same particle sizes and volatile contents. Combination of the observed d_{min} with the K_{St}-value is shown in Fig. 11. This yields a simple relation

$$d_{min} K_{St} \sim \text{Const.} \tag{2}$$

which represents a force rate with the dimension of $N \cdot s^{-1}$. The theoretical background of the relation (2) is still un-

Table 4 Experimental K_{St} and d_{min}

Dust	Size	K_{St} bar·m·s^{-1}	d_{min} m
Bituminous coal (Ref. 6)	≤ 100 μm	59 (Ref. 12)	0.60
Corn starch	10 μm	160	0.28
Anthraquinone	6 x 6 x 22 μm^3	274	0.14
Aluminum (Ref. 4)	1 x 10 x 10 μm^3	400 (Ref. 13)	0.12

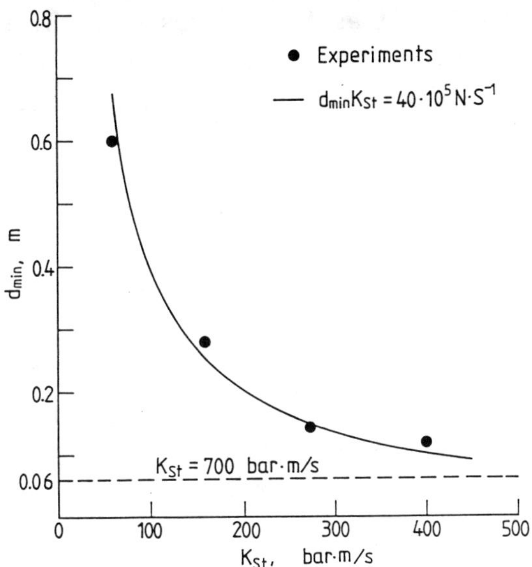

Fig. 11 K_{st}-d_{min} diagram.

clear, since it combines two parameters from two physically different combustion processes. However, it can be regarded as a "rule of thumb" for practical estimates. The majority of industrial and agricultural dusts has values of $K_{St} \leq 700$ bar · m · s^{-1}. Thus, an overall lower limit of $d_{min} \sim 60$ mm is derived. The K_{St}-values of most dusts range from 40 to 300 bar · m · s^{-1}, which corresponds to a minimum tube diameter for stable detonations of about 1 to 0.13 m. However, before any further conclusion may be drawn from relation (2), numerous detonability experiments have to be carried out to confirm its validity.

Conclusions

The experimental results demonstrate that low-mode spin detonations represent the mode of stable propagation in dust-air mixtures in round tubes. In the anthraquinone dust-air mixtures studied here at atmospheric pressure the minimum tube diameter for a stable detonation has a value of $d_{min} = 0.14$ m. For corn starch-air mixtures at atmospheric pressure $d_{min} \sim 0.28$ m is estimated, based on the assumption of $d_{min} \propto p_0^{-m}$ with $m \sim O(1)$. For both kinds of dust, the detonation possesses both the lean and the rich limit within the initial pressure range studied.

Acknowledgments

The authors would like to thank John H.S. Lee for many instructive discussions. Grateful acknowledgments go to Ciba-Geigy and BG Nahrungsmittel und Gaststätten for the supply of dusts and their explosion data for closed vessels. The authors are indebted to A. van de Ven and P. Greilich for their help during the experiments.

References

[1]Zhang, F., and Grönig, H., »Spin Detonation in Reactive Particles-Oxidizing Gas Flow,« Physics of Fluids A, Vol. 3, No. 5, 1991.

[2]Bartknecht, W., »Explosionstechnische Kennzahlen brennbarer Stäube in Rohren mit engen Querschnitten,« Internationales Symposium über Staubexplosionsgefahr in Bergbau und Industrie, Karlsbad, CSSR, 1972.

[3]Tulis, A. J., and Selman, J. R., »Detonation Tube Studies of Aluminum Particles Dispersed in Air,« 19th Symposium (International) on Combustion, The Combustion Institute, Pittsburgh, PA, 1982, pp. 655-663.

[4]Borisov, A. A., Khasainov, B. A., Veyssiere, B., Saneev, E. L., Fomin, I. B., and Khomik, S. V., »On Detonation of Aluminum Suspensions in Air and Oxygen,« 23rd Symposium (International) on Combustion, poster, Orleans, France, 1990.

[5]Lee, J. H. S., private communication, February 1991.

[6]Gardner, B. R., Winter, R. J., and Moore, M. J., »Explosion Development and Deflagration-to-Detonation Transition in Coal Dust/Air Suspensions,« 21st Symposium (International) on Combustion, The Combustion Institute, Pittsburgh, PA, 1986, pp. 335-343.

[7]Sacha, W., Wolanski, P., and Zalesinski, M., »Experimental Study of Shock Wave Initiation of Detonation in Grain Dust-Air Mixtures,« 23rd Symposium (International) on Combustion, poster, Orleans, France, 1990.

[8]Zhang, F., and Grönig, H., »Transition and Structure of Dust Detonation,« In: »Dynamic Structure of Detonation in Gaseous and Dispersed Media«, edited by Borissov A. A., Kluwer Academic Publishers, Dordrecht, 1991, pp. 157-213.

[9]Zhang, F., and Grönig, H., »Detonation Structure of Corn Starch Particles-Oxygen Mixtures,« In: Dynamics of Detonations and Explosions: Detonations, edited by Kuhl A. L., Leyer J.-C., Borisov A. A., and Sirignano W. A., Progress in Astronautics and Aeronautics, Vol. 133, 1991, pp. 342-355.

[10]Lee, J. H .S., Knystautas, R., and Chan, C. K., »Turbulent Flame Propagation in Obstacle-Filled Tubes,« 20st Symposium (International) on Combustion, The Combustion Institute, Pittsburgh, PA, 1985, pp. 1663- 1672.

[11]Zhang, F., and Grönig, H., »Unsteady Reaction Shock-Prestage of Organic Dust Detonation,« 18th International Symposium on Shock Waves, Sendai, Japan, July 1991.

[12]Fangrat, J., Glinka, W., Wolanski, P., and Wolinski, M., »Detonation Structure in Organic Dust-Oxygen Mixtures,« archivum combustionis, Vol. 7, No. 3/4, 1987, pp. 321-332.

[13]Bartknecht, W., Explosionen, Springer, Berlin, Heidelberg, FRG, 1978.

[14]Dove, J. E., and Wagner, H. Gg., »A Photographic Investigation of the Mechanisms of Spinning Detonation,« 8th Symposium (International) on Combustion, The Combustion Institute, Pittsburgh, PA, 1960, pp. 589- 600.

[15]Kogarko, S.M., and Zeldovich, Y. B., Doklady Akademii Nauk SSSR, Vol. 63, 1948, pp. 553.

[16]Lee, J. H. S., »Dynamic Parameters of Gaseous Detonations,« Annual Review of Fluid Mechanics, Vol. 16, 1984, pp. 311-336.

[17]Lewis, B., and von Elbe, G., Determination of the Speed of Flames and the Temperature Distribution in a Spherical Bomb from Time-Pressure Explosion Reconds,« Journal of Chemical Physics, Vol. 2, May 1934, pp. 283-290.

Two-Head Detonation Structure in Cornstarch-Oxygen Mixtures

F. Zhang,* P. Greilich,† A. v. d. Ven,† and H. Grönig‡
Stosswellenlabor, RWTH Aachen, Germany

Abstract

Two-head detonation waves in heterogeneous cornstarch dust-oxygen mixtures are studied in a circular tube. Pressure transducers and double-response gauges reacting on both the shock and combustion front are used in a 0.5 m test section which has 32 gauge positions arranged on four circumferences. The combined results permit the deduction of a fairly detailed description of the shock waves and induction zones existing at the detonation front. The observed structure possesses two Mach triple point configurations in the periphery. Collision of the two triple points leads to an overdriven wave followed by transient decaying, which requires continual regeneration by collisions for sustained propagation.

Introduction

In previous publications[1,2] the authors have reported the observation of single-head spin detonation waves in reac-

---

*Assistant Professor, Department of Mechanical Engineering

†Graduate Student, Department of Mechanical Engineering

‡Professor, Department of Mechanical Engineering

tive dust-oxidizing gas mixtures. The deduced structural details are in general agreement with that of gas-phase marginal detonation waves in circular tubes. The three-dimensional frontal behavior manifests itself primarily through the occurrence of a transverse shock wave, which propagates into the induction zone behind the incident shock and produces a traveling Mach stem configuration at the front. Moreover, in homogeneous gaseous mixtures it was well established that this type of Mach triple-point configuration is the basic element of the single-head spin and the cellular detonation as well. There are, however, significant differences between both kinds of detonation. In the former case, the circular tube geometry permits a progressive circumferential motion of the unique transverse wave. Therefore, the frontal shape remains unchanged and the triple point moves at a constant track angle. On the other hand, in a cellular detonation collisions between Mach triple points are necessary for sustained propagation. Thus the frontal geometry changes continuously, and the forward propagation of the triple points decays at transient track angles between two successive collisions. Clearly, an observation of the existence and the nature of a cellular detonation in dust-gas mixtures is of great importance in the study of heterogeneous detonations.

Regarding a single cell detonation involving two Mach interactions in a circular tube, Denisov and Troshin[3] observed this two-head wave structure in a gas-phase mixture of $2 H_2 + O_2$ at initial pressures of 130 Torr in a tube 16 mm in diameter. They termed it the lowest mode of some type of »pulsating« detonation. On the other hand, Duff[4] ascribed it to the spin mode by means of the superposition principle of the acoustic theory of Manson[5] and Fay.[6] In his explanation this mode corresponded to the coexistence of a right- and left-handed single spin. Some observation was also reported by Voitsekhovskii et al.[7] in a circular and annular tube formed by inserting an axial rod. Unlike the single-head spin detonation, there is a lack of publications concerning the detailed wave structure and the nature of this detonation mode.

In the present work the two-head detonation wave in corn-starch dust-oxygen mixtures is examined in order to test the cellular structure in heterogeneous mixtures and to provide further information on the wave structure and the nature of this mode.

Experimental Setup

Experiments were performed in a horizontal tube 141 mm in diameter with a test section length of 17.4 m, as described in Ref. 1. A dust-gas suspension is prepared in the test section. Its initiation is generated by a hot gas jet of a detonated stoichiometric oxyhydrogen mixture through a circular orifice 20 mm in diameter into the test section. Gauges arranged on four circumferences 166 mm apart from each other are used in the test section. In each circumference there are eight gauges at 45 degree intervals. A double-response gauge measures the arrival of both the shock and combustion front. It consists of a thin piezoelectric film of 2 mm diameter and an ion gauge separated from the film by a 2.4 mm gap.[1] This gauge is used in the study of wave front geometry and local induction times. Combination of this gauge with Kistler 603 B pressure transducers permits the deduction of a fairly detailed description of the shock waves and induction zones existing at the wave front. Signals were recorded by 48-channel transient recorders with a sampling rate of 0.5 μsec.

In the present study cornstarch dust-oxygen mixtures are used. The average particle size is 10 μm with 15 μm being the maximum value. A rich mixture having a dust concentration of 4 kg/m^3 at an initial pressure of 1.15 bar at ambient temperature is examined in order to provide a clear two-head detonation mode. The mixture has an equivalence ratio of 3.1. The propagation velocity of the detonation obtained has a mean value of D = 1940 m/s within a deviation of 0.5% in 15 experiments.

Two-Head Wave Structure

Single experiment records on the periphery are given in Figs. 1 to 3, which exhibit two Mach triple-point configurations. In Fig. 1 the records of the first and the fourth circumference are obtained by the double-response gauges, and the second and the third one by the pressure transducers. The corresponding pressure profiles are shown in Figs. 2 and 3. After the first collision at 180° the two triple points move into the opposite direction and collide again at nearly 0°. Thus there are two triple point intersections on the periphery. The transverse waves can be recognized in the pressure profiles labeled as 90° and 270° in Fig. 2. The second sharp-fronted peak in the profile labeled as 0° in Fig. 3 is caused by the collision of the two reflected shocks associated with each triple point. The time between the fronts at x = 0 and x = 49.8 cm along the tube generatrix at 0° takes 262 μsec.

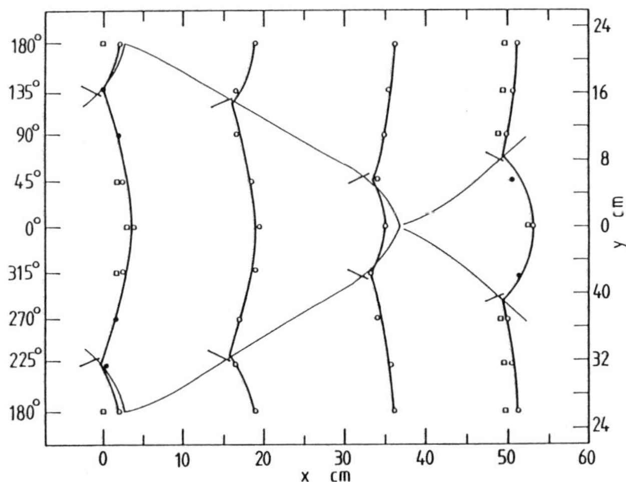

Fig. 1 Wave fronts of two-head detonation along the tube circumference in a cornstarch-oxygen mixture, σ_p = 4 kg/m^3 and p_0 = 1.15 bar. x: axial coordinate, y: circumferential coordinate. o: shock front, □: combustion front, o: unresolvable between shock and combustion front.

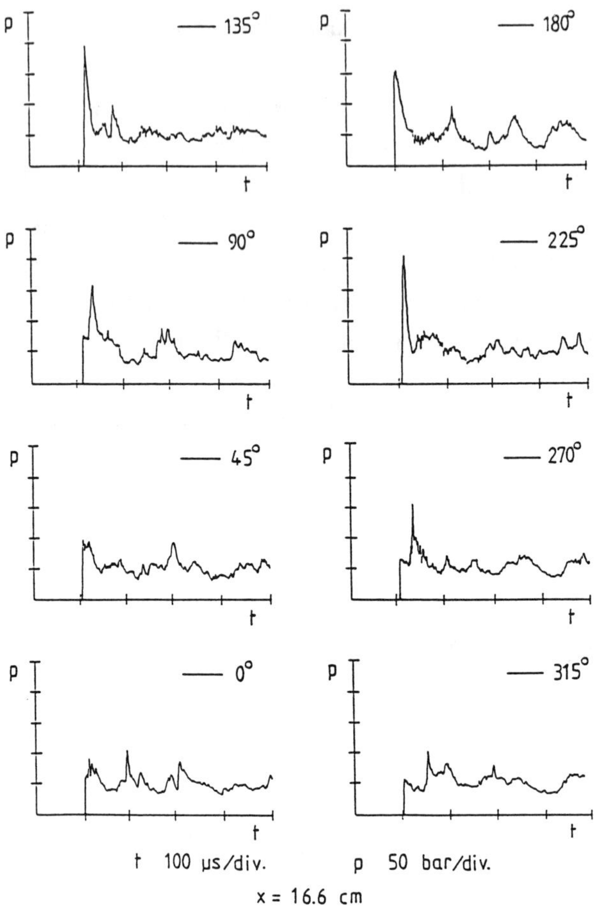

Fig. 2 Pressure records of the wave front at x = 16.6 cm (see Fig. 1).

A rather complete description of the major wave structure on the periphery is shown in Fig. 4, which is derived from four experiments. The corresponding frontal pressure peak records are given in Fig. 5. At $x = 0$ the Mach triple points collide at 0°. This event generates an explosion kernel, producing an overdriven wave with a velocity of about 1.4 D in x-direction. The structure immediately after the collision appears to be of the weak type; details cannot be resolved due to the short induction time. As the wave proceeds, the triple points move apart from each other with

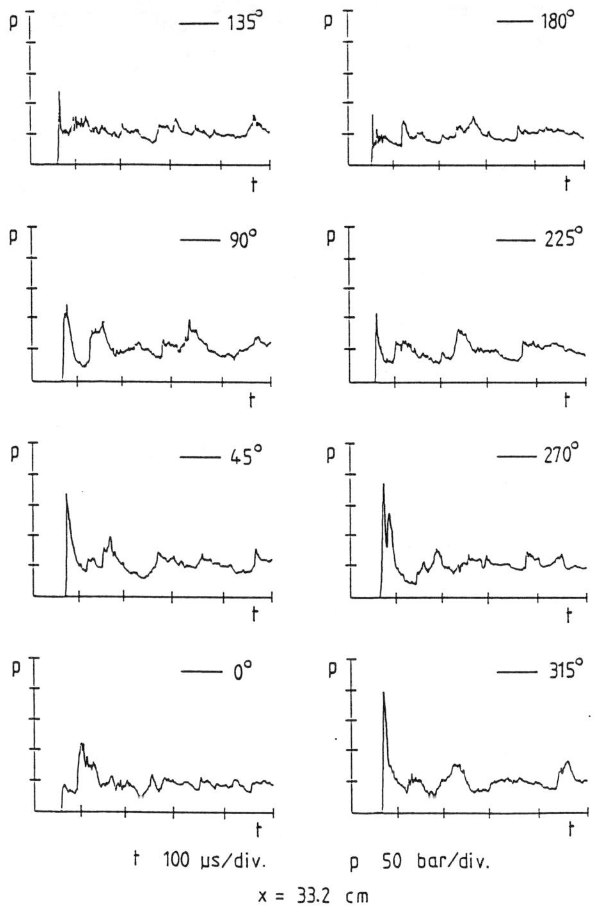

Fig. 3 Pressure records of the wave front at x = 33.2 cm (see Fig. 1).

a transverse wave velocity of about 0.62 D, while the track angle between the trace of the triple point and the generatrix of the tube increases because the front motion in x-direction is slowing down. Moreover, the induction time behind the incident shock increases. At $x \approx 34.5$ cm the collision between the triple points occurs again at the opposite position of nearly 180° to the first collision. It generates a new explosion kernel, reproducing the overdriven wave which provides, for a short time, the close coupling between the combustion front and the shock that is required

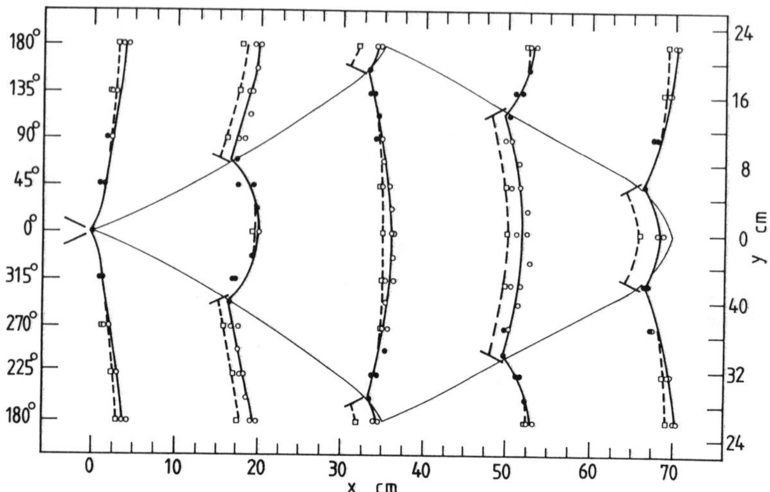

Fig. 4 Structure and triple point traces of two-head detonation wave along the tube circumference in a corn-starch-oxygen mixture, σ_p = 4 kg/m^3 and p_0 = 1.15 bar. x: axial coordinate, y: circumferential coordinate. o: shock front, □: combustion front, •: unresolvable between shock and combustion front.

for the self-sustenance. The whole process repeats. Thus the traces of two triple points form a single cell.

Figures 6 and 7 give the variation of the frontal pressure peak, induction time and axial propagation velocity along the cell-center line, i.e., the tube generatrix at 0°. There b denotes the cell length. Along the cell the pressure peak decays from 250 bar to about 75 bar, the induction time increases up to 11 μs, and the propagation velocity decreases from about 1.4 D to 0.87 D. On the other hand, the transverse shock velocity has an average value of about 0.62 D being almost constant within the cell (Fig. 7). The above results show that the transverse wave collisions are seen to be of central importance in sustaining a detonation wave. Thus, the superposition treatment of the acoustic theory by Duff[4] is not suitable for explanation of this detonation mode concerning the highly transient process between two successive collisions of the triple points.

The mean propagation velocity of D = 1940 m/s has a small deviation of only 0.5% in all of the experiments.

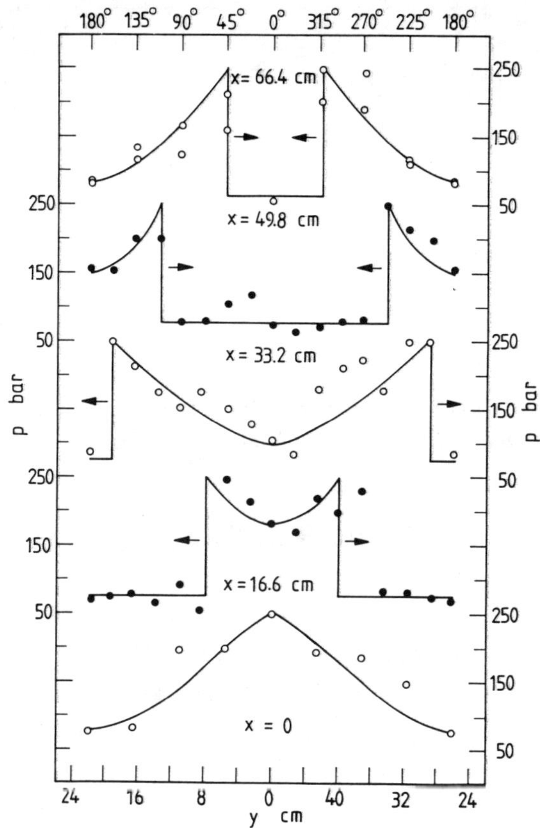

Fig. 5 Frontal pressure peak records corresponding to shock fronts shown in Fig. 4.

However, the cell length b varies between 610 mm and 690 mm. The ratio of cell width to length corresponds to $\lambda_2/b = 0.68 \pm 0.04$. Moreover, it is not possible to obtain two strictly symmetric heads, which can be seen from Figs. 1 to 3. The pressure amplitude of the oscillations behind the wave fronts is less than that of the single-head spin,[2] and the oscillations manifest themselves less regularly. It is noticeable that in the majority of the experiments, one of the two collisions on the periphery occurs near the dust dispersion tube of 12 mm in outside diameter, which is mounted with a small gap on the bottom of the whole test section. This geometry causes wave focusing effects like in a converging channel. This event probably results in the onset of the detonation mode studied here.

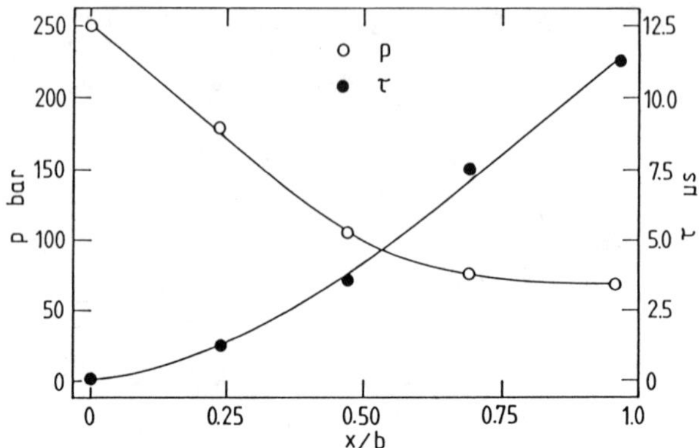

Fig. 6 Frontal pressure peak and induction time at cell center-line vs axial distance along the cell shown in Fig. 4.

Comparison Between Single-Head Spin and Two-Head Detonation

Our previous investigation[1, 2] showed that in cornstarch dust-oxygen mixtures, single-head spin detonation appears at an equivalence ratio of $\varphi \leq 1$ and an initial pressure of $p_0 \leq 1$ bar. Two-head or single-cell detonation has been observed in a certain range of rich mixtures ($\varphi \approx 2-4.5$ at $p_0 = 1$ bar), or when p_0 increases. In the case of single-head spin, the coupling of the chemical reaction and the acoustic vibration in round-tube geometry results in the

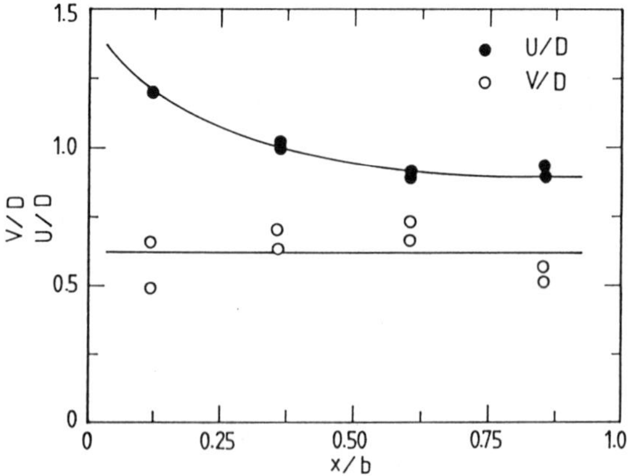

Fig. 7 Axial wave velocity at the cell center-line and transverse wave velocity vs axial distance along the cell shown in Fig. 4.

helical propagation of the unique Mach triple-point configuration, moving with a constant velocity at a constant track angle. On the other hand, the single-cell detonation involves two Mach interactions which move into opposite directions relative to each other on the circumference. In this situation the collision between the Mach triple points leads to an overdriven wave followed by decaying propagation of the new Mach triple points at transient track angles. The decaying of the waves requires continual regeneration by collisions to maintain their existence.

Phenomenally there is no cell in the single-head spin structure, since a complete cell is formed by traces of two Mach triple-point configurations, or an incomplete cell by one Mach triple point coupled to the wall reflection, for instance, in rectangular tubes. The single-head spin is considered as the limit of the lowest stable detonation mode in circular tubes. From this point of view, the sustenance of a single-head spin in a tube is regarded as a criterion for the minimum tube diameter d_{min} needed for the transition to a stable detonation. The minimum tube diameter is associated with the ordinary cell width by the empirical relation[8] $\lambda = \pi d_{min}$. This criterion is explained by Lee[9] as the result of a resonant coupling between the acoustic vibration and the periodic chemical processes in the cellular detonation. On the other hand, a two-head mode has a cell width $\lambda_2 = \pi d$. It is obvious that $\lambda_2 > \lambda$ holds because a two-head wave occurs in a larger tube than that used by a single-head spin for the same mixture and initial conditions. Thus, even though the two-head mode has the same triple-point collision mechanism as the multicellular detonation, the cell size must be greater than that corresponding to the multicellular detonation due to boundary effects.

The acoustic theory by Manson[5] and Fay[6] gives an approximate equation for the transverse wave speed on the circumference

$$V = \frac{\zeta_n a}{n}$$

thus for the tangent of the track angle

$$\tan\alpha = \frac{V}{D} \approx \frac{\zeta_n \rho_0}{n\rho}$$

Here n is the node number in the tube circumference in one rotation direction. ζ_n is the first zero of the first derivative of the Bessel function of order n, $\zeta_1 = 1.84$; a denotes the equilibrium sound speed of the detonation products; D is the detonation velocity; and ρ/ρ_0 is the mixture-density ratio across the detonation wave. For the single-head spin our previous work[1, 2] gave the experimental data of $V \approx D$ and $\alpha \approx 45°$, which are in good agreement with the acoustic theory. For the two-head mode studied here, the experiments give $V \approx 0.62\ D$ and an average value of the track angles of $\bar{\alpha} \approx 31.8°$. However, $V = D$ and $\alpha = 45°$ are predicted by the acoustic theory.

Conclusions

The propagation of a two-head or single cell detonation has been investigated in a circular tube filled with a heterogeneous cornstarch dust-oxygen mixture. The deduced detailed wave structure reveals that collisions of the two Mach triple-point configurations are of central importance in the propagation mechanism of this detonation mode. The transient decaying of the coupling between the shock and the chemical reaction waves requires continual re-ignition by successive collisions for sustained propagation.

References

[1]Zhang, F., and Grönig, H., »Detonation Structure of Corn Starch Particles-Oxygen Mixtures,« In: Dynamics of Detonations and Explosions: Detonations, edited by Kuhl, A. L., Leyer, J.-C., Borisov, A. A., and Sirignano, W. A., Vol. 103, Progress in Astronautics and Aeronautics, AIAA, New York, 1991, pp. 342 - 355.

[2]Zhang, F., and Grönig, H., »Spin Detonation in Reactive Particles-Oxidizing Gas Flow,« Physics of Fluids A, Vol. 3, No. 8, 1991, pp. 1983 - 1990.

[3]Denisov, Yu. N., and Troshin, Ya. K., »Structure of Gaseous Detonation in Tubes,« Zhurnal techniceskoj fiziki, Vol. 30, No. 4, 1960, pp. 450 - 459.

[4]Duff, R. E., »Investigation of Spinning Detonation and Detonation Stability,« Physics of Fluids, Vol. 4, No. 11, 1961, pp. 1427 - 1433.

[5]Manson, N., »Propagation des Detonations et des Deflagrations dans les Melanges Gaseaux,« L'Office National d'Etudes et des Recherches Aéronautique, Paris, 1947.

[6]Fay, J. A., »A Mechanical Theory of Spinning Detonation,« Journal of Chemical Physics, Vol. 20, June 1952, pp. 942 - 950.

[7]Voitsekhovskii, B. V., Mitrofanov, V. V., and Topchian, M. E., »Structure of a Detonation Front in Gases,« Izdatel'stvo Sibirskogo Otdeleniya AN SSSR, 1963.

[8]Kogarko, S. M., and Zeldovich, Ya. B., Doklady Akademii Nauk SSSR, Vol. 63, 1948, pp. 553.

[9]Lee, J. H., »On the Transition from Deflagration to Detonation,« In: Dynamics of Explosions, edited by Bowen, J. R., Leyer, J.-C., and Soloukhin, R. I., Vol. 106, Progress in Astronautics and Aeronautics, AIAA, New York, 1986, pp. 3 - 18.

Detonation Wave Propagation in Combustible Mixtures with Variable Particle Density Distributions

Shmuel Eidelman* and Xiaolong Yang*
Science Applications International Corporation, McLean, Virginia 22102

Abstract

A mathematical model is presented describing a physical system of detonation waves propagating in a solid particle/air mixture with a wide range of solid-phase concentrations. The mathematical model was solved numerically using the Second Order Godunov method, and numerical solutions were validated for detonation waves propagating in mixtures with concentrations of solid phase from 0.75 kg/m^3 to 1000 kg/m^3. Numerical solution was obtained for detonation waves propagating in a system consisting of clouds with a small concentration of particles and a ground layer in which solid particle densities are three orders of magnitude larger than in the cloud. Three different particle concentration distributions in the ground layer were simulated and compared in terms of detonation wave structure and parameters.

Introduction

When combustible particles are intentionally or unintentionally dispersed into the air, the resulting mixture can be detonable. Formation of this potentially explosive dust environment and the properties of its detonation are of significant practical interest in view of its destructive or creative effects.

---

*Research Scientist, Applied Physics Operation.

The experimental and theoretical study of these phenomena until now has addressed only homogenous particle/oxidizer mixtures. However, intentional or accidental processes of the explosive dust dispersion will always lead to inhomogeneous particle density distribution. Some industrial methods of explosive-forming rely on detonation of explosive powder. This powder can be deposited as a thin layer over the surface area of the forming metal, with some remaining concentration in the vicinity of the layer. The phenomenology of detonation wave initiation and propagation in this environment is the main subject of this paper.

When the detonation wave is generated in a homogeneous mixture by a "direct initiation," it starts with a strong blast wave from the initiating charge. As the blast wave decays, combustion of the reactive mixture behind its shock front starts to have a larger role in support of the shock wave motion. When the initial explosion energy exceeds some critical value, transition to steady-state detonation occurs.[1–4] In explosive dust mixtures with a nonuniform distribution of particle density, the initiation dynamics are significantly more complicated. The critical initiation energy sufficient for one of the explosive particle density strata regions is not necessarily adequate for other regions. Also, when there is a significant variation in density between the different layers (regions) of the mixture, steady detonation in one layer can result in an overdriven detonation in an adjacent layer. Our paper demonstrates that the phenomenology of these interactions is distinctly different from the classical studies of multilayer detonations in gases. This is primarily because the energy content of adjacent layers in a typical multigas layer experiment[5] varies by a factor of two or four, whereas the energy content in explosive dust/air mixtures can vary by several orders of magnitude.

In this paper we use detailed numerical simulation to study the initiation dynamics and propagation phenomenology for a general case of explosive dust dispersion. We will consider particle density variation from 1000 kg/m^3 in the ground layer to 0.5 kg/m^3 or 0 for the upper edges of the cloud. The effects of variation of the cloud density on detonation wave parameters will be examined for different cases of cloud particle density distribution. When possible, the results of computer simulations are validated in comparison with experimental and theoretical studies.

The outline of this paper is as follows. Section 2 gives a description of a mathematical model that includes governing conservation equations for two phases and the constitutive laws. We describe the model for a particle-gas interaction, combustion, and equation-of-state for gas phase. The numerical integration technique for solving the mathematical model will also be outlined. In Section 3, we present our numerical simulation results. We first validate our model by comparing one-dimensional detonation wave simulation with available experimental results. We then give the two-dimensional simulation for detonation wave propagation in combustible particles/air mixtures with variable particles density distribution. Concluding remarks are given in Section 4.

Mathematical Model and the Numerical Solution

The mathematical model consists of conservation governing equations and constitutive laws that provide closure relations for the model. The basic formulation adopted here follows the two-phase fluid dynamics model presented in the text by Kuo.[6] The approach assumes that there are two distinct continua, one for gas and one for solid particles, each moving at its own velocity through its own control volume. The sum of these two volumes represents an average mixture volume. With these assumptions, distinct equations for continuity, momentum, and energy are written for each phase. The interaction effects between the two phases are accounted for by the source terms on the right-hand side of the governing equation. The following is a short description of the two-phase flow model used in our study, with conservation equations written in Eulerian form for two-dimensional flow in Cartesian coordinates:

Continuity of gaseous phase

$$\frac{\partial \rho_1}{\partial t} + \frac{\partial(\rho_1 u_g)}{\partial x} + \frac{\partial(\rho_1 v_g)}{\partial y} = \Gamma \tag{1}$$

Continuity of solid-particle phase

$$\frac{\partial \rho_2}{\partial t} + \frac{\partial(\rho_2 u_p)}{\partial x} + \frac{\partial(\rho_2 v_p)}{\partial y} = -\Gamma \tag{2}$$

Conservation of momentum of gaseous phase in x direction

$$\frac{\partial(\rho_1 u_g)}{\partial t} + \frac{\partial(\rho_1 u_g^2 + \phi p_g)}{\partial x} + \frac{\partial(\rho_1 u_g v_g)}{\partial y} = -F_x + \Gamma u_p \tag{3}$$

Conservation of momentum of gaseous phase in y direction

$$\frac{\partial(\rho_1 v_g)}{\partial t} + \frac{\partial(\rho_1 u_g v_g)}{\partial x} + \frac{\partial(\rho_1 v_g^2 + \phi p_g)}{\partial y} = -F_y + \Gamma v_p \tag{4}$$

Conservation of momentum of solid-particle phase in x direction

$$\frac{\partial(\rho_2 u_p)}{\partial t} + \frac{\partial(\rho_2 u_p^2)}{\partial x} + \frac{\partial(\rho_2 v_p u_p)}{\partial y} = F_x - \Gamma u_p \tag{5}$$

Conservation of momentum of solid-particle phase in y direction

$$\frac{\partial(\rho_2 v_p)}{\partial t} + \frac{\partial(\rho_2 u_p v_p)}{\partial x} + \frac{\partial(\rho_2 v_p^2)}{\partial y} = F_y - \Gamma v_p \tag{6}$$

Conservation of energy of gas phase

$$\frac{\partial(\rho_1 E_{gT})}{\partial t} + \frac{\partial(\rho_1 u_g E_{gT} + u_g \phi p_g)}{\partial x} + \frac{\partial(\rho_1 v_g E_{gT} + v_g \phi p_g)}{\partial y} =$$

$$\Gamma\Big(\frac{u_p^2 + v_p^2}{2} + Echem + C_s T_p\Big) - \Big(F_x u_p + F_y v_p\Big) - \dot{Q} \tag{7}$$

Conservation of energy of solid-particle phase

$$\frac{\partial(\rho_2 E_{pT})}{\partial t} + \frac{\partial(\rho_2 E_{pT} u_p)}{\partial x} + \frac{\partial}{\partial y}(\rho_2 E_p v_p) = \dot{Q} + (F_x v_p + F_y v_p)$$

$$-\Gamma\left(\frac{u_p^2 + v_p^2}{2} + Echem + C_s T_p\right) \tag{8}$$

Conservation of number density of solid-particle

$$\frac{\partial N_p}{\partial t} + \frac{\partial(N_p u_p)}{\partial x} + \frac{\partial(N_p v_p)}{\partial y} = 0 \tag{9}$$

In the above equations, we have the following definitions and constitutive laws:

Phase densities

$$\rho_1 = \phi \rho_g, \qquad \rho_2 = (1 - \phi)\rho_p \tag{10a}$$

and fractional porosity

$$\phi = 1 - \frac{N_p M_p}{\rho_p} = \frac{\text{Volume of void}}{\text{total volume}} \tag{10b}$$

where M_p is the mass of each particle and ρ_p is the solid-particle density.

Total internal energy of gaseous phase

$$E_{gT} = E_g + \frac{1}{2}(u_g^2 + v_g^2) \quad \text{and} \quad E_g = E_g(p_g, \rho_g) \tag{11}$$

where $E_g(p_g, \rho_g)$ is the equation-of-state for gas phase, which will be discussed later.

Total internal energy of solid-particle phase

$$E_{pT} = E_p + \frac{1}{2}(v_p^2 + v_p^2) \quad \text{and} \quad E_p = \text{E}chem + C_s T_p \tag{12}$$

In order to close the above system of conservation equations, it is necessary to define certain criteria and interaction laws between the two phases, which include mass generation rate, Γ, drag force between particles and gas, F_x, F_y, and the interphase heat transfer rate $\dot{Q}$. The model for particle and gas interaction and particle combustion that results in the constitutive relation for the conservation equations is explained in detail in the next subsection.

Model for a Particle Gas Interaction and Combustion

Presently, the physics of the energy release mechanisms in solid-particles/air mixtures is not clearly understood. This can be attributed to the obvious difficulties of making a direct nonobtrusive measurement in the optically thick environment typical for this system. In the experimental and theoretical work done for the grain dust detonation conditions,[7] it was demonstrated that the volatile components released by the particle heated behind the shock front play a major role in determining the detonability limits of the mixture. Eidelman and Burcat[8] successfully applied a combination of fast evaporation and aerodynamic shattering mechanisms to simulate a two-phase detonation process.

The chemical processes of a single particle combustion, which mainly occur in the gaseous phase, are significantly faster than the physical processes of particle gasification or disintegration. Thus, in the multiphase mixtures, the rate of energy release will be mostly determined by physics of particle disintegration. It is very difficult to describe the details of particle disintegration in the complex environment prevalent behind the shock or detonation wave. For example, Reinecke and Waldman[9] defined five different disintegration regimes for a relatively simple environment of water droplets passing through a weak shock. Fortunately, in most cases of multiphase detonation, only the main features of the particle disintegration dynamics need to be captured to describe the phenomena. For example, Eidelman and Burcat[10] used simple models for particle evaporation and shattering to obtain simulation results that compared very favorably with experimental data. Because of our inability to resolve the particle disintegration problem in all its complexity, the validation of the model against known experimental data is essential.

In this paper, we consider solid particles consisting of explosive material. Explosive material contains fuel and oxidizer in a passive state at low temperature; however, when the temperature rises the fuel and oxidizer react, leading to detonation or combustion. The intiation of reaction for explosives occurs at relatively low temperature. For example, TNT will detonate when heated to the temperature[11] of 570°C. Only particles larger than a critical detonation size can detonate directly when initiated by a shock wave. Here, consider particles smaller than 4 mm in diameter that will not detonate when heated, but will burn when the temperature on the particle surface reaches a critical value. Since the heat conduction inside the explosive material is relatively slow, the process of particle heating needs to be resolved in detail. Our simulations numerically solve the temperature field in the particles at every step of numerical integration of the global conservation equations. The explosive particle combustion model examined in this paper assumes that the fraction of the particle that reaches the critical temperature will burn instantaneously.

Energy transfer by convection and conduction is simulated by solving the unsteady heat conduction equation in each computational cell at each time step. Assuming a particle's temperature to be a function of time and radial position only, the unsteady heat conduction equation may be transformed to:

$$\frac{d^2 w}{dr^2} = \frac{1}{\alpha}\frac{dw}{dt} \tag{13}$$

subject to the boundary conditions:

$$w = 0 \quad \text{at} \quad r = 0, \quad t > 0$$

$$k\frac{dw}{dr} + (h - \frac{1}{R})\, w = hRT_g \quad \text{at} \quad r = R, t > 0 \tag{14}$$

where

$w(r,t) = rT(r,t)$
r = radial position
$T(r,t)$ = temperature
R = particle radius
T_g = temperature of surrounding gas
k = thermal conductivity of particle
h = convective heat transfer coefficient

The Nusselt number, used to find h, is given by an empirical relation given by Drake.[12] The gas viscosity is derived from Sutherland's Law. The gas thermal conductivity is calculated by assuming a constant Prandtl number. Finally, the boiling temperature at a given pressure is derived from the Clapeyron-Clausius equation under the following assumptions: 1) phrasing-constant latent enthalpy of phase-change; 2) the vapor obeys the ideal equation-of-state; and 3) the specific volume of the solid/liquid is negligible compared to that of the vapor. A critical temperature is also employed to serve as an upper limit to the boiling point, regardless of pressure.

Equation 13 with boundary condition 14 can be numerically integrated using either implicit or explicit schemes.

Since the particle radius R becomes very small due to evaporation, the implicit Crank-Nicolson algorithm is used because of its stability properties and its second order temporal and spatial accuracy. Using the Crank-Nicolson scheme to predict the particle temperature profiles at times t_1 and t_2 permits easy calculation of the total energy exchange $\dot{Q}$ between t_1 and t_2, due to convection and conduction.

Knowledge of the particle temperature profile also allows the precise determination of the quantity of the mass to transfer from the particle to the gas Γ. Once any point at a radial location $0 \leq r \leq$ R has a temperature exceeding the boiling temperature, the entire mass between r and R is transferred to the gas phase in one time step. In so doing, an energy equal to the product of the mass lost and the particle intrinsic energy is transferred by the particle to the gas.

The interphase drag force Fx, Fy is determined from the experimental drag for a sphere, as presented by Schlichting.[13]

$$F_x = \left(\frac{\pi}{8}\right) N_p \rho_g C_D |\mathbf{V}_g - \mathbf{V}_p| (u_g - u_p) R^2 \qquad (15)$$

where

$$C_D = \begin{cases} \frac{24}{Re}\left(1 + \frac{Re^{2/3}}{6}\right) & \text{for Re} < 1000; \\ 0.44 & \text{for Re} > 1000 \end{cases} \qquad (16)$$

and $Re = \frac{2R|V - V_p|}{\mu_g}$, R is radius of particle, and μ_g is gas viscosity at temperature of $T_{film} = \frac{1}{2}(T_g + T_p)$.
Similarly, the formulae for Fy is

$$Fy = \frac{\pi}{8} N_p \rho_g C_D |\mathbf{V}_g - \mathbf{V}_p| (v_g - v_p) R^2 \qquad (17)$$

Equation-of-State for Detonation Products

To close the system of governing equations, one needs a constitutive relation between pressure, temperature, and energy for gas phase, which is an equation-of-state. This study uses the Becker-Kistiakowsky-Wilson (BKW) equation-of-state,[14,15] that is,

$$p_g V_g / \bar{R} T_g = 1 + x e^{bx} \qquad (18)$$

where

V_g = volume of gas phase
p_g = pressure of gas phase
T_g = temperature of gas phase
$\bar{R}$ = universal gas constant
$x = k/V_g(T + \Theta)^a$
$k = K\Sigma_j X_i k_i$

with empirical constants a, b, K, Θ, and k_i. The constants k_i, one for each molecular species, are covolumes. The covolumes are multiplied by their mole fraction of species X_i and are added to find an effective volume for a mixture. For a particular explosive, if we know the composition of detonation products, a, b, Θ, K, and all k_is can be found in Ref. 15.

The internal energy is determined by thermodynamics relation

$$\left(\frac{\partial E_g}{\partial V_g}\right)_T = T_g\left(\frac{\partial p_g}{\partial T_g}\right)_V - p_g \tag{19}$$

Integration of this equation for a fixed composition of the detonation products will allow us to calculate the energy of the detonation products as a function of temperature and volume. For each component, its thermodynamic properties as functions of temperature were calculated from the NASA tables compiled by Gordon and McBride.[16]

The BKW equation-of-state is the most commonly used and well-calibrated of those equations-of-state used to calculate the properties of detonation products. The detailed discussion and review of the BKW equation-of-state can be found in Ref. 15.

Numerical Method of Solutions

The system of partial differential equations described in the previous paragraph is integrated numerically. The Second Order Godunov method is used for the integration of the subsystem of equations describing flow of gaseous phase material and is described in Ref. 17. In the following, we will elaborate only on some specifics of its application to simulations of detonation products. The subsystem of equations describing the flow of particles is integrated using a simple upwind integration. This is done because our mathematical model neglects the pressure of interparticle interaction, and that prevents formulation of a Second Order Godunov scheme for particles.

The physical system under study will have concentrations of solid explosive powder ranging from 1000 kg/m^3 near the ground to 0.75 kg/m^3 or less in the cloud. Detonation of this mixture will create detonation products with effective γ ranging from 3 to 1.1. To describe the flow of detonation products, we use the BKW equation-of-state described above. Since the Second Order Godunov method uses primitive variables to calculate Riemann problems at the edges of the cells, its implementation for non-ideal EOS is difficult. In our simulations, we have resolved this problem by using direct and inverse equations-of-state. After integrating a system of gas conservation laws, we use the direct BKW equation-of-state to calculate pressure, gamma, and temperature as functions of thermal energy, density, and mixture composition. After this step, we have a complete set of parameters allowing calculation of the fluxes in the Second Order Godunov method as well as interaction of the multiphase processes. The "inverse" EOS calculates internal energy as a function of density, pressure, and mixture composition. In our code, we use the "inverse" EOS to calculate the fluxes of conserved variables after calculation of the flux of primitive variables.

For the multiphase system under study, dx=dy=1mm was used to allow explicit integration of the gasdynamic and physical processes of evaporation and heat release. When a mismatch occurred between the physical and gasdynamical characteristic times, the time step was adjusted by some fraction to assure stability. However, this did not result in a significantly smaller time step than the onc calculated using CFL criteria. For larger cell sizes, this approach will be impractical. Recently, we implemented a scheme in which multiphase processes are calculated implicitly; however, this will be reported elsewhere.

The numerical method is implemented in a code named MPHASE, which is fully vectorized and supported by number of graphics and diagnostics codes.

Results

Model Validation for One-Dimensional Detonation Wave Problem

The main advantage of our particle combustion model is its description of the phenomenology of detonation for a wide

Table 1 One-dimensional validation result

D[m/sec] - Detonation wave velocity,
P_{CJ}[Pa] - Pressure at Chapman-Jouguet Point
P_p[Pa] - Peak pressure; ρ_p[kg/m^3] - Peak density

RDX density (kg/m^3)	Parameters	Present calculation	Expt'l Ref. 1	Tiger calculation Ref. 2	BKW calculation Ref. 1	Soviet experiments Ref. 3
1000 kg/m^3	D	6155	5981		6128	
	P_{CJ}	1.220 × 10^{10}			1.08 × 10^{10}	1.00 × 10^{10}
	P_p	2.57 × 10^{10}				
	ρ_p	1936				
860 kg/m^3	D	6031		5900		
	P_{CJ}	0.986 × 10^{10}		0.88 × 10^{10}		0.82 × 10^{10}
	P_p	1.95 × 10^{10}				
	ρ_p	1722				
466 kg/m^3	D	4800		4500		
	P_{CJ}	0.379 × 10^{10}		0.30 × 10^{10}	0.3 × 10^{10}	
	P_p	0.625 × 10^{10}				
	ρ_p	924				
250 kg/m^3	D	4049		3600		
	P_{CJ}	0.2478 × 10^{10}		0.13 × 10^{10}		
	P_p	0.4538 × 10^{10}				
	ρ_p	552				
100 kg/m^3	D	3495				
	P_{CJ}	0.5013 × 10^9				
	P_p	0.7658 × 10^9				
	ρ_p	220				
0.75 kg/m^3	D	1622	1410*	1870*		
	P_{CJ}	0.25 × 10^7	0.284 × 10^{7*}	0.26 × 10^{7*}		
	P_p	0.484 × 10^7				
	ρ_p	8				

Ref. 1 - Mader, C., Numerical Modeling of Detonation, (University of California Press, Ltd., 1979) p. 47. Ref. 2 - Wiedermann, A., "An Evaluation of Bimodal Layer Loading Effects," IITRI Report, Feb. 1990. Ref. 3 - Stanukovitch, K. P., "Physics of Explosion" (in Russian), Nauka, 1975.

range of explosive particle sizes and densities. We will demonstrate this capability on a set of one-dimensional test problems. For these test problems, we simulated the initiation and propagation of the detonation waves in a shock tube-like setting, where the explosive particles are distributed uniformly through the shock tube volume.

Results of these simulations are summarized in Table 1, which shows detonation wave velocity, peak pressure, and peak density given as a function of the average density of the solid explosive. Here, the explosive two-phase mixture is composed from RDX particle and air, where RDX particle concentration varies from 0.75 kg/m^3 to 1000 kg/m^3. This concentration variation covers a whole range of solid explosive concentrations of interest to our problem. The simulations performed with the MPHASE code were compared with the experimental results[15,18] and calculations done with the TIGER code that are presented in Ref. 19.

From Table 1, it is clear that our simulation results compare favorably with other simulation results and experimental data. The maximum deviation between our results and referenced results is no greater than 15% for the entire range of explosives densities. Considering that our results were obtained with a single model for particle combustion applied to the extreme range of densities, our model gives an excellent prediction of the detonation wave parameters.

Two-Dimensional Simulation Results

Figure 1 shows a setup for a typical simulation with a computational domain of 25 cm × 25 cm. The explosive powder density is distributed according to the 4th power law of vertical distance, starting from the ground where the density is 1000 kg/m^3, to 1.2 cm, where the density is 0.75 kg/m^3. From this point to 25 cm height, the density is constant and equal to 0.75 kg/m^3. The density distribution in the direction of the "x" axis is uniform. The boundary conditions for the computational domain shown in Fig. 1 are specified as follows: solid wall along the "x" axis, symmetry conditions along the "y" axis, supersonic outflow for upper boundary, and at the

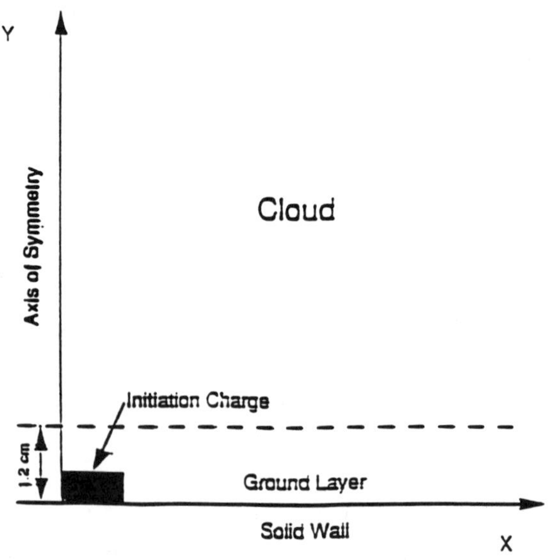

Fig. 1 Computational domain and boundary conditions.

right of the computational domain. The mixture consists of RDX powder and air at ambient conditions, and it is assumed to be quiescent at the time of initiation.

The simulation starts at t=0 when the mixture is initiated at the lower left corner of the computational domain, as shown in Fig. 1. The energy released by the initiating explosion leads to formation of the detonation wave propagating through the multiphase media. Figure 2a shows pressure contours for the propagating detonation wave at the time of t=0.012 msec after initiation. The pressure contour levels are shown on the logarithmic scale in MPa. The maximum pressure value of 7940 MPa is observed in the layer of condensed explosive located near the ground. The pressure in the layer is two to three orders of magnitude higher than pressure behind the detonation wave in the 0.75 kg/m^3 RDX cloud and air located above the distance of 1.2 cm from the ground. Figure 2a demonstrates that the detonation wave in the cloud is overdriven, since the pressure behind the shock continuously rises and reaches its maximum in the layer. From this figure, we also observe that the overdriven wave propagates faster in the cloud than in the layer. This is explained by the fact that it is easier to compress air that is very lightly loaded with particles and located above the ground layer than it is to compress air heavily loaded with a particle mixture near the ground. It is interesting to note a discontinuous pressure change between the yellow contours and the light blue and green contours behind the detonation front. This discontinuity is overemphasized by our presentation of contour lines on the logarithmic scale; however, further examination of our simulation results indicates this feature is real and is similar in nature to barrel shocks observed for strong jets.

In Fig. 2b, gas-phase density contours are shown for the time t=0.012 msec. Here the contour lines are distributed on the logarithmic scale. The main features of the shock wave structure are very similar to those observed in the pressure contours figure. We see that a jet of high-density gases reflects from the center of symmetry axis, which will create a contact discontinuity that we will observe at later times. The barrel shock is clearly visible in this figure. In Fig. 2c, the particle density contour plots are shown for t=0.012 msec. The contour levels in Fig. 2c are given on the logarithmic scale and the initial deposition of the explosive material in the ground layer of the computational domain can be clearly observed. The white contour line delineates the beginning and the end of the reaction zone in the cloud. To the left of these contours lies an area with combustion products and to the right are unburned particles in the cloud. The reaction zone length is of the order of 1 cm.

Figure 2d shows pressure contours for the same simulation for the time t=0.055 msec, just before the detonation wave leaves the computational domain. In this figure, we see that the global structure of the wave did change slightly from Fig. 2a. We observe that the barrel shock wave is fully developed and has a half-ellipse shape. The detonation wave in the cloud is still overdriven; however, part of the shock wave front that propagates vertically weakened because it gets further away from the detonation front in the layer. Another noticeable feature is the increase in distance between the detonation front in the layer and in the cloud area close to the layer. This is a result of the fact that the lightly loaded two-phase media above the layer can be compressed much more easily than the particle-heavy ground layer. In Fig. 2e, temperature contours are shown for t=0.055 msec. Comparing this figure with an early stage of the wave propagation, we observe a significant cooling of the front area propagating upwards, which indicates transition from the overdriven detonation regime to a self-sustained detonation. We also observe in Fig. 2a clear development of two detonation fronts, one moving vertically in the cloud and another moving horizontally in the layer. Because the energy density of the explosive powder in the layer is about three orders of magnitude larger than in the cloud, the vertical parts of the front represent an overdriven detonation wave in the cloud. Even though the vertical front has slowed down compared with the horizontal front, its speed and parameters far exceed those typical for detonation waves in a cloud. In fact, the self-sustained detonation regime in the cloud will

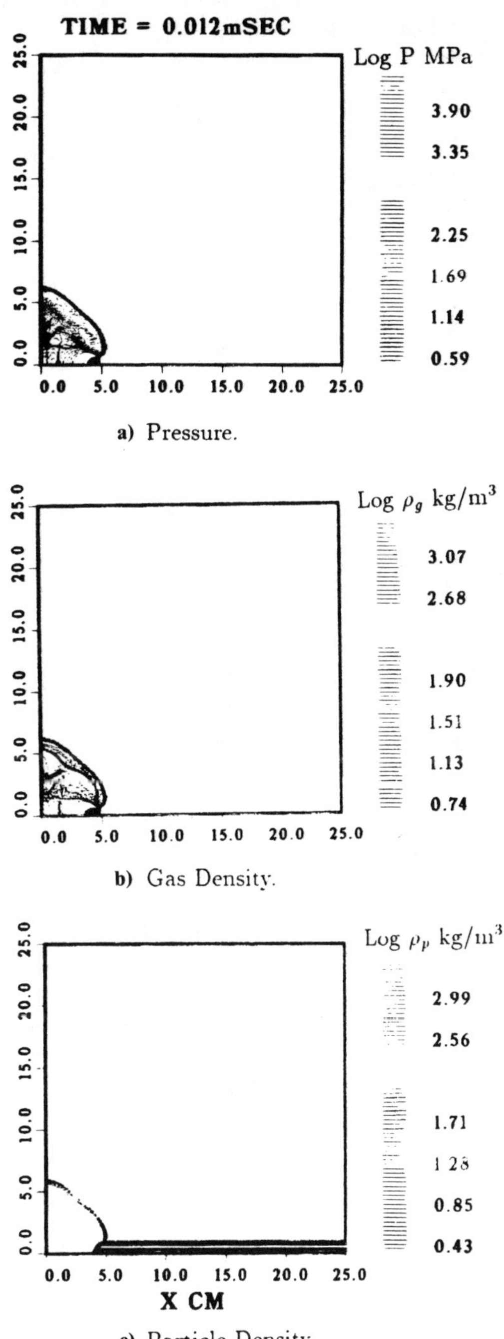

a) Pressure.

b) Gas Density.

c) Particle Density.

Fig. 2 Fourth power layer distribution; maximum density in the layer 800 kg/m^3; density in the cloud 0.75 kg/m^3; time 0.012 m/s and 0.055 m/s after initiation.

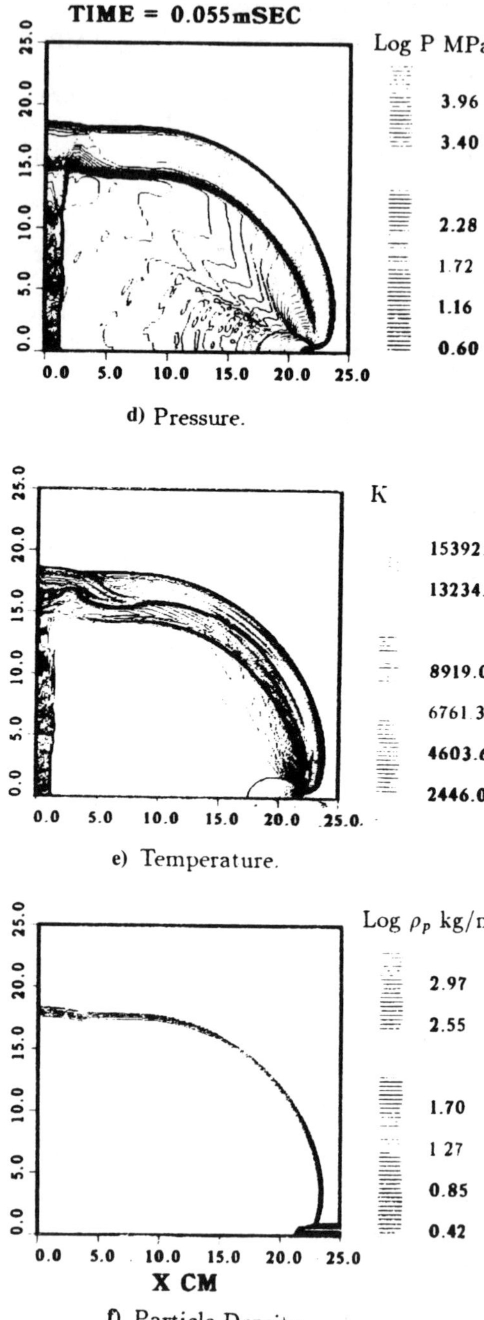

d) Pressure.

e) Temperature.

f) Particle Density.

Fig. 2 (continued) Fourth power layer distribution; maximum density in the layer 800 kg/m^3; density in the cloud 0.75 kg/m^3; time 0.012 m/s and 0.055 m/s after initiation.

develop at the distance of about 3 m from the layer. The area of the front close to the detonation wave in the layer will remain hot and overdriven, since it is located very close to the detonation front in the layer. In Fig. 2f, particle density contours are shown on a logarithmic scale. We can clearly observe the reaction zone delineated by black contour lines. In this case, the reaction zone length in the cloud is about 1 cm. Consistent with the gradual transition from overdriven to self-sustained detonation, the reaction zone length is larger for the vertical part of the detonation front. The detonation wave velocity observed in our simulation is approximately 4048 msec, which is significantly lower than the detonation wave velocity observed in RDX with a density of 860 kg/m^3 (see Table 1), the highest density in the ground layer. This can be explained by high gradient of particle density distribution in the layer, where the density drops rapidly from 860 kg/m^3 at the bottom of the layer to 1 kg/m^3 at the top strata of the layer at 12 mm above the ground.

To further explore properties and phenomenology of the detonation waves propagating in the layer/cloud systems, we simulated additional cases in which explosive powder density distribution was different from the case reported above, although total weight of fuel per unit area remained the same.

In Fig. 3, results are shown for the case of a uniform 2.5 cm-thick layer of RDX with a density of 100 kg/m^3 and a 0.75 kg/m^3 cloud initiated under the same conditions as in the previous example. Figures 3a, 3b, and 3c show pressure, gas density, and particle density contour plots at t=0.066 msec. We observe that because the layer has considerably smaller density compared to the case reported above, the precursor effect of the detonation wave in the cloud preceding the wave in the layer is less pronounced. Also, one can observe a significant difference in the shape of the strong contact discontinuity in the region of the shock front close to the layer. In Fig. 3b, we can clearly distinguish two contact surfaces. One is between condensed explosive detonation products in the layer and in the cloud, and another is between the detonation products from

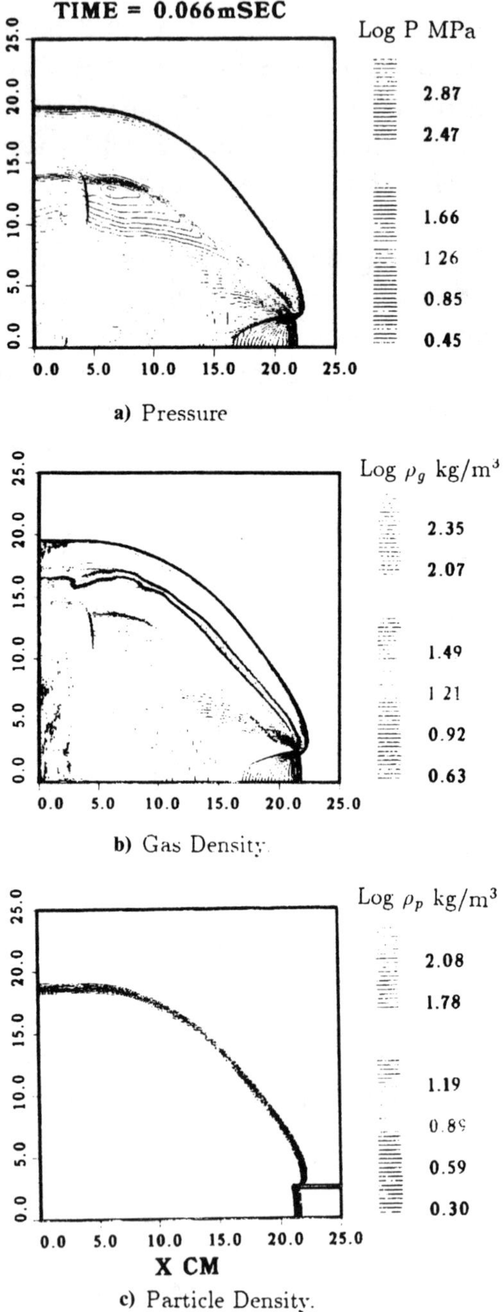

a) Pressure

b) Gas Density

c) Particle Density.

Fig. 3 Constant density 2.5-cm-thick layer; maximum density in the layer 100 kg/m^3; density in the cloud 0.75 kg/m^3; time 0.055 m/s after initiation.

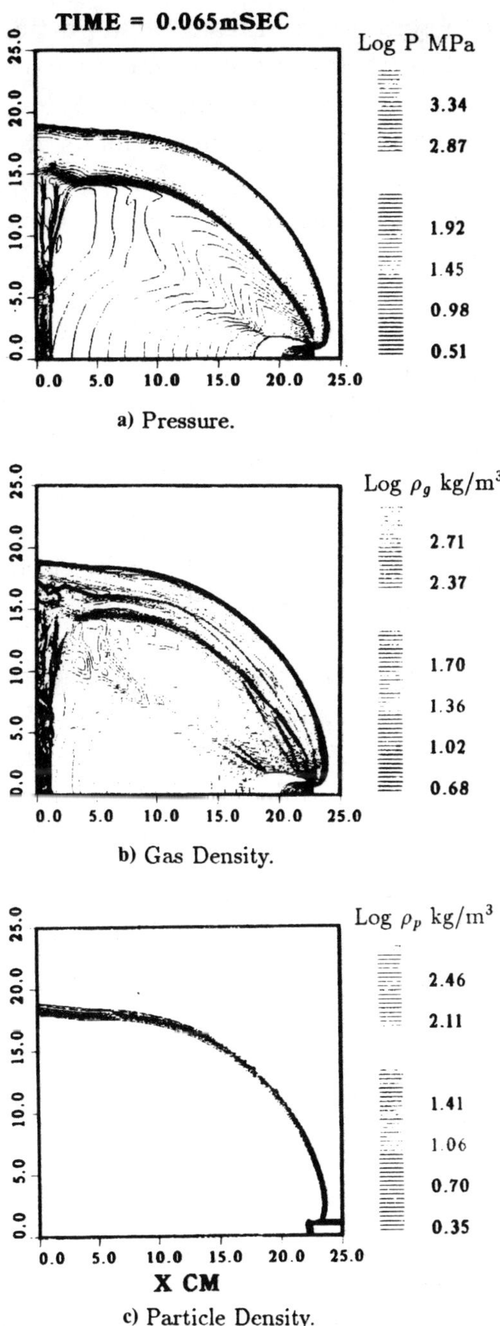

Fig. 4 Constant density 1.2-cm-thick layer; maximum density in the layer 250 kg/m^3; density in the cloud 0.75 kg/m^3.

layer explosive detonation and from cloud particle detonation. We should note that these contact surfaces are overemphasized by the logarithmic display of the contour plot levels. The maximum pressure observed in this simulation is 955 MPa, which is about one order of magnitude smaller than in previous simulation. This is consistent with the one order of magnitude difference in the maximum density of the ground layer in the two cases. The detonation wave speed in the case presented in Fig. 3 is 3407 msec. That is only slightly lower than the speed predicted by one-dimensional simulations presented in Table 1, which reflects the influence of the two-dimensional expansion on the detonation wave propagation.

Figure 4 presents results for the case of a uniform density of 250 kg/m^3 in 1.2 cm ground layer. All other parameters are the same as in the previous two cases. In Figs. 4a, 4b, and 4c, pressure, gas density, and particle density contour plots are shown at the time t=0.066 msec after initiation of the detonation wave. Here, the detonation wave propagates faster than in the previous cases U=3660 msec. This is about 400 msec slower than in the case of parabolic density distribution. Maximum pressure on the ground is 2150 MPa, which is consistent with the increase of powder density in the layer. The basic structure of the detonation front and the contact surfaces is similar to the case of parabolic density distribution.

Conclusions

We have presented a mathematical model and numerical solution for the simulation of initiation and propagation of the detonation waves in multiphase mixtures consisting of solid combustible particles and gas. Using this model, we studied detonations in mixtures of solid RDX particles and air, with the objective of examining the effects of wide variation in particle density distribution on the dynamics and structure of detonation waves. We considered a physical system of solid particle clouds in air, in which a significant amount of particles settle on the ground and the condensed-phase concentrations in the particle/air mixture range from 0 to 1000 kg/m^3. This range of solid-phase densities necessitated development of the model and its numerical implementation for a wide range of particle concentrations. Our validation study has shown good agreement between the simulations and referenced results for the whole range of particle concentrations.

Two-dimensional simulations were done for the system of low particle density concentration clouds and ground layers formed by high concentrations of the RDX powder. We examined three cases of ground layer density distribution: a fourth power distribution within 12 mm above ground with a maximum density on the ground of 860 kg/m^3; a uniform 25 mm-thick layer with a density of 100 kg/m^3; and a 12 mm-thick uniform layer with a density of 250 kg/m^3. In all these cases, the weight of condensed phase per unit area was the same, which allowed examination of the effects of the particle density distribution on detonation wave parameters.

In all examined two-dimensional cases, the detonation wave in the cloud in the computational domain was significantly overdriven and did not play an important role. We estimated that the self-sustained regime of the detonation wave in the cloud for the examined cloud concentrations can occur only at the distances of 2–3 m above ground. At the same time, the particle density distribution in the layer determines the dynamics of the detonation wave as well as pressure on the ground.

In all three two-dimensional simulations, we observed a very distinct shape of the detonation wave front in the vicinity of the layer. In this area, the overdriven detonation in the cloud is preceding the detonation wave in the ground layer. This feature of the detonation front can be explained by the fact that the energy released in the detonation wave in the ground layer produces a faster shock wave in the dilute cloud than in those heavily loaded with solid particle stratas from the ground layer. However, these structures were not observed experimentally, and more studies are needed to examine their parameters.

The maximum pressure affecting the ground was directly related to the maximum particle density in the lower strata of the layer. However, the detonation front velocity for the fourth power distribution case was considerably lower than calculated for a one-dimensional case with 860 kg/m^3 particle density, reflecting the significant effect of two-dimensional expansion. Two other cases with 250 kg/m^3 and 100 kg/m^3 maximum densities had the detonation wave velocity only slightly lower than the one-dimensional simulations of the same RDX/air concentrations. It is interesting to compare the simulation of the fourth power density distribution case and 250 kg/m^3 case. In both cases, the same amount of explosive was distributed in the same physical space; however, the parameters of developed detonations were vastly different. Existence of the high-density strata at the bottom of the ground layer in the fourth power case significantly increased the maximum pressure at the ground and produced higher detonation wave velocity.

Using a variable density layer, one can reach a combination of pressure and velocity conditions outside of Chapmen-Jougett limitations. The range of conditions that can be obtained in the variable density system and the parametrics for this range need a more systematic study. In this article, we introduced only the mathematical formulation and numerical simulation method validated for the range of conditions of interest. In addition, we have given some examples of its application for two-dimensional simulations. However, this methodology should be linked to an experimental study for a more in-depth analysis of the phenomenology discussed here.

References

[1]Eidelman, S., Timnat, Y. M., and Burcat, A., "The Problem of a Strong Point Explosion in a Combustible Medium," 6th Symposium on Detonation, Office of Naval Research, Coronado, CA, 1976, p. 590.

[2]Burcat, A., Eidelman, S., and Manheimer-Timnat, Y., "The Evolution of a Shock Wave Generated by a Point Explosion in a Combustible Medium," Symposium of High Dynamic Pressures (H.D.P.), Paris, 1978, p. 347.

[3]Oved, Y., Eidelman, S., and Burcat, A., "The Propagation of Blasts from Solid Explosives to Two-Phase Medium," *Propellants and Explosives*, Vol. 3, No. 105, 1978.

[4]Eidelman, S., and Burcat, A., "The Evolution of a Detonation Wave in a Cloud of Fuel Droplets; Part I, Influence of the Igniting Explosion," *AIAA Journal*, Vol. 18, 1980, p. 1103.

[5]Liu, J. C., Kauffman, C. W., and Sichel, M., "The Lateral Interaction of Detonating and Detonable Mixtures," Private communication, 1990.

[6]Kuo, K., *Principles of Combustion*, John Wiley and Sons, Inc., New York, NY, 1990, pp. 513-626.

[7]Kauffman, C. W., et al., "Shock Wave Initiated Combustion of Grain Dust," *Symposium on Grain Dust*, Manhattan, KS, 1979.

[8]Eidelman, S., and Burcat, A., "Numerical Solution of a Non-Steady Blast Wave Propagation in Two-Phase ('Separated Flow') Reactive Medium," *Journal of Computational Physics*, Vol. 39, 1981, p. 456.

[9]Reinecke, W. G., and Waldman, G. D., "Shock Layer Shattering of Cloud Drops in Reentry Flight," AIAA Paper 75-152, 1975.

[10]Eidelman, S., and Burcat, A., "The Mechanism of Detonation Wave Enhancement in a Two-Phase Combustible Medium," 18th Symposium on Combustion, The Combustion Institute, Waterloo, Ontario, Canada, 1980, pp. 1661-1670.

[11]*Engineering Design Handbook, Explosives Series, Properties of Explosives of Military Interest*, AMC Pamphlet, AMCP 706-7177, 1971.

[12]Drake, R. M., Jr., "Discussions on G. C. Vliet and G. Leppert: Forced Convection Heat Transfer from an Isothermal Sphere to Water," *Journal of Heat Transfer*, Vol. 83, 1961, p. 170.

[13]Schlichting, H., *Boundary Layer Theory* 7th ed., McGraw-Hill, New York, 1983.

[14]Cowan, R. D., and Fickett, W., "Calculation of the Detonation Products of Solid Explosives with the Kistiakowsky-Wilson Equation of State," *Journal of Chemical Physics*, Vol. 24, 1956, p. 932.

[15]Mader, C. L., *Numerical Modeling of Detonation*, University of California Press, Ltd. London, England, 1979.

[16]Gordon, S., and McBride, B. J., "Computer Program for Calculations of Complex Chemical Equilibrium Compositions, Rocket Performance, Incident and Reflected Shocks and C-J Detonations," NASA SP-273, 1976 (revision).

[17]Eidelman, S., Collela, P., and Shreeve, R. P., "Application of the Godunov Method and Its Second Order Extension to Cascade Flow Modelling," *AIAA Journal*, Vol. 22, 1984, p. 10.

[18]Stanukovitch, K. P., *Physics of Explosion* (in Russian), Nauka, 1975.

[19]Wiedermann, A., "An Evaluation of Bimodal Layer Loading Effects," *IITRI Report*, February 1990.

Structure of Detonation Waves in a Vacuum with Propellant Particles

Sergei A. Zhdan*

Lavrentyev Institute of Hydrodynamics, Novosibirsk, Russia

Abstract

Results of an investigation of a steady detonation wave in a vacuum with propellant particles are presented. The particles are assumed to be overheated and to burn when achieving the ignition temperature. It has been established that the detonation wave does not correspond to ZND theory, since the shock wave is absent, and the wave front represents a contact discontinuity with the gas temperature jump and continuous pressure. The relaxation wave is formed behind the contact discontinuity. Before the particles start to ignite, the two-phase flow parameters are one-parametric functions of particle temperature. The contact discontinuity in gas exists in the ignition plane, the gas temperature behind the contact discontinuity being maximum. The existence of a region of steady relaxation wave has been determined, and the profiles of parameters in the reaction zone have been calculated. Thus, it has been established that the detonation wave in vacuum with propellant particles has been a real example of C-J detonation in gas-particle mixtures with a continuous structure.

Introduction

The structure of detonation waves in propellant particle-gas mixtures was studied by Nigmatulin et al.[1,2] and Akhatov et al.[3] It was found that the structure was in qualitative agreement with the Zel'dovich-Neumann-Doring theory, i.e., the detonation complex is composed of the gas-frozen shock wave, relaxation zone, and energy-release zone. Singular and stationary points of heterogeneous equations have been investigated by Medvedev et al. [4,5]

---

*Leading Research Scientist.

In contrast to a heterogeneous medium of the gas-particle type, in which the oxidizer and the fuel are in different phases and the presence of each of them is necessary for the detonation process, the propellant particle-gas mixture possesses the unique property of the capacity to detonate in the absence of gaseous phase in the mixture, i.e., to detonate in vacuum.

This paper presents the results of an investigation of the structure of steady detonation waves in vacuum with propellant particles. It has been established that the detonation wave does not correspond to ZND theory, since the detonation complex does not involve the frozen shock wave, and wave front is a contact discontinuity. In the relaxation zone the two-phase flow parameters are one-parametric functions of particle temperature. In the ignition plane there exists a contact discontinuity in gas. The existence region of the relaxation wave has been determined. The profiles of parameters in the steady detonation wave relaxation have been calculated.

Formulation of the Problem

Let us consider a plane one-dimensional steady motion of monodispersed propellant particles in gas with combustion of particles. It is assumed that: 1) a chemical reaction of combustion of propellant particles starts after heating them up to the ignition temperature(T_{ign}) and runs at this temperature[1]; 2) the reaction products are calorically perfect gas and particles are incompressible; 3) initially the propellant particles are mixed in vacuum, i.e., the initial pressure and density of gaseous phase are equal to zero $\left(p_0 = 0,\ \rho_{10}^0 = 0\right)$.

The equations[6] describing the structure of detonation wave propagating with velocity $\mathcal{D}$ are as follows:

Laws of mass, momentum, and energy conservation of the mixture

$$\rho_1 u_1 + \rho_2 u_2 = \rho_{20}\mathcal{D} \qquad \left(\rho_i = \alpha_i \rho_i^0\right)$$

$$\rho_1 u_1^2 + \rho_2 u_2^2 + p = \rho_{20}\mathcal{D}^2, \qquad \left(n u_2 = n_0 \mathcal{D}\right)$$

$$\rho_1 u_1\left(h_1 + \left(\frac{u_1}{2}\right)^2\right) + \rho_2 u_2\left(h_2 + \left(\frac{u_2}{2}\right)^2\right) = \rho_{20}\mathcal{D}\left(h_{20} + \frac{\mathcal{D}^2}{2}\right) \tag{1}$$

Equations of mass, momentum, and energy of particles

$$\frac{d\rho_2 u_2}{dx} = -j, \quad \rho_2 u_2 \frac{du_2}{dx} + \alpha\, 2\frac{dp}{dx} = f, \quad \rho_2 u_2 \frac{de_2}{dx} = q$$

$$f = \frac{n\pi d^2}{4}\rho_1^0 C_D \frac{|u_1 - u_2|(u_1 - u_2)}{2}$$

$$C_D = \frac{24}{\mathrm{Re}} + \frac{4.4}{\sqrt{\mathrm{Re}}} + 0.42$$

$$q = n\pi d\,\frac{\lambda_1}{c_1}\,Nu\left[c_1(T_1 - T_2) + \Pr^{1/3}\frac{(u_1 - u_2)^2}{2}\right]$$

$$Nu = 2 + 0.6\,\mathrm{Re}^{1/2}\,\Pr^{1/3}$$

$$j = n\pi d\,\frac{\lambda_1}{c_1}\,Nu\,\ln\left[1 + \frac{c_1(T_1 - T_{ign})}{l_2}\right]$$

$$\mathrm{Re} = \frac{\rho_1^0 d|u_1 - u_2|}{\mu_1}$$

$$\Pr = \frac{\mu_1 c_1}{\lambda_1} \qquad (2)$$

Here $\rho_1, \rho_i^0, \alpha_i, u_i, and\ h_i \quad (i = 1.2)$ are the mean density, real density, volume fraction, velocity, and specific enthalpy of the ith phase, respectively; p is the pressure and n the number of particles per unit volume; the subscript 0 denotes the undisturbed condition; j,f, and q describe the transfer of mass, momentum, and energy from one phase to another; d is the particle diameter; and μ_1 and λ_1 are the coefficients of viscosity and heat conduction of gas, respectively. The expression for j corresponds to the gasification model by Gostintsev.[7] According to assumption 1, $j = 0$ at $T_2 < T_{ign}$ and $q = 0$ at $T_2 \geq T_{ign}$.

Let us add systems (1) and (2) using the equations of state for phases,

$$\begin{aligned} &p = \rho_1^0 RT_1, \quad \rho_2^0 = const, \qquad e_2 = h_2 - p/\rho_2^0 \\ &h_1 = c_1 T_1, \qquad h_2 = c_2 T_2 + p/\rho_2^0 + q_2 \end{aligned} \qquad (3)$$

where R is the gas constant, C_1 and T_1 are the constant pressure specific heat and the temperature of reaction products, respectively; e_2, c_2, and T_2 are the internal energy, specific heat, and temperature, of particles, respectively; and q_2 is the heat of chemical reactions per unit mass of the particles.

Define the dimensionless variables as follows:

$$R = \frac{\rho_i}{\rho_{20}}, \quad U_i = \frac{u_i}{\mathcal{D}}, \quad H_i = \frac{h_i}{q_2}, \quad \theta_i = \frac{T_i}{T_{ign}}, \quad C_i = \frac{c_i T_{ign}}{q_2}, \quad (i = 1.2)$$

$$P = \frac{p}{\rho_{20} q_2}, \quad J = \frac{j x_0}{\rho_{20}\sqrt{q_2}}, \quad F = \frac{f x_0}{\rho_{20} q_2}, \quad Q = \frac{q x_0}{\rho_{20}\sqrt{q_2}\, q_2}, \quad D = \frac{\mathcal{D}}{\sqrt{q_2}}$$

and the coordinate $\xi = x/x_0$, where x_0 is the characteristic dimension of the problem.

Statement 1: The front of the steady detonation wave of propellant particles in vacuum is a contact discontinuity in gas parameters.

Proof: Let us designate the parameters immediately after the jump as f. The discontinuity relationship following from Equation (1) take the form

$$R_{1f}U_{1f} + R_{2f}U_{2f} = 1 \tag{4a}$$

$$R_{1f}(U_{1f}D)^2 + R_{2f}(U_{2f}D)^2 + P_f = D^2 \tag{4b}$$

$$R_{1f}U_{1f}\left(H_{1f} + \frac{(U_{1f}D)^2}{2}\right) + R_{2f}U_{2f}\left(H_{2f} + \frac{(U_{2f}D)^2}{2}\right) = 1 + C_2\theta_0 + \frac{D^2}{2} \tag{4c}$$

Taking into account that in the jump zone the time is insufficient for the phases to interact, it is appropriate to use the discontinuity relations following for Equation (2)

$$R_{2f}U_{2f} = 1 \tag{5a}$$

$$R_{2f}U_{2f}E_{2f} = E_{20} = C_2\theta_0 \tag{5b}$$

The difference between Equations (4a) and (5a) provides the condition of the contact discontinuity

$$R_{1f}U_{1f} = 0 \tag{6}$$

i.e., the absence of gas flow through the discontinuity. Using Equation (6), Equation (4b) and (4c) are transformed into

$$U_{2f}D^2 + P_f = D^2, \quad E_{2f} + \alpha_{20}P_f + \frac{U_{2f}D^2}{2} = \frac{D^2}{2} + C_2\theta_0$$

Combining them and substituting Equation (5b) gives

$$(1 - 2\alpha_{20} + U_{2f})\,P_f = 0$$

The expression in parentheses is always positive, therefore, another contact discontinuity conditions is obtained

$$P_f = 0 \tag{7}$$

i.e., the pressure is continuous. Since gas temperature $\theta_{1f} > 0$, it follows from Equations (3) and (7) that $R_{1f} = 0$. Thus, at discontinuity $[P] = 0$, $[R_1] = 0$, and gas does not flow through the discontinuity . However, it is possible that $[U_1] \neq 0$ and $[H_1] \neq 0$, i.e., the detonation wave front is a contact discontinuity in gas.

Sequence 1: All of the parameters of a particle at the contact discontinuity are continuous

$$U_{2f} = 1, \quad R_{2f} = 1, \quad \theta_{2f} = \theta_0$$

Sequence 2: The frozen shock adiabatic of the mixture appearing in the analysis of wave structure in dispersed gas-particle mixtures is in degenerate form: $P_f = 0$. it should be noted that the detonation wave in vacuum with propellant particles is a unique example of detonation in gas-particle mixtures with continuous structure, i.e., without the frozen shock at the front.

Let us divide a steady detonation wave zone into two parts by the contact discontinuity placed in the plane with the coordinate $\xi = 0$: the first of them, called the zone of thermal and velocity relaxation, begins behind the contact discontinuity ($\theta_2 = \theta_0$) and ends in the plane of particle ignition ($\theta_2 = 1$); the second, called the combustion zone, begins behind the ignition plane and ends in the Chapman-Jouguet plane. Let us consider the behavior of the solution of the problem under consideration in the relaxation zone ($\theta_0 < \theta_2 \leq 1$). The dimensionless Equations (1) and (2) take the form

$$R_1 U_1 = 0, \qquad \alpha_2 U_2 = \alpha_{20}, \qquad R_2 (U_2 D)^2 + P = D^2$$

$$H_2 + \frac{(U_2 D)^2}{2} = 1 + C_2 \theta_0 + \frac{D^2}{2} \tag{8}$$

$$R_2 U_2 = 1, \qquad D \frac{dE_2}{d\xi} = Q, \qquad D^2 \frac{dU_2}{d\xi} + \alpha \frac{dP}{d\xi} = F \tag{9}$$

Since the density $R_1(\xi) \not\equiv 0$, it follows from the first Equation (8) that

$$U_1(\xi) \equiv 0 \tag{10}$$

i.e., in the relaxation zone the gaseous products are immovable. P is obtained from the law of conservation of momentum

$$P = D^2 (1 - U_2) \tag{11}$$

which, at fixed D, depends only on particle velocity. Form the law of conservation of energy it follows that U_2 depends only on particle temperature

$$U_2 = \alpha_{20} + \sqrt{(1-\alpha_{20})^2 - 2C(\theta_2 - \theta_0)/D^2} \tag{12}$$

By introducing the specific volume of particles, $V_2 = 1/R_2$ from the first Equation (9), we obtain

$$V_2 = U_2 \tag{13}$$

Thus particle velocity, volume fraction and specific volume of propellant particles, as well as the pressure of gaseous products in the detonation wave relaxation zone are the functions of only the parameter θ_2. From the last two Equations (9) it follows that

$$Q + U_2 DF = 0 \tag{14}$$

(the relation between the intensities of force and thermal interactions). Equation (14) allows the determination of gas temperature θ_1, at each point of the relaxation zone, and then the real and mean densities (R_1^0 and R1, respectively) of the first phase.

Equations(10-14) reduce the problem of detonation wave structure in the relaxation zone to the solution of the differential equation

$$DC_2 \frac{d\theta_2}{d\xi} = Q(\theta_2)$$

with the initial condition $\theta_2(0) = \theta_2$. The solution is sought up to the coordinate $\xi = \xi_*$, where $\theta_2(\xi_*) = 1$.

Existence Region of Steady Relaxation Zone

The equilibrium shock adiabatic characteristic of dispersed mixtures is not to be introduced in a standard manner when analyzing the structure of the detonation waves in vacuum with propellant particles. Actually, the equilibrium shock adiabatic should characterize the state of a medium after equalizing the relaxation parameters of the phases($U_{1e} = U_{2e}$, $\theta_{1e} = \theta_{2e}$) with no ignition of particles (J = 0). In this case, however, $U_1 \neq U_2$ everywhere in the relaxation zone, since it follows from Equations (10) and (12) that $U_1 = 0$, and U2 > α_{20}. Thus, the equilibrium shock adiabatic does not exist for the detonation wave in vacuum with propellant particles, and its detonation zone consists of the contact discontinuity relaxation compression wave, and the combustion zone adjacent to it in the particle ignition plane $\xi = \xi_*$.

To determine the necessary conditions of relaxation weave existence in the detonation wave structure, it is convenient to make an analysis in the P and V_2 variables. The next equation of straight line follows from Equation (11) and (13) :

$$P = D^2(1 - V_2) \tag{15}$$

The variation region of specific volume of particles V*2 < V2,- 1, where

$$V_2^* = \alpha_{20} + \sqrt{(1-\alpha_{20})^2 - 2\,C_2\,(1-\theta_0)/D^2} \tag{16}$$

Typical dependencies (Equations (10-13)) on V_2 in the relaxation zone are shown in Figure 1. The set of states $P(V_2^*)$, attainable at the particle ignition temperature t2 = 1, is found form Equation (15) and (16) by eliminating the velocity D (line 3, Figure 1a)

$$P(V_2^*) = 2\,C_2\,(1-\theta_0)/(1-2\alpha_{20}+V_2^*) \tag{17}$$

A subradical expression in Equation (16) must be non-negative, therefore, the lower bound upon the relaxation wave velocity is obtained

$$D \geq \sqrt{2C_2(1-\theta_0)}/(1-\alpha_{20}) = D_{\min} \tag{18}$$

Straight line Equation (15) with a minimum slope is shown in Figure 1a (line 2). Designate the C-J detonation velocity as $\mathcal{D}_J$. From Equation (18) it follows that it is necessary that $\mathcal{D}_J > \mathcal{D}_{\min}$ for a steady self-sustained detonation wave to exist. The last inequality gives the upper bound on the particle ignition temperature T_{ign}

$$T_{ign} < T_0 + (1-\alpha_{20})^2\,\mathcal{D}_J^2/2c_2 \tag{19}$$

Thus, it the ignition temperature of the propellants does not satisfy Equation (19), the steady detonation wave in vacuum is impossible.

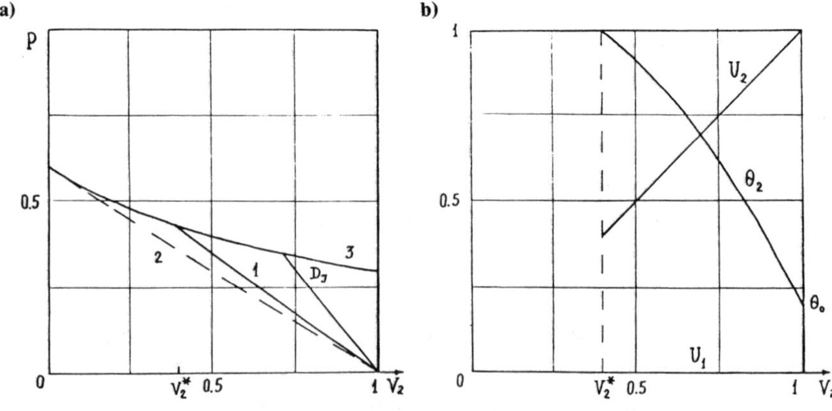

Fig. 1 Typical dependencies on V_2 in the relaxation zone.

Let us consider the behavior of the solution for $\xi \geq \xi_{*}$. Dimensionless Equations (1) and (2) in this zone is in the form

$$R_1 U_1 + R_2 U_2 = 1, \quad n U_2 = n$$

$$R_1 (U_1 D)^2 + R_2 (U_2 D)^2 + P = D^2$$

$$R_1 U_1 \left(C_1 \theta_1 + \frac{(U_1 D)^2}{2} \right) + R_2 U_2 \left(\quad H_2 + \frac{(U_2 D)^2}{2} \right) = 1 + C_2 \theta_0 + \frac{D^2}{2}$$

$$D \frac{dR_2 U_2}{d\xi} = -J, \quad D^2 R_2 U_2 \frac{dU_2}{d\xi} + \alpha_2 \frac{dP}{d\xi} = F, \quad E_2 = E_2^{*} \qquad (20)$$

Statement 2: The ignition plane of a detonation wave in vacuum with propellant particles is a contact discontinuity in gas parameters.

Proof: Denote the parameters to the right from the ignition region by the superscript (+). When $\xi \to \xi_{*}^{+}$ it is not difficult to show via the limiting transition from Equation (20) that all the parameters of the propellant are continuous when transiting through the ignition plane

$$\theta_2^{+} = \theta_2^{*} = 1, \quad U_2^{+} = U_2^{*}, \quad R_2^{+} = R_2^{*}$$

Moreover, a part of the gaseous phase parameters is also continuous

$$P^{+} = P^{*}, \quad U_1^{+} = U_2^{*}, \quad \alpha_1^{+} = \alpha_1^{*}$$

The last equalities mean that gas does not flow through the ignition plane and the pressure jump does not exist in this region. Now it is appropriate to demonstrate that $[\theta_1] \neq 0$ and $[R_1] \neq 0$. After having differentiated the equation of mixture energy and having substituted the derivatives by the intensities of interactions between particles and forces, we obtain

$$\left(C_1 \theta_1 + \frac{(U_1 D)^2}{2} - H_2 - \frac{(U_2 D)^2}{2} \right) \frac{J}{D} + F U_2 + R_1 U_1 \frac{d}{d\xi} \left(H_1 - \frac{(U_2 D)^2}{2} \right) = 0$$

Coming over to the limit when $\xi \to \xi_{*}^{+}$ and bearing in mind that $R_1 U_1 \to 0$, we find the values of gas temperature θ_1 immediately behind the ignition plane

$$C_1 \theta_1^{+} = 1 + C_2 \theta_0 + D^2/2 + U_2 D(-F)/J \qquad (21)$$

Since the gas temperature to the left of the ignition plane is determined independently of Equation (14), we come to the conclusion that $\theta_1^+ \neq \theta_1^+$ and $R_1^+ \neq R_1^+$. The statement has been proved.

Consequence: The gas temperature θ_1^+ is always higher than that at the point of complete burn out of particles. Since $F(\xi_*^+) < 0$, and $J(\xi_*^+) > 0$, from Equation (21) it follows that $C_1\,\theta_1^+ > 1 + C_2\theta_0 + D^2/2$. At the point of complete burn out of particles $R_2 = 0$, and from Equation (20) it follows that $C_1\,\theta_{1e} + (U_{1e}\,D)^2/2 = 1 + C_2\,\theta_0 + D^2/2$. Consequently, $\theta_1^+ > \theta_{1e}$.

Furthermore, the parameters referring to the plane of complete burn out will be denoted by the subscript e. The detonation wave after a complete burnout of particles ($R_{2e} = 0$). It is found from Equation (20):

$$P_e = 2\,(1 + C_2\,\theta_0)\left(\frac{\gamma+1}{\gamma-1}V_e - 1\right)^{-1} \tag{22}$$

Here γ is the specific heat ratio. It should be noted that the deflagration branch of adiabatic (22) is absent, i.e., a steady combustion wave cannot propagate in he mixture of propellant particles in vacuum. An ideal C-J detonation velocity in the detonation wave under consideration is

$$D_J^2 = 2(\gamma^2 - 1)(1 + C_2\theta_0) \tag{23}$$

It is independent of an initial mass concentration of particles.

Results and Conclusions

Let us consider the C-J detonation wave structure of the propellant particles having thermodynamical properties[8]: ρ_2^0 = 1550 kg/m^3 , q_2 = 1.993 MJ/kg, T_{ign} = 473 K, l_2 = 0.4 MJ/kg, c_1 = 1675 J/(kg K), c_2 = 1465 J/(kg K), μ_{10} = 1.73 x 10^{-5} kg/ms, λ_{10} = 3.607 x 10^{-2} kg m/(s^3 K), γ = 1.2435. The dependence of viscosity coefficients and heat conduction on temperature was taken into account in the form $\mu_1 = \mu_{10}(T_1/300)^{0.7}$, $\lambda_1 = \lambda_{10}(T_1/300)^{0.7}$. The initial pressure, density, and temperature are $p_0 = 0$, $\rho_{10}^0 = 0$, $T_0 = 0$. Substituting q_2, γ, and T_0 in Equation (23), we find the C-J ideal detonation velocity $\mathcal{D}_J$ = 1476 m/s. When solving Equations (8) and (9) in the relaxation zone, the parameter jump at the contact discontinuity was prescribed as an initial condition when $\xi = 0$.; the gas temperature being found from Equation (14) in a explicit form

$$T_{1f} = T_0 + (3\mathrm{Pr} - \mathrm{Pr}^{1/3})\mathcal{D}^2/2c_1$$

In the general case, with fixed thermodynamically properties of phases, the solution in the zone of detonation wave and its length are the functions of two independent parameters, such as an initial mass concentration of particles $m_{20} = \alpha_{20}\rho_2^0$ and their diameters d_0. However, with low volumetric concentration $(\alpha_{20} << 1)$ but finite mass concentration of particles m_{20}, it is not difficult to demonstrate, using the value $x_0 = d_0^2 p_2^0 \sqrt{q_2}/(18\mu_{10})$ as a characteristic linear dimension, that the dimensionless solution of the problem depends only on one dimensionless constant $K = m_{20} d_0 \sqrt{q_2}/(18\mu_{10})$ within the terms $0(\alpha_{20})$. Therefore, one variant of the calculation of detonation wave structure with fixed parameter K provides a continuum of solutions with different m_{20} and d_0 satisfying the condition $m_{20}d_0$ = constant.

Figure 2 illustrates the profiles of pressure (P), velocity of gas (U_1) and of particles (U_2) ; temperature of particles (θ_2) and degree of the particle burn out ($W = (d/d_0)^3$); and Mach number $M = u_1/\sqrt{\gamma p/\rho_1}$ in the zone of C-J detonation when K = 7. The characteristic dimensions of the detonation wave are of certain interest when predicting the scales of experimental facilities. A series of calculations with the parameter K being varied gives quantitative data on the structure of the solution as well as the lengths of particle heating one (ξ_*) and of detonation zone (sJ). The values ξ_* and ξ_J as functions of the parameter K (1 < K < 100) are shown in Figure 3. The theoretical data may be approximated as follows:

$$\xi_* = (3.6 + 4.8K^{1/2} + 0.07K)^{-1}$$

$$\xi_J = (2.85 + 1.86K^{1/2} - 0.02K)^{-1}$$

Thus, the problem of steady detonation wave in vacuum with propellant particles has been solved for the first time, and the peculiarities of its structure have been investigated.

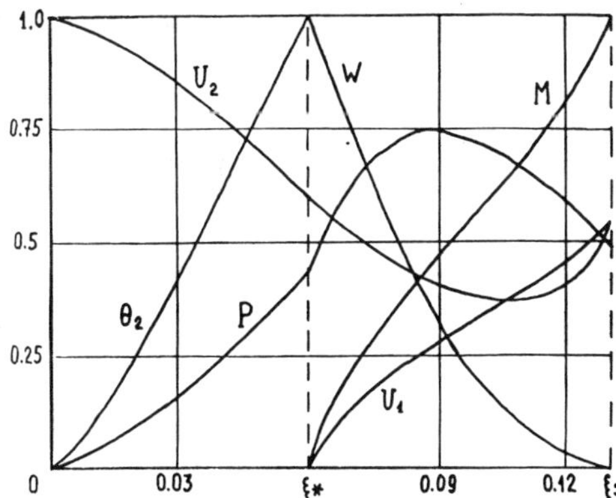

Fig. 2 Profiles of pressure P, velocity of gas U_1 and of particles U_2; temperature of particles θ_2 and degree of the particle burnout $W = (d/d_0)^3$; and Mach number $M = u_1/\sqrt{(\gamma p/\rho_1)}$ in the zone of C-J detonation when K = 7.

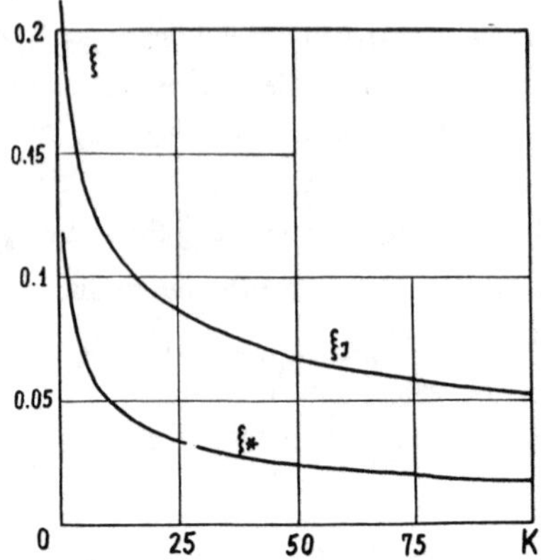

Fig. 3 Values ξ and ξ_J as functions of the parameter K (1 < K < 100).

References

1 Nigmatulin, R.I., Vainshtein, P.B., Akhatov, I.S., and Pyzh, V.A., "Structure of detonation waves in two-phase dispersed media, "*Detonation*, Chernogolovka, 1977, pp. 100-103.

2 Nigmatulin, R.I., Vainshtein, P.B., and Akhatov, I.S., "Structure of steady detonation waves in gas mixtures with propellant particles," *Detonation*, Chernogolovka, 1980, pp. 96-99.

3 Akhatov, I.S., Vainshtein, P.B., and Nigmatulin, R.I., "Structure of detonation waves in propellant particle-gas mixtures, " *Izvestiya Akademii Nauk SSSR, Mechanics of fluid and gas*, Vol. 16, No. 5, 1981, pp. 47-53.

4 Medvedev, A.E., Fedorov, A.V., and Fomin, V.M., "Heterogeneous detonation wave structure in gas-particle mixtures, " Novosibirsk, 1986, 46 pp. (Preprint of the Institute of Theoretical and Applied Mechanics, No. 36-86.

5 Medvedev, A.E., Fedorov, A.V., and Fomin, V.M., "Investigation of adiabats of heterogeneous two-phase detonation," *The Physics of Combustion and Explosion*, Vol 23, No. 2, 1987, pp. 115-121.

6 Nigmatulin, R.I., *Fundamentals of Mechanics of Heterogeneous Media*, Nauka, 1978, Moscow, Russia, 336 pp.

7 Gostintsev, Y.A., "On ignition, unsteady buring and flame extinguishment in the propellant particle combusion," *The Physics of Combustion and Explosion*, Vol. 7, No. 3, 1971, pp. 337, 344.

8 Belyaev, A.F. "On nitroglichol combustion," *Jhurnal fiziheskoi khimii*, Vol. 14, No. 8, 1940, p.1009.

Effect of Inert Particle Evaporation on the Chemical Reaction in a Combustible Medium

S. M. Frolov*
Russian Academy of Sciences, Moscow, Russia
and
J. M. Timmler† and P. Roth‡
Universität Duisburg, Duisburg, Germany

Abstract

Dispersions of various solid or liquid materials are often used as a fire safety measure in chemical processing plants or mine galleries. In this paper the effect of inert particles evaporation on the preself-ignition chemical reaction in a gaseous combustible medium is investigated. It is shown that under certain conditions the evaporation process may promote the chemical activity and may become a reason for strong secondary shock or detonation wave onset. Monodisperse particles uniformly distributed in a combustible mixture were considered first. A quenching criterion was derived indicating the amount of particles required for the suppression of the mixture self-ignition. A comparison was made between the quenching ability of H_2O, $Pl(C_2H_5)_4$ and KBr particles. It was shown that self-ignition can be promoted, if particle inertia effects in the fluid flow were taken into account. The latter is due to momentum losses during vapor acceleration and drag forces. Nonuniformly distributed evaporating particles produce temperature and dilution ratio nonuniformities in the reactive mixture. Under conditions close to self-ignition this may become a reason for strong shock or detonation wave generation in accordance with the Zeldovich mechanism. Detailed calculations were carried out indicating characteristic regions of spontaneous detonation onset. A comparison was made between the calculated results and predictions obtained on the basis of an analytical criterion.

---

* Semenov Institute of Chemical Physics. Senior Scientist
† Institut für Verbrennung und Gasdynamik. Dr.-Ing.
‡ Institut für Verbrennung und Gasdynamik. Professor

Introduction

Dispersions of various liquid or solid materials are often used as a safety expedient in chemical processing plants or mine galleries[1] for preventing high dynamic or thermal loading of construction elements. The quenching of a flame can be achieved, for example, by the formation of fog in front of it. The influence of these aerosol particles on the flame consists in chemical inhibition by loss of free radicals or in physical inhibition by gas cooling due to heat transfer during particle evaporation. Drag forces between particles and gas can also play an important role, see, for example, Ref.2. According to Ref.3, powders can be used for detonation suppression. Additionally, a whole class of detonable media exists, where chemically inert dust particles play a significant role, for example, dusty detonations. It is shown in Ref.4 that inert particles dispersed in a gaseous reactive mixture cause a widening of the detonation limits as compared to the pure gas. The influence of inert particles on the propagation of detonation waves was theoretically studied in Ref.5, in regard to heat transfer and friction between particles and gas.

In the present paper the effect of mass addition by inert particle evaporation on the chemical reaction in a combustible gas mixture was studied. It is shown that in some cases the particle evaporation process can promote the onset of strong secondary shock and detonation waves. Of particular interest is the coupling between gas phase reaction and fast evaporation of the dispersed particles. This can take place either in a flame front or in a high-temperature gas mixture, which is close to self-ignition conditions. In the present case, we restrict ourselves by considering only the problem, concerning self-igniting mixture behavior.

One-Dimensional Two-Phase Flow Conservation Equations

The flow behavior of a gas/particle mixture[4] including inter- and intra-phase rate processes can be described by the conservation equations of two interactive continua. In a simplified strategy the volume fraction of the particle phase is introduced by

$$\varepsilon = \frac{V_p}{V} = \frac{\varrho_p}{\varrho_p^0}$$

where V and V_p are the volume of the suspension and the respective volume of all particles in the suspension. The quantities ϱ_p^0 and ϱ_p are the material density of the particles and the "smeared" density of the particle phase (mass of the particles per volume of suspension). In the later equations, we only consider the limiting case of very low volume fraction of the suspended particles. This assumption facilitates the description of the two-phase-flow problem and allows the consideration of only the gas-phase properties with the respective source terms.

The conservation equations of mass, momentum and energy for the gas phase of two interactive gas/particle continua can be formulated under the assumptions of Ref.6 in the following one-dimensional way:

$$\frac{d\varrho}{dt} + \varrho\frac{\partial u}{\partial x} = J \tag{1}$$

$$\varrho\frac{du}{dt} = -\frac{\partial p}{\partial x} + F_d + J(u_p - u) \tag{2}$$

$$\varrho\frac{de}{dt} = -p\frac{\partial u}{\partial x} + J\left[(h_D - e) + \frac{(u_p - u)^2}{2}\right] + F_d(u_p - u) + Q + Q_R \tag{3}$$

In these equations d/dt is the substantial derivative and $\varrho, u, p,$ and e are the density, flow velocity, pressure, and internal energy of the gas phase, respectively. The quantities[4] J, F_d, and Q are the source terms or interphase flux terms for mass, momentum, and heat, u_p is the mean velocity of the particle phase, and Q_R is the chemical energy source term. The quantity $J\left[(h_D - e) + (u_p - u)^2/2\right]$ is the energy transferred from the particle phase to the gas phase due to mass transfer, where h_D is the specific enthalpy of the vapor when leaving the particle surface. It was introduced instead of e_D because of the possible expansion work of the vapor. Similar equations for the particle phase can also be formulated, see Ref.6.

The different source terms in Eqs. (1)-(3) are, in general, complicated functions of the gas- and particle-phase properties. In the present case we assume the particles to be spherical and monodisperse. In this case the particle phase can be described by the two properties n_p and d_p, the number concentration and the mean particle diameter. According to Ref.6, the interphase flux terms are

$$J = -n_p \cdot \frac{dm_p}{dt} = -n_p\varrho_p^0\frac{\pi}{4}d_p\frac{dd_p^2}{dt}$$

$$J = n_p \cdot \varrho_p^0\frac{\pi}{4}d_p \cdot K$$

$$F_d = n_p \cdot \frac{\pi}{2}\varrho C_D(u_p - u) \mid u_p - u \mid$$

$$Q = n_p \cdot \pi d_p \lambda Nu(T_p - T)$$

In these equations, $K = -dd_p^2/dt$ is the evaporation coefficient[7-9], C_D and λ are the drag coefficient and the heat conductivity of the gas phase, respectively, T_p and m_p are the particle surface temperature and the particle mass, and Nu the particle Nusselt number.

The chemical reaction in the gas phase is introduced in the simplest possible way by a one-step kinetics, characterized by a global rate parameter

a. The source term Q_R then results in

$$Q_R = -\varrho h_R \frac{da}{dt}$$

where h_R is the heat of reaction. The rate parameter a ranged between $a = 1$ (unreacted mixture) and $a = 0$ (combustion products).

The rate laws for d_p and a can be introduced in the following simplified way:

$$\frac{dd_p^2}{dt} = -K \quad (4)$$

$$\frac{da}{dt} = -a^n Z \; exp(-E/RT) \quad (5)$$

In Eq. (5), Z is the pre-exponential frequency factor, n the reaction order, R the gas constant, E the activation energy, and T the gas-phase temperature. Equation (4) can be integrated if K is independent of d_p and t. The result is[8,9]

$$d_p^2 = d_{p0}^2 - Kt \quad (6)$$

where d_{p0} is the particle diameter at time $t = t_0$, $K = 2D_{vg}p_v M_v/\varrho_p^0 RT_p$. Here D_{vg} is the diffusion coefficient, p_v the vapor pressure, M_v the vapor molecular weight, T_p the particle temperature. Note that the d^2-law given by Eq. (6) is a reasonable realistic model because we consider processes during the long induction period with insignificant temperature and pressure rise[9].

Uniformly Distributed Particles in a Closed Volume

In this section we consider a volume of the linear dimension L, which contains a premixed reactive gas mixture loaded with uniformly distributed suspended inert particles of initial size d_{p0}. The initial conditions $(p, T)_0$ of the gas/particle mixture are such that self explosion is possible. The effect of particle evaporation on the chemical reaction was studied under the following assumptions:

1. All thermophysical parameters of gas and vapor are identical.
2. The particle velocity is equal to the gas velocity, $u = u_p$.
3. The enthalpy of the vapor (added energy) is equal to the interphase heat flux, $J \cdot h_D = -Q$.

Under these conditions, the conservation equations of mass and energy for a quiescent mixture result in the simple form

$$\frac{d\varrho}{dt} = J \quad (7)$$

$$\varrho \frac{de}{dt} = Q_R - Je \tag{8}$$

The energy equation (8) expresses the fact that the energy e (or the temperature T) of a gas/particle mixture can be increased or decreased depending on the relative effects of gas cooling due to mass addition ($J \cdot e$) or gas heating because of chemical reaction (Q_R). It is clear from the above differential equation that the quenching criterion for the chemical reaction is

$$Q_R < J \cdot e \tag{9}$$

This criterion is fulfilled if the cooling effect exceeds the gas heating by the exothermic reaction.

The quenching criterion will now be applied to a high-temperature gas/particle mixture, in which at time $t = 0$ the reaction and evaporation processes were simultaneously started.

$$t = 0: \qquad T = T_0, \quad \varrho = \varrho_0, \quad a = a_0 = 1$$
$$J_0 = J(t=0), \quad Q_{R0} = Q_R(t=0)$$

The source terms J_0 and Q_{R0} can be expressed by the respective evaporation time τ_{ev} or chemical induction time τ_{ind},[10] which are

$$\tau_{ev} = \frac{\varrho}{J_0}$$

$$\tau_{ind} = \frac{c_p T_0}{Q_{R0}} \frac{RT_0}{E} \tag{10}$$

The quenching criterion (9) can be then rewritten in the following form:

$$\frac{\tau_{ev}}{\tau_{ind}} < \frac{E}{RT_0} \tag{11}$$

If this condition is satisfied, then the chemical reaction is quenched by the particle evaporation process.

The source term J_0 or the evaporation time τ_{ev} are mainly determined by the evaporation coefficient K, which is a complicated function of the particle as well as the gas phase properties. It includes, for example, the Sherwood number, which expresses the convective mass transfer due to the relative motion between a suspended single particle and the surrounding gas. In the present case we only assume diffusive processes of a spherical particle, that means $Sh = 2$. From the energy balance applied to a single suspended particle under stationary conditions, the difference between gas temperature T and particle temperature T_p can be determined[8]. For the particle materials KBr, H_2O, and $Pb(C_2H_5)_4$, the calculated evaporation coefficients are summarized in Table I. The Stefan flow conditions are

Table I Calculated evaporation coefficient K obtained from stationary, diffusive energy exchange between a single particle and the surrounding gas; T = gas temperature and T_p = particle surface temperature.

	KBr		Water		$Pb(C_2H_5)_4$	
T K	T_p K	K $10^{-10}m^2/s$	T_p K	K $10^{-10}m^2/s$	K K	K $10^{-10}m^2/s$
300	300	-	299	18	299.7	1.1
400	400	-	359	314	377	119
500	500	-	360	677	406	203
600	600	-	371.3	999	418	975
700	700	-	371.9	1345	425	1482
800	800	-	372.1	1724	427	2014
900	900	-	372.2	2140	432	2571
1000	1000	2.07	372.3	2594	434	3152
1100	1100	8.65	372.3	3082	436	3758
1200	1200	24.6	372.3	3605	437	4390
1300	1300	48.9	372.4	4167	438	5042
1400	1400	79.7	372.4	4763	438.7	5728
1500	1500	115.4	372.4	5395	439.4	6438

Table II Reaction quenching criterion of Eq. (11) applied to the reaction conditions behind the leading shock of methane-air detonation waves; particles of KBr with $n_p = 10^{15}$ m^{-3} and $d_{p0} = 1 \mu m$.

Φ	D m/s	$p_f p_0^{-1}$	T_f K	τ_{ind} ms	τ_{ev} ms	τ_{ev}/τ_{ind}	E/RT_f	τ_p ms
0.5	1430	19.70	1150	4.180	16.00	3.8	20.1	6.20
0.8	1660	26.40	1370	0.200	1.40	7.0	16.9	0.50
0.9	1710	28.03	1430	0.120	1.00	8.3	16.2	0.33
1.0	1750	29.30	1465	0.086	0.75	8.7	15.8	0.25
1.5	1760	29.22	1430	0.140	1.00	8.3	16.2	0.33
2.0	1680	26.24	1290	0.880	3.00	3.4	17.9	1.00
2.5	1590	23.25	1170	6.320	15.00	2.4	19.7	5.00

included in the given set of data. It is clear from Table I that at a gas-phase temperature of $T = 1000K$ the evaporation coefficient of H_2O and $Pb(C_2H_5)_4$ is two to three orders of magnitude larger than that of KBr.

The quenching condition of Eq. (11) is now applied to the properties behind the leading shock (index f) of a stationary methane-air detonation. The induction time of the gas phase reactions[11] was compared to the evaporation time of KBr particles with $n_p = 10^{15}$ m^{-3} and $d_{p0} = 1$ μm. The suspension density was assumed to be $\varrho_p = 1$ kg/m^3. The result of the calculations are shown in Table II. The gas-phase detonation is characterized by the equivalence ratio Φ, the detonation velocity D, the pressure ratio p_f/p_0 of the leading shock, and the temperature T_f in the induction zone. The reaction induction time τ_{ind} and the calculated evaporation time τ_{ev} are given. The comparison of τ_{ev}/τ_{ind} with the dimensionless activation energy E/RT_f indicates that according to Eq. (11) the detonation is always suppressed. The property in the last column of Table II is the particle life time, τ_p, according to Eq. (6). The condition $\tau_{ind} < \tau_p$ is always fulfilled (except for $\Phi = 2.5$). Note that in this simplified theory, heat consumption by the particles and the influence of mixture dilution was not considered. It is also clear from Table I that the reaction quenching by H_2O or $Pb(C_2H_5)_4$ particles is much more effective than that of KBr, because of the considerably higher evaporation coefficients.

Relative Motion Between Particles and Gas

We now consider the case of relative motion between the particles and the combustible gas mixture. To simplify Eq. (4) we assume the following:

1. The particle velocity is zero, $u_p = 0$.
2. The drag force F_d is of minimum importance, $F_d = 0$.

For flow conditions the energy equation results in

$$\varrho \frac{de}{dt} = Q_R - J\left(e - \frac{u^2}{2}\right) \tag{12}$$

It can be seen from the right hand side of Eq. (12) that because of the explosion suppression criterion

$$Q_R < J\left(e - \frac{u^2}{2}\right)$$

the gas cooling effect due to particle evaporation is qualitatively less efficient than in the case considered before. This is a result of momentum loss due to

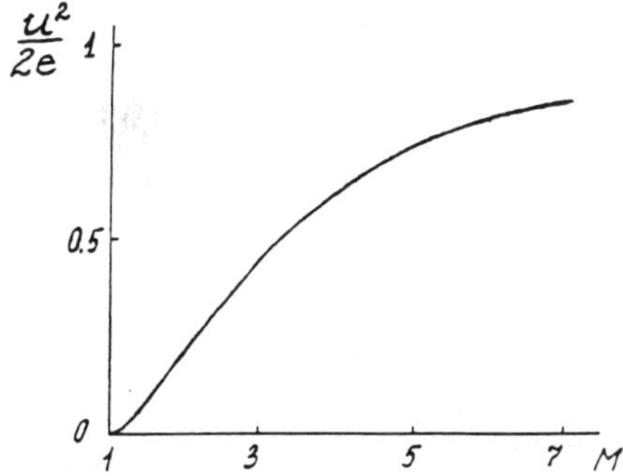

Fig. 1 Dependence of the kinetic to thermal energy ratio behind a shock wave on the shock Mach number.

vapor acceleration. When making quantitative analysis one should take into account the dependence of the evaporation coefficient on the flow Reynolds number[7]. At $u^2/2 > e$ the evaporation process promotes self-explosion. Figure 1 shows the dependence of the quanty $u^2/2e$ behind the shock wave on the shock Mach number for fuel-air mixtures. It appears that particle evaporation behind the shock wave always leads to temperature decrease ($u^2/2e < 1$). The effect of reaction promotion may come into play in nozzle flows. If the process of kinetic energy dissipation due to drag force is also considered, the full energy equation (3) must be discussed. Remember, the term $F_d(u_p - u)$ is always positive and contributes to the effect of ignition promotion.

Nonuniformly Distributed Particles

In the case of nonuniformly distributed particles $n_p = n_p(x)$, the source term J of the gas/particle mixture no longer depends only on t but also on x, $J = J(x,t)$. This means that the gas cooling effect by the particles is nonuniform resulting in pressure, temperature, and velocity gradients, as can be seen from the conservation equations (1)-(3) with the simplified assumptions $F_d = 0$ and $Jh_D = -Q$:

$$
\begin{aligned}
\frac{d\varrho}{dt} + \varrho\frac{\partial u}{\partial x} &= J(x,t) \\
\varrho\frac{du}{dt} &= -\frac{\partial p}{\partial x} + J(u_p - u) \\
\varrho\frac{de}{dt} &= -p\frac{\partial u}{\partial x} + Q_R + J(x,t)[-e + \frac{(u_p-u)^2}{2}]
\end{aligned}
\tag{13}
$$

Stationary conditions are no longer possible, and waves will be generated by the particle nonuniformity. Spatially nonuniform chemical energy release can be expected in this case. According to the Zel'dovich mechanism[12,13] under conditions close to self-ignition this may become a reason of strong shock or detonation wave generation.

The physical meaning of the Zel'dovich coupling criterion[13] consists in a requirement for the reaction wave to propagate at the local sound velocity. Now we consider conditions when particle evaporation promotes spontaneous detonation onset in the preheated combustible mixture. Evaporation of particles produces the temperature gradient of the order

$$
\omega(t) \approx \Delta T(t)/L \tag{14}
$$

where ΔT is the temperature increment at the length L. According to Eqs. (6)-(8) and using Taylor series

$$
\Delta T = T_0\delta/(1+\delta)
$$

with $\delta = (\pi/6)(d_{p0}^3 - d_p^3(t))n_p(L)\varrho_p^0/\varrho_0$ for particles distributed initially as $n_p = n_p^0 x/L$. The parameter δ is approximately equal to $\tau_{ind}\ \tau_{ev}$ to first order for $k\tau_{ind}\ d_{p0}^2 \ll 1$. A further simplification gives $\Delta T = T_0\delta$ because $\tau_{ind}/\tau_{ev} \ll 1$ due to the occurence of self-ignition.

Here n_p^0 is the reference value of particle number concentration. The temperature gradient given by Eq. (14) should be compared with the characteristic temperature gradient for the Zel'dovich mechanism to come into effect[13]:

$$\omega_* \approx \frac{RT_0^2/E}{c_0\tau_{ind}(e-1)} \tag{15}$$

where c_0 is the sound velocity in the initial mixture. Eq. (15) can be derived from Eq. (10) using high activation energy asymptotics. Equating Eq. (14) to Eq. (15) and taking into account Eqs. (10) gives

$$\frac{\tau_{ev}}{\tau_{ind}} \approx \frac{E}{RT_0}\,\frac{c_0\tau_{ind}(e-1)}{L} \tag{16}$$

A comparison between criteria (11) and (16) shows that at

$$\frac{c_0\tau_{ind}(e-1)}{L} < 1$$

reaction quenching by evaporating particles may produce conditions for spontaneous onset of strong shock waves. This may become a case if particles are nonuniformly distributed in the volume.

To verify the criterion (16) we integrated numerically Eqs. (13) for initially quiescent methane-air mixture. The following set of parameters was used for the computational example, which corresponds closely to Eq. (16): $\varrho = 0.3$ kg/m^3, $T_0 = 1200$ K, $p_0 = 0.1$ MPa, $C_p = 8.67$ kcal/kmole K, $\mu = 30$ kg/kmole, $E = 44.5$ kcal/mole, $n = 1$, $Q = 15$ kcal/mole s, $Z = 10^{10}\,\mathrm{s}^{-1}$, $L = 0.1$ m, $n_p^0 = 10^{11}\,\mathrm{m}^{-1}$, $\varrho_p^0 = 2225$ kg/m^3, $d_{p0} = 20\ \mu$m (KBr particles)

The evaporation coefficient for KBr particles was calculated by the use of the dependence[8]

$$k = 2.126 \cdot 10^{(-1.7618-8665/T)}(T/T_e)^{0.75}(p_e/p)$$

where $T_e = 300$ K, $p_e = 1$ bar.

The kinetic parameters Z, n, and E of the mixture have been chosen by comparison of an adiabatic induction period with an empirical correlation for ignition delay[14]. The ratio of specific heats is assumed to be constant and equal to the average value between the corresponding quantities for the intial mixture and combustion products, the composition of the latter being calculated through the use of a stoichiometric formula neglecting dissociation. The heat release Q is then determined from the experimental value of the ideal detonation velocity.

Initial conditions for Eqs. (13) are as follows:

$$t = 0, \quad 0 \le x \le L: \quad p = p_0,\ \varrho = \varrho_0,\ u = 0,\ a = 1$$

Bounday conditions:

$$u(t > 0, x = 0) = u(t > 0, x = L) = 0$$

Two limiting cases were considered:

1. Particles were assumed taken in motion immediately, i.e., $u - u_p = 0$
2. The particle velocity was assumed zero, i.e., $u - u_p = u$

The integration procedure of Eqs. (13) was the same as proposed in Ref. 15.

Shown in Fig. 2 is the calculated temporal evolution of various quantities. Results of the calculations are similar whether or not the relative motion of the phases is taken into account. This is due to the fact that the most important stage of the process is the pre-ignition stage. Figure 2a shows the evolution of the parameter $n_p d_p (n_p^0 d_{p0})^{-1}$ representing a modified particle diameter. At time $t \approx 400 \mu s$ self-ignition occurs near the boundary $x = 0$ where the particle number concentration is minimum. In the further evolution, particles disappear behind the reaction wave.

Figure 2b shows the dimensionless temperature history. At time $t = 300 - 400 \mu s$ one can see the formation of nonuniform temperature distribution induced by nonuniform cooling effect of the particle ensemble. After the self-ignition event at $t \approx 400 \mu s$ spontaneous detonation onset is evident. The fully developed detonation wave originates at $t \approx 420 \mu s$. The detonation wave forms as a result of spatially nonuniform chemical energy release[12,13]. Also evident is the reaction proceeding at the far end of the vessel.

Figure 2c shows the dimensionless pressure history. Relatively low-pressure values in the detonation wave are due to the high initial temperature T_0. Unstable character of the detonation wave propagation is due to one-dimensional instability of the detonation front at given E and Q values.

Figure 2d shows dimensionless gas-phase velocity history. The detonation velocity is of about 1750 m/s which is somewhat less than the CJ value ($\approx$ 1810 m/s). The difference should be attributed to momentum losses on vapor acceleration[16].

Thus the computational example confirms the criterion given by Eq. (16). It should be mentioned that experiments on detonation suppression by water-based foams reveal a possibility of strong secondary shock formation behind a decaying precursor blast wave[16]. The secondary explosions may be attributed to reaction onset in nonuniformly preheated zones in a two-phase combustible mixture.

To verify the criterion in general, we made calculations for the whole variety of fuel-air and fuel-oxygen mixtures with different quenching agents.

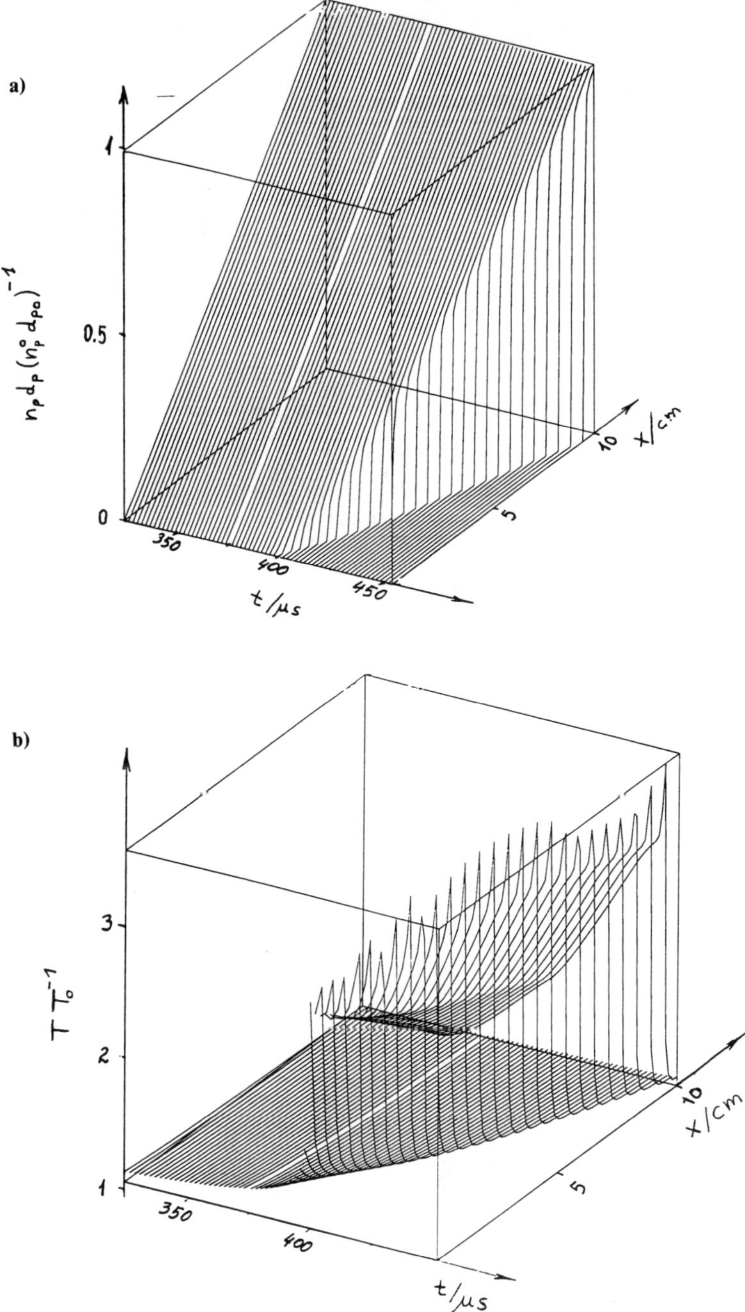

Fig. 2 Calculated temporal evolution of reaction in an initially quiescent methane-air mixture loaded with nonuniformly distributed inert evaporating particles: a) modified particle diameter, b) gas-phase temperature, c) pressure, and d) gas-phase velocity.

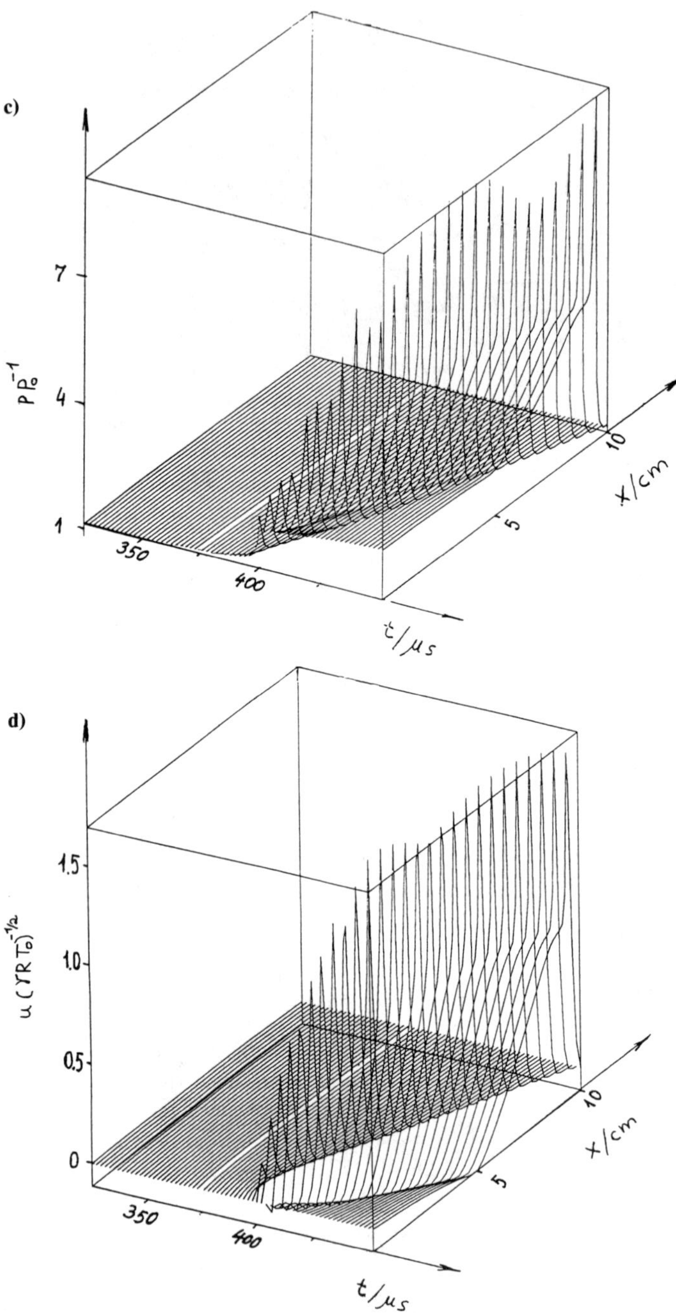

Fig. 2 (continued) Calculated temporal evolution of reaction in an initially quiescent methane-air mixture loaded with nonuniformly distributed inert evaporating particles: a) modified particle diameter, b) gas-phase temperature, c) pressure, and d) gas-phase velocity.

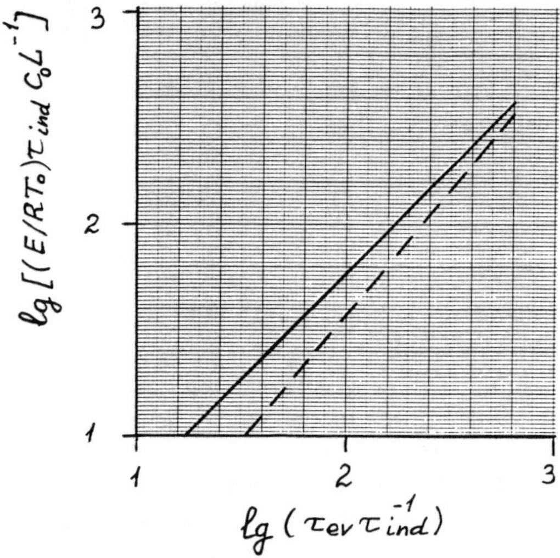

Fig. 3 Resultant dimensionless diagram indicating parametric domains with and without spontaneous shock/detonation formation: solid line corresponds with the approximate coupling criterion of Eq. (16). Dashed line is the numerically calculated boundary between reaction regimes with (upper region) and without (lower region) shock/detonation formation.

Presented in Fig. 3 is the resultant dimensionless diagram. It shows parametric domains with and without spontaneous shock or detonation formation. The dashed line is the numerically calculated boundary separating the two characteristic regions. In the upper region the maximum pressure during system evolution is more than 10% larger than the pressure of constant volume explosion at $T = T_0$. In the lower region the maximum pressure differs less than by 10% from it. The 10% criterion is somewhat conditional here. It is used to indicate only a possibility of considerable local pressure jumps (shocks) during system evolution. The solid line in Fig. 3 represents the approximate criterion given by Eq. (16). If one takes into account strong dependence of τ_{ev} and τ_{ind} on temperature, the agreement of results is satisfactory.

Conclusion

A simplified theory developed in this paper shows that evaporating inert particles may quench or promote chemical activity in the combustible mixture. Physically clear criteria are proposed to predict critical conditions for gaseous detonation quenching. It is shown that suppression of chemical activity in the combustible gas may be accompanied by strong secondary shock formation and even detonation onset. It is a matter of great concern now to study a combined effect of particle evaporation and drag due to relative motion of phases.

Acknowledgments

The authors gratefully acknowledge the financial support of the Deutsche Forschungsgemeinschaft.

References

[1]Nettleton, M. A. *Gaseous Detonations, Their Nature, Effects and Control*, Chapman and Hall, London - New York, 1987, pp. 191-207.

[2]Frolov, S. M., Gelfand, B. E., Tsyganov, S. A., and Medvedev, S. P. "Quenching of Shock Waves by Barriers and Screens", *Proceedings of the 17th Symposium (International) on Shock Tubes and Waves*, A.I.P., Ann-Arbor, MI, 1989, pp. 314-320.

[3]Laffitte, P., and Bouchet, R., "Suppression of Explosion Waves in Gaseous Mixtures by Means of Fine Powders", *Proceedings of the 7th Sysmposium (International) on Combustion*, The Combustion Institute, Pittsburgh, PA, 1959, pp. 504-508.

[4]Kauffman, C. W., Wolanski, P., and Arisoy, A., "Dust, Hybrid and Dusty Detonations", *Dynamics of Shock Waves, Explosions and Detonations*, Vol. 94, Progress in Astronautics and Aeronautics, AIAA, New York, 1984, pp. 227-237.

[5]Borisov, A. A., Gelfand, B. E., and Gubin, S. A., "Effect of Inert Particles on Detonation of a Gaseous Combustible Mixture", *Soviet Journal Fizika Goreniya Vzriva*, Vol. 11, No.6, 1975, pp. 909-916.

[6]Reichelt, R., Roth, P., and Wang, L., "Stoßwellen in Gas-Partikel-Gemischen mit Massenaustausch zwischen den Phasen", *Wärme- und Stoffübertragung*, Vol. 19, 1985, pp. 101-109.

[7]Williams, F. A., *Combustion Theory*, Addison Wesley, Reading, MA, 1965, Chapter 3.

[8]Timmler, J., and Roth, P., "Measurements of High Temperature Evaporation Rates of Solid and Liquid Aerosol Particles", *International Journal of Heat and Mass Transfer*, Vol. 32, No.10, 1989, pp. 1887-1892.

[9]Law, C. K., "Recent Advances in Droplet Vaporization and Combustion", *Progress in Energy and Combustion Science*, Vol. 8, 1982, pp. 171-218.

[10]Frank-Kamenetskii, D. A., *Diffusion and Heat Transfer in Chemical Kinetics*, Plenum Press, New York, 1969, Chapter 6.

[11]Westbrook, C. K., "Chemical Kinetics of Hydrocarbon Oxidation in Gaseous Detonations", *Combustion and Flame*, Vol. 46, 1982, pp. 191-216.

[12]Zel'dovich, Y. B., Librovich, V. B., Makhviladze, G. M., and Sivashinski, G. I., "On the Development of Detonation in a Nonuniformly Preheated Gas", *Astronautica Acta*, Vol. 15, 1970, pp. 313-319.

[13]Zel'dovich, Y. B., Gelfand, B. E., Tsyganov, S. A., Frolov, S. M., and Polenov, A. N., "Concentration and Temperature Nonuniformities (CTN) of Combustible Mixtures as a Reason of Pressure Waves Generation", *Dy-*

namics of Explosions, Vol. 114, Progress in Astronautics and Aeronautics, edited by A. L. Kuhl, J. R. Bowen, J.-C. Leyer, A. Borisov, AIAA, New York, 1988, pp. 99-123.

[14]Gelfand, B. E., Frolov, S. M., and Nettleton, M., "Gaseous Detonations - A Selective Review", *Progress in Energy and Combustion Science*, Vol. 17, 1991, pp. 327-371.

[15]Growley, B. K., "PUFL, An Almost Lagrangian Gasdynamic Calculation for Pipe Flows with Mass Entrainment", *Journal of Computational Physics*, Vol. 2, 1967, pp. 61-86.

[16]Frolov, S. M., and Gelfand, B. E., "On Detonation Suppression by Fogs and Foams", *Soviet Journal Fizika Goreniya Vzriva*, Vol. 27, No.6, 1991, pp. 116-124.

Ignition Mechanism of Coal Suspension in Shock Waves

V. M. Boiko,* A. N. Papyrin,† and S. V. Poplavski‡
Russian Academy of Sciences, Novosibirsk, Russia

Abstract

Results of experimental study of the ignition process of dust-gas mixture in shock waves are presented. The experiments were carried out in a shock tube at Mach numbers ranging from 2.5 to 4.2. Three types of coals with volatiles $V_0 = 9$, 26, and 55% and dispersivity $d < 40$ μm were studied. The stage of formation of the dust cloud was studied in detail, and its dimensions and the dust concentration were evaluated. The ignition delays of coal dust dispersed in air and in pure oxygen were obtained at gas temperatures ranging from 1000 to 2500 K and at constant pressure equal to 2.3 MPa. It was discovered that coal particle ignition is controlled by the simultaneous action of several different mechanisms, including the heating particle, devolatilization kinetics, and the kinetics of volatile ignition.

Introduction

In recent years, there has been considerable interest in studies of explosion hazard problems because of increased production, storage, and use of fuel-dust materials. The explosive organic dusts, such as coal and wheat, are the most widespread in practice, and that is why

*Senior Research Associate, Institute of Theoretical and Applied Mechanics, Siberian Division.

†Professor, Institute of Theoretical and Applied Mechanics, Siberian Division.

‡Research Associate, Institute of Theoretical and Applied Mechanics, Siberian Division.

an intensification of such study as dust ignition and detonation behind the shock waves has occurred (see, e.g., Refs. 1-4).

It is known that the ignition characteristics and explosion properties of organic dusts essentially depend on the quantity and composition of volatiles generated during organic thermal decay. This is especially true of coal dusts because their volatile content is changed considerably from one type of coal to another. Therefore it seems that investigation of coal dust ignition mechanism and, in particular, of the effect of volatiles on ignition delays in shock waves is very important.

Experiment

The experiments were carried out in a shock tube consisting of driver and driven sections, respectively, 1.5 and 5 m long, with a duct having a cross section 52x52 mm [5]. The pressures of the driver gas (helium) and the driven gas (air, oxygen) were, respectively, $P_4 = 2.5-9.0$ MPa and $P_1 = 0.01 - 0.1$ MPa. The Mach numbers of the incident shock waves (ISW) ranged from 2.5 to 4.2. The gas parameters behind the shock wave were determined from the tables, calculated by Lapworth[6] for the real gas with reference to the temperature dependence of adiabat index. Of special note is that the pressure of driven P_1 and driving P_4 gases were selected in such a way as to provide a constant pressure in the reflected shock wave (RSW) $P_5 = 2.3$ MPa irrespective of the Mach number.

In this work, a most frequently used method for producing gas suspension in the experiments conducted on shock tubes was used, i.e., the sputtering of powder by means of gas flow following the front of incident shock wave. A sample of coal dust with a mass m = 10-20 mg in the form of compact charge was placed on a flat substrate having 10 mm in diameter installed at the level of duct axis at a distance 70-150 mm from the reflecting wall.

A fast multiframe shadow laser visualization technique[5,7] recorded the dynamics of the dust-gas mixture formation. The exposure duration ($\simeq$ 30 ns), the number of frames, and the time interval Δt between them were set by a laser stroboscopic light source, and the spatial separation of the frames was realized with the help of a camera with a rotating mirror prism.

A synchronization system provided the necessary sequence in time for triggering the elements of the shock tube as well as diagnostic equipment that permitted us to coordinate exactly the starting of light pulse generation with the moment the shock front passes through studied region. A series of 15-20 frames obtained in each experiment reflected the dynamics of the process under study during the time 300-800 μs, including the stage of dust suspension formation in the ISW and the stage of interaction between the RSW and the cloud of suspension.

A streak camera was used to study the ignition and combustion processes in a coal dust suspension. The combination of these techniques, reliable recording of all the necessary spatial-temporal characteristics, of the process, including the ignition delay of every point of the cloud, i.e., the time interval between the RSW passing some point in the region under study and the moment of luminescence at the same point.

Results

Coal dust with volatile content $V_0 = 9$, 26 and 55% was analyzed in the present work. Coal dust was scattered into the fractions with the help of sieves, and a $0 - 40\mu$m fraction was used. Figures 1 and 2 are series of shadow photographs illustrating the dynamics of interaction of a coal dust sample with ISW and RSW for Mach number M=4.0 and 2.6, respectively. The ISW front is moving from left to right (frame 1) and reflecting from the wall that is located on the right border of the frames (successive locations of the RSW are clearly visible in frames 2-6).

From two photographs similar to Figs. 1 and 2, it is possible to study the dynamics of dust-gas mixture formation, particularly to determine the time required for all the dust in the sample to turn into suspension, and to estimate the dust cloud dimensions and the average concentration of the mixture. Thus, according to Fig. 2, complete dispersion of sample m=20 mg corresponds to frame 7, at which moment the dusty cloud occupied about 20 cm^3, and, the average dust concentration was $\simeq 1$ kg/m^3.

It was easy to change an average concentration by varying the sample mass and the distance l from the reflecting wall; however, in this method of mixture formation the distribution of concentration was essentially inhomogeneous.

Using the shadow photographs it is also possible to study ignition and combustion processes to find the location of the ignition site and the flame propagation velocity in the fuel mixture cloud. In Fig. 1 ignition appears on frame 4 and the site location is behind the RSW

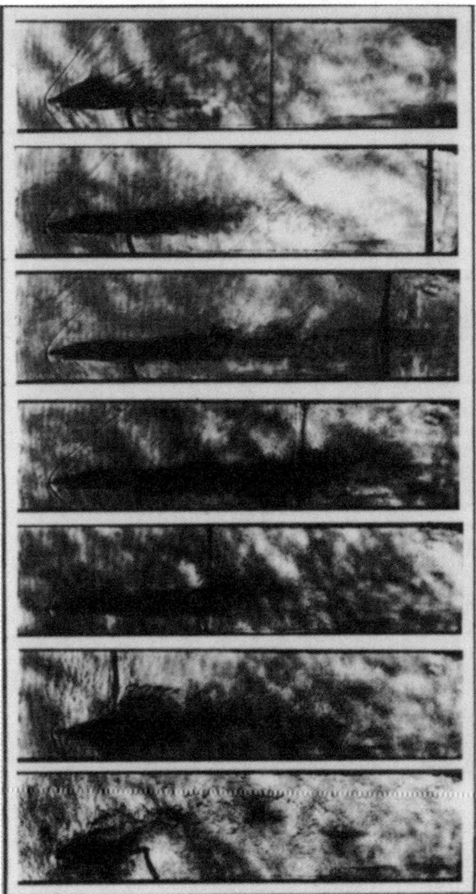

Fig. 1. Multiframe shadow photographs of the interaction of reflected shock with a coal dust cloud in oxygen. M = 4.0; Δt = 40 μs.

30 mm from the reflecting wall. The combustion of coal suspension was registered, first, by the lightening of dusty cloud caused by fast burning of small particles and, second, by the changes in shadow picture outside the cloud that were connected with the propagation of thermal disturbances (frames 5-7).

Figure 3 represents typical streak-camera records of ignition and combustion processes of coal dust suspension with $V_0 = 55\%$ in pure oxygen behind RSW. In Fig. 3 are images of light impulses, which are used for recording the moment of SW reflection and for the introduction of the time scale ($\Delta t = 100 \pm 0.2\mu$s). The lighter line denotes

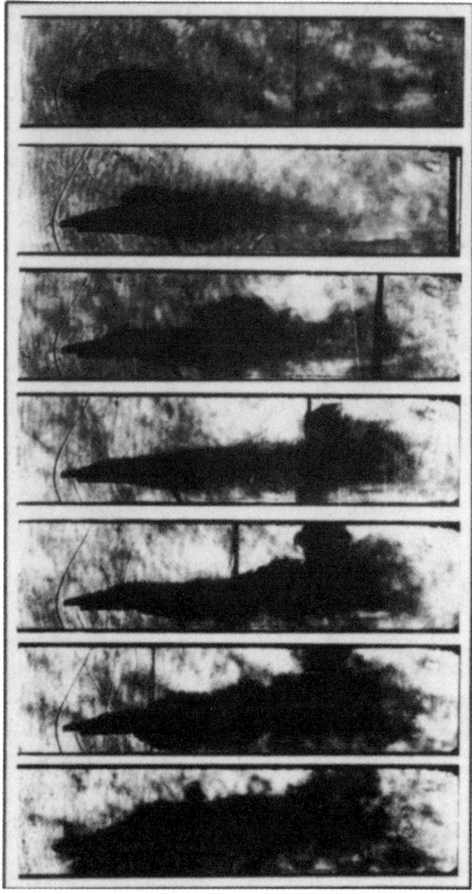

Fig. 2. Multiframe shadow photographs of coal dust cloud formation in a shock wave. M = 2.6; Δt = 40 μs.

the trajectory of RSW determined by shadow photographs. Clearly visible are the moment of ignition, the spatial location of the ignition site, the time and character of the mixture combustion, and also separate tracks corresponding to the largest particles of this fraction or the particle agglomeration.

As numerous experiments have shown, at lower temperatures, ignition typically occurs on separate sites and the flame front is moving in essentially irregular manner and is appreciably behind the RSW (Fig. 3b). With increase in temperature, the flame front begins to move at constant velocity similar to the velocity of RSW (Fig. 3a).

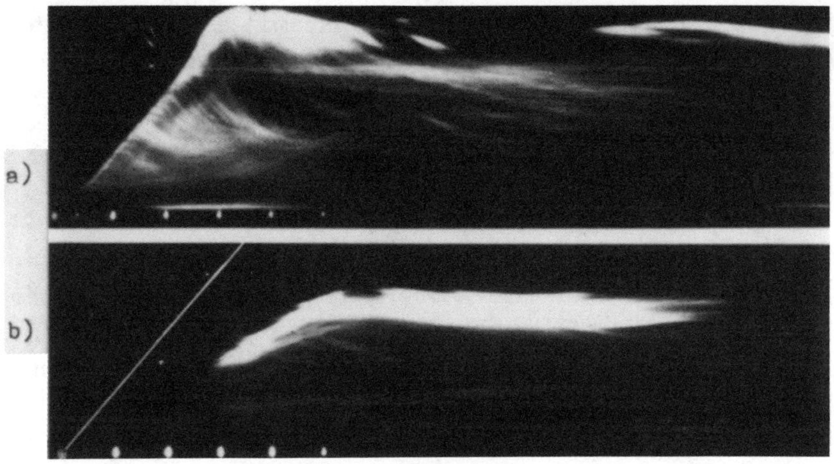

Fig. 3. Streak-camera records of ignition of coal dust suspension in oxygen. $V_0 = 0.55$: a) $M = 4.0$; b) $M = 2.6$

The obtained data demonstrate that ignition starts from the small particles, because at the initial stage the short tracks prevailed. The ignition site was some distance from the reflecting wall, and it was displaced to the area with high dust concentration as the gas temperature decreased. While the ISW Mach number decreased, the ignition delays increased, and the number of igniting particles and their combustion time decreased. It is also substantially true that, in the studied range of Mach numbers, the time of coal dust presence at ISW had no practical influence on its nature and on the ignition delay in the RSW. (In the present series of experiments, l was changed from 30 to 600 mm).

Since the ignition site is some distance from the reflecting wall, the ignition delay is defined as a minimal time interval between the RSW passing the ignition site location point and the appearance of solid particle luminescence.

These observations show that complex use of the highly informative methods of multiframe laser shadow visualization and the streak camera allows us to interpret data on dust mixtures ignition more precisely, increasing their authenticity.

Figure 4 presents the ignition delays of coal dusts with different volatile content depending on air temperature behind the RSW. Also shown is the dependence $\ln(\tau_{ig}) = f(1/T)$ for methane-air mixture (2% CH_4) behind the RSW[8]. The data on ignition delay of the same samples of coal dust in pure oxygen are illustrated in Fig. 5. Here one can also find the data for a methane-oxygen mixture[9], which were obtained by calculation for oxygen partial pressure behind the RSW $P_5 = 2.3$ MPa according to the following equation:

$$\tau_{ig} = 1.19 \quad 10^{-12}[CH_4]^{0.48}[O_2]^{-1.94} \quad \exp(194/RT)$$

proposed by Cheng and Oppenheim[9].

From Figs. 4 and 5 it is clear that volatile content strongly influences the ignition delay of coal dust. It is evident that the dependence $\ln(\tau_{ig}) = f(1/T)$ has a pronounced nonlinear nature for both dust-

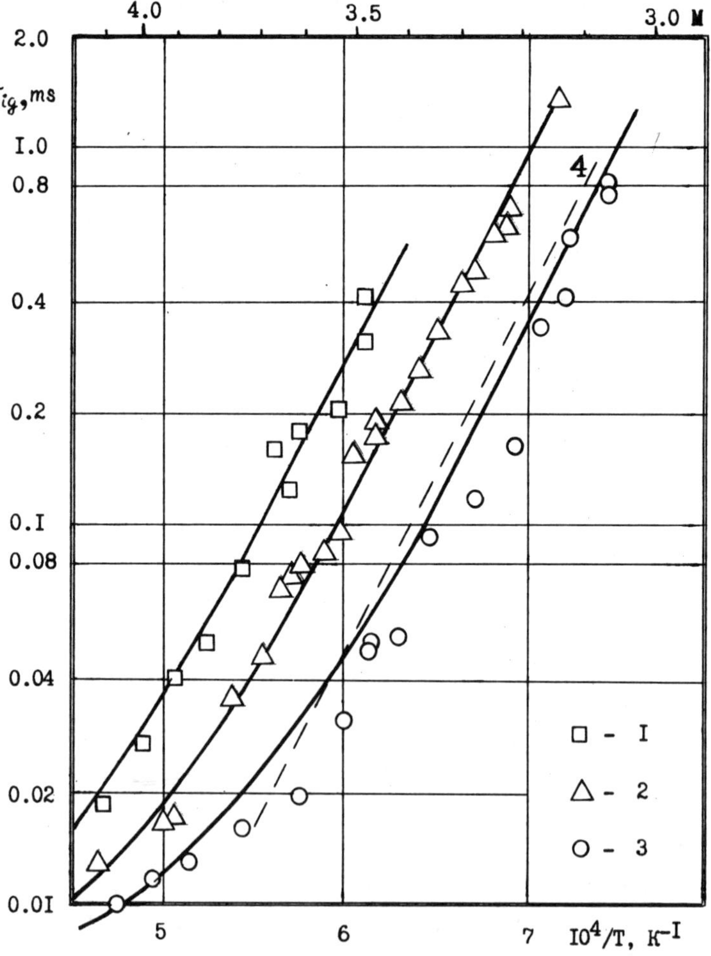

Fig. 4. Ignition delays of coal dust behind a reflected shock in air. 1, $V_0 = 0.09$; 2, $V_0 = 0.26$; 3, $V_0 = 0.55$; 4, $V = 1\%$ CH_4

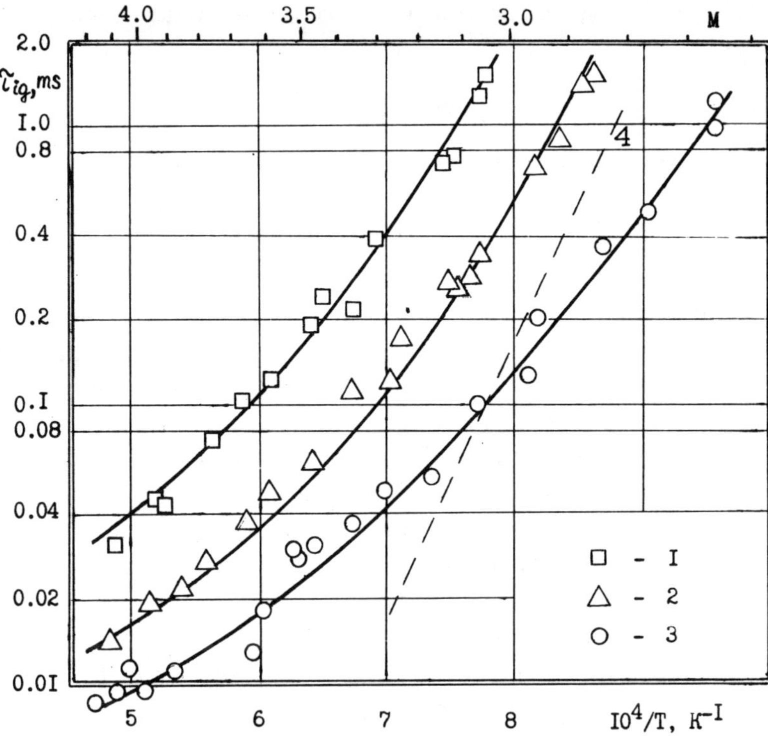

Fig. 5. Ignition delays of coal dust behind a reflected shock in oxygen. 1, $V_0 = 0.09$; 2, $V_0 = 0.26$; 3, $V_0 = 0.55$; 4, $V = 1\%$ CH_4

oxygen and dust-air mixtures. It indicates that coal particle ignition is controlled by the simultaneous action of several different mechanisms, and it is seen that the effect of those mechanisms is different and depends on gas composition and temperature. Thus, for a coal-air mixture at temperature $T_5 < 1700$ K ($1/T > 6\ 10^{-4}$ K^{-1}), the slope of dependence $\ln(\tau_{ig}) = f(1/T)$ is the same as that for methane-air mixtures. On this basis, one can suppose that the ignition of coal particles in air at temperatures 1300-1700 K is controlled by the volatile ignition kinetics (CH_4). For the coal-oxygen mixture, the slope of dependence $\ln(\tau_{ig}) = f(1/T)$ is different from that of the methane-oxygen mixtures in practically all temperature ranges being studied, although, at low temperatures there is a tendency toward rapproachment of their slopes. Hence the volatile ignition kinetics in pure oxygen is not the main mechanism of coal particle ignition. As will be shown later, for pure oxygen in all temperature ranges and for air at high temperature ($T > 1700$ K), it is necessary to take into account devolatilization kinetics.

Discussion

Investigation of dust organic fuel ignition, performed as a rule in stationary conditions at the heating rates $\simeq 10^2$ K/s (see, e.g., Ref.10) has shown that during fuel particle heating there exists thermal decay of organic matter with volatile exuding. These volatiles contain a considerable number of fuel components, which were identified at a low heating rate. Among them are H_2, CO, C_nH_m and, in particular, methane. Combustion of these components promotes the heating of particles to the temperature of particle ignition.

The process of organic particle ignition can be separated into relatively independent stages: particles heating up to the start of devolatilization, the exude of volatiles itself, volatile ignition, and volatile combustion with further heating of particles to ignition. These stages have a different rate, duration, and extent of influence on particle ignition depending on such factors as temperature, pressure, oxidizer composition, volatile content, and dispersivity of fuel dust. If we suppose that preflame stages happen in consecutive order, the particle ignition delay can be regarded as a sum of the following characteristic periods:

τ_1 : A period of the particle inert heating up to the temperature of intensive organic matter decay (> 1000 K).

τ_2 : Devolatilization delay, controlled by pyrolysis reaction rate.

τ_3 : Volatile ignition delay.

Therefore, particle ignition delay is $\tau_{ig} = \tau_1 + \tau_2 + \tau_3$.

There are three characteristic stages of particle inert heating: first, in the ISW, heating due to convective heat transmission during particle acceleration; second, the analogous heating mechanism during particle braking in the RSW; and, finally, heat transmission to particles in a motionless gas after particle braking. To estimate the maximum value of τ_1, it is enough to take into account only the last of the inert heating stages, assuming that the main factor limiting particle heating rate is the heat conductivity of the gas. Thus, the time required for particles to heat up to the temperature T_5 will be

$$\tau_1 = c\rho d^2 \ln[(T_5 - T_0)/(T_5 - T_s)]/12\lambda \qquad (1)$$

where T_5, λ are the gas temperature and heat conductivity coefficients; c, ρ, and d are the particle specific heat, density and diameter; and T_0 is the initial temperature of the particles.

The choice of d value for this estimation is not connected to some average sized coal particle. Value d was chosen as a minimal particle diameter from the sample with $d < 40\mu$m because τ_{ig} was measured as a minimal time delay of flash appearance. Microscopic analysis of samples has shown that dust contains a great number of particles with diameters of about 1 μm. Thus the time of inert heating for such particles, even up to the temperature $T_s \simeq 0.9T_5$ does not exceed 5 μs. From Figs. 4 and 5 it follows that the addition of an inert heating stage become appreciable only at temperature $T_5 > 2000$ K.

We suppose that coal dust particle ignition happens as a result of absorption of thermal energy released in the time of burndown of fixed volatile quantity $V = a$. Total mass devolatilizing to the moment τ_2 is[10]

$$a = V_0 \sum_{i=1}^{n} C_{0i} \left\{ 1 - \exp\left[- \int_0^{\tau_2} k_{0i} \exp\Big(-E_{vi}/RT_s(t) \Big) \right] dt \right\} \qquad (2)$$

where V_0 is the volatile content, n is the number of pyrolytic reaction groups, C_{0i} is the quantitative characteristic of a particular reaction group, k_{0i} and E_{vi} are the kinetic constants of such groups, and $T_s(t)$ is the particle temperature history. If the one-component scheme of pyrolysis ($n = 1; C_{0i} = 1$) at the constant temperature $T_s = 0.9T_5$ is considered, then,

$$a = V_0\{1 - \exp[-\tau_2 k_0 \ \exp(-E_v/RT_5)]\}$$

and devolatilization delay is

$$\tau_2 = A \ln[V_0/(V_0 - a)] \ \exp(E_v/RT_5) \qquad (3)$$

The one-component scheme of pyrolysis is highly approximated and the kinetic constants take different values at very different heating rates. Since authors have no information on pyrolysis kinetic constants at heating rates of about $10^6 - 10^8$ K/s, the values of A, a, E_v have been arrived from the experimental curves $\ln(\tau_{ig}) = f(1/T_5)$ which were obtained at high temperatures in pure oxygen.

At gas temperatures below 1500 K, total ignition delay is governed mainly by the volatile ignition kinetics. It is well known that methane is a basic component of volatiles, which are formed under high-temperature pyrolysis of coals. The most reliable data on its ignition both in oxygen and in air were obtained from the shock tubes (see, e.g., Ref.11). Investigations were carried out in a wide range parameters ($800 < T_5 < 2600$ K; P_5 up to 0.6 MPa). To describe the methane ignition delay, the following kind of expression is often useful: $\tau_{ig} \approx C_F^m C_0^n \exp(E_a/RT)$, where C_F and C_0 are the fuel and oxygen concentrations, and E_a is the activation energy. In spite of a great number of papers on methane oxidation, the kinetic high temperature range (> 1500 K) has scarcely been explored. Thus, values of E_a, as obtained by different authors, range from 86 to 230 kJ/mol, m from 0.3 to 0.48 and n - from -1.0 to -1.94.

For the more complete hydrocarbons (from C_3H_8 to kerosene) at the concentration of fuel or pyrolysis products over 3-5%, the change in C_F either does not practically influence the ignition delay (m = 0)[12] or it influences the delay in the opposite direction ($m < 0$)[13]. The influence of C_0 remains strong (n from -0.6 to -2.0), and the value of E_a lies in the same range as methane.

From Figs. 4 and 5 one can see that the volatile content increases, accordingly, the methane concentration, too, leads to a decrease in the ignition delay of a coal-air (or coal-oxygen) mixture, but the nature of dependence $\ln(\tau_{ig}) = f(1/T_5)$ is not changed. It can be supposed that the local methane concentration in the ignition seats is more than 5%. Then the influence of volatile content tells upon the particle ignition delays through the heat transmission features from the fuel-oxygen reaction zone to the coal particles and also by the way of the total heat energy released in the reaction zone. When gas fuel is plentiful, then the influence of volatile content can be expressed by the reaction order for oxygen, $n = \varphi(V_0)$:

$$\tau_3 = BV_0^{0.3}(P/P_0)^{-(1+bV_0)} \exp(E_a/RT) \tag{4}$$

The final expression for ignition delay of coal particles about 1 μm in diameter is obtained, according to the present analysis, by approximation of experimental data:

$$\begin{aligned}\tau_{ig} &= 3 \cdot 10^{-6} + 5.5 \cdot 10^{-6} \ln[V_0/(V_0 - 0.04)] \exp(83/RT_5) + \\ &+ 2 \cdot 10^{-9} V_0^{0.3}(P/P_0)^{-(1+3.7V_0)} \exp(190/RT_5), \ s\end{aligned} \tag{5}$$

where P is the partial pressure of oxygen (MPa), $P_0 = 0.1$ MPa, T_5 is the gas temperature behind RSW (K), and R is the gas constant in kJ/K·mol.

The curves in Figs. 4 and 5 were obtained by means of expression (5). One can see that Eq. (5) describes experimental data on ignition delays for different coals in the air as well as in pure oxygen.

Although expression (5) was obtained for particles $\approx 1\mu$m, we can recommend it for the coal dusts of a wide range of dispersivity because, in reality, every coal sample always contains fine fractions. To estimate τ_{ig} for the larger particles, we cannot neglect the inert heating time of the particle, Eq. (1); also, the particle temperature history has to be taken into account, Eq. (2).

References

[1] Nettleton, M.A., and Stirling, R., "The Ignition of Clouds of Particles in Shock-Heated Oxygen", Proceedings of the Royal Society, London, Vol.300, No.1460, 1967.

[2] Wolanski P., "Deflagration and Detonation Combustion of Dust Mixtures", in: Dynamics of Deflagrations and Reactive Systems: Heterogeneous Combustion. Ed. by A.L.Kuhl, J.-C. Leyer, A.A.Borisov, and W.A.Sirignano, Vol.132, Progress in Astronautics and Aeronautics, AIAA, New York, 1991, p.3

[3] Borisov,A.A., Gelfand,B.E., Timofeev, E.I., Tsyganov, S.A., and Khomic, S.V., "Ignition of Dust Suspensions Behind Shock Wav es", in: Dynamics of Shock Waves, Explosions and Detonations. Ed. by J.R.Bowen, N.Manson, A.K.Oppenheim, and R.I.Soloukhin, Vol.94, Progress in Astronautics and Aeronautics, AIAA, New York, 1984, p.332

[4] Lee, J.H.S., "Dust Explosion: An Overview", in: Shock Tubes and Waves. Proceedings of the 16th International Symposium on Shock Tubes and Waves, ed. by H.Gronig, VCH Verlagsgesellschaft, Weinheim, Aahen, 1987.

[5] Boiko, V.M., Fedorov, A.V., Fomin, V.M., Papyrin, A.N., Soloukhin R.I., "Ignition of Small Particles Behind Shock Waves", in: Shock Waves, Explosions and Detonations, Ed. by J.R.Bowen, A.K.Oppenheim, N.Manson, and R.I.Soloukhin, Vol.87, Progress in Astronautics and Aeronautics, AIAA, New York, 1983, p.71.

[6] Lapworth K.C., "Normal Shock Wave Tables for Air, ... and Oxygen", ARS Current Papers, 1970.

[7] Boiko V.M., Papyrin A.N., "The Quick-Acting Laser Visualization of Processes Arising by Interaction of Shock and Detonation Waves and Small Particles", Proceedings of the 17th International Symposium on Shock Waves and Shock Tubes, ed. by Y.W.Kim, AIP Conference Proceedings No.208, New York, 1990, p.512

[8] Zellner R.N., Niemitz K.J., Warnatz J., Gardiner, N.C. Jr., Eubank, C.S., and Simmie, J.M., "Hydrocarbon Induced Acceleration of Methane-Air Ignition", in: Flames, Lasers and Reactive Systems, ed. by J.R.Bowen, N.Manson, A.K.Oppenheim, and R.I.Soloukhin, Vol.88, Progress in Astronautics and Aeronautics, AIAA, New York, 1983, p.252

[9] Cheng, R.K., and Oppenheim, A.K., Autoignition in Methane-Hydrogen Mixture, Combustion and Flame, Vol.58, 1984, p.125.

[10] Pomerancev V.V. (editor), Fundamentals of Practical Theory of Combustion, Leningrad, Energoatomizdat, 1986, (In Russian).

[11] Tsuboi T., and Wagner H., "Homogeneous Thermal Oxidation of Methane in Reflected Shock Waves", Proceedings of the 15th International Symposium on Combustion, The Combustion Society, Pittsburgh, 1974.

[12] Zimont V.L., and Trushin Yu.N., On Ignition Delays in Hydrocarbon Fuels at High Temperatures, in: Fizika Goreniya i Vzryva, Vol.3, No.1, 1967, p.86, (in Russsian).

[13] Freeman, G., and Lefebvre, A.N., "Spontaneous Ignition Characteristics of Gaseous Hydrocarbon-Air Mixtures", Combustion and Flame, Vol.58, 1984, p.153.

Chapter III. Vapor Explosions

Developments of the CULDESAC Physical Explosion Model

D. F. Fletcher*
Atomic Energy Authority Technology, Oxfordshire, United Kingdom

Abstract

This paper contains a description of an improved physical explosion model developed to simulate the explosive interaction of a hot liquid with a cold volatile liquid. The modeling presented in this paper represents a significant extension of an earlier version of the CULDESAC code.[1] After describing the physics of this problem, a complete description of the conservation equations and constitutive relations which form the model is given. Particular attention is paid to the recent modifications to the model. The solution procedure is then described and some results from simulations are presented. The model has been used to make calculations of relevance to other modeling and experimental work being performed world wide.

---

*Research Scientist, Culham Laboratory.

Nomenclature

A	= area factor
b_0	= coefficient in the artificial viscosity formula
c_d	= drag coefficient
c_{frag}	= constant in the fragmentation model
c_v	= specific heat at constant volume
e	= internal energy
e_s	= stagnation energy
h	= heat transfer coefficient
h_s	= stagnation enthalpy
K	= momentum exchange function
L	= length-scale
m	= mass stripped from a single droplet
p	= pressure
R	= energy exchange function
T	= temperature
t	= time
V	= velocity
x	= spatial coordinate
α	= volume fraction
β	= void fraction
Γ_f	= mass exchange rate between melt droplets and fragments
Γ_{frag}	= length-scale source term
Δt	= time step
Δx	= space step
μ_a	= artificial viscosity
ρ	= density

Subscripts

e	= effective fluid (water plus fragments)
f	= fragments
m	= melt droplets
s	= two-phase zone separating the mixture and the wall
w	= water

I. Introduction

If a hot liquid (melt) contacts a cooler volatile liquid, in some circumstances the energy transfer rate can be so rapid and coherent that an explosion results. An explosion of this type derives its energy from the stored thermal energy of the hot fluid. Such explosions are a well-known hazard in the metal casting industry,[2] in the transportation of liquefied natural gas over water,[3] in the paper industry where a molten salt (called smelt) may contact water,[4] and it is postulated that they may occur in submarine volcanisms.[5] They are also studied in the nuclear industry, to assess the consequences of the unlikely event that in a severe accident molten material contacts residual coolant and such an explosion results.[6]

It is postulated that explosions of this type progress through a number of distinct phases.[6] Initially, the melt and water (a coolant of water is assumed in the remainder of the paper) mix on a relatively slow time-scale ($\sim$1 s). During this stage the melt and water zones have a characteristic dimension of the order of 10 mm. Because of the high temperature of the melt, a vapor blanket insulates the melt from the water and there is relatively little heat transfer. If this vapor blanket is collapsed in some small region of the mixture, high heat transfer rates result and there is a rapid rise in the pressure locally. In some circumstances this pressure pulse can cause further vapor film collapse, so that it escalates and propagates through the mixture, causing coherent energy release. The propagating pressure pulse (which steepens to form a shock wave) has two main effects. Firstly, it collapses the vapor blankets around the melt droplets initiating rapid heat transfer. Secondly, it causes differential acceleration of the melt and water, which in turn leads to relative velocity breakup of the melt and a large increase in the melt surface area. As the energy of the melt is rapidly transferred to the water, high-pressure steam is produced, which expands, with the potential to cause damage to any surrounding structures.

The history of the development of models of the propagation process has been reviewed by Fletcher and Anderson.[7] They concluded that although modeling in this area has been pursued for almost 20 years there is still much to be done before such models can become truly predictive tools. There are a number of reasons for this. The physical mechanisms of fragmentation and explosive heat transfer are still poorly understood. There are no data which are sufficiently detailed to validate the existing models. In addition, all of the existing models have many shortcomings and gross approximations. Indeed, there is no suitable mathematical framework available yet which can cope with the large spatial and temporal thermal gradients which arise in the explosion process.

At Culham Laboratory, an attempt is being made to improve the understanding of the propagation process via model development. This model development is being performed in stages and is based on use of the CULDESAC code. This model was originally developed by Fletcher and Thyagaraja[1, 8] and has subsequently been extended by Fletcher.[9, 10] This paper contains an update on the status of the model and the results from a number of calculations performed to address issues of current interest in the vapor explosion community.

Section II contains a description of the model and the constitutive physics used to close it. The solution procedure is outlined in Sec. III and results from the simulations are presented in Sec. IV. Finally, Sec. V contains the conclusions.

II. Description of the Model

In this section the partial differential equations and constitutive relations which constitute the model are described. The model is similar to that described in earlier papers,[1, 8] except that it is now possible to use artificial viscosity to capture strong shocks. The model is transient and one dimensional, although this may be planar, cylindrical, spherical, or any user-specified slowly varying shape. The mixture is assumed to consist of melt droplets, water and steam. Behind the pressure front the droplets are fragmented and the water is heated by energy transfer from the fragments. This situation is shown schematically in Fig. 1.

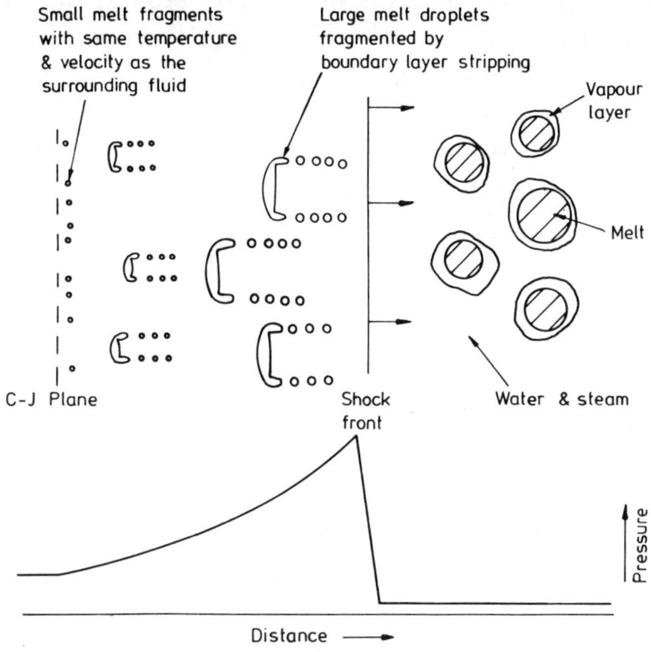

Fig. 1 Schematic representation of physical picture underlying the model.

In the model this situation is represented using three different components, namely, melt droplets, melt fragments and water. (The presence of a permanent gas is no longer allowed in the modeling, since calculations performed using the earlier model showed that small amounts of such gas had a very small effect on the propagation behavior.[1]) To restrict the number of constitutive relations required to close the model, the simplifying assumption that steam and water are always in mechanical and thermal equilibrium is made. This assumption is justified once supercritical temperatures and pressures occur and provides a first approximation for subcritical conditions. Each species is assumed to have its own velocity field (in previous calculations the water and fragments were assumed to be in mechanical equilibrium). The fragmentation process is assumed to be boundary-layer stripping,[11] although the model is sufficiently general to apply to any fragmentation process. Heat is transferred from the fragments to *all* of the water at a finite rate. Thus no allowance is made for local thermal disequilibrium, around the fragments, in the water phase, or within the fragments themselves.

This problem has been formulated mathematically using the usual multiphase flow equations,[12] where the presence of each species is specified by a volume fraction and all of the species are at a common pressure. The simplifying assumption that the melt and fragments are incompressible is made, so that $\rho_f = \rho_m = \text{const}$.

A. Conservation Equations

Conservation of mass applied to the water, melt, and fragments gives

$$\frac{\partial}{\partial t}(\alpha_w \rho_w) + \frac{1}{A}\frac{\partial}{\partial x}(A\alpha_w \rho_w V_w) = 0 \tag{1}$$

$$\frac{\partial}{\partial t}(\alpha_m \rho_m) + \frac{1}{A}\frac{\partial}{\partial x}(A\alpha_m \rho_m V_m) = -\Gamma_f \tag{2}$$

and

$$\frac{\partial}{\partial t}(\alpha_f \rho_m) + \frac{1}{A}\frac{\partial}{\partial x}(A\alpha_f \rho_m V_f) = \Gamma_f \tag{3}$$

In Eqs. (2) and (3) Γ_f is the mass exchange rate between the droplet and fragment species due to fragmentation and is specified later.

Conservation of momentum for the water, melt, and fragments gives

$$\begin{aligned}\frac{\partial}{\partial t}(\alpha_w \rho_w V_w) &+ \frac{1}{A}\frac{\partial}{\partial x}(A\alpha_w \rho_w V_w^2)\\ = -\alpha_w \frac{\partial p}{\partial x} &+ \frac{\partial}{\partial x}\left(\alpha_w \mu_a \frac{1}{A}\frac{\partial}{\partial x}(AV_w)\right)\\ &+K_{mw}(V_m - V_w) + K_{fw}(V_f - V_w)\end{aligned} \tag{4}$$

$$\frac{\partial}{\partial t}(\alpha_m \rho_m V_m) + \frac{1}{A}\frac{\partial}{\partial x}(A\alpha_m \rho_m V_m^2)$$
$$= -\alpha_m \frac{\partial p}{\partial x} + K_{mw}(V_w - V_m) - \Gamma_f V_m \tag{5}$$

and

$$\frac{\partial}{\partial t}(\alpha_f \rho_m V_f) + \frac{1}{A}\frac{\partial}{\partial x}(A\alpha_f \rho_m V_f^2)$$
$$= -\alpha_f \frac{\partial p}{\partial x} + K_{fw}(V_w - V_f) + \Gamma_f V_m \tag{6}$$

The terms on the right-hand side of Eqs. (4–6) represent the effect of the pressure gradient force, drag between the various species (the fragments and the melt droplets are assumed to be surrounded by water so that they do not interact, i.e., $K_{fm} \equiv 0$), and momentum transfer between the large droplets and the fragments due to fragmentation. The fragments are assumed to be moving with the melt droplet velocity at the instant they are formed.

There are also energy conservation equations for each species. It is convenient to work in terms of the stagnation energy, defined by $e_s = e + 1/2V^2$ and the stagnation enthalpy defined by $h_s = e_s + p/\rho$. Conservation of energy for the water, melt, and fragments gives

$$\frac{\partial}{\partial t}(\alpha_w \rho_w e_{sw}) + \frac{1}{A}\frac{\partial}{\partial x}\left[A\alpha_w V_w \left(\rho_w h_{sw} - \mu_a \frac{\partial V_w}{\partial x}\right)\right]$$
$$= -p\frac{\partial \alpha_w}{\partial t} + R_{mw}(T_m - T_w) + R_{fw}(T_f - T_w)$$
$$+ V_w K_{mw}(V_m - V_w) + K_{mw}(V_w - V_m)^2$$
$$+ V_w K_{fw}(V_f - V_w) + K_{fw}(V_w - V_f)^2 \tag{7}$$

$$\frac{\partial}{\partial t}(\alpha_m \rho_m e_{sm}) + \frac{1}{A}\frac{\partial}{\partial x}(A\alpha_m \rho_m V_m h_{sm})$$
$$= -p\frac{\partial \alpha_m}{\partial t} + R_{mw}(T_w - T_m) + V_m K_{mw}(V_w - V_m) - \Gamma_f h_{sm} \tag{8}$$

and

$$\frac{\partial}{\partial t}(\alpha_f \rho_m e_{sf}) + \frac{1}{A}\frac{\partial}{\partial x}(A\alpha_f \rho_m V_f h_{sf})$$

$$= -p\frac{\partial \alpha_f}{\partial t} + R_{fw}(T_w - T_f) + V_f K_{fw}(V_w - V_f) + \Gamma_f h_{sm} \quad (9)$$

In the above equations the terms involving R_{mw} etc., represent thermal equilibration and the terms containing K_{mw} etc., are the drag work. It has been assumed that all of the irreversible drag work heats up the water. For completeness, the viscous work due to the artificial viscosity term is included in the water energy equation, but this is found to be insignificant in practice.

In multiphase flow the surface area per unit volume of the interpenetrating phases plays an important role in determining the interaction between the different phases, i.e., in determining interfacial drag and heat transfer. A transport equation for this quantity was developed by Ishii[12] in his fundamental review of the equations of multiphase flow. In this work we assume that the melt is composed of spherical droplets, so that it is a simple matter to convert this equation to one for the droplet diameter. The resulting equation for the lengthscale of the melt droplets is given by

$$\frac{\partial}{\partial t}(\alpha_m \rho_m L_m) + \frac{1}{A}\frac{\partial}{\partial x}(A\alpha_m \rho_m V_m L_m) = -\Gamma_f L_m - \Gamma_{\mathbf{frag}} \quad (10)$$

In the above equation the term involving Γ_f arises as a consequence of writing the transport equation in conservation form and the term $-\Gamma_{\mathbf{frag}}$ models the chosen fragmentation process.

In addition to the above equations there is the constraint that

$$\alpha_m + \alpha_f + \alpha_w = 1 \quad (11)$$

This completes the specification of the differential equations. The next subsections contain a description of the constitutive relations used to close the model, the equations of state (EOS), the form of the artificial viscosity term used, and the boundary and initial conditions.

B. Constitutive Relations

In this section the constitutive relations for drag, heat transfer, and fragmentation employed in the model are described.

1. Momentum Exchange

If the volume fraction of droplets is α_m, there are $6\alpha_m/\pi L_m^3$ spherical droplets per unit volume. The drag force on a single droplet may be written as

$$F_D = \frac{1}{2} c_{d,mw} \rho_e \pi \frac{L_m^2}{4} |V_w - V_m|(V_w - V_m) \tag{12}$$

where $\rho_e = (\alpha_w \rho_w + \alpha_f \rho_f)/(\alpha_w + \alpha_f)$ is the effective density of the fluid dragging the melt droplets. The effective density is used because there is assumed to be no direct interaction between the fragments and the unfragmented droplets; the presence of the fragments is assumed to increase the inertia of the fluid surrounding the droplets. Thus the total drag force is

$$F_T = \frac{3}{4} \frac{c_{d,mw}}{L_m} \rho_e \alpha_m |V_w - V_m|(V_w - V_m) \tag{13}$$

and comparison of Eq. (13) with Eq. (5) shows that

$$K_{mw} = \frac{3}{4} c_{d,mw} \alpha_m \frac{\rho_e}{L_m} |V_w - V_m| \tag{14}$$

In the present work a constant value of $c_{d,mw} = 2.5$ has been used. This value is higher than the usual value of 0.4 to account for the increased drag when droplets are fragmenting.[13]

Similarly, the drag between the melt fragments and the water is given by

$$K_{fw} = \frac{3}{4} c_{d,fw} \alpha_f \frac{\rho_w}{L_f} |V_w - V_f| \tag{15}$$

where $c_{d,fw} = 0.4$. (The present model can be used to simulate the situation of no slip between the water and fragments by setting $c_{d,fw}$ to a large value.)

2. *Heat Transfer*

If the heat transfer rate is specified as the product of a heat transfer coefficient and the temperature difference, it is easily shown that

$$R_{mw} = 6\alpha_m \frac{h_{mw}}{L_m} \tag{16}$$

and

$$R_{fw} = 6\alpha_f \frac{h_{fw}}{L_f} \tag{17}$$

where h_{mw} is the heat transfer coefficient between the melt and water, h_{fw} the heat transfer coefficient between the fragments and water, and L_f the fragment size.

The heat transfer mechanisms between the melt and water are very complex and depend on the time history of each particle. The only experimental data available is rather crude and consists of time-averaged heat transfer coefficients for the duration of the fragmentation process.[7] Thus at present, only crude models of this process can be formulated. In the present version of CULDESAC it is possible to specify either constant heat transfer coefficients for the fragments and droplets or to make the fragment to water heat transfer rate a function of the local water density. This is achieved by using the following simple formula[10]:

$$h_{fw} = h_{\text{steam}} + (h_{\text{liquid}} - h_{\text{steam}}) \frac{(\rho_w - \rho_{\text{steam}})}{(\rho_{\text{liquid}} - \rho_{\text{steam}})} \tag{18}$$

where h_{steam} and h_{liquid} are appropriate heat transfer rates to steam and liquid water respectively. The end-point densities were given the following values: $\rho_{\text{liquid}} = 1000 \text{ kg m}^{-3}$ and $\rho_{\text{steam}} = 0.6 \text{ kg m}^{-3}$. Values of $h_{\text{steam}} = 10^3 \text{ Wm}^{-2}\text{K}^{-1}$ and $h_{\text{liquid}} = 10^5 \text{ Wm}^{-2}\text{K}^{-1}$ were used in the present study[7, 10]. The enhancement of heat transfer from the large droplets following vapor film collapse was exactly the same as that used in the earlier model.[9]

3. *Fragmentation Model*

A boundary-layer stripping model for fragmentation is used, as this is thought to be the most appropriate for the study of vapor explosions.[11] The model proposed by Carachalios et al.[14] is used. This gives the following stripping rate from a single droplet

$$\frac{\mathrm{d}m}{\mathrm{d}t} = c_{\mathbf{frag}}|V_m - V_w|\pi L_m^2\sqrt{\rho_m\rho_e} \tag{19}$$

where the empirical constant $c_{\mathbf{frag}}$ takes a value of approximately 1/6. This model was used to generate expressions for Γ_f and $\Gamma_{\mathbf{frag}}$ in exactly the same manner as in the earlier model.[1]

The fragment size L_f is not determined by the fragmentation model currently employed and was specified by reference to experimental data. A typical value of $L_f = 100\ \mu$m was used.[15]

C. Equations of State

The melt equation of state is very simple. The melt is assumed to be incompressible so that $\rho_m = \rho_f =$ const, and to have a simple caloric equation of state so that $E_m = c_{vm}T_m$ and $E_f = c_{vm}T_f$.

The EOS for water is more complicated and there is virtually no thermodynamic data on the properties of water at high temperatures and pressures in a suitable form for use in a model such as this one. In this study an approximate EOS, based on the Grüneisen approximation, has been used. The equation of state is fully described elsewhere[8] and has been shown to agree well with steam table data in the region where they can be compared.[1]

D. Artificial Viscosity

In an earlier version of the CULDESAC code[1, 8] the conservation equations were solved without the inclusion of an artificial viscosity term. This procedure was found to work well, provided that the water density behind the front was not too great. For example, a series of calculations performed to examine the effect of varying the void fraction (the fraction of the water phase in vapor form) ahead of the front, showed the increased presence of high-frequency oscillations in the pressure at the front with falling void fraction.[8] Simulations also showed that decreasing the heat transfer rate from the fragments to the water led to these oscillations.[1] In some cases a solution could not be obtained beyond a certain time because these oscillations caused unphysical compression of the water (they led to densities outside the liquid region of the phase plane).

In this version of the model an artificial viscosity term is included in the water species momentum equation to remove this instability. The chosen form is that advocated by von Neumann and Richtmyer[16] and discussed by Richtmyer and Morton.[17] The artificial viscosity μ_a is assumed to be of the following form:

$$\begin{aligned} \mu_a &= \rho_w (b_0 \Delta x)^2 \left| \frac{\partial V_w}{\partial x} \right| \qquad \text{if } \frac{\partial V_w}{\partial x} < 0 \\ &= 0 \qquad \text{otherwise} \end{aligned} \tag{20}$$

This form for μ_a has the property that viscosity is only added in regions of flow compression, thus it does not smear out expansion fans and it is only added where there are large velocity gradients, i.e., at the shock front itself.

In the original work of von Neumann and Richtmyer, it was shown that if the constant b_0 was chosen to be of order unity, then the shock would be spread over approximately three grid cells. This conclusion was based on analysis for an *ideal gas* equation of state. In this application values of b_0 of order 5–15 were used, with higher values required for the less compressible initial condition, i.e., for low initial void fractions or low heat input rates. Higher values of b_0 are required because the equation of state of water is much "stiffer" than that of an ideal gas.

E. Boundary and Initial Conditions

The above equations are solved in a solution domain which represents an initial mixture contained in a closed vessel. Thus the only boundary condition needed is to set the velocities to zero at the vessel walls.

Initially, the volume fractions, void fraction, velocity, and particle size distribution are specified. To simulate triggering a fraction of the melt is fragmented in a small region of the solution domain. This causes a high heat transfer rate in these cells, the pressure rises locally, and a propagating wave may develop.

III. Solution Procedure

The partial differential equations are solved using a finite difference method which employs the usual staggered grid arrangement. All convective terms are modeled using upwind differencing for stability. An implicit scheme is used, with all physical terms, except for the rates obtained from constitutive laws, being evaluated using midtime values. The fragmentation, heat transfer, and drag rates are evaluated using old time values, once and for all, at the beginning of each time step. The artificial viscosity term in the water momentum equation is treated fully implicitly for stability. The same basic strategy as described in earlier work[8, 18] is employed, except that iteration must now be carried out at each time step. Superscripts n and $n+1$ are used to denote old and new time values, respectively. At any time during the calculation new time values of the field variables are obtained using the following procedure.

1) Determine the fragmentation, heat transfer, and drag rates using old time values.

2) Equations (1), (2), and (3) are used to obtain $(\alpha_w \rho_w)^{n+1}$, α_m^{n+1} and α_f^{n+1}, respectively. Equation (11) is then used to determine α_w^{n+1} and hence ρ_w^{n+1} can be determined.

3) Equations (4–6) are used to obtain the new velocity fields. A tri-diagonal matrix is solved for each species, with the momentum flux, pressure gradient and drag terms evaluated using the latest estimates of the midtime values.

4) The new stagnation energies are found by time advancing the three energy equations (7–9). Using the stagnation form ensures that the Rankine-Hugoniot equation can be built into the solution scheme to give it good shock-capturing properties. The new velocity field is then used to determine the new internal energies.

5) The caloric equations are used to determine the new melt and fragment temperature. The water EOS is then used to determine the new pressure field and the water temperature.

6) A check is made to determine the largest change in the pressure in each cell, from that calculated at the previous iteration step, normalized by the pressure in that cell. If this is less than 10^{-3}, Eq. (10) is solved to obtain the new melt lengthscale and the iteration cycle is finished. If it is not, new midtime values are calculated using the latest estimate of the values at the time level $n+1$ and the iteration process starts again at step 2.

This completes a brief description of the numerical scheme, which has proved to be very robust and stable. Typically, two or three iterations are required at each time step. Because an implicit method is being used, the time step can exceed the value used in the previous version of the code by about two orders of magnitude. In addition, the use of an implicit scheme removes the need to use a small time step to prevent the acoustic instability which occurred in stagnant regions when an explicit scheme was used.[19] The main restriction on the time step is that imposed by the need to model rate processes, such as fragmentation, accurately.

The sensitivity of the model to solution parameters, such as the time step, grid size, choice of b_0, and pressure iteration tolerance has been considered in detail elsewhere.[9] Figure 2 shows

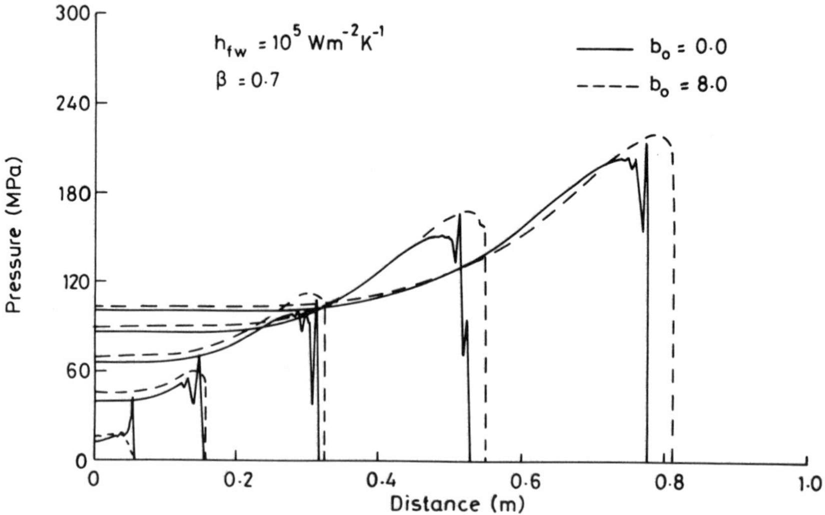

Fig. 2 Comparison of the development of the pressure profile for simulations with and without artificial viscosity; initial void fraction 0.7 and pressure profiles shown every 0.5 ms.

Table 1 Parameters used in the simulations

Parameter	Value
Time step, s	10^{-6}
Space step, m	0.01
Coefficient in artificial viscosity term (b_0)	8.0
Initial particle size, m	0.005
Initial pressure, MPa	0.1
Heat transfer rate: vapor blanketed, $\mathrm{Wm^{-2}K^{-1}}$	10^3
Heat transfer rate: liquid-liquid contact, $\mathrm{Wm^{-2}K^{-1}}$	10^5
Pressure increase required to collapse vapor blanket, MPa	1.0
Initial melt temperature, K	3400
Melt density, $\mathrm{kg\,m^{-3}}$	8400
Melt heat capacity, $\mathrm{J\,kg^{-1}K^{-1}}$	500
Melt surface tension, $\mathrm{N\,m^{-1}}$	0.4
Initial melt volume fraction	0.1
Initial void fraction	0.7
Length of triggered zone, m	0.02
Fraction of melt fragmented as a trigger	0.9

how the artificial viscosity term removes the dispersive instability which occurs at the shock front (see Table 1 for the conditions used in the simulation).

IV. Results from Simulations

The next five subsections contain results of simulations which have been performed to answer questions of direct relevance to on-going experimental or computational studies being performed world wide, or to answer specific questions raised by previous calculations. Unless otherwise stated the conditions shown in

Table 1 are used in the calculations. All of the simulations are for a planar geometry, i.e., $A \equiv 1$. The simulations are relevant to the interaction of molten core material with water. (An initial melt temperature above expected values is used to allow for the latent heat which would be released on freezing.)

A. Effect of Trigger Size

Experiments performed at Sandia National Laboratories[20] have shown that the work yield from single-droplet fragmentation experiments increases with the trigger strength. A series of numerical simulations were performed to determine what the present mathematical model predicts will occur as the trigger strength is increased. This was achieved by increasing the length of the triggered zone from 5 mm to 50 mm. In these calculations the non density-dependent heat transfer model was used to ensure that a strong detonation developed.

The results are shown in Fig. 3. The calculations show that increasing the trigger size increases the escalation rate but that all of the solutions are contained within a common envelope. This is the expected result and suggests that a model which can explain the experimentally observed threshold would have to contain a different fragmentation model. It is possible that both hydrodynamic and thermal fragmentation occur in some circumstances and that thermal fragmentation may need a high trigger to initiate it. Further developments in this area will depend on the availability of new experimental data.

B. Role of Particle Motion in Restarting Propagation

In an earlier paper, calculations were performed to determine the effect of mixture inhomogeneity.[21] A geometry was set up in which regions of mixture were separated by regions of highly voided water. It was found that as the pressure wave left the mixture region it would decay very rapidly but that it would start a propagating wave again very easily once it reached more mixture. Even a steam-filled zone a meter long was not sufficient to prevent a propagating wave redeveloping. Calculations using

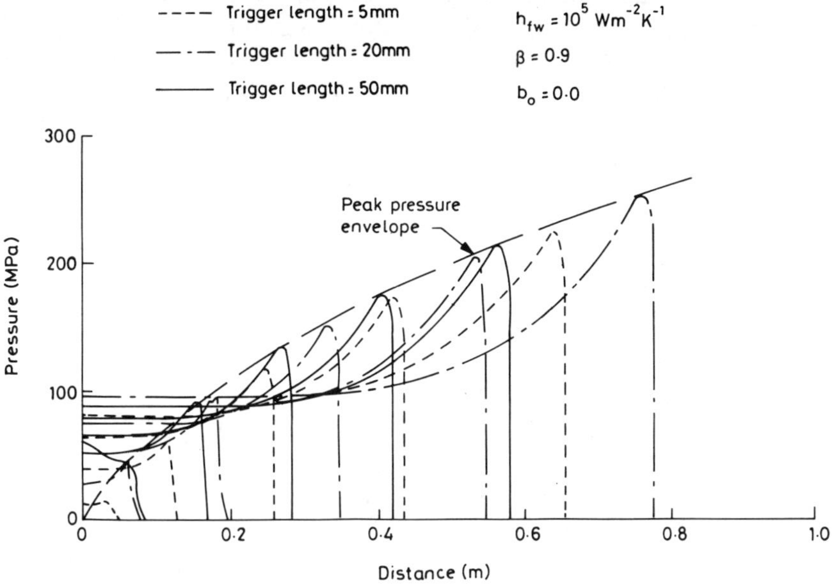

Fig. 3 Comparison of the development of the pressure profile for three different trigger pressures; note that all of the solutions are enclosed within a common envelope. (Pressure profiles are shown every 0.4 ms for the 5- and 20 mm trigger-length cases and every 0.3 ms for the 50 mm trigger-length case.)

the present model, in which the fragments can slip relative to the water phase, exhibit the same behavior, but there is some reduction in the ease with which a secondary propagation can be initiated.[10]

Further calculations have been performed to examine this effect, from which it has become clear that the fragments cross the voided region and initiate a propagation by a "hammer" pressure effect. Separating the two regions by a low void fraction region reduces the distance the fragments can travel before coming to rest but allows the shock wave to propagate with less attenuation. Figure 4 shows the results from these computations and illustrates the very different behavior of the pressure wave in the low and high void fraction cases.

Thus although mixture inhomogeneities do inhibit the propagation process, their effect is not as large as one might expect. Their effect is likely to be much more important in two-

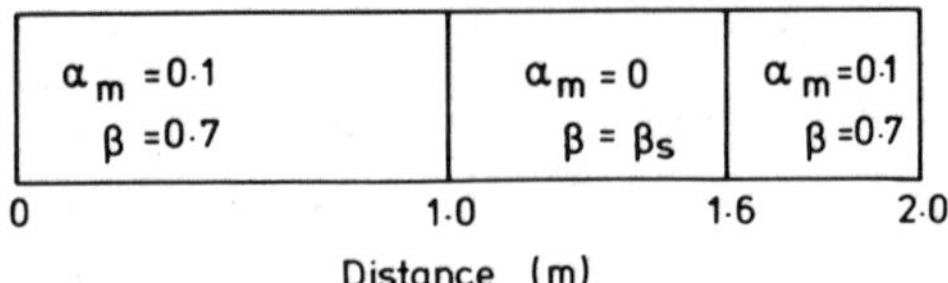

a) Geometry used in the simulations

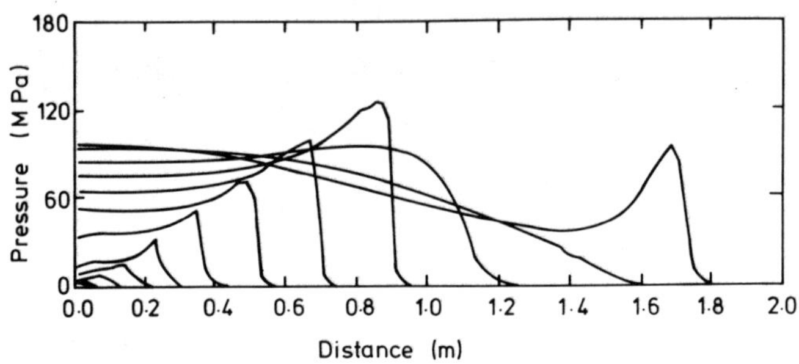

b) Pressure profile every 0·5ms for β_s =0·99 case

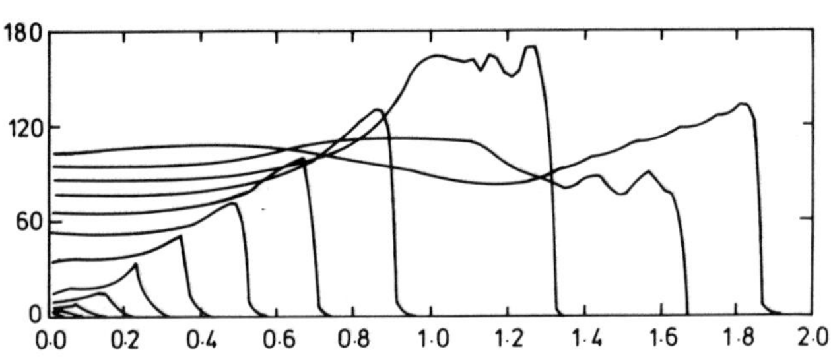

c) Pressure profile every 0·5 ms for β_s =0·2 case

Fig. 4 Results from simulations showing how the void fraction of the buffer region controls the manner in which a propagation is restarted.

dimensions because the flow will not be focused. However, it is clear that experimental data, especially for highly voided mixtures, would be very useful to guide future modeling attempts.

C. Effect of Computational Grid Size

To perform computation in two space dimensions there is a temptation to use a coarse computational mesh. Previous work[22] has shown that for high heat input rates this practice can lead to numerical solutions which bear no resemblance to the true

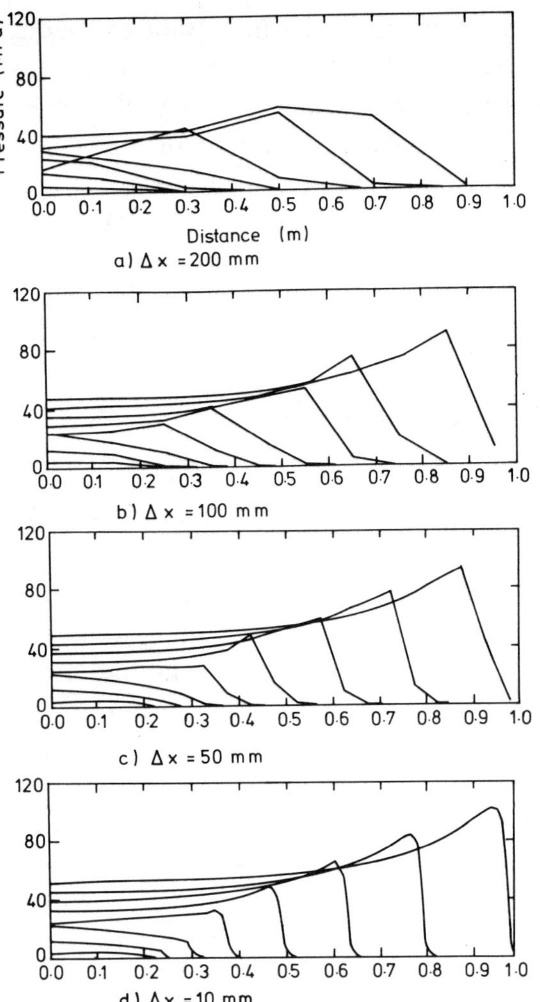

Fig. 5 Comparison of the development of the pressure profile for various grid sizes (pressure profiles are shown every 0.5 ms).

solution. A study was performed to examine the effect of grid size for a case where the heat transfer rate was lower. In these simulations the initial void fraction was 0.9, $b_0 = 0$, and the density-dependent heat transfer rate was used. Figure 5 shows the development of the pressure profile for four different grid sizes, ranging from 10 to 200 mm. In all cases a trigger length of 200 mm was used so that all of the solutions are for exactly the same problem. Once a cell size of 50 mm is reached, the essential features of the solution are predicted but the shock front is very smeared. A cell size of 200 mm gives a very poor solution.

It is clear from this and previous grid refinement studies[22] that the level of resolution required depends very strongly on the heat transfer and fragmentation rates. Using a coarse grid necessitates applying a trigger over a large volume of the solution domain. In addition, when moving to two dimensional simulations the shock wave will not always be aligned with the grid, a problem which is known to degrade the solution further.

D. Effect of Ambient Pressure

In studies of explosion propagation of relevance to light water reactor safety it is of interest to examine the effect of ambient pressure on the propagation process. Figure 6 shows the results from computations at pressures of 0.1, 6, and 15 MPa. In each case the the same trigger was used and the initial void fraction was 0.7. Simulations showed that as the pressure was increased less artificial viscosity was required to obtain a solution, presumably because there is a smaller density change when vapor is condensed at the shock front at higher ambient pressures.

The figure also shows that the peak pressure at the front does not vary significantly with ambient pressure and that the pressure increase due to the passage of the shock wave is about the same in each case. The propagation velocity increases with pressure, with most of this effect being due to the faster escalation rate. This is because the density-dependent heat transfer rate was used which causes the heat transfer rate to increase with pressure. The fragmentation rate is also changed in a complex manner. The density difference between the melt and steam/water mixture falls with increased pressure, so that the relative velocity causing fragmentation is reduced, but the density of the fluid causing fragmentation increases, so the net effect on fragmentation is not easy to predict.

E. Vapor Explosion Pressures

In most experiments performed to study vapor explosions the principal diagnostic is an array of pressure transducers. Thus it

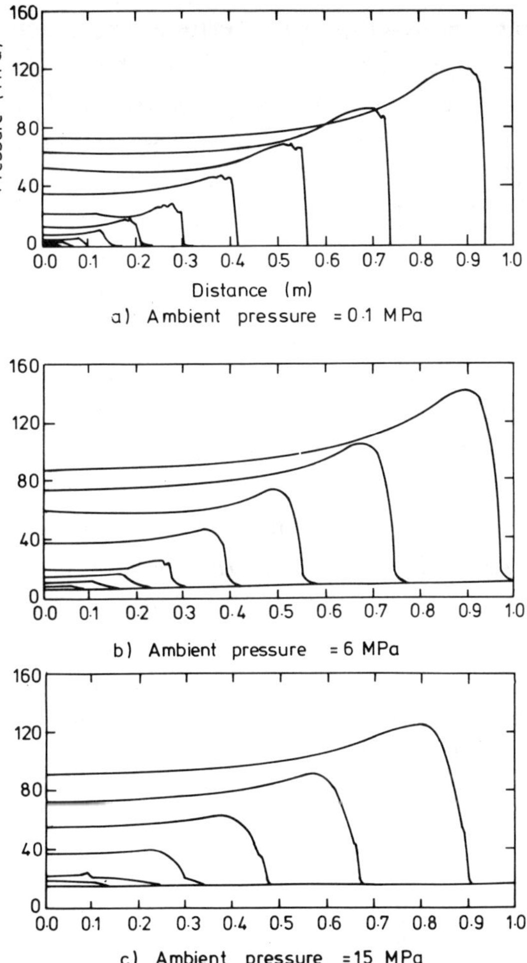

Fig. 6 Comparison of the development of the pressure profile for three different ambient pressures (pressure profiles are shown every 0.5 ms).

is important to gain an idea of how pressure transducer location affects the measured pressure. The transient pressure produced by such an explosion is also of crucial importance in determining the damage it causes. To investigate both of these features a number of different simulations were performed in which the transient pressure was recorded at a number of locations in the solution domain. Simulations were performed for a 1 m-long tube containing mixture, and for situations where only the first 0.8 m contained mixture and the last 0.2 m contained a water/steam mixture with a void fraction denoted by β_s. These simulations were performed to attempt to determine the effect of measuring the pressure a short distance away from the explosion zone.

Figure 7 shows the calculated pressure transient at the right-hand wall (at the opposite end from the trigger location) for a case where the mixture void fraction is 0.7. The figure shows results for a tube filled with mixture and for cases where the steam/water region has a void fraction of 0.2 or 0.9. In all of the cases a shock impacts the wall giving very high pressures, of order 300–600 MPa. This shock loading is relatively short lived, typically less than 0.5 ms. The peak pressure is reduced in the 0.9 void fraction case because the material close to the wall has a much higher compressibility. The pressure in the tube equilibrates in approximately 12 ms, with the final pressure being a function of the melt/water mass ratio in the region where melt is present (discussed later). This equilibration is achieved by the passage of shock and rarefaction waves up and down the tube.

It must be remembered that these calculations are very idealized, in that no expansion of the mixture is allowed, so that very high pressures are achieved. Such pressures would be reduced by acoustic relief, especially if the mixture region were surrounded by a region of highly voided water, as is likely for

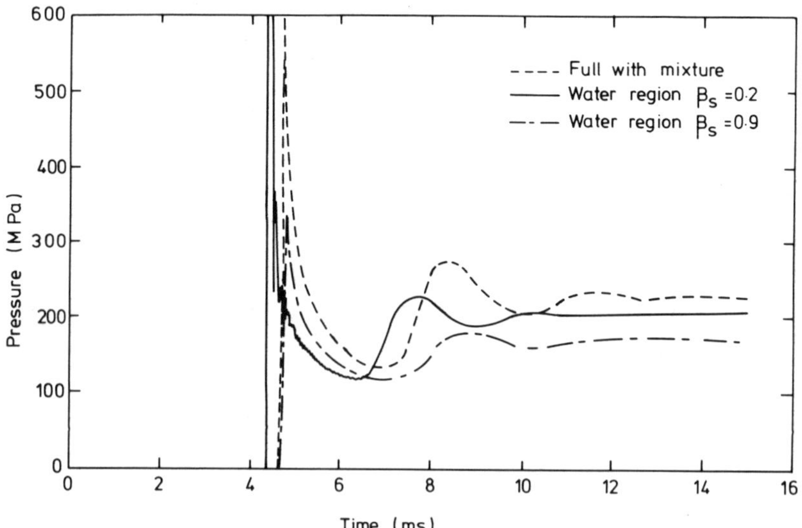

Fig. 7 Predicted wall pressure transients for a mixture void fraction of 0.7.

the high-temperature melt used in this simulation.[23] The exact shape and height of the pressure pulse is determined in these calculations by the amount of artificial viscosity used in the simulation. In reality, three dimensional effects and dispersion of the shock front in multicomponent mixtures will reduce the peak pressure.[7] The values given here are indicative and should not be taken as precise estimates.

Figure 8 shows a repeat of the above simulation but for a mixture void fraction of 0.9. In this case the final pressures are lower and the equilibrium pressure is only achieved after several oscillations in the pressure. The figure shows that even for a highly voided mixture a large shock pressure can be generated at the wall.

The variable area option of CULDESAC was used to investigate pressure relief by coupling the explosion zone to a steam-filled region with ten times the volume of the explosion zone. In this case the equilibrium pressure was reduced considerably and the equilibration time was lengthened to many tens of milliseconds. It is clear from this scoping calculation that to model the quasistatic pressure generated in the explosion accurately it is necessary to have a detailed model of the region surrounding the explosion

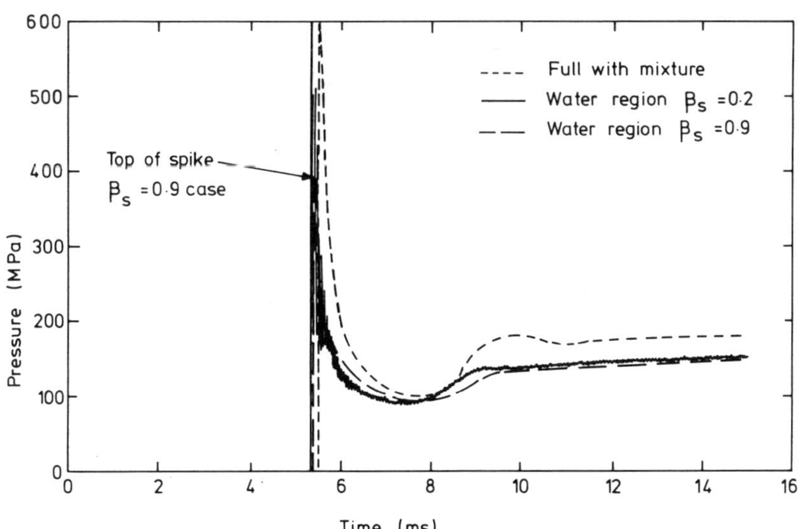

Fig. 8 Predicted wall pressure transients for a mixture void fraction of 0.9.

zone. If a large steam chimney is located above the explosion zone, as is predicted by mixing calculations,[23] pressures can be readily relived.

For the case where the tube was filled with mixture, the pressure was recorded at three locations: the two ends and the middle of the tube. Figure 9 shows these results superposed on one plot. The oscillatory behavior of the pressure equilibration process becomes very apparent. In addition, the figure shows the very different transient pressure loads at the three locations, with a very slow (8 ms) ramp loading at the trigger end, compared with a shock loading of 350 MPa decaying to 150 MPa over 0.5 ms at the opposite end of the tube. Half-way along the tube the pressure rises by ~100 MPa as the initial shock wave passes and then steps up again as the reflected shock passes back down the tube. These calculations illustrate why pressure-transducer traces recorded at various locations in an interaction vessel are often impossible to interpret. They also highlight the difficulty in assessing the possible damage which could be caused by such an explosion.

Volume fraction plots were produced to determine the location of various components at pressure equilibration. Figure 10 shows the volume fraction distribution at various times for the

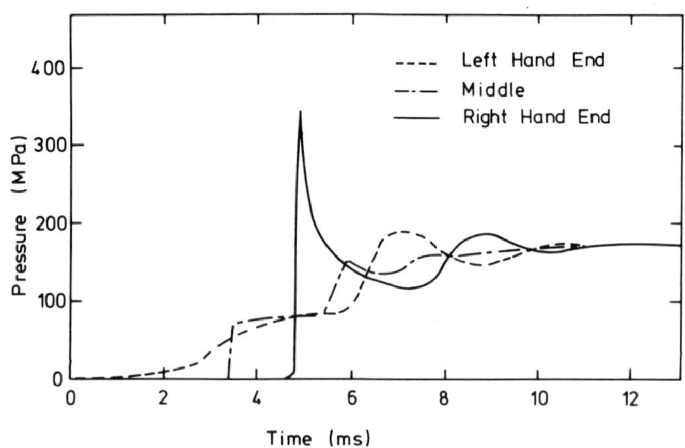

Fig. 9 Predicted pressure transients at various locations in the tube for a mixture void fraction of 0.9.

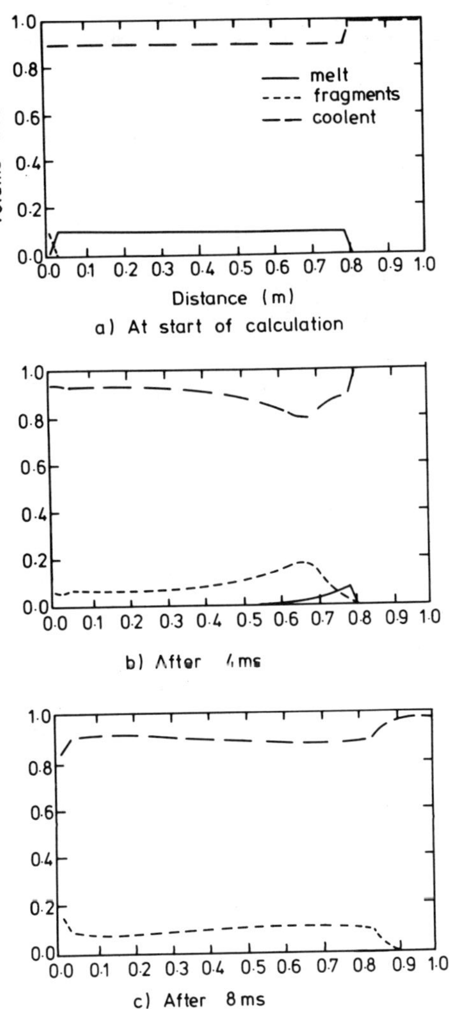

Fig. 10 Predicted volume fraction distributions for a case where the mixture zone is separated from the right-hand wall by a steam/water zone with a void fraction of 0.2.

case of a mixture void fraction of 0.7 and a void fraction of 0.2 in the steam/water zone. The figure shows that the propagating front generates a zone of approximately 0.2 m long over which fragmentation occurs. The final material distribution is achieved after about 8 ms and shows that the fragment volume fraction is not uniform throughout the tube. Fragments have only mixed about 70 mm into the surrounding water and some of this will have been caused by the artificial viscosity used in the calculation. For the case of steam/water region with a void fraction of 0.9 the fragments were spread almost uniformly throughout the tube (with material collected at the two ends due to the zero velocity boundary condition). This result is interesting, since it provides an estimate of the possible amount of additional water than could be included in the mixture on a short time scale.

V. Conclusions

An improved mathematical model of physical explosions has been described. This model is applicable to a much wider range of situations than an earlier version of the model.[1] The constitutive relations used to close the model have been described, together with the solution scheme used to solve the equations which form the model. A series of computational results of relevance to a number of experimental and theoretical programs being performed world wide have been presented.

It is clear from the assumptions required to formulate the model, and the lack of data available for model validation, that much work still remains to be done in this area. High quality data are needed to determine the appropriate fragmentation model and heat transfer rates. Experimental investigation of the physics of propagation as a function of void fraction would be very helpful, since it is still not clear whether propagation can occur at high void fractions. At high void fractions both the hot and cold liquids would be in droplet form ahead of the shock and the key question is whether shock fragmentation of these two components could produce sufficiently good heat transfer to sustain a propagating wave.

On the theoretical side, more calculations are required to study the effects of propagation in two and three dimensions, the influence of obstructions in the flow domain, and to explore other heat transfer and fragmentation assumptions. In addition, a framework which allows local thermal disequilibrium to be properly represented within a set of conservation equations is required.

Acknowledgment

This model described in this paper was developed as part of the General Nuclear Safety Research Programme funded by the Department of Energy.

References

[1]Fletcher, D. F., and Thyagaraja, A., "Multiphase Detonation Modeling using the CULDESAC Code," Progress in Astronautics and Aeronautics, Vol. 134, edited by A. L. Kuhl, J.-C. Leyer, A. A. Borisov and W.A. Sirigano, AIAA, Washington, D.C., 1991, pp. 387–407.

[2]Long, G., "Explosions of Molten Metal in Water — Causes and Prevention," Metal Progress, Vol. 71, May 1957, pp. 107–112.

[3]Katz, D. L., and Sliepcevich, C. M., "Liquefied Natural Gas/water Explosions: Cause and Effect," Hydrocarbon Process, Vol. 50, Nov. 1971, pp. 240–244.

[4]Reid, R. C., "Rapid Phase Transitions from Liquid to Vapor," Advances in Chemical Engineering, Vol. 12, 1983, pp. 105–208.

[5]Colgate, S. A., and Sigurgeirsson, T., "Dynamic Mixing of Water and Lava," Nature, Vol. 244, July/Aug. 1973, pp. 552–555.

[6]Corradini, M. L., Kim, B. J., and Oh, M. D., "Vapor Explosions in Light Water Reactors: A Review of Theory and Modelling," Progress in Nuclear Energy, Vol. 22, No. 1, 1988, pp. 1–117.

[7]Fletcher, D. F., and Anderson, R. P., "A Review of Pressure-Induced Propagation Models of the Vapour Explosion Process," Progress in Nuclear Energy, Vol. 23, No. 2, 1990, pp. 137–179.

[8]Fletcher, D. F., and Thyagaraja, A., "A Mathematical Model of Melt/Water Detonations," Applied Mathematical Modelling, Vol. 13, No. 6, 1989, pp. 339–347.

[9]Fletcher, D. F., "An Improved Model of Melt/Water Detonations, Part I: Model Formulation and Example Results," International Journal of Heat and Mass Transfer, Vol. 34, No. 10, 1991, pp. 2435–2448.

[10]Fletcher, D. F., "An Improved Model of Melt/Water Detonations, Part II: A Study of Escalation," International Journal of Heat and Mass Transfer, Vol. 34, No. 10, 1991, pp. 2449–2459.

[11]Baines, M., Board, S. J., Buttery, N. E., and R. W. Hall, "The Hydrodynamics of Large-Scale Fuel-Coolant Interactions," Nuclear Technology, Vol. 49, June 1980, pp. 27–39.

[12]Ishii, M., Thermo-fluid Dynamic Theory of Two-phase Flow, Eyrolles, Paris, France, 1975, Ch. 1–3.

[13]Pilch, M., and Erdman, C. A., "Use of Breakup Time Data and Velocity History Data to Predict the Maximum Size of Stable Fragments for Acceleration-Induced Breakup of a Liquid Drop," International Journal of Multiphase Flow, Vol. 13, No. 6, 1987, pp. 741–757.

[14]Carachalios, C., Bürger, M., and Unger, H., "A Transient Two-Phase Model to Describe Thermal Detonations Based on Hydrodynamic Fragmentation," Proceedings of the International Meeting on LWR Severe Accident Evaluation, Cambridge, MA, Aug. 28 — Sept. 1, American Nuclear Society, LaGrange Park, IL, USA, 1983, paper 6.8.

[15]Fletcher, D. F., "The Particle Size Distribution of Solidified Melt Debris from Molten Fuel-Coolant Interaction Experiments," Nuclear Engineering and Design, Vol. 105, No. 3, 1988, pp. 313–319.

[16]von Neumann, J., and Richtmyer, R. D., "A Method for the Numerical Calculation of Hydrodynamical Shocks," Journal Applied Physics, Vol. 21, March 1950, pp. 232–237.

[17]Richtmyer, R. D., and Morton, K. W., Difference Methods for Initial-Value Problems, Interscience, New York, 1967, pp. 311–317.

[18]Fletcher, D. F., and Thyagaraja, A., "Some Calculations of Shocks and Detonations for Gas Mixtures," Computer & Fluids, Vol. 17, No. 2, 1989, pp. 333–350.

[19]Thyagaraja, A. and Fletcher, D. F., "Low Mach Number Instability of an Explicit Numerical Scheme," Applied Mathematical Modelling, Vol. 15, No. 1, 1991, pp. 40–45.

[20]Beck, D. F., Berman, M., and Nelson, L. S., "Steam Explosion Studies with Molten Iron-Alumina Generated by Thermite Reactions," Progress in Astronautics and Aeronautics, Vol. 134, edited by A. L. Kuhl, J.-C. Leyer, A. A. Borisov and W.A. Sirigano, AIAA, Washington, D.C., 1991, pp. 326–355.

[21]Fletcher, D. F., "Some Computations of Detonations Using the CULDESAC Code," Culham Lab. Rept. CLM–R291, Abingdon, U.K., 1989.

[22]Thyagaraja, A., and Fletcher, D. F., "Numerical Aspects of Multiphase Detonation Modelling," Proceedings 6th International Conference on Numerical Methods in Laminar and Turbulent Flow, Vol. 6, 1753–1763, Pineridge, Swansea, U.K., 1989, pp. 1753–1763.

[23]Fletcher, D. F., and Thyagaraja, A., "A Mathematical Model of Premixing," Proceedings 25th National Heat Transfer Conference, (Houston, Tex.), American Nuclear Society, LaGrange Park, IL, USA, HTC–3, 1988, pp. 184–194.

Behavior of Free-Falling Boiling Spheres with Relation to Vapor Explosion Phenomena

F. S. Gunnerson* and P. R. Chappidi†
University of Central Florida, Orlando, Florida 32816

Abstract

An experimental and analytical research program has been conducted to qualitatively illustrate the influence of boiling on the settling or falling behavior of hot spherical particles in subcooled water. The influence of the boiling mode; nucleate, transition or film boiling on the settling velocity and corresponding drag coefficient were of particular interest. Results indicate that hot particles fall differently than particles in thermal equilibrium with the surrounding fluid. Boiling produces transient forces that vary as the hot particulate settles. Thermal energy exchange gives rise to local variation in viscosity, fluid shear stresses, projected form area, and, consequently, the settling behavior of the hot mass. Warm spheres at temperatures below the onset of boiling have been shown to fall slightly faster than colder spheres with reductions up to approximately 7% in the corresponding drag coefficient. Nucleate and transition boiling generally increase the drag coefficient and result in terminal settling velocities that are comparable to the thermal equilibrium state. Film boiling, however, gives rise to a significant increase in the settling velocity. The maximum average velocity experimentally recorded within the film boiling region was about 18-20% greater than the cold, single-phase velocity. At elevated sphere temperatures, well into the film boiling regime, there is a distinct turnaround in the drag and settling behavior. Very hot spheres experience "a drag bucket phenomenon" and can actually fall slower than their counterparts in the thermal equilibrium. Material properties, geometry, and particularly, surface conditions were found to greatly influence the onset of boiling and the transition from one boiling regime to another.

---

*Professor, Department of Mechanical & Aerospace Engineering.
†CFD Research Corporation, Huntsville, Alabama.

Introduction

Boiling, like any change in phase, can influence the motion of objects as they pass through a liquid. Changes in the buoyancy, skin friction, and pressure drag forces result from the heat exchange and the vapor production processes. Nuclear reactor safety analyses, metal forming and extrusion processes, and underwater projectile studies exemplify where boiling induced forces can be of fundamental importance.

Over the years, several experimental and theoretical studies have been conducted to assess the influence of boiling on hydrodynamic drag forces. It has been shown, for example, that film boiling can significantly reduce skin friction [1-4] and overall (pressure + skin friction + buoyancy) drag forces [5-10]. Unfortunately, many aspects of boiling behavior itself and its influence on hydrodynamic drag remain beyond the scope of current modeling capabilities.

Several recent experimental efforts have been directed toward boiling spheres free-falling through a liquid [5-11]. The spherical geometry is a favorite choice of experimental investigators in part because of the applications to nuclear reactor safety assessments.

Certain nuclear reactor accidents are postulated to result in extensive core melting and subsequent molten fuel motion. When molten fuel contacts coolant, core internals, or the reactor vessel, multiphase thermal interactions can occur which may jeopardize the control, structural integrity and safe shutdown of the reactor system. When molten fuel comes in contact with the coolant, several interaction scenarios can be hypothesized. A worst case scenario is a highly energetic, coherent fuel-coolant interaction where a fraction of the fuel's thermal energy is rapidly converted to mechanical energy, overpressurization, and subsequent reactor pressure vessel failure. Although highly unlikely, a vapor explosion has been postulated as a potential source of containment building breach or rupture [12,13]. Failure of a reactor pressure vessel or containment would significantly increase the potential for environmental radioactive release. Fortunately, such a scenario has not been realized within a commercial light water reactor. Another possible scenario, that is supported by previous studies, is that of a nonviolent, nonexplosive nature where the molten fuel undergoes rapid breakup and continues to boil and solidify while settling into the coolant.

Two key elements are generally attributed with the onset of vapor explosions [14-16]:

(1) An initial period of film boiling with hot molten fuel within the liquid coolant.

(2) Fragmentation and coarse intermixing of the fuel within the coolant.

The second element requires basic knowledge of boiling particulate hydrodynamics to estimate the extent and degree of fuel-coolant intermixing.

Because of the surface tension force, small molten fuel masses tend toward spherical shapes and undergo boiling and single-phase convective cooling while they settle and solidify within the coolant. The rate at which hot particulate, undergoing boiling and subsequent cooling, settle or fall through a coolant has not been extensively investigated. To model the effects of molten fuel motion it is of fundamental interest to determine if hot masses settle faster, slower or at the same rate as masses in thermal equilibrium with the coolant.

Experiments were conducted with subcooled water and cryogenic liquid nitrogen to evaluate the effects of boiling on the settling behavior of hot, spherical masses. Results indicate that hot solid particulate fall differently than particulate in thermal equilibrium with the surrounding fluid. Boiling produces transient forces that vary as the hot particulate settles. Thermal energy exchange during the cooling process gives rise to local variations in viscosity, fluid shear stresses, the presence of vapor and consequently, the settling behavior of the hot mass.

This paper provides an update on experimental and theoretical studies being conducted at the University of Central Florida (UCF), Orlando. The results from these studies support and compliment those of recent investigators [5-11] and provide additional insight into boiling induced forces. Emphasis herein is focussed on experimental efforts with spherical geometries.

Theoretical Considerations

The motion of a sphere moving through a continuous fluid can be described by Newton's second law of motion

$$m_s \frac{dV}{dt} = \rho_s \left(\frac{4}{3} \pi R^3 \right) g - \rho_f \left(\frac{4}{3} \pi R^3 \right) g - \frac{1}{2} \rho_f C_D (\pi R^2) V^2 \quad (1)$$

were V is velocity, m mass, t time, ρ density, g the gravitational acceleration constant and R is the sphere radius. Subscripts s and f refer to the sphere and fluid, respectively.

The lefthand side of Eq. (1) represents the rate of momentum change and the right hand side the weight, buoyancy, and drag forces, respectively. For nonboiling conditions, the coefficient of drag (C_D) includes the skin friction and pressure drag contributions and is traditionally evaluated graphically by iteration with the Reynolds number.

Boiling can affect the buoyancy force through vapor production and the drag force by modifying the frictional and pressure distribution characteristics. Intuitively, vapor production yields an additional upward component to the buoyancy force. The influence of boiling on the drag force is not so obvious. The skin friction component of the drag force is altered by the rate of thermal energy exchange and the boiling mode (nucleate, transition or film) as is the fluid viscosity. Simple heating (without an interfacial phase change) alters the local fluid viscosity. Generally, for gases the viscosity and drag forces increase with surface heating whereas for most liquids viscosity and drag decrease with surface heating [7].

The experimental and theoretical studies of several investigators [1-4] have shown that the skin friction component of the drag force can be significantly altered in the presence of film boiling. Boiling increases the projected area of a settling sphere and hence can influence the form or pressure drag contribution. Unfortunately, separate effects experiments have not been conducted; thus the skin friction, pressure drag and buoyancy contributions to the total drag force have yet to be independently isolated.

Under certain conditions, the settling velocity of hot, spherical geometries becomes approximately constant and the left side of the Eq. (1) becomes zero. Such a terminal velocity condition has been shown experimentally [5,7] for brief settling periods and in the absence of boiling mode changes.

The total drag coefficient (C_D) as given in the Eq. (1) for two-phase flow conditions is interpreted by

$$C_D = C_{D,2\theta} + C_{D,BF} \tag{2}$$

Where $C_{D,2\theta}$ is a two-phase pressure and skin friction drag coefficient, and $C_{D,BF}$ is the vapor buoyancy force contribution to the total drag coefficient.

Experiment

A series of experiments (approximately 120) was conducted to evaluate the settling behavior of heated spheres within an aquarium of subcooled water. The aquarium tank was constructed from 1.5 cm thick transparent plexiglass, 31 x 27 cm wide and 182 cm high as shown in Fig. 1. The tank is fitted with a vertical calibrated scale for position reference. The tank is also equipped with a mirror so that a third dimension of sphere movement could be added to the total displacement. This mirror was attached to the side of the tank and positioned in such a manner that from one view the sphere could be seen as it changed position from top to bottom, side to side, and front to back. The tank was filled with water to the level of 174 cm. The tank also contained a removable catch screen which is used in the recovery of the tested spheres.

A furnace was constructed for attachment to the top of the tank. The furnace was required to heat the sphere for a particular test and also to provide a release mechanism for dropping the sphere (Fig. 2). The furnace heated the sphere by an electrical resistance method. The furnace was controlled by a variable potentiometer which adjusts the current that is delivered to the heating element. Sphere temperatures were measured with

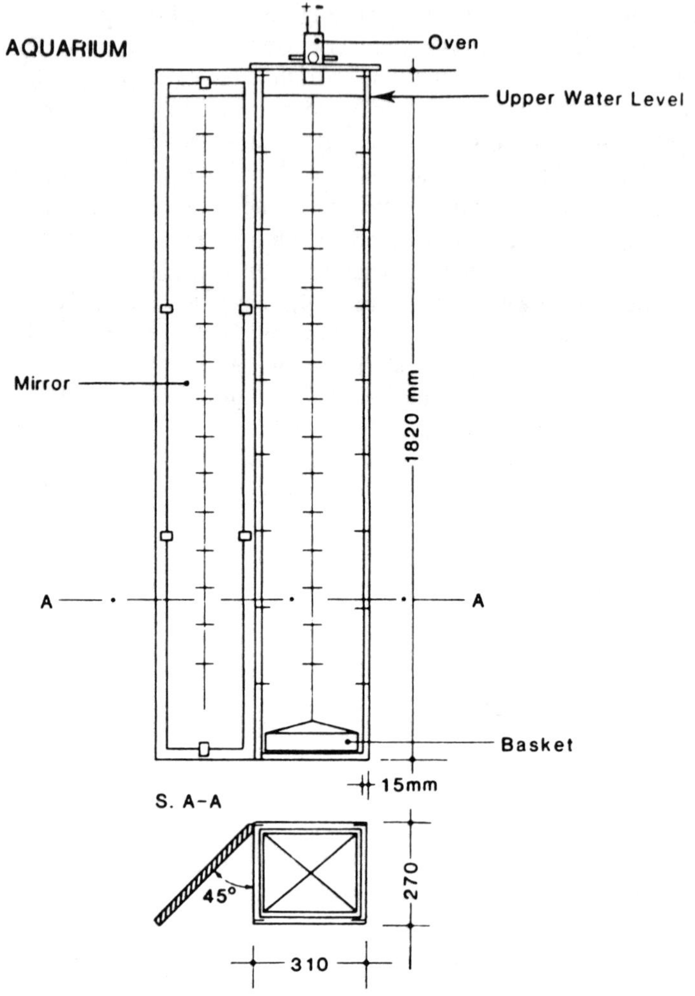

Figure 1. Diagram of aquarium tank.

a thermocouple located in the furnace. The temperature range for sphere testing was from approximately 20 to 500 °C. A device was provided as a release mechanism for the sphere which consisted of a metal plunger driven by an electric solenoid. When the sphere is in the heating position it rests on the plunger. As the electric solenoid is activated the plunger is retracted and the sphere is permitted to fall approximately 8 cm to the water surface. The sphere must pass through a guide tube which aids in keeping the sphere on its directed path. Guide tubes were produced with varying inside diameters which correspond to the diameter of the sphere being tested. The positioning of the furnace on the tank is partially adjustable. The position may be changed from front to back, however, only very minor

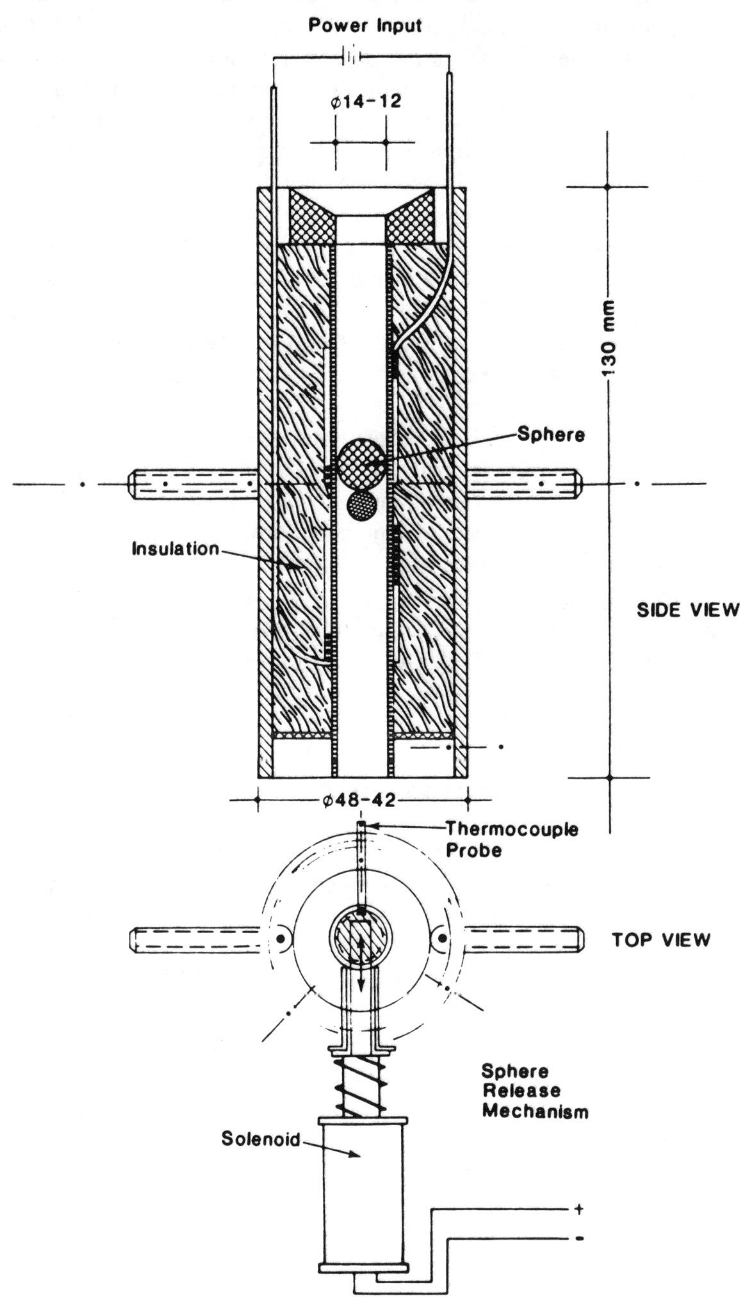

Figure 2. Aquarium sphere heater.

changes may be made from side to side. The range of sphere diameters tested corresponds to theoretical predictions of maximum particle sizes from molten UO_2 fuel-coolant interactions. The upper limit was determined to be approximately 10 mm. The lower limit, set by equipment limitations, was about 5 mm. Lead, bronze, brass, steel and ceramic spheres were selected in various diameters. Another component of the aquarium experiment is the video camera equipment. The video equipment consists of a high speed camera (200f ps), recorder, monitor, strobe light and required hardware for positioning the instruments. The collected data was reduced using a computer technique to digitize each frame of the high-speed VHS tape. A cursor was manipulated to the position of the sphere in that frame and the coordinates were recorded. These coordinates were converted to position at time intervals and graphed to provide an indication of the sphere's falling behavior.

The boiling mode (nucleate, transition or film) was estimated visually from the VHS tape. Film boiling, particularly near the minimum film boiling point, is difficult to see due to the thinness and transparency of the vapor film. Subsequent vapor bubble detachment, following cool down into the transition mode, however, can be used as a positive indication of previous boiling. Transition boiling was characterized by vapor patches and regions of partial film boiling. Frequently (in approximately 10% of the experiments), a sphere would exhibit a significant deflection ("shoot-off phenomena") from the straight-line path during it's passage through the transition boiling mode.

Results and Discussion

Fig. 3 illustrates typical average settling velocity behaviors for 10 mm bronze and steel spheres and 5.5 mm lead spheres in highly subcooled (20 °C) water. Distinct variations in the settling velocity behaviors with temperature are noticeable. Each sphere exhibits an increased velocity at elevated temperature and, for the steel and lead spheres, a turnaround decreased velocity at still higher temperatures. The lead sphere (with the greatest thermal diffusivity) reaches a maximum velocity at a temperature considerably less than that of the bronze or steel. Maximum velocity behaviors could be visually correlated to initial film boiling conditions. Fig. 4 relates boiling modes (nucleate, transition and film) to buoyant force, average velocity, and average drag coefficient. Fig. 4a illustrates the classical pool boiling curve for highly subcooled water and defines the no boiling, nucleate/transition and film boiling regions. Also shown is a representative quench path followed during the settling and cooling of a hot sphere.

Fig. 4b illustrates the buoyancy force contribution due to vapor production for a 8-mm-diam lead sphere suspended from a sensitive dynamic balance scale in liquid nitrogen. Although the temperature scale is quantitatively different for this figure, it does qualitatively illustrate the

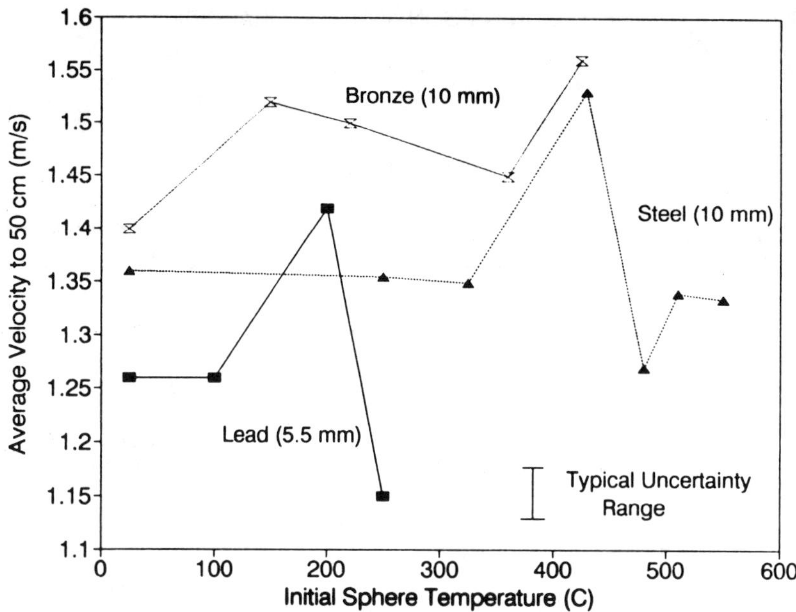

Figure 3. Typical settling behavior for spheres.

possible variation of the vapor buoyancy force with changes in the boiling mode. Notice that the maximum vapor induced buoyancy force occurs as the sphere transitions from film to nucleate boiling.

Figs. 4c and 4d show the velocity and drag characteristics for a 6.35 mm ceramic sphere in 20 °C water. Within the low temperature, no boiling, regime (T < 100 °C), the falling sphere is cooled by single phase natural and forced convection heat transfer. There is no additional buoyancy force from vapor production. As a result of local heating of the boundary-layer fluid, the local fluid viscosity and density decreases allowing the sphere to accelerate to a slightly higher velocity (Fig. 4c). The drag coefficient (Fig. 4d) is estimated from classical theory (Eq. 1) using the fluid viscosity at the initial sphere temperature.

At the onset of nucleate boiling (ONB) vapor buoyancy forces commence, the boundary layer is perturbed and the sphere velocity starts to decrease. The ONB is highly dependent on surface conditions and liquid superheat behavior. Generally, a rough, oxidized microsurface structure will permit the ONB at temperatures only a few degrees above saturation. For clean, polished spheres, higher degrees of excess surface temperature can be required for the ONB.

Within the nucleate and transition boiling regimes, an elevated heat flux gives rise to rapid vapor production. This results in a disturbed two-phase boundary layer, additional vapor buoyancy forces, and relatively low settling velocities; velocities comparable to the single-phase no boiling conditions.

At elevated temperatures, the sphere becomes vapor blanketed and the velocity significantly increases (Fig. 4c). Because of the insulating effect on the surrounding vapor layer, the heat flux is reduced on the vapor

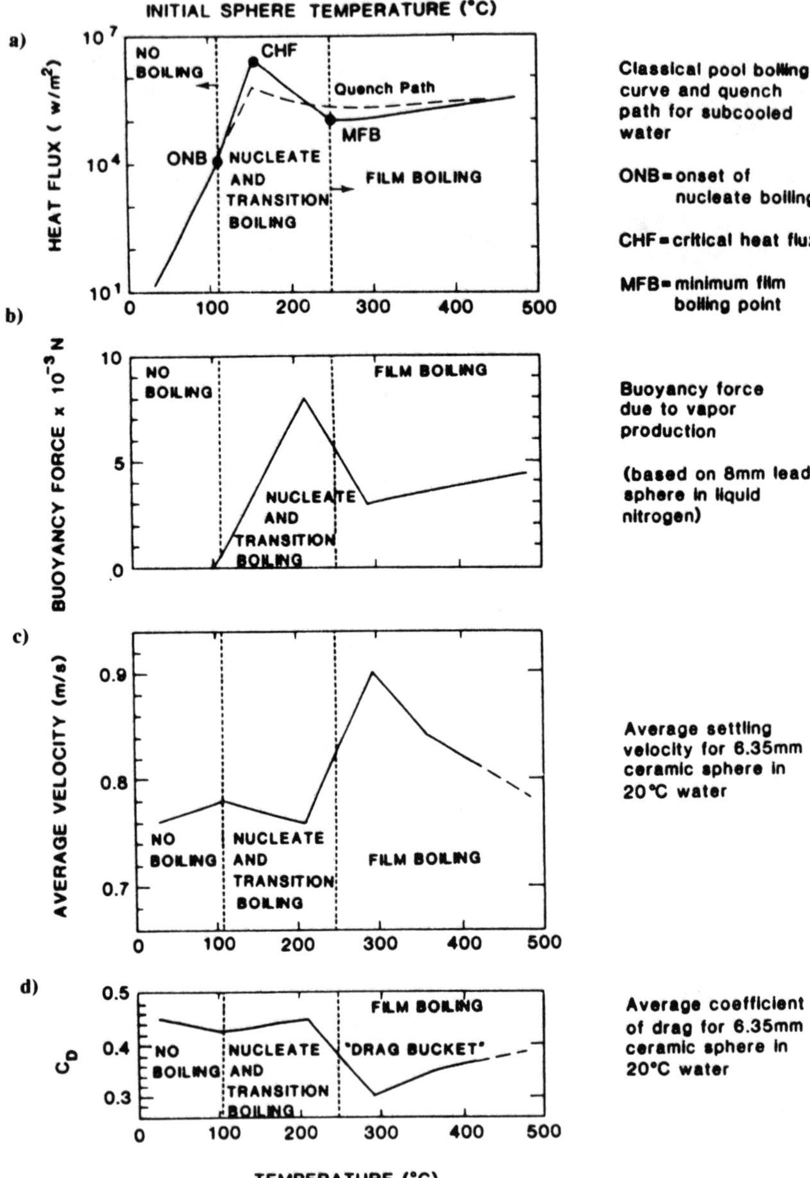

Figure 4. A description of boiling and settling behaviors.

production rate and buoyancy force decrease. Early in the film boiling region the drag coefficient likewise becomes minimum (Fig. 4d).

At higher film boiling temperatures ($T \approx 300$ °C) unusual settling behavior occurs. The sphere experiences an upward acceleration and velocity begins to decrease as shown in Fig. 4c. At such temperatures heat transfer by radiation may become apparent and increased vapor production results. It is believed that under such circumstances the sphere no longer 'feels', the liquid, but rather is affected by a high-velocity vapor film. The greater the temperature the greater the vapor film thickness and form drag, and the greater the resistance to falling. This slowing down behavior of the sphere is manifested as a minimum drag coefficient or "drag bucket", as shown in Fig. 4d. At very high temperatures, beyond the drag bucket behavior, the sphere can actually fall slower than in the cold single-phase condition. Very small hot particulate may actually float upward prior to cooling. The maximum average velocity experimentally recorded within the film boiling regime was about 18-20% greater than cold, single-phase velocity. The minimum average velocity experimentally recorded within the film boiling regime was approximately 11% less than the cold, single-phase terminal velocity.

The viscosity of most gases increases with increasing temperature whereas the viscosity of most liquids decreases with increasing temperature. The single-phase drag characteristics likewise follow the viscosity trends. A heated object moving in air, for example, has a greater drag force than the same object in thermal equilibrium with the air. In contrast, a heated object moving through a liquid has less drag than the same object in thermal equilibrium with the liquid. When a local change in phase (boiling) occurs, the drag force may be significantly altered as was illustrated in Figs. 3 and 4. The smallest drag force has been experimentally demonstrated to occur when the surface heat flux is sufficient to produce a condition of film boiling.

Zvirin, et al. [5,10,11] conducted similar free-fall boiling experiments. Their test spheres were instrumented with centerline thermocouples inserted into the spheres so that the thermal history during free fall could be monitored. The warm trailing thermocouple wire, however, inherently provides additional drag and thus may be expected to reduce the falling velocity somewhat below that of a free sphere. Additionally, the attached thermocouple wire may disrupt the vapor removal mechanism and the resultant buoyancy force acting on the sphere. Fig. 5 compares the results of Zvirin et al.[11] for a 20 mm copper sphere with those of the present authors for ceramic, steel and bronze spheres.

In most cases, the experimental results show that the falling velocity of the hot sphere is higher with boiling than the velocity of an unheated sphere. The free-fall velocity also peaks in the elevated temperature range (indicative of film boiling) then decreases at even higher temperatures. At higher sphere temperatures a turnaround or decreased velocity may occur. Subsequent to film boiling collapse, liquid-solid interactions can frequently occur (approximately 10% of the author's experiments) producing lateral

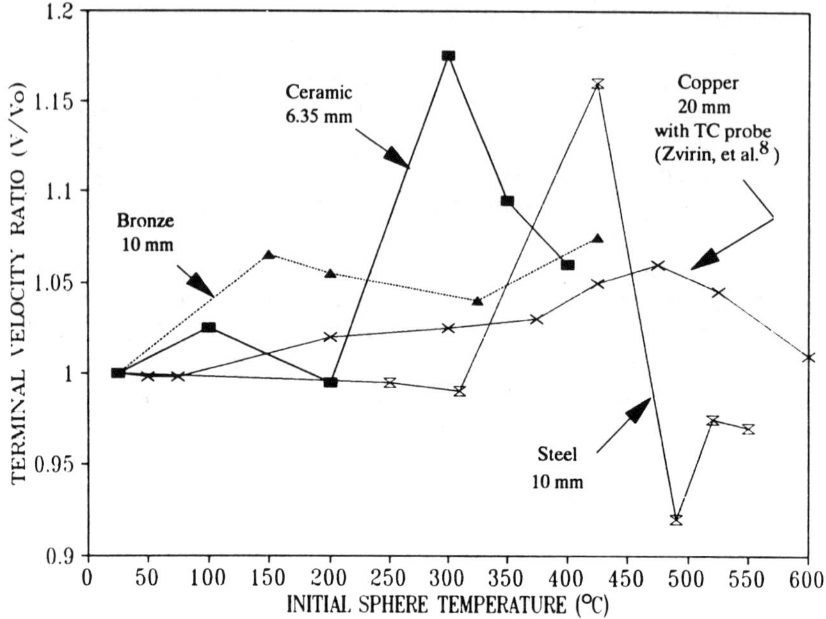

Figure 5. Comparison of settling velocity behaviors in highly subcooled water.

motion of the sphere "shoot-off phenomena"; such data is not included. Uncertainties in the authors' terminal velocity ratio data is estimated to be +/- 3%.

Any vapor explosion model which seeks to characterize the coarse intermixing behavior of boiling fuel within coolant should consider the effects of boiling. In general, it has been shown that film boiling enhances the intermixing capabilities by promoting lower particulate drag and greater distribution velocities.

Acknowledgments

The authors thank J. Meyer and R. Zeyen for their valuable efforts in conducting portions of this experimental program. Experimental support was provided by programs at the JRC, Italy and the University of Central Florida.

References

1. Bradfield, W. S., Barkdol, R. O. and Byrne, J. T. "Some Effects of Boiling on Hydrodynamic Drag", Int. J. Heat Mass Transfer Vol. 5, pp. 614-622 1962.

2. Cess, R. D. and Sparrow, E. M., Film Boiling in a Forced-Convection Boundary-Layer Flow, ASME Journal of Heat Transfer Vol. 83, pp. 370-376 1961.

3. Chappidi, P. R. and Gunnerson, F. S. "A Numerical Study of the Drag and Heat Transfer Characteristics of a Wedge in Film Boiling Flow", ASME Proceedings of the 1988 National Heat Transfer Conference, Vol. 2, Houston, TX, pp. 475-486 July 1988.
4. Chappidi, P. R. and Gunnerson, F. S. "Skin Friction and Heat Transfer Characteristics of a Vertical Flat Plate in Film Boiling Flow", Proceedings of the 5th Miami International Symposium on Multi-Phase Transport and Particulate Phenomena, Miami Beach, FL, pp. 337-344 December 1988.
5. Zvirin, Y., Aziz, S., Hewitt, G. F., and D. B. R. Kenning, "Experimental Investigations of the Transition from Film Boiling on Free Falling Spheres", ANS Proceedings, 23rd National Heat Transfer Conference, Houston, TX, pp. 209-216 July 1988.
6. Cho, D. H. and Lambert, G. A. "Effect of Boiling on Particle Drag", Nuclear Engineering and Design, Vol. 108, pp. 523-524 1988.
7. Gunnerson, F. S., Meyer, J. A. and Zeyen, R. "An Investigation of Boiling Behavior with Applications of Nuclear Reactor Safety", University of Central Florida Rep. 16-26-805, Orlando, FL, Aug. 1987.
8. Zvirin, Y., Hewitt, G. F. and Kenning, D. B. R. "Boiling on Free-Falling Spheres: Drag and Heat Transfer Coefficients", Experimental Heat Transfer, Vol. 3, No. 3, pp. 185-215 1990.
9. Aziz, S. "Forced Convection Film Boiling on Spheres", Ph.D. Thesis, Dept. of Engineering Science., Univ. of Oxford, U.K. 1986.
10. Zvirin, Y., Hewitt, G. F. and Kenning, D. B. R. "Drag and Heat Transfer Coefficients in Boiling on Free Falling Spheres", Rep. No. R13289, Harwell UKAEA, 1988.
11. Aziz, S., Hewitt, G. F. and Kenning, D. B. R. "Heat Transfer Regimes in Forced-Convection Film Boiling on Spheres", Proceedings 8th International Heat Transfer Conference, Vol. 5, San Francisco, CA, pp. 2149-2154 1986.
12. Theofanous, T. G., Najafi, B. and Rumble, E. "An assessment of Steam Explosion Induced Containment Failure, Part I: Probabilistic Aspects", Nuclear Engineering Design, Vol. 97, pp. 259-281 1987.
13. Abofadl M. A. and Theofanous, T. G. "An Assessment of Steam Explosion Induced Containment Failure, Part II: Premixing Limits", Nuclear Engineering Design, Vol. 97, pp. 282-295 1987.
14. Corradini, M. L. "Molten Fuel/Coolant Interactions: Recent Analysis of Experiments", Nuclear Science and Engineering, Vol. 86, pp. 372-387 1984.
15. Fletcher, D. "Buoyancy Driven Transient, Two-Dimensional Thermo-Hydrodynamics of a Melt-Water-Steam Mixture, Rep. CLM-P790, Culham UKAEA, 1986.
16. Schins, H. and Gunnerson, F. S. "Boiling and Fragmentation Behavior During Fuel-Sodium Interactions", Nuclear Engineering Design, Vol. 91, pp. 221-235 1986.

Effect of Fluid Flow Velocity on the Fragmentation Mechanism of a Hot Melt Drop

G. Ciccarelli* and D. L. Frost†
McGill University, Montreal, Quebec, Canada

Abstract

The effect of flow velocity on the fragmentation of hot molten drops (tin and a low melting point alloy have been studied) immersed in water has been investigated experimentally with high-speed photography and flash X-ray radiography. For low flow velocities (i.e., $V \sim 5$ m/s or weber no. ~ 150) the drop interaction results in the formation of a vapor bubble which undergoes cyclic expansion/collapse phases. In this case the drop fragmentation is governed by thermal effects and hydrodynamic fragmentation due to relative velocity between the drop and coolant plays no role. For high flow velocities (i.e., $V \sim 50$ m/s, weber no. ~ 15,000) the fragmentation of a *cold* liquid metal drop is a result of hydrodynamic stripping of the upstream surface of the drop and the penetration of long wavelength waves due to the incident water flow. For hot drops subject to high flow velocities, the growth rate of the *projected area* of the dispersed drop fragments is similar to that of a cold drop. However, X-ray photography reveals that thermal effects still play a role at high velocities, i.e., a vapor bubble is generated and radial jets of melt are visible protruding from the drop surface in the first several hundred microseconds. However, the high ambient flow velocity inhibits the bubble growth and high-speed vapor flow within the bubble enhances the fragmentation rate.

I. Introduction

The direct contact of a volatile liquid (or coolant) such as water with a hot liquid (e.g., molten metal) whose temperature is significantly above the boiling

* Graduate Student, Mechanical Engineering Department.

† Assistant Professor, Mechanical Engineering Department.

temperature of the coolant can result in a vapor explosion. A vapor explosion involves the rapid production of vapor as a result of rapid heat transfer from the melt to the coolant and subsequent phase transition of the superheated coolant. Enhanced heat transfer is achieved through the fragmentation of the melt and hence an enhancement of the surface area available for heat transfer. The expansion of the high-pressure vapor generated can produce strong compression waves and accelerate the coolant ahead placing any surrounding structures under hydrodynamic loading. Such explosions occur frequently in metal foundries as a result of the accidental mixing of molten metal and cooling water. The nuclear industry has also studied the possibility of a strong vapor explosion causing failure of the reactor containment in the unlikely event of a core meltdown. An extensive summary of the occurrence of vapor explosions in different industries can be found in Reid.[1]

Perhaps the most important process in a vapor explosion is the fine fragmentation of the melt which governs the overall heat transfer rate. Since the rate of fragmentation is generally much slower than the heat transfer or evaporation rates, it is therefore the rate determining step in a vapor explosion. In the past there have been many fragmentation models proposed to describe the fine fragmentation of the melt in vapor explosions (see the recent comprehensive review by Fletcher and Anderson[2]). These fragmentation models can be grouped into two large classes, based on either *thermal* or *hydrodynamic* effects. In thermal-fragmentation mechanisms the energy dissipated in the melt breakup is derived from the internal energy of the melt itself. In many thermal-fragmentation models it is assumed that the coolant is introduced into the melt by the formation of re-entrant jets during film collapse[3,4] which are subsequently superheated. The evaporation of the encapsulated coolant droplets produces high-pressure vapor which upon expansion generates a hydrodynamic flowfield which tears the drop apart. Drumheller[5] has proposed that fragmentation may be caused directly by the symmetrical impact of the collapsing film on the drop surface. In other thermal-fragmentation models, the local production of high-pressure vapor following film collapse causes fragmentation by the production of annular jets ("splash theory" of Ochiai and Bankoff[6]) or by Rayleigh-Taylor instability of the melt surface (Corradini[7]). Thermal-fragmentation processes are often cited as the source of the fine fragmentation in spontaneous or self-triggered vapor explosions where the effective flow velocity is the drop terminal velocity which is typically of the order of 1 m/s.

In so-called hydrodynamic-fragmentation mechanisms, the breakup occurs as a result of the acceleration of the melt drops due to the relative velocity of the coolant. The relative velocity causes fine melt fragments to be stripped from the surface of the drop. A boundary-layer stripping model was first proposed by Taylor.[8] In this model the flow of the coolant induces a boundary-layer flow in the vicinity of the upstream surface of the melt drop. When the melt mass in the drop boundary layer is convected to the equator of the drop it breaks away from the surface, causing a continuous stripping of the melt drop surface. Other hydrodynamic-fragmentation models are based on the stripping of wave crests generated on the upstream surface of the drop due to Rayleigh-Taylor and Kelvin-Helmholtz instabilities.[9] The breakup of rain drops, for example, in a strong air flow (e.g., behind a shock wave) is usually attributed to this type of hydrodynamic fragmentation. A large body of literature exists on the differential velocity fragmentation of drops (see Pilch and Erdman[10] for a review of experimental data). To apply one of the hydrodynamic-fragmentation mechanisms to the propagation of a vapor explosion it is necessary to have a detailed knowledge of the coolant flowfield.

In a typical large-scale accident scenario a mass of melt is released into a tank of water. As the melt falls through the coolant it breaks up into fragments of the order of centimeters in size creating a "coarse mixture" of melt and coolant. During this mixing process the melt is surrounded by a vapor film which limits the heat transfer to the coolant. If the vapor film surrounding one or more fragments is somehow destabilized causing liquid-liquid contact, a localized vapor explosion can occur involving only a small region of the coarse mixture. This local interaction will generate a high pressure vapor region within the mixture. The expansion of this local high-pressure region will generate a pressure wave and convective flow which will cause the collapse of the vapor film around adjacent melt fragments. In this way the sequential explosion of the fragments can lead to propagation of the interaction throughout the remainder of the mixture. If this propagation were to escalate with the pressure wave steepening to form a shock wave, a coherent energy release wave may develop. In this escalation process, the melt drops at the explosion front would be subjected to a spectrum of coolant flow velocities as the front accelerates. During the early stage of initiation coolant flow velocities of the order of meters per second would be encountered while at the other extreme of the escalation process flow velocities of hundreds of meters per second would exist behind the leading shock wave of a fully developed coherent propagating wave.

Board and Hall[11] have proposed that the propagation of a coherent vapor explosion wave is analogous to that of a classical chemical detonation wave. Existence of such a thermal detonation wave has not been confirmed experimentally. If such a thermal detonation can exist, an escalation process is required because, in general, a trigger strong enough for direct initiation does not exist in accident scenarios. However, to model the escalation process it is necessary to provide detailed models for the complex processes that occur within the transient multiphase vapor explosion.

A number of researchers have proposed models of the *escalation* process using transient detonation codes.[12-16] In the past, with the exception of Oh and Corradini[12] who developed a thermal-fragmentation model for their code, these numerical codes relied solely on hydrodynamic-stripping-type-fragmentation models. It is likely during the escalation process that thermal-fragmentation mechanisms will play an important role, particularly for low energy melt-coolant mixtures. Burger et al.[14] attempted to model tin/water experiments performed on the KROTOS facility at ISPRA using their transient code with only a hydrodynamic-fragmentation mechanism. The numerical simulation produced much lower pressures compared with the pressures recorded during the experiments. They postulated that this disagreement was due to the lack of a thermal-fragmentation model in their code. By including a simple ad hoc thermal-fragmentation model in their code better agreement was obtained with the experimental results.

Although several investigators have looked at the hydrodynamic breakup of a cold liquid drop in a liquid medium at elevated flow conditions, no experimental results are available showing the effect of flow velocity on the fragmentation of a *hot* molten drop where thermal effects will also play a role. In particular, there exists no experimental results to describe the transition from a thermal- to a hydrodynamic-type-fragmentation mechanism as the flow velocity is increased. None of the existing thermal-fragmentation models incorporate the effect of coolant flow, and therefore according to these models, thermal-fragmentation effects are not inhibited at high flow velocities. Whether thermal fragmentation acts independently of the hydrodynamic fragmentation and whether they mutually reinforce or suppress one another have not been investigated.

The present experiments investigate how the fragmentation process of a hot drop is influenced by coolant flow velocity. In particular, emphasis is placed on determining if there is a transition from thermal fragmentation at low flow velocities to hydrodynamic fragmentation at high flow velocities, and if so, what is the nature of this transition? The coolant flow is generated by the piston action of the vapor bubble generated by an underwater spark discharge or electric detonator. The interaction between the melt drop and the coolant is observed through high-speed photography and X-ray radiography.

II. Experimental

The experimental facility, shown in Fig. 1, consists of a 10-cm-wide, 25.5-cm-long and 12.5-cm-high aluminum tank. Windows are located on all four sides so as to allow simultaneous flash X-ray and regular high-speed photography of the drop interaction. The width of the tank was chosen to provide the best contrast on the X-ray film, and Lexan windows were used for the X-ray radiographs to minimize absorption of the x rays. The tank has a heavy walled vertical cylinder (3.8 cm i.d., 18 cm long) welded onto the base which houses the triggering system. Two drop materials were used: tin and a low melting point alloy ($T_{m.pt.}$ = 49 °C) consisting of 45% bismuth, 23% lead, 19% indium, 8% tin, and 5% cadmium (commercially called "cerrolow"). The use of the alloy material was convenient in that both hot- and cold-drop experiments (cold refers to the condition when the drop is at the same temperature as the water) could be carried out with the same drop material. The drop is heated in a small furnace located on the top of the tank. The furnace consists of a hollow graphite cylinder lined with a nichrome heating coil. The graphite cylinder has a bottle neck at the bottom which is plugged by a vertical stainless-steel rod. The temperature of the oven is monitored by a thermocouple which is placed inside the central rod. The drop is released into the water by the raising of the rod. The interaction is triggered when the drop intercepts a laser beam directed through the tank by a low-power He-Ne laser. The laser beam is focused onto a photodiode, and with a simple electrical circuit the triggering system is activated when a drop in light intensity is measured by the photodiode. The trigger system consists of a high-voltage capacitor discharge circuit in which either an exploding wire or blasting cap is used to generate the blast wave and flowfield. The blasting cap contains a 0.1-g lead azide primary charge and a 0.25-g-PETN base charge. An 8-μf (30-kV maximum charging voltage) capacitor is used in conjunction with the exploding wire while a smaller 2-μf capacitor charged to 3 kV is used to initiate the blasting cap. The exploding wire and blasting cap are located at the bottom of the cylinder, as shown in Fig. 1. The interaction is triggered when the drop is 2 cm above the tube exit.

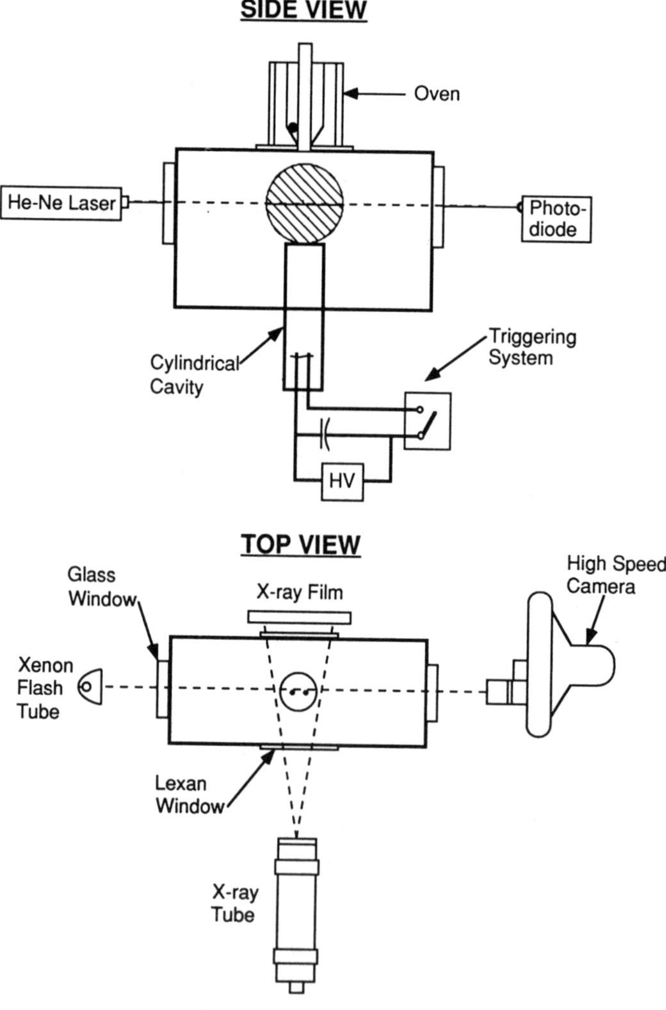

Fig. 1 Schematic of experimental apparatus

The initiation of the exploding wire or blasting cap generates a high-pressure vapor bubble which upon expansion produces a shock wave that runs ahead of the bubble surface up the tube. Since the inner diameter of the tube is fairly small, the bubble interface rapidly becomes planar as it propagates up the tube. High-speed photographs show that the bubble interface is planar after it emerges from the tube. If we assume that evaporation and condensation of the bubble interface is negligible during this time it can be treated as a massless piston. In this case due to the planar nature of the interface and the proximity of the drop to the tube exit, the flow velocity which the drop would be subjected to will be very close to the bubble velocity. By varying the capacitor voltage one can obtain a range of flow velocities with a maximum of about 60 m/s for the blasting cap. As the bubble surface emerges from the tube it decelerates reaching a steady-state velocity. The time required to achieve a steady-state velocity varies depending on the strength of the trigger. The bubble surface position as a function of time is shown in Fig. 2. In each case, a straight line is fit to the steady-state portion of the curves. For the strongest trigger (i.e., the blasting cap), a steady-state velocity of 53.9 m/s is achieved almost immediately after the bubble emerges from the tube. In the case of a 3-kV discharge across an exploding wire (i.e., the weakest trigger), the final steady-state velocity of 0.53 m/s is achieved after 1.5 ms. For the intermediate triggers the deceleration times lie between

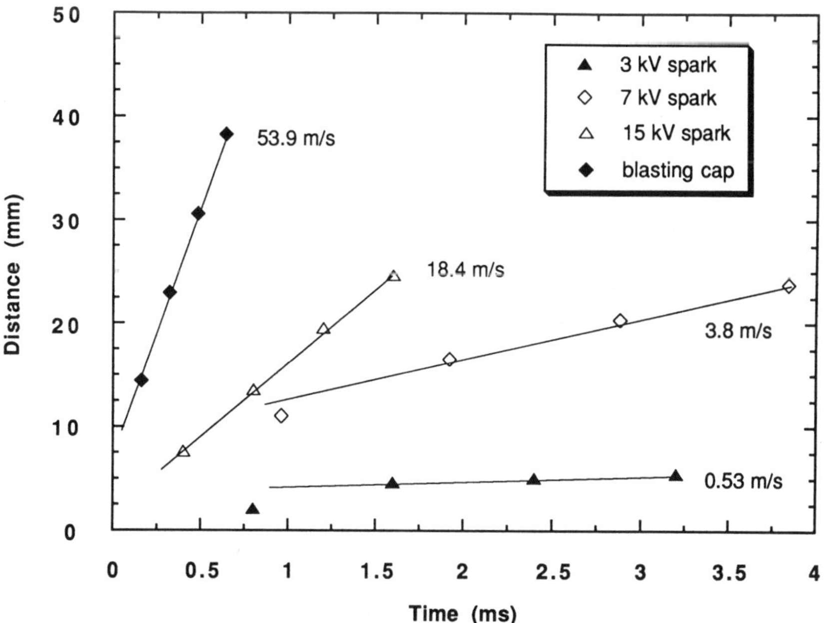

Fig. 2 Time history of bubble surface position relative to the outlet of the vertical triggering tube.

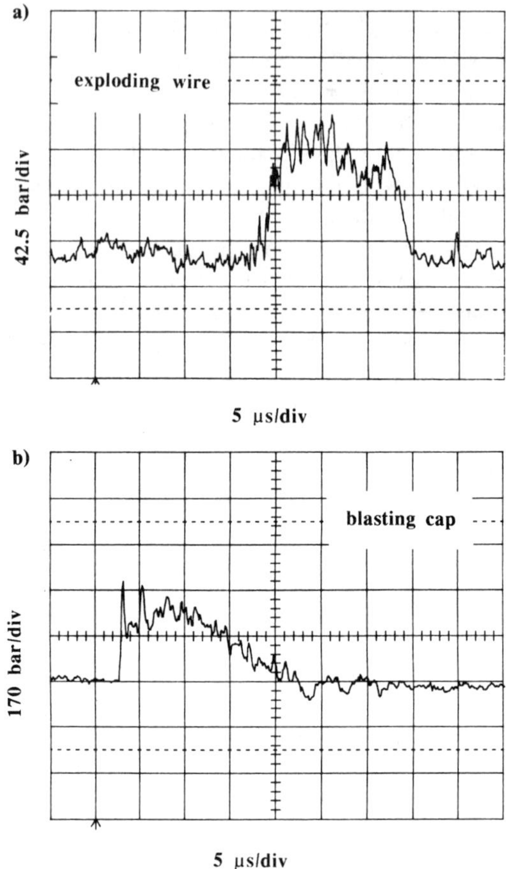

Fig. 3 Pressure traces obtained at drop location for a) exploding wire discharge, 8 µf, 15 kV and b) blasting cap.

these two limits. The steady-state value will be used to characterize the flow velocity in the remainder of the paper unless otherwise stated.

A Cordin 16-mm rotating drum camera was used to record the drop/water interaction at a framing rate of 25,000 fps. Back lighting is provided by a 6-ms duration xenon flash tube. A single head 150-kV Scandiflash X-ray system with a 35-ns X-ray pulse duration was also used in conjunction with the high speed camera for a series of hot- and cold-drop experiments at a flow velocity of about 40 m/s. Kodak XAR-5 high-speed X-ray film was used in conjunction with Kodak Min-R intensifying screens. The shock pressure is recorded using a IMOTEC polyvinyldifluoride (pvdf) pencil-type pressure transducer (90-ns rise-time). Figure 3 shows a typical pressure time history at the drop position for the case of an exploding wire (15-kV capacitor charge) and a blasting cap. The peak pressure and positive phase duration are 12.8 MPa and 16 µs for the exploding wire and 37.5 MPa and 18 μs for the blasting cap. Because of the short duration of the flow velocity associated with the blast wave, the hydrodynamic breakup of a drop is governed primarily by the flow generated ahead of the expanding trigger bubble.

III. Experimental Results

A. Cold-Drop Experiments

Before considering the effect of flow velocity on a hot drop, it is useful to first characterize the influence of flow velocity on the fragmentation of a cold drop. Since the same drop material is used in both the hot- and cold-drop experiments (i.e., cerrolow alloy) a direct comparison of the two can give insight on the relative importance of thermal- and hydrodynamic-fragmentation mechanisms during the explosion of a hot drop. For the cold-drop experiments the drop is released from below the water surface so that the drop and water are in thermal equilibrium (the water temperature is maintained at 60 °C). This method of drop release eliminates the possibility of a trailing gas bag which is characteristic of above-surface drop release. Nevertheless, it was observed that a trailing gas bag had no noticeable effect on drop breakup. In general, the drop takes on an ellipsoidal shape as it descends in the water. Since the drop is usually slightly tilted to one side as it falls, its shape appears different in the photographs from one trial to the next, depending on the drop orientation. When describing drop fragmentation it is common to characterize the flow disturbance with the weber number, or the ratio of inertial to surface tension forces. Here the weber number is defined as $\rho u^2 r_0/\sigma$, where ρ is the density of the fluid medium, u the flow velocity, r_0 a characteristic length scale for the drop (taken to be the initial major radius of the ellipsoid), and σ the surface tension of the drop. Since the precise surface tension of the cerrolow alloy is not available, a value of 0.455 N/m will be used which is based on the mass-weighted average of the surface tension of the components. The terminal velocity of the drop falling in the water is about 1 m/s, and this value is added to the flow velocity to calculate the weber number.

Figures 4a-4d show a sequence of photographs illustrating the fragmentation process of a 0.5-g cold alloy drop at four different water flow velocities. The protrusion which is visible in the lower portion of the first photograph in each sequence is the head of a bolt which is attached to the top of the vertical tube. The dark region at the bottom of the photographs corresponds to the bubble generated by the trigger system. The photographs are arranged to show the time history of both the drop and bubble surface position. The variation of bubble position with time shown in Fig. 2 corresponds to the trials shown in Figs. 4a-4d. For the lowest flow velocity of 0.5 m/s (we = 14.5) obtained using an exploding wire with a 3-kV capacitor charge (Fig. 4a) it is apparent that on the timescale of milliseconds the drop undergoes little fragmentation. Since the weber number is just above the critical weber number of about 17 (Tan and Bankoff[17]), the drop undergoes simple oscillatory motion. Figure 4b shows the

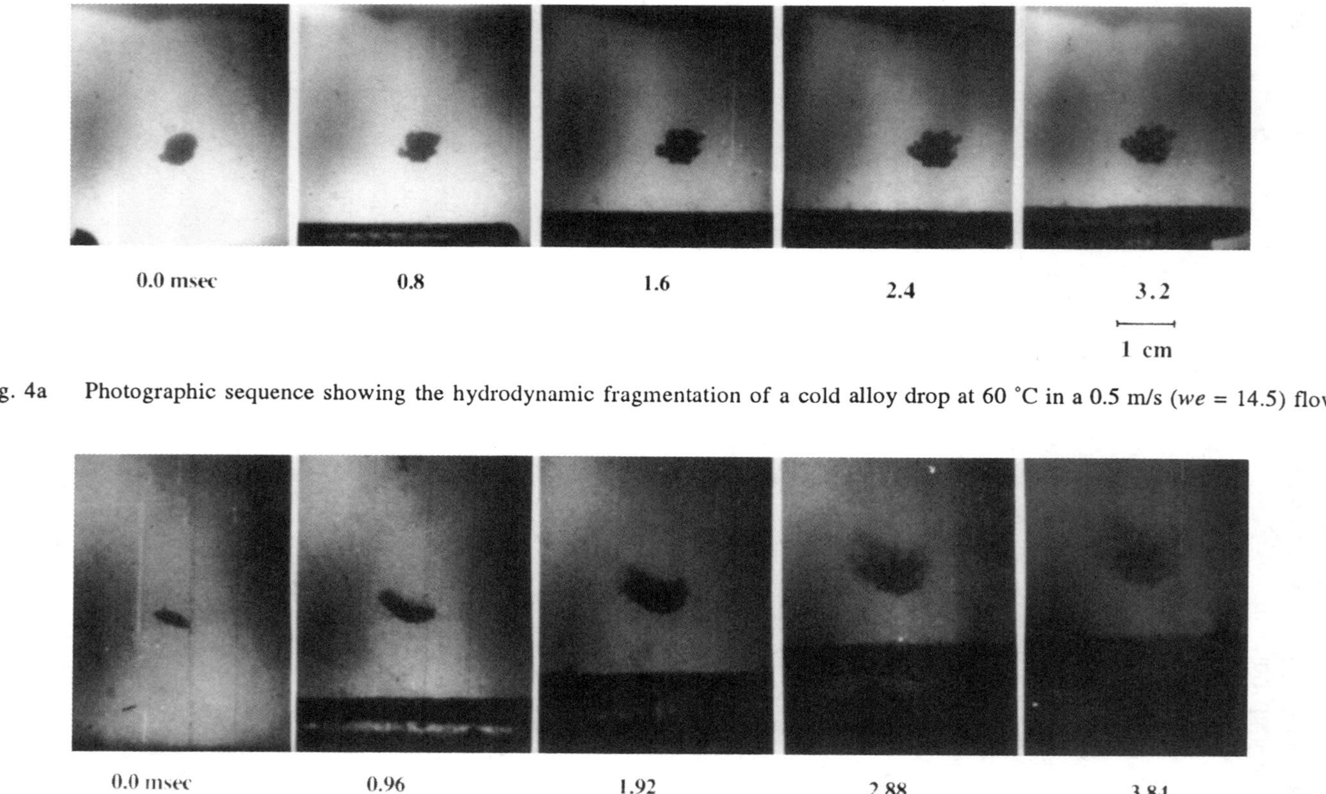

Fig. 4a Photographic sequence showing the hydrodynamic fragmentation of a cold alloy drop at 60 °C in a 0.5 m/s (we = 14.5) flow.

Fig. 4b Photographic sequence showing the hydrodynamic fragmentation of a cold alloy drop at 60 °C in a 3.8 m/s (we = 94) flow.

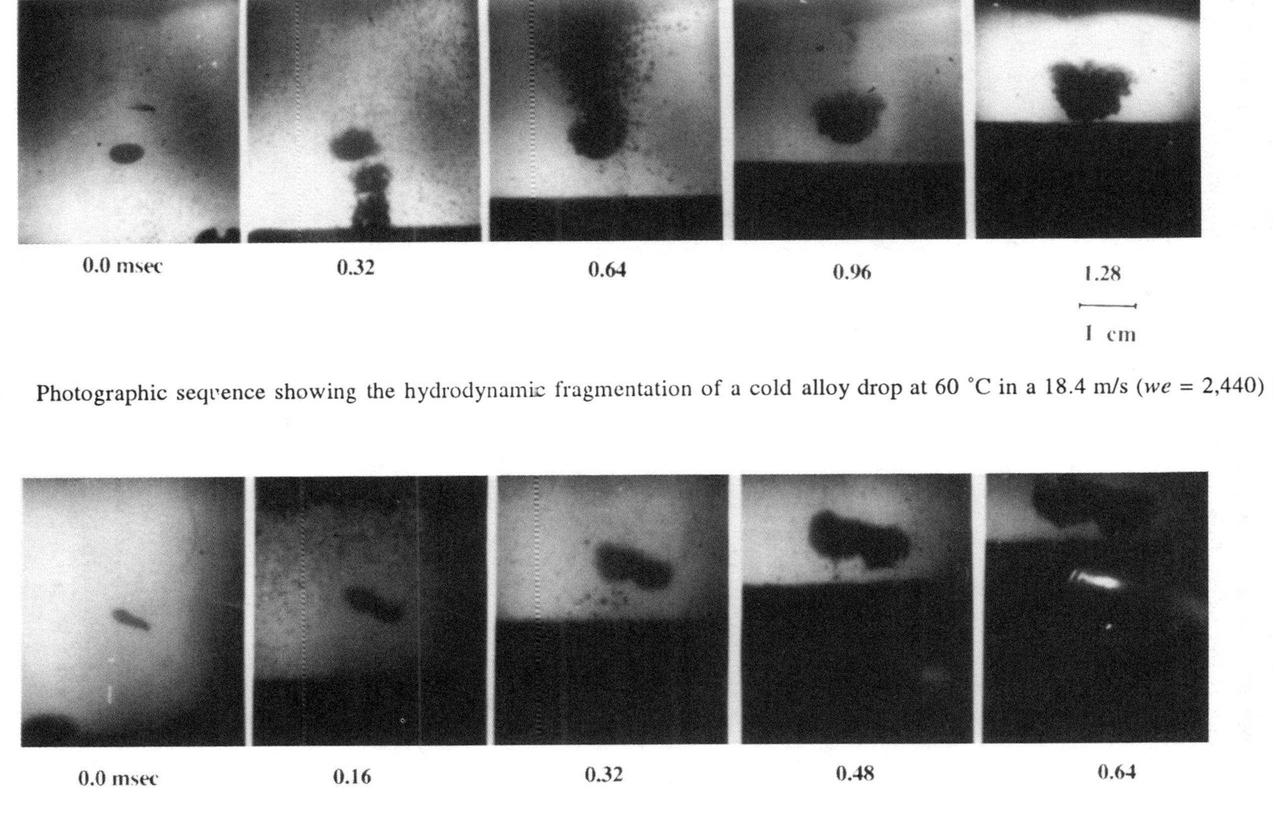

Fig. 4c Photographic sequence showing the hydrodynamic fragmentation of a cold alloy drop at 60 °C in a 18.4 m/s (we = 2,440) flow.

Fig. 4d Photographic sequence showing the hydrodynamic fragmentation of a cold alloy drop at 60 °C in a 53.9 m/s (we = 13,442) flow.

fragmentation process for a flow velocity of 3.8 m/s (*we* = 94). After 3 ms the projected area of the fragmentation zone has nearly quadrupled compared to the initial drop area. During the first 2 ms the drop surface is drawn off producing long filaments extending from the drop on the downstream side of the drop. After about 3 ms fine particles can be seen to form downstream of the drop as a result of the breakup of the long filaments. This type of behavior was observed by Tan and Bankoff[17] using mercury drops in water and Kim et al.[21] with gallium drops in a water flow. The upstream surface of the drop shows only minimal signs of roughness caused by the growth of instabilities. There is also very little lateral expansion of the drop as a result of drop deformation. Figure 4c shows the drop breakup in a 18.4 m/s (*we* = 2440) flow. Shortly after the initiation of the exploding wire (up to 0.640 ms), cavitation bubbles can be seen to form throughout the bulk of the water as a result of the sharp expansion behind the blast wave generated by the exploding wire. The breakup is again characterized by a drawing off of the drop surface at the equator, only in this case there is little evidence of the filament structure downstream of the drop. Instead the drawn off surface material quickly breaks up into small fragments. At this weber number the upstream surface roughening is much more pronounced.

For the maximum flow velocity in this series of experiments of 53.9 m/s (*we* = 13,442), obtained using a blasting cap (Fig. 4d), the breakup is much faster and more violent. Fine fragments are immediately torn off the drop surface. The fragmentation rate is so high that the fragment mist behind the drop appears opaque on the photographs. Because of the initial flat shape of the drop, the lateral growth observed in this trial is a result of fragments from the upstream surface being convected parallel to the surface rather than due to the deformation of the parent drop. From the high-speed regular photography for the high weber number breakup, an indentation appears to grow on the upstream surface of the drop near the center of the drop dividing the parent drop into two sections. This behavior is usually referred to as "catastrophic breakup" and was also observed by Reinecke and Waldman[18] for weber numbers above 30,000 for water droplets in an air stream. The catastrophic breakup can be attributed to the growth of large wavelength waves on the upstream surface of the drop. The drop is severed when the amplitude of these waves becomes comparable to the drop width.

1. Estimation of Drag Coefficient

A second series of cold-drop experiments was carried out using smaller 0.25-g drops of gallium to have a more spherical and reproducible initial drop shape. For this series of experiments only high-speed regular photography was used to record the fragmentation process. The photographic results for the gallium drops are similar to those obtained for the alloy drops shown in Figs. 4a-4d. It is of interest to calculate the effective drag coefficient C_D exhibited by the drop. The drag coefficient will be different from that of a solid since the drop continually deforms and fragments as a result of the water flow. The Reynolds number in all gallium drop experiments varied from 60,000 to 200,000. The drag coefficient of a sphere in this Reynolds number range is about 0.5. Figure 5 shows the displacement of the drop x, normalized with the initial drop diameter perpendicular to the flow d, as a function of a nondimensional time T for several experiments. Here the nondimensional time is defined as

$$T = \frac{\bar{U}\,t}{d}\sqrt{\frac{\rho_f}{\rho_d}} \tag{1}$$

where ρ_d and ρ_f are the density of the water and drop, respectively, and $\bar{U}$ is the average flow velocity. The graph in Fig. 5 shows that all of the curves for the different experiments collapse approximately to a single curve with the nondimensional parameters chosen. From Newton's Law one can readily derive the following expression for the drag coefficient for a solid object in a flow in terms of the displacement time history:

$$\frac{x}{d} = \frac{T}{\sqrt{\rho_f/\rho_d}} - \frac{4}{3C_D\rho_f/\rho_d}\ln\left\{1 + \frac{3C_D}{4}T\sqrt{\frac{\rho_f}{\rho_d}}\right\} \tag{2}$$

The experimental data can be fit roughly using the above expression with a drag coefficient of 4 (see Fig. 5). Using early times (up to a nondimensional time of

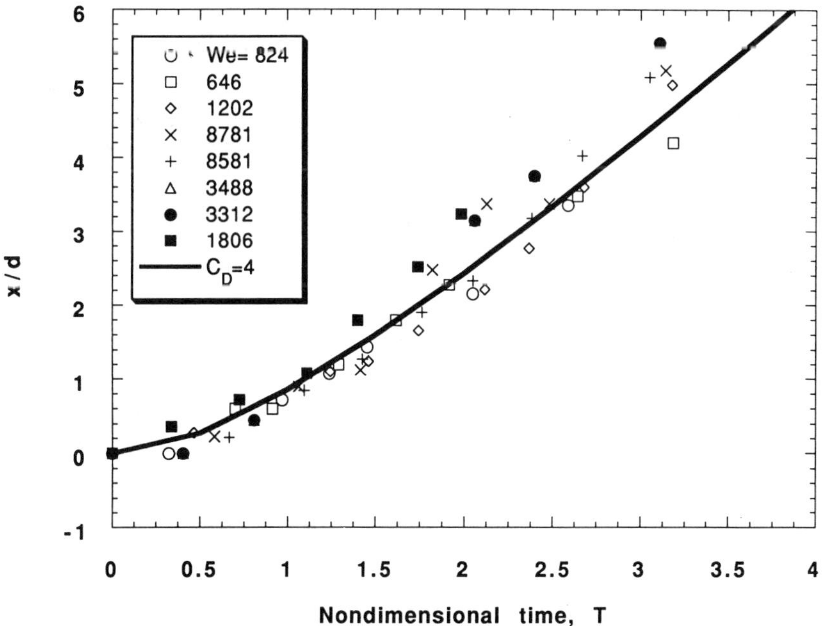

Fig. 5 Dimensionless drop displacement vs nondimensional time for cold-drop gallium experiments

1) when drop deformation and breakup are not as advanced, a constant drag coefficient between 2 and 3 is more representative. This value is higher than that of a solid sphere because the drop more closely resembles an oblate spheroid as a result of aerodynamic drop flattening and the effective projected area of the drop increases with time as a result of drop fattening. This value of the drag coefficient for early times is in agreement with that reported by Simpkins and Bales[19] who fit their water drop/air data with a drag coefficient of 2.5 and Patel and Theofanous[20] who fit their mercury drop/water data with a drag coefficient slightly less than 2.5.

2. Estimation of Nondimensional Breakup Time

Since regular photography was used to film the drop breakup, direct observation of complete drop fragmentation was not possible. The only information which can be obtained from the high-speed film concerns the size and shape of the silhouette or profile of the cloud of fine fragments stripped from the drop surface. In the past there have been various criteria used to arbitrarily define the size of the silhouette corresponding to the time of total drop breakup. For example, Patel and Theofanous[20] used the doubling of the initial drop diameter and Kim et al.[21] used the quadrupling of the silhouette area along with several other criteria to define the total fragmentation time of the drop. To compare with past results,

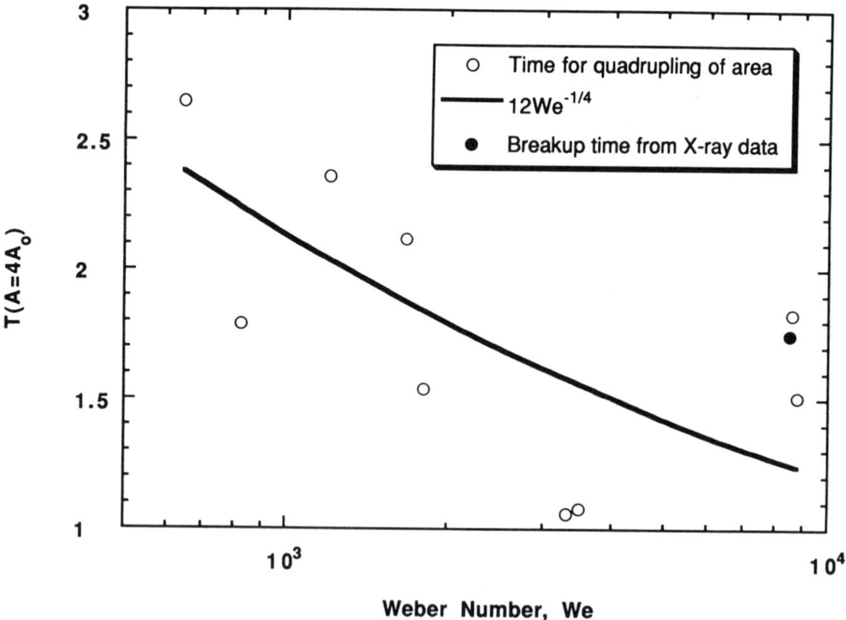

Fig. 6 Nondimensional time for the quadrupling of the fragmentation cloud projected area from the photographic results of the cold-drop gallium experiments

we have calculated the time history of the silhouette area for each trial by digitizing the photographs and using image analysis software to calculate the area. Figure 6 shows the nondimensional time for quadrupling of the initial drop projected area as a function of the weber number for each experiment. The single solid circle represents the time for total fragmentation obtained from the X-ray radiographs, which will be discussed later, using 0.4-g alloy drops from Fig. 9. This point fits well with the trend of the quadrupling of the projected area obtained from the gallium experiments. On the average the nondimensional time for quadrupling of the area is between 1.5 and 2 but there does appear to be a decreasing trend with weber number. The data can be well correlated with

$$T_{A} = 4A_{0} = 12\ \mathrm{We}^{-1/4} \tag{3}$$

This dependence of nondimensional breakup time with weber number has also been observed experimentally in Refs. 18-20. Reinecke and Waldman[18] pointed out that a similar weber number dependence results from the characteristic amplification time for the fastest growing Rayleigh-Taylor instability wave. Consequently, they attributed their observed drop breakup to catastrophic drop breakup, caused by drop piercing of long wavelength Rayleigh-Taylor instability waves. Different proportionality constants have been obtained by the different investigators due to the different criteria used to define total breakup. Patel and Theofanous[20] obtained a constant of proportionality of 1.66 based on the doubling of the initial drop diameter, Simpkins and Bales[19] obtained a value of 22 based on the first appearance of waves on the upwind surface of the drop and 65 for total fragmentation, and Reinecke and Waldman[18] fit their X-ray data for total fragmentation with a constant of 45. However, other investigations of drop fragmentation have found no clear dependence of breakup time with Weber number. For example, Kim et al.[21] found $T_A = 4A_0 \sim 2$ for gallium drops and Baines and Buttery[22] reported $T_{\mathrm{breakup}} \sim 4$.

B. Hot-Drop Experiments

A series of experiments was carried out to investigate the interaction of a hot alloy drop, initially at 700 °C, immersed in water at 60 °C and subject to different flow velocities. Figures 7a-7d show a sequence of photographs taken during four such experiments. For a flow velocity of 0.5 m/s (we = 14.5) the interaction results in the formation of a vapor bubble centered at the drop. The vapor bubble expands to a maximum size after about 1 ms. The dispersed fragments from the initial interaction obscure the bubble collapse. The degree of fragmentation is extensive at 2 ms when the vapor bubble has collapsed leaving fine scale fragments dispersed in the surrounding liquid. From Fig. 4a it is apparent that for a cold drop under similar flow conditions at the same time, that fragmentation is minimal. Based on this comparison, it is evident that the fragmentation mechanism in a hot-drop interaction, under this flow condition, is domi-

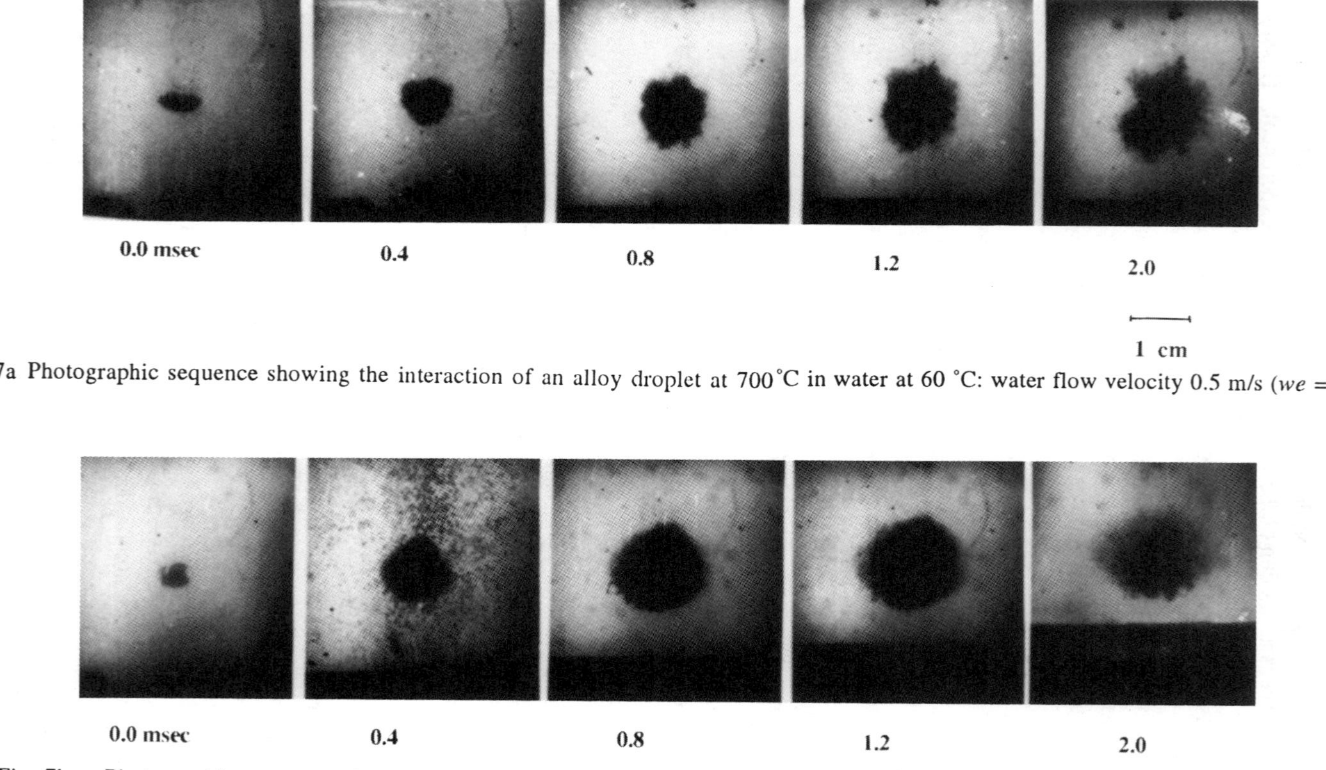

Fig. 7a Photographic sequence showing the interaction of an alloy droplet at 700°C in water at 60 °C: water flow velocity 0.5 m/s (we = 14.5).

Fig. 7b Photographic sequence showing the interaction of an alloy droplet at 700 °C in water at 60 °C: water flow velocity 4.8 m/s (we = 186).

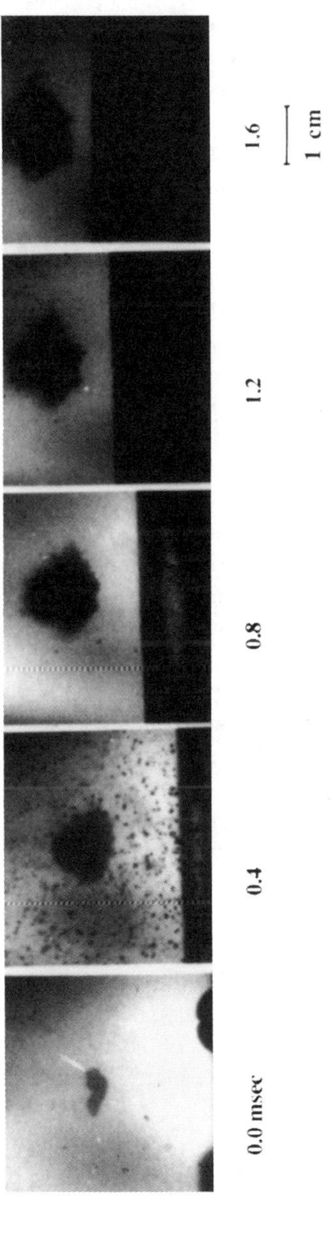

Fig. 7c Photographic sequence showing the interaction of an alloy droplet at 700 °C in water at 60 °C: water flow velocity 17 m/s (*we* = 2654).

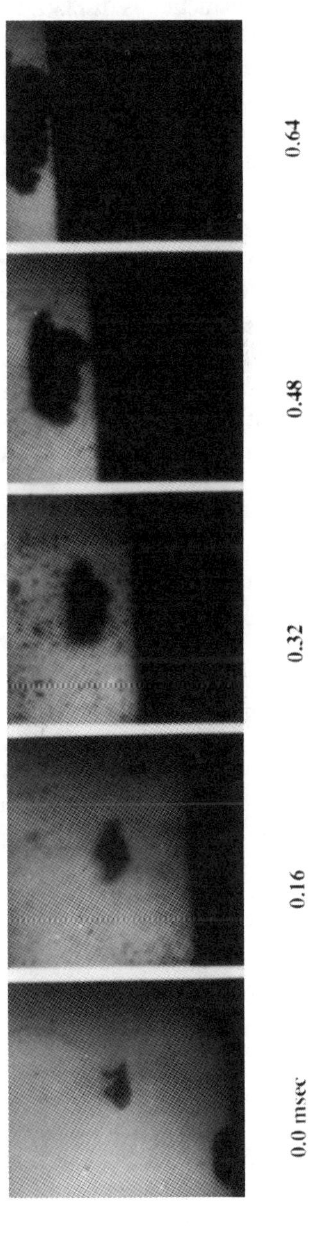

Fig. 7d Photographic sequence showing the interaction of an alloy droplet at 700 °C in water at 60 °C: water flow velocity 56.7 m/s (*we* = 14,760).

nated by thermal effects with little contribution from hydrodynamic fragmentation resulting from the incident flow.

For a higher flow velocity the vapor bubble formed around the drop as a result of the interaction is displaced upward relative to the drop in the direction of the flow. This bubble migration can be seen in the interaction shown in Fig. 7b for a flow velocity of 4.8 m/s (*we* = 186). The vapor bubble initially expands symmetrically ($t < 0.240$ ms), but at later times the bubble expands preferentially upward. As a result of the deflection and separation of the flow around the vapor bubble there exists a higher pressure on the upstream surface of the bubble. At early times the bubble can expand fairly symmetrically because the vapor pressure inside the bubble is higher than the pressure in the water just ahead of the lower bubble surface. As the vapor pressure drops, due to bubble expansion, the advancement of the bottom bubble surface is impeded due to the elevated pressure. A comparison of the time scales for hydrodynamic fragmentation of a cold drop (see Fig. 4b) with the hot drop shows that hydrodynamic fragmentation is too slow to account for the breakup of the hot drop. Even at this moderate velocity the presence of the vapor bubble surrounding the drop negates any contribution from the hydrodynamic-stripping process which is present for the cold alloy drop at the same flow velocity.

Figure 7c shows the interaction of a hot alloy drop with a flow velocity of 17 m/s (*we* = 2654). From the photographs there is no evidence of vapor formation on the upstream surface of the drop, however, a vapor bubble does form in the wake region. The vapor bubble grows to a maximum size after about 800 μs and then collapses. The asymmetric collapse of this bubble causes further fragmentation at the downstream surface of the drop. From the sequence of photographs it is not clear if the evaporation is completely suppressed at the bottom of the drop or if the high-pressure vapor is restricted from expanding as a result of the high flow velocity. At 0.4 ms the bottom of the dark cloud is fairly smooth but by 0.8 ms the bottom surface becomes highly corrugated indicating stripping of the drop surface. A comparison of the hot-drop interaction in Fig. 7c with the cold-drop breakup at a similar flow velocity reveals that the fragmentation of the upstream surface of the drop is very similar to that in the cold drop (see Fig. 4c) for times later than 0.4 ms. The growth of the vapor bubble above the drop also disperses the fragments stripped from the front of the drop farther downstream than would be possible without a vapor bubble due to the low viscosity of vapor compared to water. The combined fragmentation of the drop due to the stripping of the upstream surface and the fragmentation of the back surface due to rapid evaporation and bubble collapse leaves the wake region littered with fragments at later times not shown in the figure.

Figure 7d shows the interaction of a hot drop with a flow velocity of 57 m/s (*we* = 14,760). The behavior of the drop is very similar to the breakup of the cold alloy drop at the same flow conditions, as shown in Fig. 4d. There is strong lateral expansion due to the stripping of fragments from the upstream surface of the drop and the drop appears to be in the process of dividing into several smaller drops.

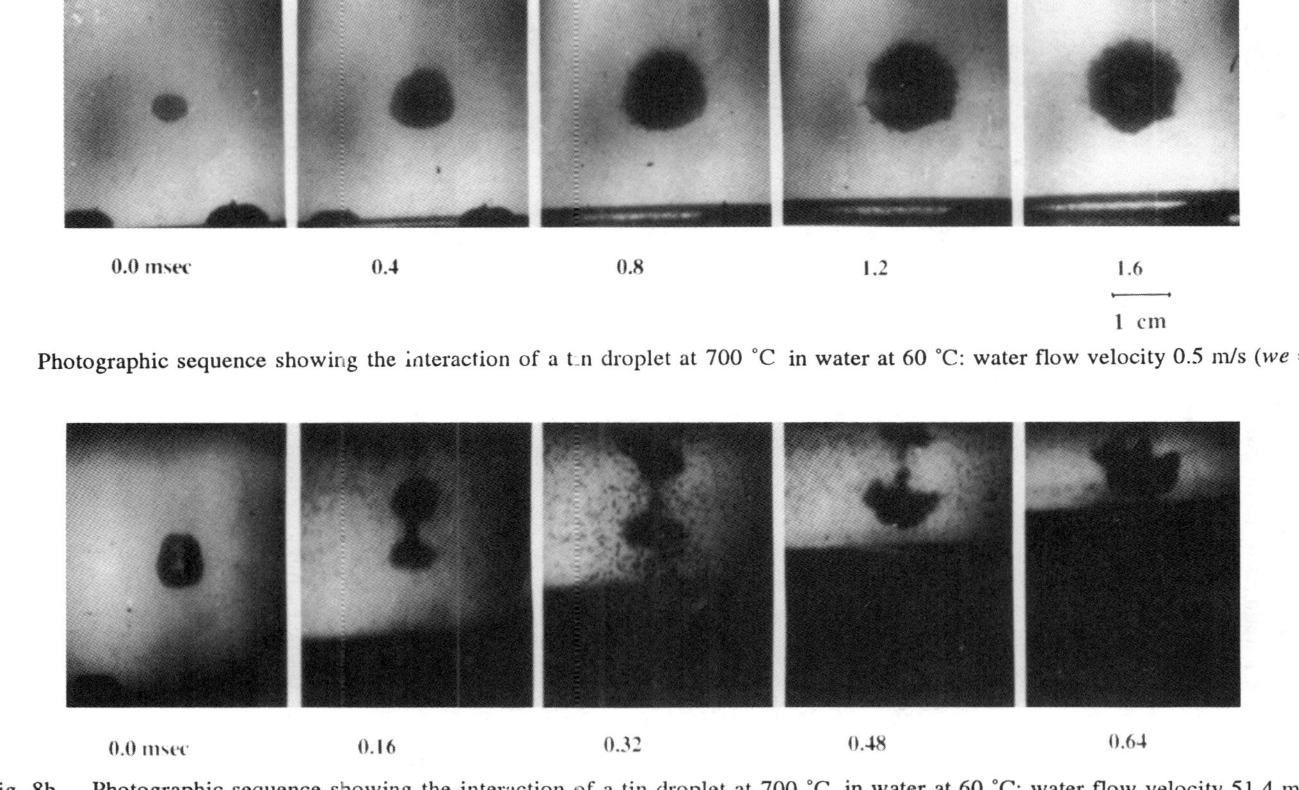

Fig. 8a Photographic sequence showing the interaction of a tin droplet at 700 °C in water at 60 °C: water flow velocity 0.5 m/s (we = 14.5).

Fig. 8b Photographic sequence showing the interaction of a tin droplet at 700 °C in water at 60 °C: water flow velocity 51.4 m/s (we = 18,500).

A similar series of hot-drop experiments was carried out using tin as the drop material. Figures 8a and 8b show the results for the two extreme flow velocity cases. For a flow velocity of 0.5 m/s the tin-drop interaction is unaffected by the flow. Characteristically a vapor bubble is formed which expands to a maximum and then collapses. This bubble expansion and collapse is very similar to that observed in the hot-alloy experiments. For the case of a 51-m/s flow velocity, shown in Fig. 8b, a large gas bag is dragged down behind the drop. However, the flow strips the bag away and the drop fragmentation appears to be unaffected by its presence. The tin drop rapidly fragments with no obvious physical sign of vapor generation, in a fashion similar to the hot and cold alloy drops at this flow velocity.

C. X-ray Visualization of Fragmentation Process

Flash X-ray visualization has been used to study the fragmentation of 0.5-g tin drops at an initial temperature 700 °C in a water flow of about 4 m/s (Frost et al.[23]). From the radiographs it was observed that following film collapse and during the first bubble expansion spikes of tin formed extending radially out from the drop surface. This surface distortion continues until the melt and water come back into contact at the time of bubble collapse. The hydrodynamic forces generated at the drop surface by the inrushing water and explosive evaporation lead to the complete fragmentation of the already highly distorted drop. Under similar flow conditions the fragmentation of a cold alloy drop was characterized by surface stripping. The heating of the alloy drop to 700 °C resulted in a similar fragmentation behavior as the heated tin drops. Based on these observations it was concluded that under these drop and flow conditions the fragmentation process was solely driven by thermal effects.

In the present study flash X-ray photography was used in a series of experiments using blasting caps to generate flows of the order of 40 m/s. Both hot- and cold-drop conditions were studied to determine if the fragmentation mechanism in both cases was the same, as suggested by the high-speed regular photographs presented in Figs. 4d and 7d. Shown in Fig. 9 is a series of radiographs taken at different times in the fragmentation process of a 0.4-g cold alloy drop. Shown below each photograph is the dimensional and nondimensional (in parentheses) time after the arrival of the triggering shock wave at the drop location. The nondimensional time (see Eq. (1)) is calculated using the average velocity of the flow up to the time at which the X-ray photograph was taken. The average weber number for these trials is about 8500.

At time zero the drop takes on an ellipsoidal shape which is characteristic of the cold drop alloy trials. The initial drop shape can vary from experiment to experiment and thus the resulting shape of the fragment cloud behind the drop will vary accordingly. The sequence of photographs shows that the fragmentation of the drop is a result of mass being stripped from the drop surface and carried downstream. Typically there is a darker region upstream which contains the parent drop and a lighter region downstream which is comprised of the stripped

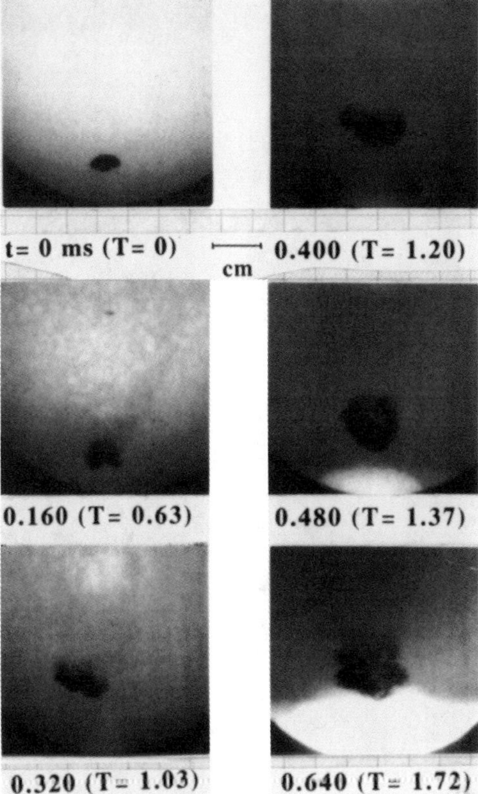

Fig. 9 Sequence of radiographs showing the interaction of a cold alloy drop at 60 °C in a 40 m/s (*we* = 8500) flow.

fragments. Using regular photography it is not possible to distinguish between these two regions. By 0.64 ms there is no distinguishable sign of a parent drop upstream and thus fragmentation can be assumed to be complete. An insufficient number of experiments were carried out using X-ray photography to obtain an accurate nondimensional total breakup time. However, based on the limited data a value between 1.5 and 2 is most representative and was included in Fig. 5. Recall, the main objective of these cold-drop experiments was to characterize, qualitatively, the hydrodynamic-fragmentation mechanism at elevated weber numbers so as to compare with the hot-drop experiments.

Figure 10 shows a sequence of X-ray radiographs taken of a 0.5-g tin drop initially at 700 °C interacting in a water flow velocity of about 40 m/s (similar to the cold-drop experiments shown in Fig. 9). The photographs show that up to a time of 0.185 ms a vapor bubble is present which forms around the drop following the initial contact between the drop and the water. However, the vapor bubble does not undergo the characteristic growth and collapse cycle which is

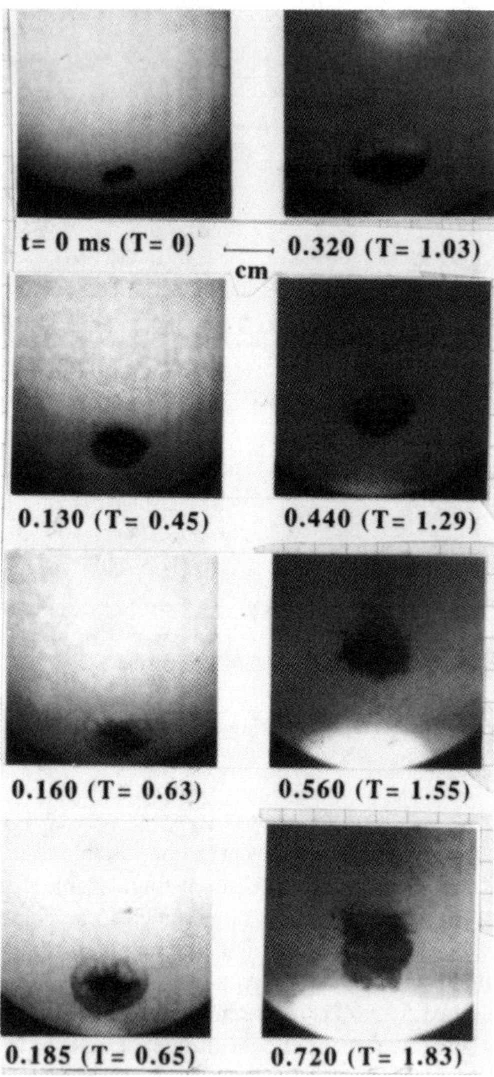

Fig. 10 Sequence of radiographs showing the interaction of a tin droplet at 700 °C in water at 60 °C: water flow velocity 40 m/s (*we* = 8500).

observed under low flow conditions (e.g., see Figs. 7a and 7b). As was observed in the X-ray records taken for low flow velocity hot-drop experiments (see Frost et al.[23]), melt spikes protruding radially outward from the drop surface are evident. There appears to be a substantial number of fine fragments located at the surface of the vapor bubble creating a ring around the drop. At later times the vapor bubble is displaced upward exposing the drop to the oncoming water flow, as observed at 0.320 ms. At this time the vapor bubble region above the drop is fairly free of fragments. At later times, 0.440 ms, the bubble region behind the drop becomes increasingly dark as more and more melt is stripped by the oncoming water flow. By 0.560 ms there is no sign of vapor with only a cloud of fine fragments remaining. The subsequent dispersion of these fragments due to the water flow can be seen at 0.720 ms. Several X-ray photographs were taken at each time and the phenomena described above was found to be quite repeatable from one experiment to the next.

IV. Discussion

A. Estimation of Breakup Time

A brief discussion on the estimation of breakup time using past and present results using regular photography was given earlier. There is considerable scatter in the breakup times arising from the different criteria used by the various investigators. The present investigation highlights the limitations of using regular backlit photography to estimate fragmentation rates. In the low flow cold-drop experiments, regular photography proved to be adequate and X-ray radiographs did not provide any additional information. In this case the fragmentation process was relatively slow and was characterized by the formation of melt filaments which could be resolved using regular photography. For the high flow cold-drop experiments the fragmentation was much faster producing a cloud of fine fragments that obscure the visibility of the parent drop. Regular photography proved to be inadequate in this case since it is impossible to resolve the extent of fragmentation of the parent drop. As a result investigators using regular photography techniques are forced to analyze the fragmentation process in terms of the growth of this opaque cloud of fragments. Ultimately any criterion for drop breakup based on the size of the cloud is inaccurate since the size is directly proportional to the extent of the dispersion of the fragments once they detach from the parent drop and not on the degree of fragmentation. To obtain quantitative information concerning the fragmentation rate or time for complete fragmentation, visualization with X-ray radiography is necessary. From the X-ray photographs the fragmentation rate could be determined by measuring the rate of change in the volume of the parent drop. To obtain such information the X rays must be of sufficient strength to penetrate the fragment mist and not the parent drop. This technique was effectively used by Reinecke and Waldman.[18] The energy of the soft X rays used in the present experiments was not ideally suited for this purpose. In the radiographs, the parent drop could be discerned from the fragment mist as a darker area superimposed on a larger grayer region. In some cases one could interpret the disappearance of this darker area as the time of complete fragmentation but in most cases a certain amount of subjectiveness in the interpretation is inevitable.

In the hot-drop experiments regular photography is not useful for observing the fragmentation process since, in general, a vapor bubble forms which obscures the view of the parent drop. For the high flow velocity the regular photography proved to be misleading since the rate of growth of the cloud appeared similar for both the hot- and cold-drop experiments. X-ray radiographs showed that the phenomenon was quite different at early times with vapor being formed in the hot-drop experiment which could not be observed using regular photography. In conclusion, caution must be used in interpreting data on hot- and cold-drop fragmentation when using regular photography.

B. Transition from Thermal- to Hydrodynamic-Fragmentation Mechanisms

The present experiments have shown at low flow velocities, that the drop fragmentation is governed by thermal effects and direct hydrodynamic fragmentation by the surrounding flow of coolant does not play a role. The production of vapor during the interaction has two effects on the overall heat transfer from the drop to the water. The low thermal conductivity of the vapor, relative to that of the water, greatly reduces the heat transfer coefficient and thus the overall heat transfer from the drop. The vapor bubble which forms around the drop also acts as a physical barrier against any additional fragmentation due to direct stripping of the drop surface by the water flow around the drop surface. For flows as high as 40 m/s ($we \sim 8500$), X-ray photographs have demonstrated that thermal effects are still important and a vapor bubble is generated during the interaction. However, at high flow velocities the flow serves to displace and condense the vapor bubble formed after a short time, exposing the drop to the ambient flow and hydrodynamic fragmentation. As the flow velocity increases still further, the relative contribution to the drop fragmentation from relative velocity hydrodynamic effects will become more important. It seems reasonable above some critical flow velocity, that the formation of vapor will be suppressed and drop breakup will be governed solely by hydrodynamic-fragmentation mechanisms.

To determine a measure of the time during which thermal effects play a role in the fragmentation process, it is necessary to obtain a rough estimate of the time which the vapor bubble surrounds the drop and to determine if there is an upper limit for the flow velocity above which vapor production is suppressed. The formation of vapor will be inhibited if the pressure in the water surrounding the drop is above the critical pressure (22 MPa). Therefore, an estimate of the flow velocity which will suppress boiling at the stagnation point of the flow can be obtained by equating the stagnation pressure $1/2\, \rho u^2$ to the critical pressure, giving a flow velocity of 214 m/s. This must be considered a minimum since the pressure drops as one moves away from the stagnation point and thus boiling can still occur away frcm the stagnation point. The flow velocities in the present experiments are considerably below this value, so the generation of vapor following film collapse is expected, as indeed is observed.

Once a vapor bubble is formed the only way it can grow is if the bubble pressure is higher than the pressure in the water at the bubble surface. As the bubble expands the bubble pressure decreases. For a given initial bubble pressure and radius, the dynamics of the bubble growth can be estimated using the Rayleigh-Plesset equation,

$$R\ddot{R} + \frac{3}{2}\dot{R}^2 = \frac{1}{\rho}\left\{p_B - p_\infty - \frac{2\sigma}{R} - \frac{4\mu\dot{R}}{R}\right\} \tag{4}$$

where R, $\dot{R}$, $\ddot{R}$, ρ, p_B, p_∞, σ, and μ are the bubble radius, radial velocity, radial acceleration, liquid density, pressure in the bubble, pressure at infinity, liquid surface tension, and viscosity, respectively. If we assume isentropic expansion of the vapor within the bubble and ideal gas behavior, then an estimate of the dependence of the lifetime of the vapor bubble (i.e., the time for bubble collapse) on flow velocity can be obtained by setting the liquid pressure to be the stagnation pressure. A higher flow velocity implies a higher stagnation pressure and hence a more rapid collapse of the vapor bubble generated. For example, if we set the initial bubble radius to be the effective drop radius (2.5 mm), the initial bubble pressure equal to the critical pressure, and the ambient pressure equal to the stagnation pressure of the flow, then the variation of bubble collapse time as a function of flow velocity is shown in Fig. 11. The time over which thermal effects will play an important role in the fragmentation process decreases rapidly with increasing flow velocity. For a flow velocity of 40 m/s the calculation predicts a bubble collapse time of 0.7 ms which is on the order of the experimentally observed time of 0.56 ms. The bubble collapse time then decreases slowly for higher flow velocity reaching a minimum of zero time for 22 MPa. In summary, the transition from thermal- to hydrodynamic-fragmentation mechanisms occurs over a range of flow velocities. Hydrodynamic effects will dominate at higher flow rates and at some point the contribution from thermal effects will be negligible.

C. Implications for the Propagation of Vapor Explosion Waves

In vapor explosion waves observed experimentally, the propagation of the coherent interaction is driven by the production of high-pressure vapor in the reaction zone which compresses the mixture ahead of the front. For example, Frost and Ciccarelli[24] used high-speed photography to study the propagation of a steam explosion through a linear array of molten tin drops confined within a narrow channel. They observed that the explosion propagates (at a speed of about 60 m/s) as a result of the expansion of the high-pressure vapor generated by a single drop causing the sequential explosion of adjacent drops. The present results indicate that as the flow generated ahead of the propagating front increases,

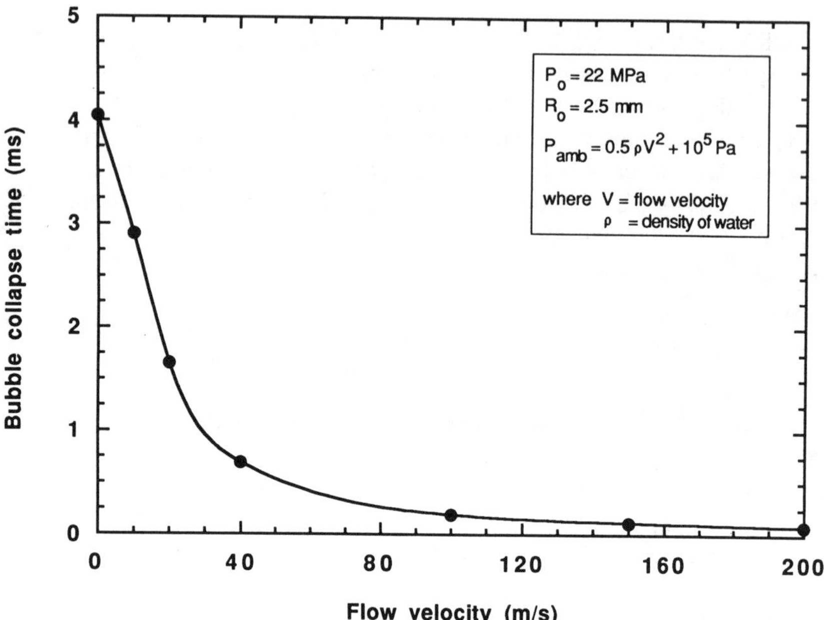

Fig. 11 Vapor bubble collapse time as a function of flow velocity.

the production of vapor will be inhibited and possibly suppressed altogether for high enough flow velocities. In the absence of vapor generation, the work required to sustain the propagation of the wave must come from the expansion of the water heated within the reaction zone. If there is insufficient expansion due to the limited heat transfer to the water (e.g., in the case of low temperature melts and small melt/coolant volume ratios) then the explosion front would be retarded and the propagation velocity would decay. As a result there would exist a maximum velocity for the propagation of a self-sustained vapor explosion wave. This is a possible explanation why in most intermediate-scale experiments (e.g., Refs. 14, 25, and 26) there appears to be an upper limit on the explosion propagation velocity (of the order of 200 m/s), well below the velocities predicted by Hugoniot calculations based solely on energetics considerations.[27,28] It is likely for the low temperature melt mixtures studied extensively experimentally, that a transition mechanism (analogous to deflagration-to-detonation transition in a combustible gas) does not exist by which a local interaction will escalate to a self-sustained super-critical thermal "detonation" wave. In vapor explosions, the escalation mechanism is intimately linked to the fragmentation mechanism which determines the rate of energy release that drives the propagation of the explosion.

V. Conclusions

From the experimental results it is clear that the vapor explosion of a hot drop is influenced by the magnitude of the flow velocity of the surrounding water. For the lower flow velocity (<5 m/s) the interaction of the drop takes on the characteristic vapor bubble growth and collapse cycle. The global fragmentation time for the hot and cold drops under high flow conditions (>45 m/s) are the same, however, the mechanism of fragmentation is very different in each case. In the cold-drop experiments the fragmentation was through stripping of the drop surface. In the hot-drop interactions there was no evidence of any vapor bubble oscillations and there was no noticeable difference in the cold- and hot-drop interactions using regular high-speed photography. However, X-ray radiographs taken during the hot-drop interactions showed that at early times a vapor bubble forms around the entire drop and is later displaced downstream exposing the drop to the oncoming flow. High-speed flow of vapor within the bubble enhances the rate of fragmentation of the drop. The vapor then condenses leaving a cloud of fragments which are subsequently dispersed by the flow. In this case both thermal and hydrodynamic mechanisms play a role in the drop breakup. As the flow velocity increases, a transition in relative importance from thermal to hydrodynamic effects occurs. It is proposed that the transition from thermal to hydrodynamic fragmentation may explain the maximum propagation velocities observed in intermediate-scale experiments to date. It is clear that any attempt at numerical modeling of the escalation of a vapor explosion must consider the effects of flow velocity on the fragmentation behavior of the melt drops within the reaction zone.

References

[1]Reid, R. C., "Rapid Phase Transitions From Liquid to Vapor," *Advances in Chemical Engineering*, Vol. 12, 1983, pp. 105-208.

[2]Fletcher, D. F., and Anderson, R. P., "A Review of Pressure-Induced Propagation Models of the Vapour Explosion Process," Progress in Nuclear Energy, Vol. 23, No. 2, 1990, pp. 137-179.

[3]Kim, B. J., "Heat Transfer and Fluid Flow Aspects of a Small-Scale Single Droplet Fuel-coolant Interaction," PhD Thesis, Department of Mechanical Engineering, University of Wisconsin, Madison, WI, USA, 1985.

[4]Buchanan, D. J., "A Model for Fuel-Coolant Interactions," *Journal of Physics D: Applied Physics*, Vol. 7, 1974, pp. 1441-1457.

[5]Drumheller, D. S., "The Initiation of Melt Fragmentation in Fuel-Coolant Interactions," *Nuclear Science and Engineering*, Vol. 72, 1979, pp. 347-356.

[6]Ochiai, M., and Bankoff, S. G., "Liquid-Liquid Contact in Vapor Explosions," *Proceedings of the International Conference on Fast Reactor Safety*, American Nuclear Society, Chicago, IL, Oct., 1976.

[7]Corradini, M. L., "Analysis and Modelling of Steam Explosion Experiments," Sandia National Lab. Rep., SAND80-2131, Albuquerque, NM, Apr., 1981.

[8]Taylor, G.I., "The Shape and Acceleration of a Drop in a High Speed Air Stream," *The Scientific Papers of G.I. Taylor*, edited by G.K. Batchelor, Vol. III, Cambridge University Press, Cambredge, England, UK, 1963, pp. 457-464.

[9]Burger, M., Kim, D.S., Schwalbe, W., Unger, H., Hohmann, H., Schins, H., "Two Phase Description of Hydrodynamic Fragmentation Process Within Thermal Detonation Waves," *ASME Journal of Heat Transfer*, Vol 106, 1984, pp. 728-734.

[10]Pilch, M., and Erdman, C.A., "Use of Break Up Time Data and Velocity History Data to Predict the Maximum Size of Stable Fragment for Acceleration-Induced Break Up of a Liquid Drop," *International Journal of Multiphase Flow*, Vol. 13, No. 6, 1987, pp. 741-757.

[11]Board, S.J., and Hall, R.W., "Propagation of a Thermal Explosion, Part 2: A Theoretical Model," Central Electricity Generating Board, Berkley Nuclear Laboratories, Berkeley, Gloucestershire, England, UK, RD/B/N 3249, 1974.

[12]Oh, M. D., and Corradini, M. L., "A Propagation/Expansion Model for Large-Scale Vapor Explosions," *Nuclear Science and Engineering*, Vol. 95, 1987, pp. 225-240.

[13]Medhekar, M., Abolfadl, M., and Theofanous T.G., "Triggering and Propagation of Steam Explosions," *Proceeding of the American Nuclear Society National Heat Transfer Conference* (Houston, TX), Vol. 2, July, 1989, pp. 244-251.

[14]Burger, M., Miller, K., Buck, M., Cho, S.H., Schatz, A., Schins, H., Zeyens, R., and Hohmann, H., "Analysis of Thermal Detonation Experiments by Means of a Transient Multiphase Detonation Code," *Proceedings of the 4th International Topical Meeting on Nuclear Reactor Thermal-Hydraulics* (Karlsruhe, Germany), Oct., 1989, Vol. 1., pp. 304-311.

[15]Fletcher, D.F., and Thyagaraja, A., "A Mathematical Model of Melt/Water Detonations," *Applied Mathematical Modeling*, Vol. 13, 1989, pp. 339-347.

[16]Young M.F., "The TEXAS Code for Fuel-Coolant Interaction Analysis," *Proceedings of the Light Metal Fast Breeder Reactor Safety Topical Meeting* (Lyon, France), Vol. III, July, 1989, pp. 567-575.

[17]Tan, M.J., and Bankoff, S.G., "On the Fragmentation of Drops," *Transactions of the ASME Journal of Fluids Engineering*, Vol. 108, 1986, pp. 109-114.

[18]Reinecke, W. G., and Waldman, G. D., "An Investigation of Water Drop Disintegration in the Region Behind Strong Shock Waves," *Proceedings of the 3rd International Conference on Rain Erosion and Associated Phenomena*, Vol. 2., 1970, pp. 629-667.

[19]Simpkins, P.G., and Bales, E.L., "Water Drop Response to Sudden Acceleration" *Journal of Fluid Mechanics*, Vol. 55, 1972, p. 629.

[20]Patel, P.D., and Theofanous, T.G., "Hydrodynamic Fragmentation of Drops," *Journal of Fluid Mechanics*, Vol. 103, 1981, pp. 207-223.

[21]Kim, D. S., Bürger, M., Fröhlich, G., and Unger, H., "Experimental Investigation of Hydrodynamic Fragmentation of Gallium Drops in Water Flows," *Proceedings of the International Meeting on LWR Severe Accident Evaluation* (Cambridge, MA), Aug., 1983, Vol. 1, TS-6.4.

[22]Baines, M. and Buttery, N. E., "Differential Velocity Fragmentation in Liquid-Liquid Systems," Central Electricity Generating Board, Berkeley Nuclear Laboratories, Berkeley, Gloucestershire, England, UK, Rep. RD/B/N4643, 1979.

[23]Frost, D.L., Ciccarelli, G., and Watts, P., "Flash X-ray Visualization of the Steam Explosion of a Molten Metal Drop," To appear in *Progress in Astronautics and Aeronautics*, AIAA Press, Washington, D.C., 1993.

[24]Frost D.L., and Ciccarelli, G., "Propagation of Explosive Boiling in Molten Tn-Water Mixtures," *Proceedings of the ASME/AIChE National Heat Transfer Conference* (Houston, TX), July, 1988, HTD-96, Vol. 2, pp. 539-548.

[25]Fry C. J., and Robinson C. H., "Experimental Observations of Propagating Thermal Interactions in Metal/Water Systems," *Proceedings of the 4th Committee on the Safety of Nuclear Installations Specialist Meeting on Fuel-Coolant Interaction in Nuclear Reactor Safety* (Bournemouth, UK), 1979.

[26]Baines, M., "Preliminary Measurements of Steam Explosion Work Yields in a Constrained System," First U.K. National Conference on Heat Transfer, *Institute of Chemical Engineers Symposium Series*, No. 86, 1984, pp. 97-108.

[27]Frost, D.L., and Ciccarelli, G., "Implications for the Existence of Thermal Detonations from Equilibrium Hugoniot Analysis," To appear in *Progress in Astronautics and Aeronautics*, AIAA Press, Washington, D.C., 1993.

[28]McCahan, S., and Shepherd, J. E., "Models of Rapid Evaporation in Nonequilibrium Mixtures of Tin and Water," To appear in *Progress in Astronautics and Aeronautics*, AIAA Press, Washington, D.C., 1993.

Implications for the Existence of Thermal Detonations from Equilibrium Hugoniot Analysis

D. L. Frost* and G. Ciccarelli†
McGill University, Montreal, Quebec, Canada

Abstract

Experimental studies have shown that a mixture of molten metal and water can support the propagation of a quasisteady vapor explosion wave. Analysis of steadily propagating vapor explosion waves has been carried out by applying the one-dimensional conservation laws of mass, momentum, and energy and appropriate equations of state to a homogeneous mixture of molten tin, water, and steam. For low-temperature melts, the Hugoniot curve lies partially inside the saturation dome and a classical Chapman-Jouguet (CJ) detonation point occurs only for low void fractions. For high melt temperatures, the downstream states lie entirely outside of the saturation region and high-pressure CJ detonation states can be found. To address the question of the existence of a thermal detonation, it is necessary to consider other factors, such as the dynamic processes within the reaction zone (including the momentum transfer, melt fragmentation, and heat transfer mechanisms), the influence of boundary conditions and the ability to initiate a thermal detonation. Consideration of the reaction zone length of a thermal detonation suggests that for conventional melt-water mixtures, large mixture volumes (> 1 m^3) will be required to support a self-sustained thermal detonation. The generation of vapor at subcritical pressures places limits on the rate of fragmentation and heat transfer from melt to water and, hence, direct initiation of a thermal detonation will likely require a trigger with a supercritical peak pressure and a long duration. The existence of a CJ detonation (and CJ deflagration) for a multiphase melt-coolant mixture has yet to be substantiated experimentally. Propagating vapor explosion waves observed experimentally for low energy melt-coolant mixtures are poorly modeled by equilibrium Hugoniot analysis. Incomplete fragmentation and

* Assistant Professor, Mechanical Engineering Department.

† Graduate Student, Mechanical Engineering Department.

nonequilibrium mechanical and thermal effects play a role in the divergence between theory and experiments.

I. Introduction

Numerous industrial accidents and laboratory experiments have demonstrated that a violent vapor explosion can result when a hot and cold liquid are rapidly mixed together (see, for example, Reid[1] and Corradini et al.[2] for reviews of accidents and experimental work, respectively). Experiments using high-speed photography have demonstrated the existence of a propagation phenomenon in the so-called "coarse mixture" formed by the fragments of the hot liquid distributed in the cold liquid. Board et al.[3] first proposed the analogy between the propagation of an explosion wave through a fuel-coolant mixture and a detonation wave in a chemically reactive medium. Propagation of a thermal detonation wave is supported by the thermal energy transfer from the hot fuel fragments to the coolant resulting in a rapid phase transition of the superheated liquid coolant to the vapor state. In recent years, numerous steady-state and transient models have been developed to describe the propagation of steam explosions (see Fletcher and Anderson[4] for a recent comprehensive review of propagation models).

In analogy to a chemically reactive medium, a spectrum of propagation speeds for vapor explosions should be possible. Conclusive experimental demonstration of steadily propagating thermal detonations has not been realized to date. However "quasisteady" propagation of vapor explosions has been observed in a variety of small-scale experiments. For example, the propagation of vapor explosions in a line of single droplets of molten tin in water has been demonstrated by Frost and Ciccarelli[5] and in a planar array of molten tin droplets from a series of jets has been reported by Fröhlich.[6] Board and Hall,[7] Fry and Robinson,[8] Baines,[9] and Bürger et al.[10] have measured the propagation of a pressure wave through confined mixtures of molten tin and water. Quasisteady propagation of a vapor explosion in a stratified geometry between molten tin and water has also been observed by Anderson et al.[11] and Ciccarelli et al.[12] Because of the complexity of the multiphase phenomena and the difficulty in carrying out definitive experiments, the precise initial and boundary conditions required for initiation and propagation of such vapor explosion waves are not known. The propagation mechanisms are also poorly understood and the propagation speeds of such quasisteady waves have not been correlated with theoretical models.

If steadily propagating vapor explosion waves do exist, they must satisfy the one-dimensional conservation laws of mass, momentum, and energy. Without any knowledge of the nonequilibrium mechanical and thermodynamical processes in the "reaction" zone of the thermal explosion wave, it is possible to determine the possible downstream equilibrium states on the basis of the conservation laws and given equations of state. Such calculations are referred to as Hugoniot analysis and since the calculations are essentially based on the energetics and not the kinetics of the energy release process, they can be made even with total ignorance of the mechanism of propagation of the wave. Valuable insights into the properties of vapor explosion waves can be deduced from these energetics considerations. Hugoniot calculations for a fuel-coolant mixture were first carried out by Board et al.[3] They assumed that the structure of a thermal detonation was analogous to a chemical detonation, and performed Hugoniot calculations to obtain the characteristic wave velocity and pressure which would result from such a propagating wave. They assumed during the interaction that the melt is fragmented to such a small size that the fragments can oe assumed to instantaneously come into mechanical and thermal equilibrium with the surrounding coolant, forming a homogeneous mixture. Board et al. solved the Hugoniot relation and obtained the downstream state by using the classical construction of the Rayleigh line tangent to the Hugoniot curve. In their analysis it was implicitly assumed that complete fragmentation of the melt exists at the CJ plane (equivalent to total energy release from the melt to the coolant in the reaction zone). Later contributors[10, 13-16] considered a model for the structure of the reaction zone and relaxed the requirement of complete fragmentation, i.e., they considered the possibility that all of the melt may not be fragmented by the end of the reaction zone. As a result, they obtained a family of "partial" Hugoniot curves for a given initial condition. For a given fraction E, of the initial melt that is fragmented, the CJ point for the resulting Hugoniot curve was determined by applying a sonic condition (the various expressions used for the speed of sound will be described later). In general, no attempt was made by the authors to determine which of the partial Hugoniot solutions represented the most probable endstate.

Application of Hugoniot analysis to vapor explosion waves assumes that the wave is steady and that equilibrium conditions (i.e., thermal, mechanical, and chemical equilibrium) prevail at the two planes bounding the energy release zone. There is no apparent restriction that the medium must be single phase. Solving the steady one-dimensional conservation laws for mass, momentum and energy across the wave yields a relationship, referred to as the Hugoniot equation, between the change of energy function of the medium (i.e., enthalpy or internal energy) and the state variables (i.e., pressure and specific volume). With the caloric equation of state (i.e., $h(P,V)$ or $e(P,V)$) and the equation of state (i.e., f,$(P,V,T) = 0$) for the medium specified, the Hugoniot equation defines the locus of equilibrium downstream states corresponding to various propagation speeds of the wave itself. An additional criterion is required to determine a unique wave speed for given initial conditions. In chemical reactive media, the detonation solution evokes the so-called Chapman-Jouguet condition that the flow is sonic (relative to the wave) at the downstream equilibrium plane. The CJ condition can be justified from stability arguments which may or may not be applicable to multiphase vapor detonations. The CJ solution also corresponds to the minimum of the wave velocity and entropy change. However, no physical principles can be used to justify the use of the minimum velocity or entropy change condition for determining the solution. Even if a criterion (such as the

CJ criterion) can be used to determine uniquely a solution, there is no guarantee such a wave exists in reality. Nevertheless, Hugoniot calculations permit some insight into the phenomenon by providing an estimate of the detonation state based on the energetics of the processes.

It appears worthwhile to re-examine the problem of Hugoniot calculations for thermal detonations with emphasis on the implications of the equilibrium calculations for the existence of a thermal detonation in practice. In the following section, the basic Hugoniot relations will first be given, with a discussion of the assumptions and the methods of closing these equations. This is followed by Hugoniot calculations for mixtures of tin, water, and steam to illustrate the characteristic properties of the detonation state. Finally, the properties of experimentally observed vapor explosion waves are compared with the Hugoniot calculations and the possibility of achieving a thermal detonation is discussed.

II. Governing Relations

Shown in Fig. 1 is a schematic of a steadily propagating energy release zone in the reference frame moving with this steady zone. Note that the details of the reaction zone are left out. Knowledge of the processes inside this zone is of no consequence as long as this zone is steady and equilibrium conditions prevail at the boundary planes of the zone itself. With subscript o denoting upstream conditions and subscript 1 denoting downstream conditions, application of mass, momentum, and energy conservation across the zone yields

$$\rho_1 U_1 = \rho_o D \tag{1}$$

$$P_1 + \rho_1 U_1^2 = P_o + \rho_o D^2 \tag{2}$$

$$h_1 + \frac{1}{2} U_1^2 = h_o + \frac{1}{2} D^2 \tag{3}$$

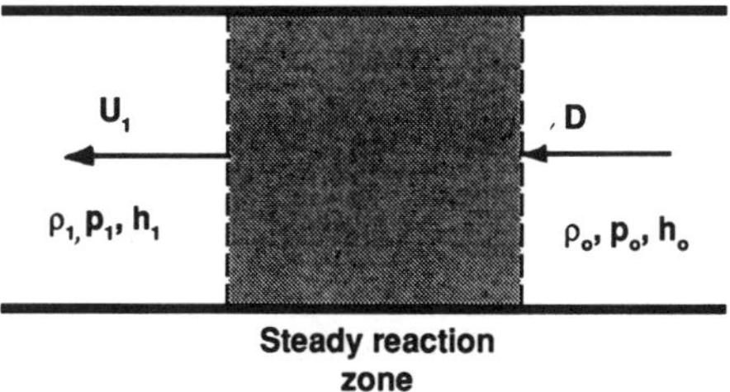

Fig. 1 Schematic of a steadily propagating energy release zone in a moving reference frame.

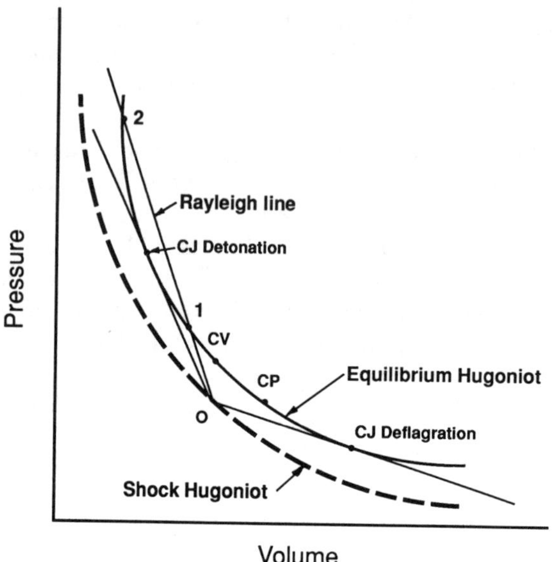

Fig. 2 Schematic of shock and equilibrium Hugoniot curves and Rayleigh line on the P-V plane. Points denoted CV and CP refer to constant volume and constant pressure processes.

Combining Eqs. (1) and (2) we obtain the so-called Rayleigh line

$$D = V_o \sqrt{\left(\frac{P_1 - P_o}{V_o - V_1}\right)} \tag{4}$$

which relates the velocity D of this propagating zone (or wave) to the change in pressure and specific volume V across the wave. Equation (4) gives a straight line on a P-V diagram where the slope is proportional to the wave velocity. Therefore, for a given initial condition (P_o, V_o) and wave velocity, the downstream state (P_1, V_1) must lie on this line. If we eliminate the wave velocity from Eq. (4) using Eqs. (3) and (1) we get the Hugoniot equation as follows.

$$h_1 - h_o = \frac{1}{2}(P_1 - P_o)(V_o + V_1) \tag{5}$$

The Hugoniot curve together with the Rayleigh line are shown schematically in Fig. 2.

If no chemical changes occur, then the reactant species and product species are identical and Eq. (5) denotes the so-called *shock Hugoniot* or *shock adiabat*. It should be emphasized again that the validity of Eqs. (4) and (5) places no restriction on the thickness of the transition zone between the upstream and downstream boundaries of the zone. A normal shock may or may not be present. The flow may even be transient and three dimensional inside the zone. Equations (4) and (5) apply to the steady one-dimensional flow and equilibrium conditions that must prevail upstream and downstream of the boundaries of the zone only. If we assume the existence of a normal shock as in the classical ZDN model for detonation structure, then since transition across a shock takes a few molecular collisions and chemical equilibration takes 10^4 to 10^5 collisions, it is reasonable to assume that in a chemical detonation the transition to the equilibrium Hugoniot must go via the shock Hugoniot; i.e., from the initial state to the shock adiabat first via the Rayleigh line prior to any reaction, then proceed from the shock adiabat to the equilibrium Hugoniot. Furthermore, if it can be assumed that intermediate states are quasiequilibrium and the one-dimensional conservation laws of mass and momentum are valid, then the intermediate states from the shock to the final equilibrium Hugoniot must also follow the Rayleigh line. However, it should be emphasized, on the basis of the one-dimensional conservation laws valid for a steadily propagating wave, that although the initial state and the final state on the Hugoniot are connected by the Rayleigh line, intermediate states cannot be defined and it is not necessary that a shock must be present at the transition zone.

The intersection of the Rayleigh line and the Hugoniot defines the downstream state. As we can see from Fig. 2, for a given wave velocity (slope of Rayleigh line), in general, there exist two solutions; the strong solution given by point 2 and the weak solution given by point 1 on the diagram. According to *classical theory* in chemical detonation, the choice for the detonation velocity is the unique minimum velocity corresponding to the tangency of the Rayleigh line to the Hugoniot curve. The criterion for choosing this tangency solution in classical detonation theory is based on the fact that the weak detonation solution violates the second law of thermodynamics, if we adopt the ZDN model for this structure of a normal shock transition prior to energy release by chemical reactions. The overdriven detonation solution is ruled out on the basis of stability. The increase in density across the wave necessitates the formation of an expansion wave behind the detonation. Since the flow behind the overdriven detonation is subsonic, the expansion can overtake the reaction zone and quench the reaction. However, we must emphasize, within the framework of the Hugoniot calculations, that *there exists, a priori, no criterion for the choice of the appropriate detonation solution for given initial conditions.*

The unique points at which the Rayleigh lines are tangent to the Hugoniot curve are denoted as CJ points and several additional conditions are associated with these points. Consider the detonation (or upper) branch of the Hugoniot curve. From Eq. (4) we see that the slope of the Rayleigh line is proportional to the square of the mass flux j ($j = \rho_0 D$) through the wave. The point of tangency of the Rayleigh line and Hugoniot curve therefore represents a minimum in the mass flux. The following expression can be derived giving the change in entropy along the Hugoniot curve:

$$\frac{dS}{d(j^2)} = \frac{(V_0 - V_1)^2}{2T_1} > 0 \tag{6}$$

Since the CJ point represents a minimum in the mass flux, it follows from Eq. (6) that the entropy must also be stationary (i.e., $dS = 0$) at this point and, in fact, the entropy is a minimum at this point. Therefore, at the CJ point, the Rayleigh line is tangent to both the Hugoniot and the isentrope going through that point. As a result it can be shown that the downstream flow velocity is sonic relative to the shock wave, which is denoted as the *CJ condition*. Note that there is a second CJ point in which the Rayleigh line is tangent to the Hugoniot curve, which is referred to as the lower-branch solution, or the deflagration mode of propagation. In the present report we will be concerned mainly with the upper-branch solution where the wave is supersonic relative to the mixture ahead of the wave.

One can eliminate the complexities of two-component mixtures by treating the mixture as being homogeneous. Homogeneous refers to a mixture which can be treated as a pseudofluid (with all the appropriate averaging of properties) that obeys all the equations for single-component fluids. For the homogeneous approximation to hold there can be no slip between the components, hence mechanical equilibrium. The *initial* temperatures of the components may be different, so that the mixture may not be initially in thermal equilibrium. For a homogeneous mixture the properties are obtained by mass averaging of the properties of the components. Using the Hugoniot equation given by Eq. (5) in conjunction with the mass-averaged specific volume and enthalpy with appropriate equations of state for the melt and the coolant one can obtain the Hugoniot curve on a P-V diagram.

As mentioned earlier, the Hugoniot curve gives a *locus* of possible downstream states. For example, different slopes of the Rayleigh lines intersect the Hugoniot curve at different points, giving different downstream states. Many previous investigators have assumed that when a thermal detonation occurs, it is a CJ detonation corresponding to the unique solution of the tangency of the Rayleigh line to the Hugoniot (or the minimum velocity solution). The tangency solution is obtained, like in the single-component case, by drawing the tangent to the Hugoniot curve from the initial state. At the CJ point, the entropy along the Hugoniot curve will be a minimum, and so the isentrope as well as the Rayleigh line will be tangent at this point. This ensures that the velocity relative to CJ plane will equal the sonic velocity (as determined from the slope of the isentrope). As in single-component Hugoniot theory, without some knowledge of the structure of the reaction zone (e.g., existence of a leading shock wave, etc.), we cannot use stability or entropy arguments to determine if this solution is the appropriate choice. However, as in the case of chemical detonations, it appears reasonable that nature would choose a solution in which a choking or sonic condition applies as energy addition always drives a flow toward the sonic condition.

Tangency of the Rayleigh line with the Hugoniot gives the unique CJ solution which has the property of sonic conditions at the downstream plane. However, other authors have used a separate equation for computing the sound speed and obtain the CJ solution formally by requiring that the particle velocity be equal to the sound speed at the CJ plane. In a two-component mixture it is difficult to define a unique sonic velocity. Wallis[17] points out that transient drag forces between the components would probably be both amplitude and frequency dependent and, therefore, the wave velocity would be a function of these variables as well as the properties of the components. Nevertheless, if we assume homogeneous flow (no slip between the components) one can derive an expression for the homogeneous mixture speed of sound (c_H) based on the void fractions of the components[17]

$$\frac{1}{\rho_m c_H^2} = \frac{\alpha}{\rho_1 c_1^2} + \frac{1 - \alpha}{\rho_2 c_2^2} \tag{7}$$

where α is the void fraction of component 1, c the speed of sound, and ρ_m the mixture density, where $\rho_m = \alpha \rho_1 + (1 - \alpha)\rho_2$. Note that since the adiabatic compressibility of a substance is given by $\beta_s \equiv -(1/V)(\partial V/\partial p)_s = 1/\rho c^2$ we can see that the above expression (7) for the homogeneous sound speed just expresses the adiabatic compressibility (and hence the sound speed) of the mixture as a (volume fraction) weighted sum of the adiabatic compressibility of the components. One can use this expression to find the point on the Hugoniot curve where the downstream flow velocity equals this homogeneous speed of sound, thus satisfying the sonic condition $U_1 = c_H$. This criterion is *not* equivalent to drawing a tangent to the Hugoniot (this will be shown later by example) because of the incompatibility of the assumptions used in Eq. (7) and in obtaining the Hugoniot itself. So this method of determining the solution is not self-consistent thermodynamically in the sense that the CJ point determined in this manner does not correspond to a minimum in entropy (so that the Rayleigh line is *not* tangent to the isentrope at this point).

Another method of choosing the solution, assumed by Sharon and Bankoff,[14] is based on a *separated flow* choking condition. The CJ solution is determined by requiring that the particle velocity is equal to the stratified flow sonic velocity c_S, given by (e.g., see Wallis[17])

$$c_S = \frac{\alpha_d \rho_f + (1-\alpha_d)\rho_d}{\alpha_d \rho_f c_f^2 + (1-\alpha_d)\rho_d c_d^2} \tag{8}$$

Here subscript d refers to the unfragmented drop component and the subscript f refers to the homogeneous mixture consisting of the fragments plus the coolant (vapor and liquid). It can be easily shown that the stratified sonic velocity is higher than the homogeneous speed of sound calculated using Eq. (8).

So in summary, there are three possible methods which can be used to determine the final CJ state on the Hugoniot:
1) Find tangency of the Rayleigh line to the mass-averaged Hugoniot curve (analogous to what is done for single-component mixtures).
2) Set the downstream flow velocity equal to the homogeneous speed of sound given by Eq. (7).
3) Set the downstream flow velocity equal to the stratified speed of sound obtained from by Eq. (8).

In principle, all three methods should yield the same solution. However, since the expressions for the sound speed imposed differ, each of the three conditions above gives a different solution for the CJ point in practice. A priori, from the basic Hugoniot theory, it is not possible to choose which method is correct. However, only the first method gives a solution that is thermodynamically self-consistent in that at the determined CJ point the Rayleigh line is tangent to the Hugoniot curve as well as the isentrope. In addition, without a detailed knowledge about the structure of the reaction zone in a thermal detonation, there does not appear to be any compelling reason to choose the latter two conditions over the classical tangency solution. Therefore, in the present calculations, the CJ point solution was obtained by constructing the tangent to the mass-weighted Hugoniot curve (which corresponds to the point of minimum entropy change).

III. Results

A. Calculation Procedure

As noted above, some earlier investigators have varied the fraction E of the melt that fragments and, hence, transfers its heat to the coolant. To determine E, it is necessary to invoke a specific model for fragmentation within the reaction zone. Given the present uncertainty in modeling the complex energy release processes, we feel that the *addition* of an extra parameter E to the Hugoniot analysis represents an unnecessary complication, given the number of assumptions that are already inherent in the basic Hugoniot analysis. Therefore, as the conservative case, in the present Hugoniot calculations we have assumed that *all* the thermal energy of the melt is transferred to the coolant and that thermal equilibrium is attained at the CJ plane.

The Hugoniot equation (5) was solved by an iterative technique, assuming homogeneous flow so that the specific volume and enthalpy can be expressed as a weighted sum of the properties of the individual components. The equation of state for water was based on the fitted equations given by Reynolds[18] which are based on the steam-table data from Keenan et al.[19] To test the validity of the equation of state for water, we calculated the shock Hugoniot for water (i.e., setting the volume fraction of melt to zero) and compared the results with calculated values from Kirkwood and Montroll[20] which are based on the Tait equation of state for water. Figure 3 shows that the shock adiabat calculated using the steam-table-based equations agrees well with the calculations of

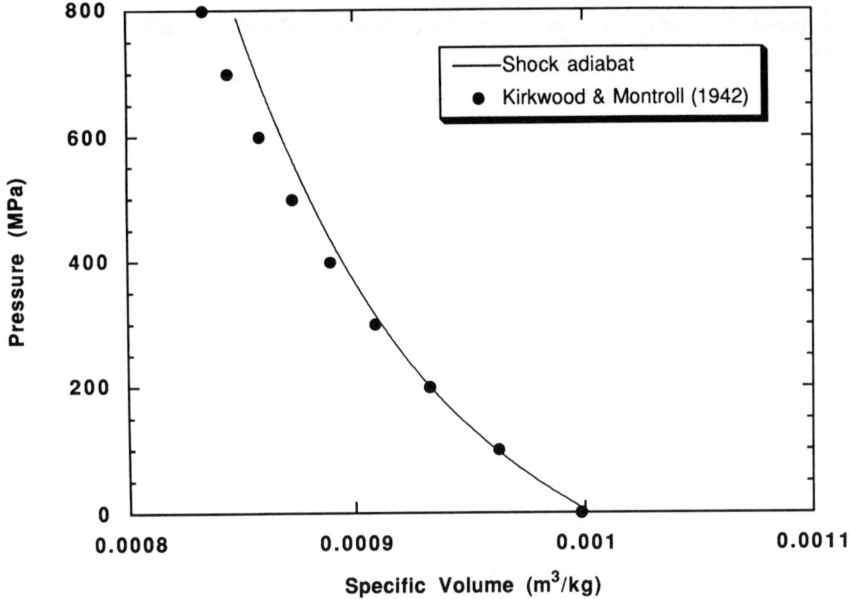

Fig. 3 Comparison of the shock adiabat for water using the equation of state formulation of Reynolds[18] with calculated values from Kirkwood and Montroll[20] using the Tait equation of state.

Kirkwood and Montroll for pressures up to about 400 MPa. The melt was taken to be incompressible. Since the isothermal compressibility of water is about an order of magnitude larger than that of tin, this assumption is reasonable except in the case of very high pressures (or for mixtures with high initial melt/water volume fraction). Thermodynamic properties for molten tin used in the calculations were obtained from Iida and Guthrie.[21]

B. Comparison with Previous Hugoniot Calculations

To illustrate the characteristic features of the Hugoniot analysis results, calculations were carried out for mixtures of molten tin, water, and steam. Many of the earlier Hugoniot calculations (e.g., by Board et al.[3]) also considered molten tin and water, so repeating the calculations for this mixture is important for comparison with earlier work. Figure 4 shows Hugoniot curves for a mixture consisting of equal volumes of molten tin, water, and steam. Results for an initial tin temperature of 1000 °C are shown from the present investigation together with the corresponding curve from Board et al.[3] The results are in poor agreement, but without additional information on the details of the earlier calculation method, it is not possible to determine the cause of the difference in results. Sharon and Bankoff[13] also computed Hugoniot curves for the case of tin at 1000 °C with equal initial volumes. The Hugoniot curves have the same general shape as the present calculation, but a direct comparison with the present results is not possible since they calculated Hugoniot curves for different values of E, or fraction of melt that fragments.

C. Results for Tin/Water/Steam Mixtures

The Hugoniot curve for 1000 °C tin in Fig. 4 illustrates several features that are common to Hugoniot curves for low energy melts. The Hugoniot curve has two distinct sections separated by a sharp discontinuity or "kink" at the saturation boundary: a lower region, inside the saturation dome, that is nearly flat, and an upper region, in the compressed liquid region, that is quite steep. For the case in which equal volumes of tin, water, and steam are present initially, the Rayleigh line shown in the figure contacts the Hugoniot curve at the kink in the Hugoniot curve at the phase boundary (at a pressure of 14.8 MPa). This point represents a minimum in the mass flux, but it is not a true CJ point in the sense that the slope of the Hugoniot as well as the isentrope at this point is indeterminate.

In Fig. 5, the Hugoniot curve for 1000 °C tin (with equal volumes of tin, water, and steam) is replotted on a semilog scale to show the other thermodynamic points of interest. The horizontal displacement of the Hugoniot curve at a pressure of about 3 MPa inside the saturation boundary is a result of the contribution of the latent heat of the melt when the melt freezes. Although a CJ detonation does not exist for this Hugoniot curve (since the Rayleigh line

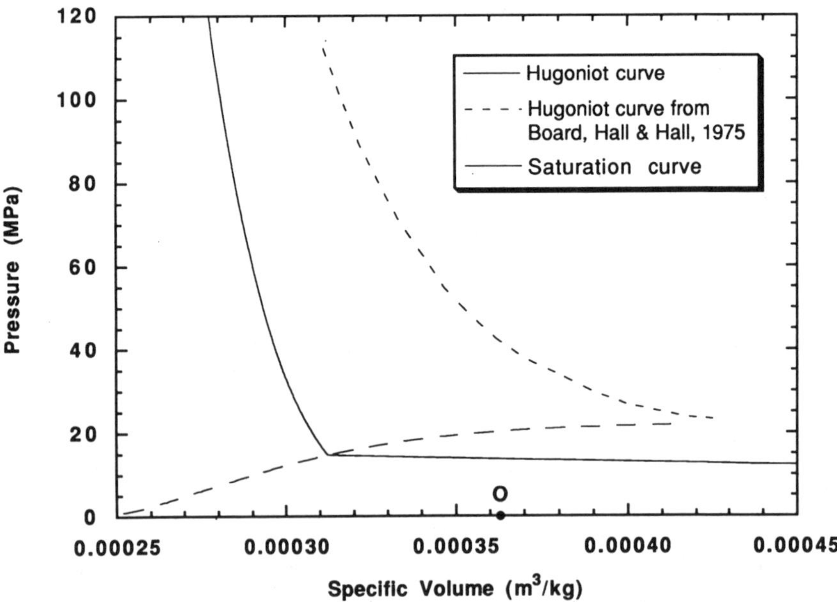

Fig. 4 Hugoniot curves for a mixture containing equal initial volumes of tin, water, and steam: tin temperature is 1000 °C.

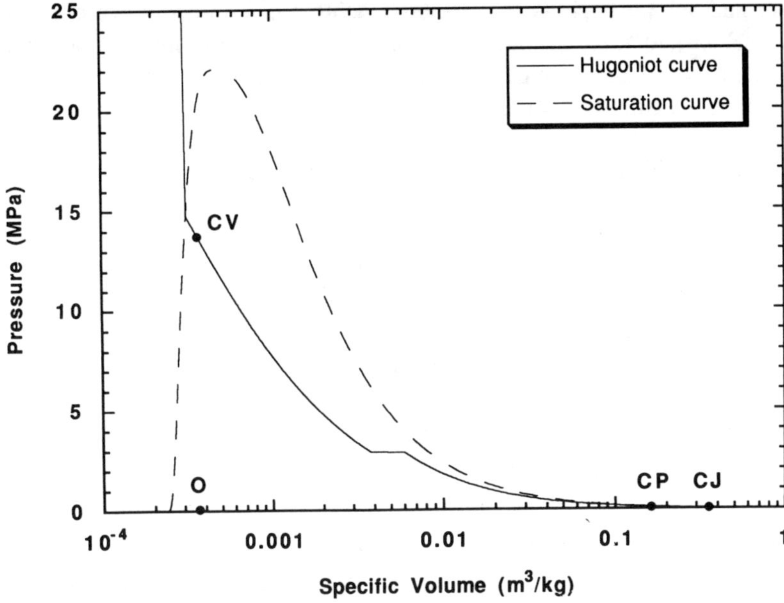

Fig. 5 Hugoniot curve for tin at 1000 °C (equal initial volumes of tin, water and steam): points denoted O, CV, CP, and CJ refer to initial conditions, constant volume process, constant pressure process, and Chapman-Jouguet deflagration, respectively.

intersects the Hugoniot at the kink), a CJ deflagration point exists which is located within the two-phase saturation region. The constant volume and pressure points are also located inside the saturation region.

As the thermal energy of the melt increases, the Hugoniot curve shifts up. This is illustrated in Fig. 6 which shows the Hugoniot curve for 1500 °C tin (equal volumes of tin, water, and steam). The Hugoniot just intersects the saturated vapor curve, but in this case both CJ solutions exist and are located outside the saturation region. The entropy change in the supercritical region near the CJ detonation point is very small, so that the Hugoniot and the isentrope are virtually indistinguishable in this region. For the Hugoniot curve corresponding to tin at 1500 °C, we can now compare the various sonic velocities at the CJ point. From the detonation wave speed (647 m/s), we find (using the continuity equation) that the flow velocity at the CJ plane is U_1 = 546 m/s. Since the Hugoniot curve is tangent to the isentrope at this point, this velocity will be the same as the velocity determined by taking the tangent to the isentrope. Now, if we calculate the homogeneous and stratified sound speeds for the two component mixture, as defined in Eqs. (7) and (8), we find that c_H = 627 m/s and c_S = 1161 m/s. Note that the homogeneous sonic velocity is not too different from the velocity obtained by using the tangency condition, therefore, the first two criteria for determining the CJ point described earlier will give similar results for the CJ pressure and velocity. However, if the CJ point is determined by forcing the particle velocity to equal the stratified sound speed, the resulting CJ point will be considerably different than that obtained using the tangency criterion.

Hugoniot calculations are useful for studying the effect of varying different thermodynamic parameters on the predicted CJ detonation state. When the initial void fraction is changed, keeping the relative volume fractions of tin and water the same, the Hugoniot curve changes only slightly. However, the initial mixture specific volume changes considerably when the void fraction changes. As a result, the CJ solution changes substantially with a change in void fraction. For example, for the case of tin at 1000 °C for void fractions above 26%, a CJ detonation state does not exist since the Rayleigh line intersects the Hugoniot curve at the discontinuity at the saturation boundary (see, for example, Fig. 5). For void fractions below 26%, the detonation pressure and velocity increase rapidly with decreasing void fraction, ranging from 20 MPa and 370 m/s at 25% void to 370 MPa and 1360 m/s at 10% void, respectively.

D. Comparison with Transient Numerical Models

A number of investigators have proposed transient models for the escalation and propagation phases of a vapor explosion, including Medhekar et al.,[22] Bürger et al.,[23] and Fletcher and Thyagaraja.[24] In these models, the differential conservation laws for mass, momentum, and energy are solved together with

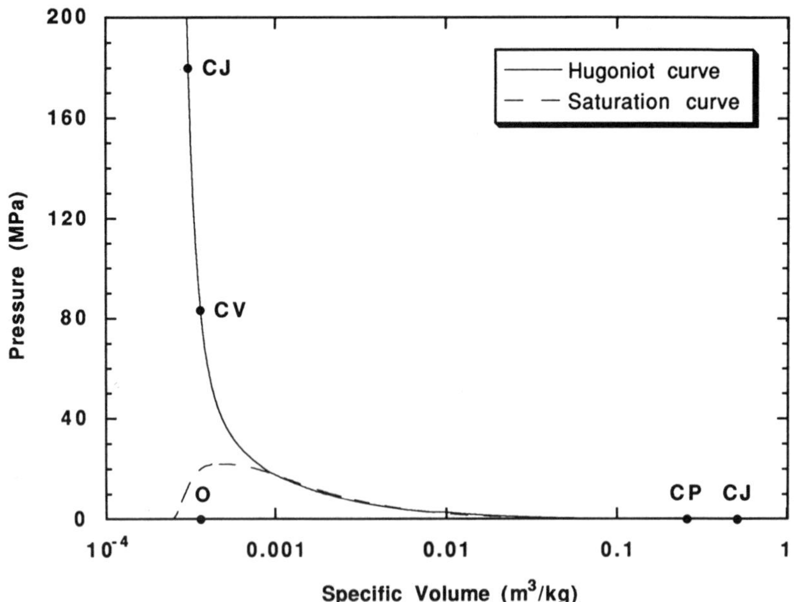

Fig. 6 Hugoniot curve for tin at 1500 °C (equal initial volumes of tin, water, and steam).

constitutive relations for drag, heat transfer, and fragmentation. There is considerable uncertainty regarding the choice of appropriate models for the highly transient heat transfer and fragmentation processes due to a scarcity of relevant experimental data. Nevertheless, it is of interest to compare the results of the transient models with the predictions of the Hugoniot analysis. To make such a comparison, CJ detonation states were calculated using the conditions (i.e., initial melt temperature = 2500 K, melt density = 7000 kg/m^3, melt heat capacity = 500 J/kgK, initial melt volume fraction = 0.1) used by Fletcher and Thyagaraja[24] in their transient model calculations. Fletcher and Thyagaraja[24] show the variation of peak pressure and "detonation" speed for void fractions of 0.5, 0.7, and 0.9. The transient calculations do not directly predict a sonic CJ plane, so a comparison with the CJ pressure cannot be made. Since the Hugoniot analysis does not give the peak shock pressure, we calculated the shock adiabat assuming negligible heat transfer from the melt to the coolant through the shock. By choosing a shock velocity equal to the CJ detonation speed, the shock pressure was estimated. A comparison between the Hugoniot predictions and the results of the transient model is given in Table 1. The wave velocities from the transient model calculations agree with the CJ detonation velocities from the Hugoniot analysis within about 20% and in both cases the largest pressures are predicted for the 0.7 void fraction case. The differences between the velocities calculated from the numerical code and the CJ detonation velocities suggests that the propagating "waves" calculated with the transient model should not be described as "detonation" waves to avoid confusion with the classical definition of a CJ detonation wave.

IV. Discussion

A. Equilibrium Assumptions

Application of equilibrium Hugoniot analysis implies at the downstream state that the mixture is in mechanical (i.e., velocity) and thermal equilibrium. The validity of the assumption of velocity equilibration between the melt fragments and the coolant depends on the fragmentation mechanism. If the coolant velocity within the reaction zone is high and fragmentation of the melt is assumed to occur solely by hydrodynamic stripping of small fragments, then the fragments will be rapidly accelerated to the coolant velocity. In this case the assumption of mechanical equilibrium would be well founded. However, if the fragmentation is dominated by thermal effects (see Ciccarelli and Frost[25] for a comparison of thermal- and hydrodynamic- fragmentation mechanisms) and vapor is present in the reaction zone, then velocity equilibration is unlikely. Thermal equilibrium implies that all of the heat from the melt is transferred to *all* of the coolant. However, during rapid heat transfer from the melt to the coolant, thin thermal boundary layers (and, hence, sharp temperature gradients) are present and nonequilibrium effects are important. For example, consider the case of a mixture of molten tin and water. After vapor film collapse, if we consider only diffusive heat transfer, then the thermal boundary-layer thickness grows with time as $\delta \sim (\kappa t)^{1/2}$, where κ is the thermal diffusivity. For the tin/water case, the ratio of thermal boundary thicknesses is, therefore, $\delta_{water}/\delta_{tin} = (\kappa_{water}/\kappa_{tin})^{1/2} \sim 0.06$, i.e., only a very small fraction of the water actually participates in the interaction. The production of vapor within the reaction zone will also inhibit heat transfer between the melt and liquid coolant and prevent thermal equilibration. Even with high-temperature melt-rich mixtures, the slow diffusivity of heat in liquids implies that the assumption of thermal equilibrium will not be satisfied for any vapor explosion waves that have been observed experimentally. The importance of nonuniform heating of water has been pointed out in the analysis of tin-water experiments by Baines[9] and Bürger et al.[23]

B. Reaction Zone Length

To obtain a conservative estimate of the thickness of the reaction zone for a thermal detonation, we will assume that the melt is completely fragmented at the downstream CJ plane. To estimate the time required for total fragmentation of the melt, we will first briefly review the correlations that have been used in earlier work for drop breakup based on actual experimental results for drop fragmentation.

In the majority of models for vapor explosion propagation, drop breakup is assumed to occur due to the relative velocity between the melt and coolant. The fragmentation models used are usually based on either boundary-layer stripping or Rayleigh-Taylor instability models. A large quantity of experimental data exists on the fragmentation of liquid drops in a liquid or gas medium following the passage of a shock. However, only recently (e.g., Ciccarelli and Frost[25]) has the fragmentation of a *hot* molten drop subject to a relative flow been studied where the effects of the vapor film and thermal energy of the drop play a role in the fragmentation process. The time for drop breakup t is usually expressed as a dimensionless time T as follows:

$$T = t \frac{U}{d} \sqrt{\rho_f/\rho_d} \tag{9}$$

where subscripts d and f refer to the drop and the surrounding fluid, respectively; and U is the initial relative velocity between the drop and the fluid. Many different definitions have been proposed for breakup time. The variations in breakup time definitions and the difficulty in interpretation of experimental data make it difficult to compare results from different studies. Pilch and Erdman[26] have reviewed the literature and found three characteristic times of interest: initiation of breakup, primary breakup and total breakup. Baines and Buttery[27] and Kim et al.[28] have presented results on total breakup times for liquid-liquid systems based on regular photography of the fragmentation process. The dimensionless complete breakup times range from 2 to 5 and show no clear correlation with weber number ($we = \rho_f U^2 R/\sigma$, where R is the drop radius and σ is the surface tension of the drop). The average nondimensional breakup time of about 3.5 is the same as the nondimensional time for catastrophic breakup of a liquid drop in a gas medium observed by Waldman et al.[29] In experiments with hot drops with high flow rates (weber numbers up to 30,000), Ciccarelli and Frost[25] observed total breakup times of about 1.5 using *x-ray radiography*. They also pointed out the difficulty in determining total breakup times solely with the use of regular photography.

Since the correlations for drop breakup are based on the fragmentation of cold liquid metal drops (i.e., at the same temperature as the surrounding liquid), the effects of freezing on the fragmentation have not been incorporated into existing numerical models for propagation. Yang and Bankoff[30] have carried out experiments to investigate the effect of solidification on the fragmentation of drops of Wood's metal subject to a weak shock wave. They observed that crust formation inhibited fragmentation in some cases although the sensible enthalpy of the drops was low (initial drop temperatures ranged from 104 to 125 °C) and the weber number range was small (*we* ranged from 35 to 270). They applied a stability theory accounting for elastic crust stiffness and obtained good agreement with the threshold for fragmentation inhibition. Sharon and Bankoff[13] applied the theory of Cooper and Dienes[31] for the effect of freezing on the development of Rayleigh-Taylor instabilities to a system of UO_2 and Na. They showed that the growth of interfacial waves, and hence hydrodynamic fragmentation, is inhibited if the mixing time scale is larger than 100 μs. However, the conclusions of the theoretical models will remain speculative until validated with experiments with high-temperature melts and large coolant flow velocities.

With the assumption of complete fragmentation at the downstream CJ plane, the time for complete fragmentation τ_{frag} can be used to estimate the reaction zone thickness L, of a thermal detonation wave, i.e., $L \sim D\tau_{frag}$. For example, consider the case shown in Table 1 for a void fraction of 0.7. For a CJ detonation speed of 690 m/s, from the continuity equation, the particle velocity after the shock is 439 m/s. If we take a nondimensional total breakup time of 1.5, and assume the initial drop/coolant relative velocity is the flow velocity behind the shock, then from Eq. (9) the breakup time for a 5-mm drop is 45 μs. In this case the corresponding reaction zone length is about 3 cm. In their calculations with the same initial conditions, Fletcher and Thyagaraja[24] used a fragmentation model in which the e-folding time for fragmentation is equal to a nondimensional time of unity. From their plot of the volume fraction profiles for a void fraction of 0.7, they observed that complete fragmentation of the melt occurs also over a distance of about 3 cm.

During the propagation of a steam explosion wave in an unconfined melt/coolant mixture, the propagation of lateral expansion waves due to sideways expansion of the reaction zone will decrease the pressure within the reaction zone and attenuate the wave velocity if the scale of the mixture is not substantially larger than the thickness of the reaction zone. As a conservative estimate, to minimize the influence of boundary conditions on the steady propagation of a thermal detonation, the scale of the melt/coolant mixture should be at least an order of magnitude larger than the reaction zone thickness. This suggests that only large-scale mixtures (with a scale ~ 1 m or larger) can support propagation of a thermal detonation.

C. Experimental Evidence

The propagation of a vapor explosion wave has been observed with high-speed photography in intermediate-scale experiments with high-temperature melts (Mitchell et al.[32]). However, due to the difficulties in obtaining quantitative data in large-scale experiments, the propagation of a pressure wave in a melt-coolant mixture has only been measured in mixtures containing relatively low-temperature melts (e.g., molten tin) confined within a tube (typically with a length of 1-2 m). In experiments with 0.5-1 kg of molten tin and water confined within a vertical tube, Baines[9] observed the propagation of a pressure pulse with propagation velocities ranging from 50 to 250 m/s. The pressure pulses had an initial peak pressure of 3-4 MPa before falling to a pressure plateau of about 2 MPa. In experiments in the KROTOS facility[23], 7.5 kg of tin at 1350 K were released into water contained within a vertical tube. In test #21, a propagating pressure wave was observed to decelerate from 270 m/s to 150 m/s near the end of the tube. The pressure profiles consisted of initial peak pressures or 5-6 MPa followed by a long pressure plateau of about 2-2.5 MPa. In both the above experiments, the premixture geometry cannot be directly observed, and varies vertically.

To compare with experiments, the Hugoniot curve was calculated for the initial conditions corresponding to experiment CT16 of Baines,[9] i.e., T_{melt} = 800 °C, T_{water} = 85 °C, melt/water/steam volume fractions 0.14/0.76/0.10. The Hugoniot is plotted in Fig. 7 on a log-log scale and a classical CJ detonation state does not exist because of the discontinuity in the Hugoniot curve at the saturation boundary. The discontinuous CJ solution shown (corresponding to minimum wave velocity) occurs at a pressure of about 0.5 MPa corresponding to a velocity of about 70 m/s, both much less than the experimentally observed values. Nonuniform heating of the coolant appears to play an important role in the divergence between the experimental results and the equilibrium Hugoniot predictions. The propagation of the "non-Hugoniot" vapor explosion waves observed experimentally is not governed solely by energetics considerations but also on nonequilibrium dynamical processes. The propagation velocity depends on the details of the complex thermo/hydrodynamic processes within the reaction zone and the boundary conditions.

D. Deflagration Propagation

Calculation of the equilibrium state that satisfies the CJ criterion on the deflagration branch of the Hugoniot has recently been carried out by Frost et al.[33] and McCann and Shepherd[34] for tin/water/steam mixtures. McCann and Shepherd[34] have developed a model for a propagating wave in a melt/coolant

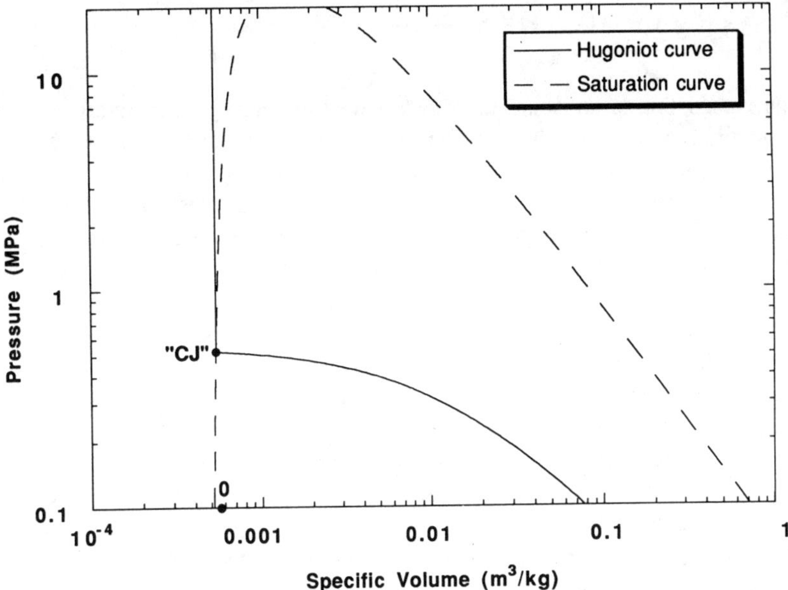

Fig. 7 Hugoniot curve corresponding to conditions from experiment CT16 from Baines[9]: initial conditions are denoted by O,. CJ represents the discontinuous CJ detonation solution.

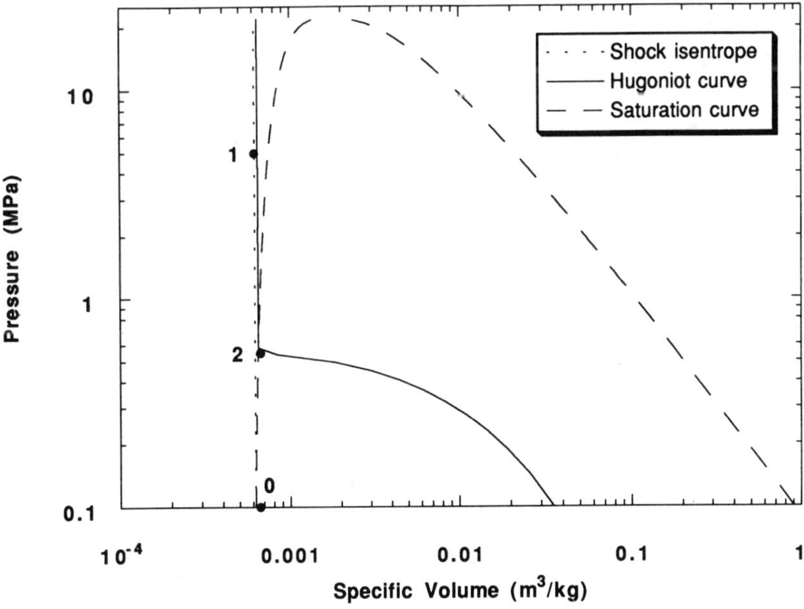

Fig. 8 Model for KROTOS test #21 (Bürger et al.[23]) consisting of isentropic shock followed by deflagration wave bringing flow to rest; upstream conditions denoted by 0, state after an shock denoted by 1, and final state after evaporation wave denoted by 2.

mixture consisting of an isentropic shock wave followed by a deflagration wave. The precursor supersonic shock collapses the vapor present and compresses the mixture. The pressurized water then expands through an evaporation (or deflagration) wave producing a multiphase mixture downstream. The flow downstream of the deflagration wave is brought to rest to satisfy the downstream boundary condition. The expansion of the mixture generates the flow that drives the shock wave. To compare with experiments, the model can be evaluated using the initial conditions of the KROTOS test #21 described above. From Bürger et al.,[23] 6.5 kg of tin at 1350 K are assumed to mix with 7.4 kg of water with a void fraction of 0.08. If we choose a shock pressure of 5 MPa to match the experimental conditions, then the thermodynamic states predicted by the model are shown in Fig. 8. State 0 corresponds to the initial conditions and state 1 falls on the shock isentrope and refers to the conditions just after the shock. The downstream state 2 is the point on the Hugoniot curve that satisfies the condition of zero velocity downstream. For the points shown, the shock velocity is 220 m/s, the evaporation wave velocity is 187 m/s and the flow velocity between the two waves is 15 m/s. However, note that because of the discontinuity in the Hugoniot curve, state 2 does *not* correspond to the classical CJ deflagration solution. In fact to obtain a CJ deflagration solution that satisfies the downstream boundary condition, the shock pressure (at point 1) must be less than about 0.5 MPa. Although the shock-deflagration model yields pressures and velocities that are closer to experimental values (than the CJ detonation solution), the CJ deflagration solution does not appear to be relevant to the experimental results.

Berman and Beck[35] have suggested that the complete spectrum of propagation mechanisms in combustible gases have analogs with propagation mechanisms for melt-coolant mixtures. Although melt-coolant mixtures can also support a range of propagation mechanisms and velocities, direct analogies with the detailed phenomena cannot be made. For chemical detonations, observed wave speeds are quite close to the calculated CJ detonation wave speed. However, for vapor explosion waves observed experimentally, nonequilibrium effects such as nonuniform heating of the coolant, partial fragmentation of the melt, and relative velocities between phases suggest that the waves are poorly modeled with equilibrium Hugoniot analysis. Observed vapor explosion waves are *superficially* more like fast chemical deflagrations, in that in both cases the propagation mechanism depends on boundary conditions and the details of the processes in the reaction zone and not purely on the energetics of the mixture.

E. Dynamic Properties of Thermal Detonations

For chemical detonation waves, considerable information on dynamic parameters such as the critical tube diameter for transmission, initiation energy and propagation limits is available. However, the existence and behavior of corresponding dynamic properties for thermal detonations are not established. Nevertheless, with a considerations of the characteristics of a thermal detonation, some estimates of the dynamic properties can be made.

Table 1 Comparison of Hugoniot analysis results with the transient model calculations of Fletcher and Thyagaraja[24]

	Results from Fletcher and Thyagaraja		Hugoniot Analysis Results		
Void fraction	Detonation speed, m/s	Peak pressure, MPa	CJ detonation speed, m/s	CJ pressure, MPa	Shock pressure, MPa
0.5	580	160	654	75	228
0.7	725	285	690	139	291
0.9	770	210	652	133	270

1. *Direct Initiation*

As in the case of chemical detonations, it is likely that *direct* initiation of a thermal detonation will require a trigger pressure on the order of the CJ pressure with a duration on the order of L/D, where L is the reaction zone length and D is the detonation velocity. For a numerical estimate of the trigger strength required, consider the case calculated by Fletcher and Thyagaraja[24] and shown in Table 1 for a void fraction of 0.7. In this case, the CJ detonation pressure is about 139 MPa, and the above estimate of the reaction zone thickness gives a value of about 45 μs for L/D. So the impulse of the triggering system must be on the order of 0.006 MPa-s. To obtain a triggering shock wave with this amplitude and duration, conventional solid explosives can be used. Since the effect of the triggering shock will be felt for some distance, to establish if a *steady* thermal detonation is possible, large-scale mixtures will probably be required (with dimensions > 1 m).

2. *Sensitivity*

The *sensitivity* of a melt-coolant mixture, or the ability of the mixture to support the initiation and self-sustained propagation of a detonation wave, is not known. In gaseous detonations, insensitive mixtures require a large mixture volume and a strong initiation source to form a detonation wave. To gain insight into the sensitivity of a melt-coolant mixture, consider the sensitivity of another heterogeneous system in which a detonation wave phenomenon has been observed. In particular, dust-air mixtures that involve dust particles with little or no volatile content are particularly insensitive to detonation. In this case, very large surface area to mass (or volume) ratios are necessary to support a detonation. For example, Tulis and Selman[36] found that for flake aluminum-air mixtures, detonation could only be achieved if the surface area to mass ratio was about 3-4 m^2/g which corresponds to spherical particles of diameter less than 1 μm. Steam explosions typically generate fragments with sizes on the order of 100 μm. Since heat transfer rates depend on surface area, it is likely that fuel-coolant mixtures are even *less* sensitive to detonation than dust air mixtures. An additional factor influencing the detonability of fuel-coolant mixtures is the fact that the fuel is not finely dispersed *prior* to the explosion (as in a dust detonation), but rather the fine fragmentation occurs in the reaction zone itself. If the fine fragmentation mechanism can give rise to very fine particles, then the energy required to disperse these particles in the coolant increases sharply with decreasing particle size. Thus it appears that fine fragmentation and the subsequent dispersion are of opposite effect and high energy transfer rates must balance the energy requirement of dispersing the fine particles. This could pose a limit on the sensitivity of a thermal detonation.

3. Transition to Detonation

In general, a trigger strong enough for direct initiation of a thermal detonation does not exist in most melt/water accident scenarios, so to achieve a thermal detonation a transition process is required in which a local explosion escalates to a coherent self-sustained detonation wave. In chemically reactive mixtures, if certain critical conditions occur during the propagation of a deflagration wave (i.e., a turbulent flame), transition from a deflagration to a detonation (DDT) can occur. The transition process involves the rapid amplification of an instability within the flow and leads to a resonant coupling between the fluid flow-field and chemical energy release. The existence of a transition phenomenon similar to DDT for steam explosions has not been established. The lack of experimental evidence for a transition process for steam explosions suggests that such an amplification mechanism may not exist. Recent experiments[25] on the effect of flow velocity on drop fragmentation suggest during the escalation stage of a steam explosion that thermal fragmentation mechanisms play an important role. For modest flow velocities (< 50 m/s), the rapid generation of vapor after film collapse shields the drop from the flow, limiting the fragmentation and heat transfer rates and moderating the escalation process. Therefore intrinsic limits to the rate of energy release may exist that limit the ability of a melt-coolant mixture to support the propagation of a classical thermal detonation wave which depends only on the energetics of the mixture and is independent of the rate processes that occur within the reaction zone.

4. Explosion Yield

The pressure generated during a thermal detonation will be a function of the initial melt/water volume ratio. For a numerical example, consider the effect of varying the melt/water volume ratio for tin at 1000 °C, with the initial void fraction equal to 33.3% in each case. Both the CJ detonation velocity and pressure increase with tin/water volume ratio and reach a maximum at some point. In this case, the detonation pressure reaches a maximum (of about 160 MPa) when the tin/water volume ratio is about four and decreases with increasing tin/water volume ratio. A plot of detonation pressure vs melt/water volume ratio can be used to define an effective *stoichiometry* for a melt-coolant mixture, where a stoichiometric mixture yields the largest predicted detonation pressure.

To estimate the work done during the expansion phase of a steam explosion, a standard practice is to calculate the work done in the equilibrium Hicks and Menzies process,[37] i.e., constant volume mixing of the fuel and coolant until thermal equilibrium is reached, followed by an isentropic expansion to atmospheric pressure. For a fuel/coolant mixture undergoing a CJ detonation, an alternative method of estimating the work yield can be obtained by expanding the high-pressure steam from the *detonation* state (rather than from the Hicks and Menzies pressure) to atmospheric pressure. The work can be visualized as the area under the isentrope drawn on a pressure-volume diagram. In the compressed liquid region the isentrope follows the Hugoniot closely and is quite steep. The contribution to the expansion work in the compressed liquid region is small (relative to the contribution within the two-phase region), and so the total work yield will not be very sensitive to the point on the Hugoniot chosen to start the expansion. Therefore, the work calculated by expanding from the Hicks and Menzies pressure will not be substantially different from the corresponding value in expanding from the CJ detonation state. To illustrate the calculation of the work yields, a numerical example is given below.

In calculating the expansion work, the expansion is assumed to be sufficiently rapid so that negligible heat transfer from the melt occurs, i.e., the melt temperature remains constant during the isentropic expansion. Figure 9

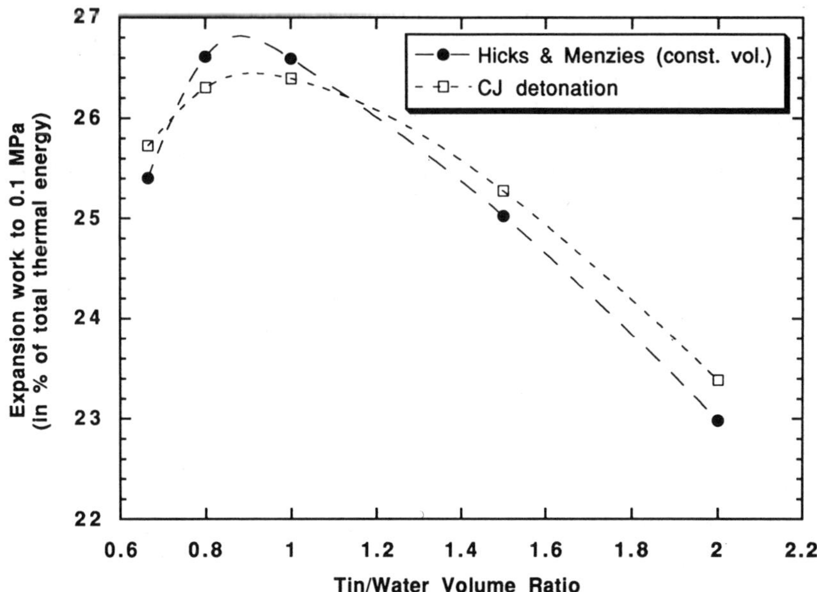

Fig. 9 Comparison of expansion work following a constant volume process (Hicks and Menzies[37] process) and a CJ detonation.

shows the work yield during isentropic expansion from the CJ detonation state as well as the corresponding Hicks and Menzies value, as a function of the ratio of the initial volume fractions of the melt and water. The work yields are normalized with the total sensible enthalpy of the melt. In each case the mixture consists of molten tin at 1500 °C with an initial void fraction of 50%. Note that the work yields calculated using the two methods give similar results and in each case the maximum work yield occurs for about equal initial volume fractions of melt and water.

V. Conclusions

Analysis of a steadily propagating vapor explosion waves can be carried out by applying the one- dimensional conservation laws of mass, momentum, and energy. Solution of the Hugoniot equation does not require any information regarding the structure of the reaction zone. However, even if a criterion (such as the CJ criterion) can be used to uniquely determine a solution, there is no guarantee that such a wave exists in reality. Nevertheless, Hugoniot calculations permit some insight into vapor explosion waves by providing an estimate of the detonation (or deflagration) state based on the energetics of the processes.

Hugoniot calculations have been carried out for a mixture consisting of equal volumes of molten tin, water, and steam to illustrate the features of the Hugoniot curve for a multiphase mixture. For low energy melts (e.g., tin at 1000 °C) and water, the Hugoniot has a "kink" at the saturation boundary, and therefore classical CJ detonation solutions only exist if the initial mixture has a low initial void fraction (e.g., less than about 26% void for tin at 1000 °C). As the energy of the melt increases, the Hugoniot curve shifts upward and CJ solutions can be obtained for a wide range of initial void fractions. Eventually the Hugoniot curve lies completely above the saturation curve.

Hugoniot calculations give a prediction of the pressure and velocity associated with a thermal detonation, but do not give any information on whether or not such a thermal detonation can be realized in practice. Conclusive experimental demonstration of a steadily propagating thermal detonation has not been realized to date. To address the question of the existence of a thermal detonation, it is necessary to consider other factors, such as the structure of the thermal detonation (including the momentum transfer, melt fragmentation, and heat transfer mechanisms), the influence of boundary conditions and the ability to initiate a thermal detonation. Consideration of the reaction zone length of a thermal detonation suggests that for conventional melt-water mixtures, large mixture volumes (> 1 m^3) will be required to support a thermal detonation. Direct initiation of a thermal detonation will likely require a strong trigger, with a trigger pressure on the order of the CJ pressure and duration on the order of the time associated with the reaction zone. Predicted CJ detonation pressures increase with increasing thermal energy of the melt and decreasing initial void fraction of the mixture.

The properties of propagating vapor explosion waves observed experimentally do not coincide with the predictions for a CJ detonation. The pressures and velocities observed are more similar to that predicted by a coupled shock-deflagration system although the CJ deflagration solution does not appear relevant. Nonequilibrium effects due to the low thermal conductivity of water (and hence long thermal equilibration times) imply that the propagation waves observed experimentally in low energy melt-coolant mixtures are poorly modeled using equilibrium Hugoniot analysis. Therefore the velocity of the propagating vapor explosion waves observed are not governed solely by energetics, but also depend on the boundary conditions and the details of the complex processes that occur within the reaction zone.

Acknowledgments

The work was supported by the Natural Sciences and Engineering Research Council of Canada and Sandia National Laboratories. The authors thank John Lee for many stimulating discussions and acknowledge the useful comments provided by Joe Shepherd and Michael Cowperthwaite regarding the formulation of the Hugoniot analysis for multicomponent systems and for information concerning appropriate equations of state.

References

[1]Reid, R. C., "Rapid Phase Transitions from Liquid to Vapor," *Advances in Chemical Engineering*, Vol. 12, 1983, pp. 105 208.

[2]Corradini, M. L., Kim, B. J., and Oh, M. D., "Vapour Explosions in Light Water Reactors: a Review of Theory and Modelling," *Progress in Nuclear Energy*, Vol. 22, 1988, pp. 1-117.

[3]Board, S. J., Hall, R. W., and Hall, R. S., "Detonation of Fuel Coolant Explosions," *Nature*, Vol. 254, 1975, pp. 319-325.

[4]Fletcher, D. F., and Anderson, R. P., "A Review of Pressure-Induced Propagation Models of the Vapour Explosion Process," *Progress in Nuclear Energy*, Vol. 23, 1990, pp. 137-179.

[5]Frost, D. L., and Ciccarelli, G., "Dynamics of Explosive Interactions between Multiple Drops of Tin and Water," *Progress in Astronautics and Aeronautics*, edited by A.L. Kuhl, J.R. Bowen, J.-C. Leyer, and A. Borisov, AIAA, Washington, D.C. 1988, pp. 451-473.

[6]Fröhlich, G., "Propagation of Fuel Coolant Interactions in Multi-Jet Experiments," *Proceedings of the 4th International Topical Meeting on Nuclear Reactor Thermal Hydraulics* (Karlsruhe, Germany), Vol. 1, 1989, pp. 282-289.

[7]Board, S. J., and Hall, R. W., "Recent Advances in Understanding Large Scale Vapour Explosions," *Proceedings of the 3rd Specialist Meeting on Sodium Fuel Interactions in Fast Reactors* (Tokyo, Japan), SNI 6/7, 1977.

[8]Fry, C. J., and Robinson, C. H., "Experimental Observations of Propagating Thermal Interactions in Metal/Water Systems," *Proceedings of the 4th Committee*

on the Safety of Nuclear Installations Specialist Meeting on Fuel-Coolant Interactions in Nuclear Reactor Safety, Bournemouth, UK, 1979.

[9]Baines, M., "Preliminary Measurements of Steam Explosion Work Yields in a Constrained System," First U.K. National Conference on Heat Transfer, *Institution of Chemical Engineers Symposium Series*, No. 86, 1984, pp. 97-108.

[10]Bürger, M., Carachalios, C., Kim, D. S., and Unger, H., "Theoretical Investigations on the Fragmentation of Drops of Melt with Respect to the Description of Thermal Detonations (Vapor Explosions) and their Application in the Code FRADEMO," Commission of the European Communities Rept. EUR 10660 EN, 1986.

[11]Anderson, R., Armstrong, D., Cho, D., and Kras, A., "Experimental and Analytical Study of Vapor Explosions in Stratified Geometries," *American Nuclear Society Proceedings of the 1988 National Heat Transfer Conference*, Vol. 3, 1988, pp. 236-243.

[12]Ciccarelli, G., Frost, D. L., and Zarafonitis, C., "Dynamics of Explosive Interactions between Molten Tin and Water in Stratified Geometry," *Progress in Astronautics and Aeronautics*, edited by A. L. Kuhl, J.-C. Leyer, A. A. Borisov, and W. A. Sirignano, AIAA, Washington, D.C., Vol. 134, 1991, pp. 307-325.

[13]Sharon, A., and Bankoff, S. G., "Propagation of Shock Waves in a Fuel-Coolant Mixture," In: *Topics in Two-phase Heat Transfer and Flow,* edited by S. G. Bankoff, ASME, 1978.

[14]Sharon, A., and Bankoff, S. G., "On the Existence of Steady Supercritical Plane Thermal Explosions," *International Journal of Heat and Mass Transfer*, Vol. 24, 1981, pp. 1561-1572.

[15]Scott, E., and Berthoud, G. J., "Multiphase Thermal Detonation," In: *Topics in Two-phase Heat Transfer and Flow*, edited by S.G. Bankoff, ASME, 1978.

[16]Condiff, D. W., "Contributions Concerning Quasi-Steady Propagation of Thermal Detonations through Dispersions of Hot Liquid Fuel in Cooler Volatile Liquid Coolants," *International Journal of Heat and Mass Transfer,* Vol. 25, 1982, pp. 87-98.

[17]Wallis, G. B., *One-Dimensional Two-Phase Flow,* McGraw-Hill, Toronto, Canada, 1969.

[18]Reynolds, W. C., *Thermodynamic Properties in SI*, Stanford University Press, Stanford, CA., 1979.

[19]Keenan, J. H., Keyes, F. G., Hill, P. C., and Moore, J. G., *Steam Tables*, John Wiley, New York, 1969.

[20]Kirkwood, J. G., and Montroll, E. W., "The Pressure Wave Produced by an Underwater Explosion II," *OSRD Rept* . 676, 1942.

[21]Iida, T., and Guthrie, R. I. L., *The Physical Properties of Liquid Metals,* Clarendon Press, Oxford, England, UK, 1988.

[22]Medhekar, S., Abolfadl, M., and Theofanous, T. G., "Triggering and Propagation of Seam Explosions," Proceedings of the American Nuclear Society National Heat Transfer Conference (Houston, TX), July 1988, Vol. 2, 1988, pp. 244-251.

[23]Bürger, M., Miller, K., Buck, M., Cho, S. H., and Schatz, A., "Analysis of Thermal Detonation Experiments by means of a Transient Multiphase Detonation Code," *Proceedings of the 4th International Topical Meeting on Nuclear Reactor Thermal-Hydraulics* Karlsruhe, Germany, Oct. 1989, pp. 304-311.

[24]Fletcher, D. F. and Thyagaraja, A., "A Mathematical Model of Melt/Water Detonations," *Applied Mathematical Modelling*, Vol. 13, 1989, pp. 339-347.

[25]Ciccarelli, G., and Frost, D. L., "The Effect of Fluid Flow Velocity on the Fragmentation Mechanism of a Hot Melt Drop," To appear in *Progress in Astronautics and Aeronautics*, AIAA Press, Washington, D.C., 1993.

[26]Pilch, M., and Erdman, C. A., "Use of Breakup Time Data and Velocity History Data to Predict the Maximum Size of Stable Fragments for Acceleration-Induced Breakup of a Liquid Drop," *International Journal of Multiphase Flow*, Vol. 13, 1987, pp. 741-757.

[27]Baines, M., and Buttery, N. E., "Differential Velocity Fragmentation in Liquid-Liquid Systems," Berkeley Nuclear Labs, UK, Rept. RD/B/N4643, 1979.

[28]Kim, D. S., Bürger, M., Fröhlich, G., and Unger, H., "Experimental Investigation of Hydrodynamic Fragmentation of Gallium Drops in Water Flows," *Proceedings of the International Meeting on Light Water Reactor Severe Accident Evaluation,* (Cambridge, MA), Aug. - Sept. 1983, Vol. 1, TS-6.4.

[29]Waldman, G. D., Reinecke, W. G., and Glenn, D. C., "Raindrop Breakup in the Shock Layer of a High-Speed Vehicle," *AIAA Journal* , Vol. 10, 1972, pp. 1200-1204.

[30]Yang, J. W. and Bankoff, S. G., "Solidification Effects on the Fragmentation of Molten Metal Drops behind a Pressure Shock Wave," *Journal of Heat Transfer*, Vol. 109, 1987, pp. 226-231.

[31]Cooper, F., and Dienes, J., "The Role of Rayleigh-Taylor Instabilities in Fuel-Coolant Interactions," *Nuclear Science and Engineering*, Vol. 68, 1978, pp. 308-321.

[32]Mitchell, D. E., Corradini, M. L., and Tarbell, W. W., "Intermediate Scale Steam Explosion Phenomena: Experiments and Analysis," Sandia National Lab. SAND81-0124, Albuquerque, N.M., 1981.

[33]Frost, D. L., Lee, J. H. S., and Ciccarelli, G., "The use of Hugoniot Analysis for the Propagation of Vapor Explosion Waves," *Shock Waves,* Vol. 1, 1991, pp. 99-110.

[34]McCann, S., and Shepherd, J. E., "Models of Rapid Evaporation in Nonequilibrium Mixtures of Tin and Water," To appear in *Progress in Astronautics and Aeronautics*, AIAA Press, Washington, D.C., 1993.

[35]Berman, M., and Beck, D., "Steam Explosion Triggering and Propagation: Hypothesis and Evidence," *Proceedings of the 3rd International Seminar on Containment of Nuclear Reactors,* Univ. of California, Los Angeles, CA, Aug. 1989.

[36]Tulis, A., and Selman, R., *19th Symposium (International) on Combustion,* The Combustion Institute, Pittsburgh, PA, 1982, p. 655.

[37]Hicks, E. P., and Menzies, D. C., "Theoretical Study of the Fast Reactor Maximum Accident," *Proceedings of the Conference on Safety, Fuels and Core Design in Large Fast Power Reactors*, Argonne National Lab. Rept: ANL-7120, 1965.

Flash X-Ray Visualization of the Steam Explosion of a Molten Metal Drop

D. L. Frost,* G. Ciccarelli,† and P. Watts†
McGill University, Montreal, Quebec, Canada

Abstract

The fragmentation processes that occur during the vapor explosion of a single molten metal drop in water have been studied experimentally with the use of flash X-ray radiography and high-speed photography. For a cold liquid metal drop the hydrodynamic flow associated with the triggering system causes direct fragmentation due to relative velocity, leading to complete breakup of the drop in about 5–10 ms. In contrast, for a hot molten drop, drop breakup is caused by the action of the hydrodynamic flow field generated by the rapid production of vapor. After collapse of the initial vapor film, growth of a high-pressure vapor bubble is first visible within about 80 μs and complete drop breakup occurs after about 1 ms. The melt-water interaction consists of several cycles involving bubble growth and collapse. The X-ray photographs show at early times that thin metal filaments are ejected radially at high speed from the drop surface. Considerable elongation and distortion of the drop surface occurs during the growth of the first vapor bubble. Collapse of the bubble leads to complete drop breakup and dispersion. The radial bubble growth rates observed experimentally are modeled reasonably well with a simple theoretical model, and a new phenomenological model for the drop fragmentation mechanism is proposed.

I. Introduction

When a hot liquid such as molten metal is brought into contact with a cold volatile liquid a violent vapor explosion can occur. The explosion is a result of rapid heat transfer between the liquids and the subsequent "explosive" production

---

*Assistant Professor, Mechanical Engineering Department.

†Graduate Student, Mechanical Engineering Department.

of vapor. Explosions can occur for homogeneous mixtures consisting of melt fragments dispersed in a coolant or for relatively inhomogeneous stratified mixtures of melt and coolant. In each case, explosions that initiate locally can escalate to form a coherent wave that collapses the vapor films separating the melt and coolant and propagates through the mixture. The propagating pressure wave also causes fine fragmentation of the melt from 1) hydrodynamic effects due to the differential velocities of the liquids, 2) thermal effects that occur following film collapse. Expansion of the high-pressure vapor formed can generate shock waves and do mechanical damage to surrounding structures. The hazards associated with accidental vapor explosions in industry are well-documented (e.g., see the review by Reid[1]) and provide the motivation for the study of vapor explosions at laboratory scale.

In the absence of an initial strong hydrodynamic flow of the coolant relative to the drop, the energy required for the fragmentation of a drop during a vapor explosion must be derived from the thermal energy of the drop itself. Fragmentation due to thermal effects plays an important role in spontaneous or self-triggered vapor explosions of single drops immersed in a coolant. Cronenberg and Benz[2] have reviewed the variety of thermal fragmentation models that have been proposed for the breakup of a drop. Many of the models involve the formation of coolant jets following film collapse (analogous to the jets formed during cavitation bubble collapse) or the asymmetric collapse of vapor bubbles that are formed following the contact between melt and coolant. The jets penetrate the melt surface and lead to encapsulation of small coolant droplets within the melt. Vaporization of the trapped droplets then tears the melt drop apart. Buchanan[3] and Kim[4] have developed detailed models of the above process to explain the explosion of single melt droplets. Drumheller[5] proposed a model in which fragmentation is caused by the impact of the coolant on the melt following symmetrical film collapse. Other models are based on the effect of the high-pressure vapor generated at local melt-coolant contact points. For example, Ochiai and Bankoff[6] developed a "splash theory" for vapor explosions in which random contacts cause local regions of high pressure that result in the formation of annular jets. The liquid jets formed then cause mixing of the liquids. Corradini[7] also considered the fragmentation of the drop due to Rayleigh-Taylor instability resulting from the acceleration of the drop by high-pressure steam generation. Many of the details of the above models are highly speculative and conclusive experimental evidence supporting one model over another does not exist.

Thermal fragmentation mechanisms have also been incorporated into models for the propagation phase of a vapor explosion (see Fletcher and Anderson[8] for a comprehensive review of propagation models). For example, Oh and Corradini[9] developed a parametric model of propagation and expansion based on a thermal fragmentation model involving film collapse and coolant jet impingement and entrapment. They used the model to interpret results from Sandia National Laboratory FITS tests. However, the initial melt-coolant mixture geometry is poorly characterized in the tests, and agreement with experimental peak pressure and conversion ratio cannot be considered to validate the model. Burger et al.[10] used a transient multiphase detonation code to model tin/water experiments performed at the KROTOS facility. Initial simulations, using only a hydrodynamic fragmentation model, were unable to simulate the propagating pressure wave observed experimentally. To obtain better agreement with experiments, they incorporated a simple ad hoc thermal fragmentation model. However, they concluded that the relative contributions of thermal and hydrodynamic fragmentation mechanisms as well as the relative importance of the triggering process cannot be determined based on the existing experimental evidence. Additional well-characterized experimental data on the fragmentation mechanisms involved are required.

Many investigators have photographically studied the important fine fragmentation process in experiments with single molten metal drops immersed in water or another coolant (see Appendix A of Corradini et al.[11] for a comprehensive summary of small-scale melt-coolant experiments). For example, Nelson et al.[12] obtained high-speed and high-resolution photographs of interactions of drops of molten tin and iron oxide triggered with a shock wave. Although they documented the details of the growth and collapse of the steam explosion bubble generated during the interaction, it is not possible with regular high-speed photography to directly observe the fragmentation process because the production of vapor obscures the visibility of the drop. To view the breakup of the melt within the vapor bubble it is necessary to use flash X-ray radiography to "see" through the vapor.

The present experiments were carried out to investigate the fragmentation process during the vapor explosion of a melt drop with the use of simultaneous flash X-ray and regular photography. To illustrate the importance of thermal effects on the drop breakup, the fragmentation of a cold liquid metal drop is contrasted with that of a hot melt drop. The effect of melt properties on the explosion process is illustrated by comparing the explosion of hot drops of tin and a low melting point alloy. On the basis of the X-ray photographs of the explosion of the hot drops, a phenomenological model of the thermal fragmentation process is proposed. Finally, results from a simple model for the dynamics of the steam explosion bubble generated during an interaction are compared with the experimental results.

II. Experimental Facility

The experimental facility shown in Fig. 1 consists of a rectangular aluminum tank (10 cm wide, 25.5 cm long, 12.5 cm high) together with a cylindrical tube (3.8-cm inner diameter) that protrudes into the tank through the base. The high-voltage discharge system used for initiating the drop interaction is mounted at the base of the tube. Lexan windows were used for the X-ray photographs to minimize absorption of the X-rays. A Scandiflash 150 kV single head X-ray system generated a dose of 25 mR of soft X-ray radiation with an exposure duration of 35 ns. Kodak XAR-5 high-speed X-ray film was used in

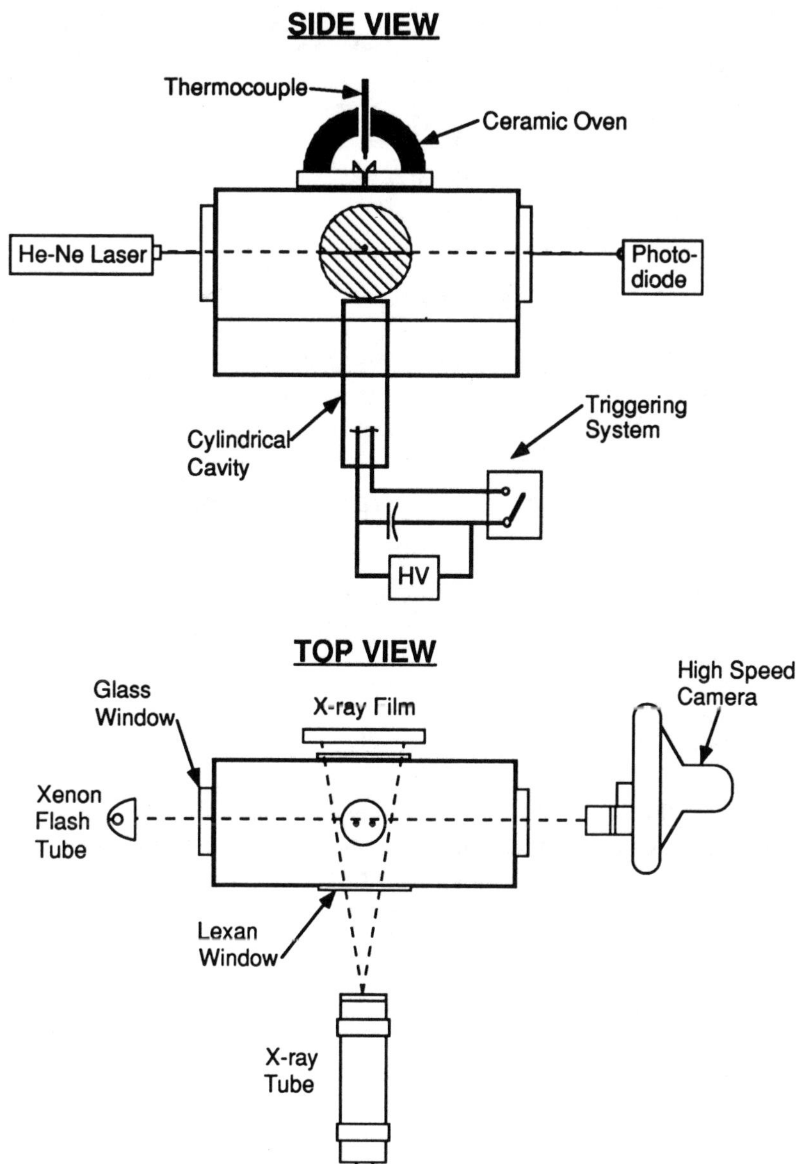

Fig. 1 Schematic diagram of the experimental apparatus.

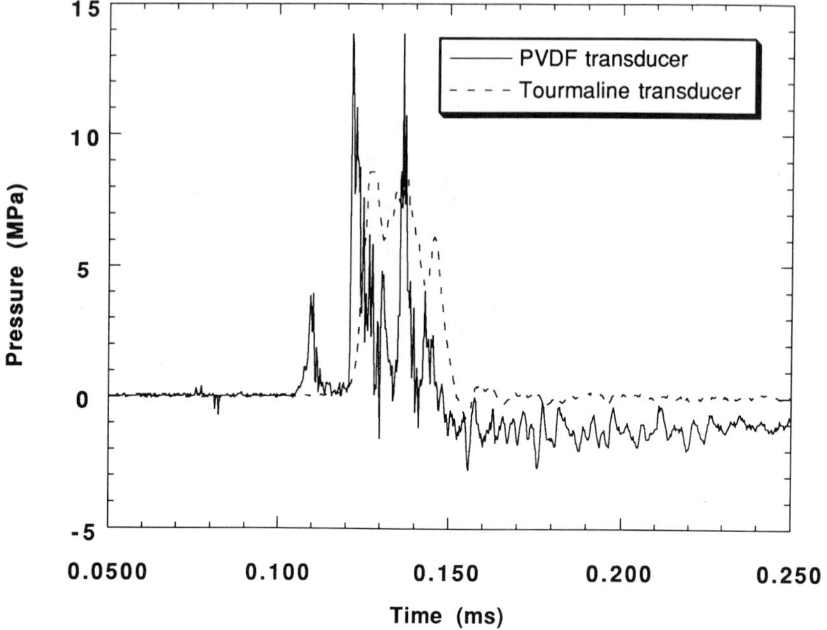

Fig. 2 Pressure near drop location from exploding wire trigger system alone, recorded simultaneously with two types of pressure transducers. At time zero the trigger system is fired.

conjunction with Kodak Min-R intensifying screens and a Kodak Min-R cassette. The width of the tank and the X-ray head–drop distance (55 cm) were chosen to provide the best contrast on the X-ray film. Regular photography was taken in a direction perpendicular to the X-ray radiographs through pyrex windows and an opal glass diffuser. High-speed photography utilized either a Hycam 16-mm camera operating at 4,000 frames/s or a Cordin 16-mm rotating drum camera operating at 25,000 frames/s. Backlighting was provided by discharging about 500 J stored in a capacitor bank through a xenon flash-tube producing a flash with a duration of 5 ms. Kodak TMAX 100 or 400 ASA film was used. High- resolution single-frame photographs were also taken with a 35- mm camera together with a spark light source. The oven used for heating the drop consists of a semicylindrical ceramic oven and a plunger system is used for dropping the drop. The fall distance from the drop to the surface of the water is about 4 cm.

The trigger system consists of an 8 μF capacitor, which is charged to 7 kV and discharged through a low-inductance circuit. During a trial, the discharge is initiated when the drop breaks a light beam from a low power He-Ne laser, causing a photodiode circuit to generate an electrical triggering pulse. To ensure that the drop hits the laser beam, a plastic funnel is submerged in the water to direct the path of the drop. The triggering pulse is then used, together with appropriate delay units, to fire the flash, exploding wire, and flash X-ray sequentially. The trigger pressure is measured with the use of underwater pressure transducers placed in the water near the drop location. Two types of pressure transducers are used: 1) an underwater tourmaline gauge (PCB Piezotronics 138A05, sensitivity 140 mV/MPa, quoted rise-time of 1.5μs), and 2) an underwater PVDF pressure transducer (Imotec, sensitivity 8.5 mV/MPa, rise-time 70 ns, sensitive diameter 0.5 mm). Signals from the pressure transducers are recorded with a LeCroy 9400A oscilloscope recording at 30 MHz. An example of the trigger pressure recorded with the two transducers at different locations in the test section in the *absence* of a drop is shown in Fig. 2. At t = 0 the exploding wire discharge system is initiated and about 135 μs later (corresponding to an exploding wire–drop distance of 20 cm) the shock arrives at the drop location. The pressure recorded consists of several pressure pulses due to reflections from the base and sides of the cylinder. The PVDF transducer captures the peak pressure transients accurately, but because of the small size of the sensing element, the charge (and hence the pressure) measured rapidly decays. The tourmaline transducer records a more integrated measure of the pressure transients and gives a more accurate estimate of the pressure-pulse duration, which is typically about 30 μs.

Two different drop materials were used in the experiments. Tin was chosen because of its low melting point, well-known properties, and its widespread use in steam explosion studies. For a direct comparison of the fragmentation of hot and cold drops, a low melting point alloy ($T_{m.p.}$ = 49°C) consisting of 45% bismuth, 23% lead, 19% indium, 8% tin, and 5% cadmium (commercially called *Cerrolow*) was also used.

III. Results

A. Fragmentation of a Cold Alloy Drop

To determine the relative importance of hydrodynamic and thermal fragmentation effects, experiments were first carried out to subject a cold liquid metal drop (i.e., at the same temperature as the surrounding water) to a blast wave. In this case, since the drop and coolant form an isothermal system, only hydrodynamic effects can play a role in the breakup of the drop. The fragmentation process is best illustrated by considering a series of X-ray photographs at different times. Figure 3 shows a series of X-ray photographs (each photograph corresponds to a different trial) showing the breakup of a 0.5 g drop of cerrolow alloy. The water temperature is 60°C so that the alloy drop remains a liquid at all times. The times shown are with respect to the time that the shock arrives at the drop location.

Initially the drop has an ellipsoidal shape. Mass is stripped continuously from the drop surface and is convected downstream of the drop into the wake region (the convective flow is from bottom to top). During the first several milliseconds, a surface layer of mass is drawn off, producing long distorted filaments extending from the drop downstream. At later times these filaments break up into fine particles as they are accelerated by the flow. The radiograph at 5 ms in

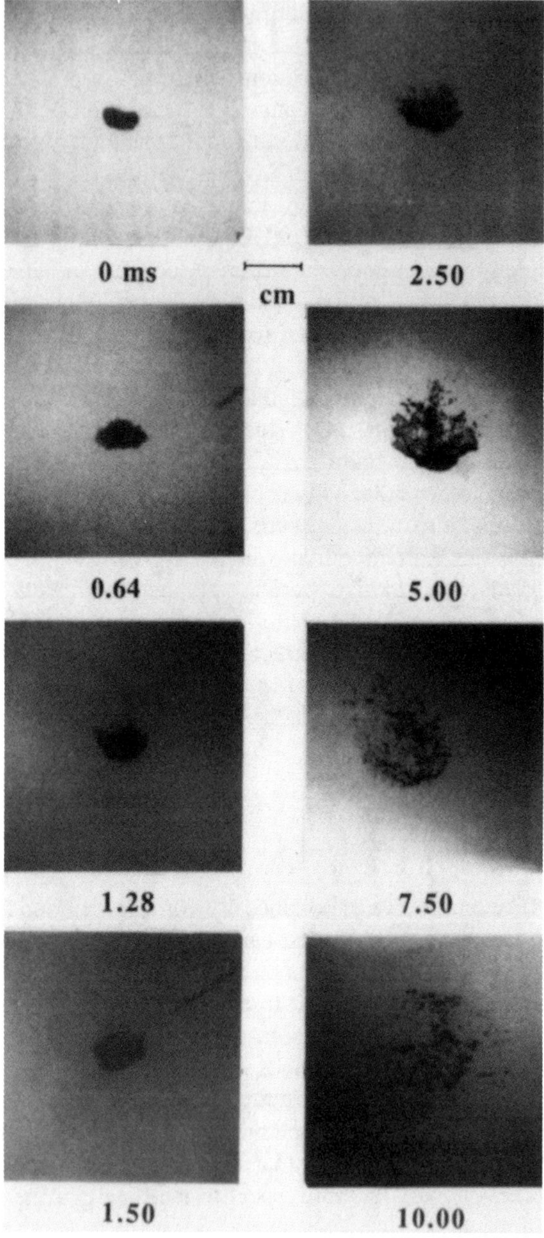

Fig. 3 X-ray photographs of the hydrodynamic fragmentation of 0.5 g drops of cerrolow alloy in thermal equilibrium with surrounding water (at 65°C). Each photograph corresponds to a different drop. Times shown denote the time after the triggering shock wave arrives at the drop location. Flow of water from the expansion of the steam bubble from exploding wire trigger at bottom of tank is from bottom to top.

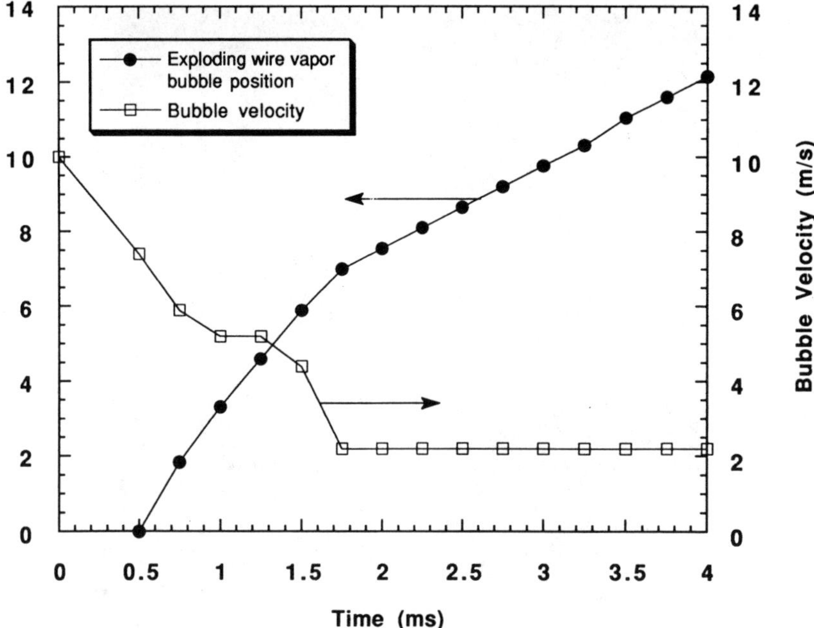

Fig. 4 Position and velocity of vapor bubble generated by trigger system.

Fig. 3 shows an example of the thin filaments extending from the surface of the drop and the micromist of fine particles that form in the wake region. This process continues until at a time of about 7.5 ms the drop appears to be completely fragmented. The drop remains liquid during the drop breakup, so after the interaction the fragments tend to coalesce at the bottom of the tank. This makes it impossible to determine an accurate distribution of sizes of the fragments generated during the interaction using the post-trial debris.

Since the duration of the pressure waves from the exploding wire trigger is only about 30 μs, it is clear that the fragmentation is a result of the convective flow generated by the expansion of the exploding wire vapor bubble. An estimate of the ambient flow velocity can be obtained by measuring the velocity of the vapor bubble generated by the exploding wire trigger. From regular high-speed photography, the bubble is visible within the test section after about 500 μs (for example, see Fig. 5). The bubble position and velocity are shown in Fig. 4 (taken from the trial shown in Fig. 5) as a function of time. The bubble rapidly decelerates and eventually reaches a constant velocity of about 2 m/s. Although the convective flow velocity is not constant and it is difficult to determine when the drop is "completely fragmented," an order of magnitude estimate for the nondimensional total breakup time can be made to compare with previous results. For example, if we take the average convective flow velocity of 4 m/s, a total breakup time of 7.5 ms with an initial drop diameter of 6 mm,

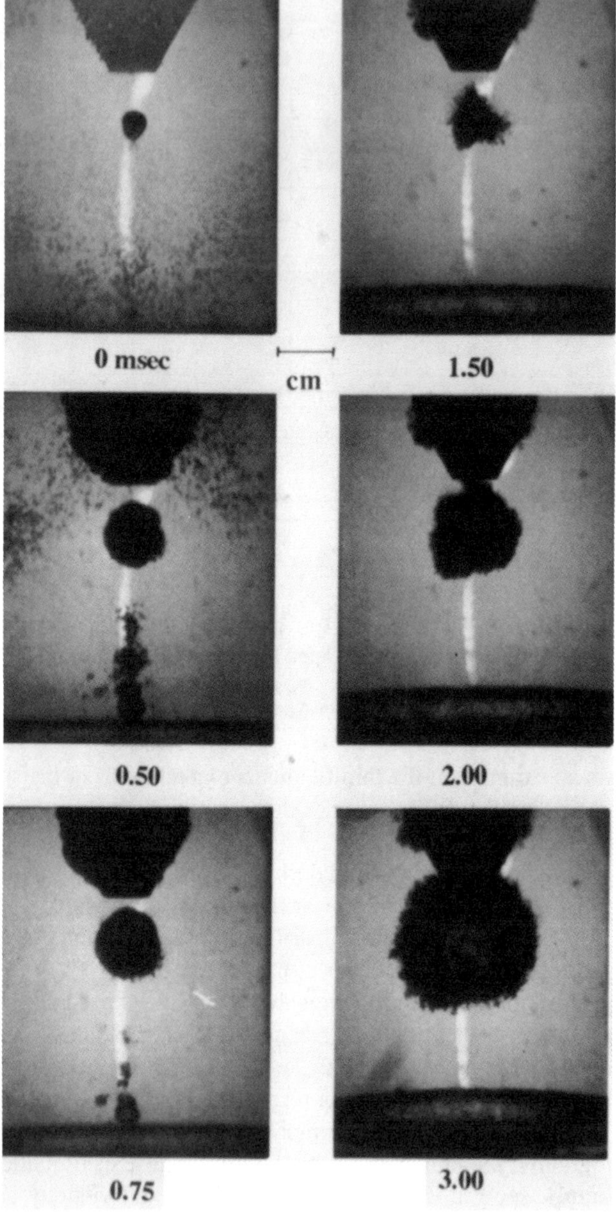

Fig. 5 High-speed photographs of the fragmentation of a 0.5 g drop of tin (at 700°C). Dark rectangle at top of each frame is a funnel immersed within the water to guide the drop. Bubble that comes into view at bottom is generated by exploding wire trigger.

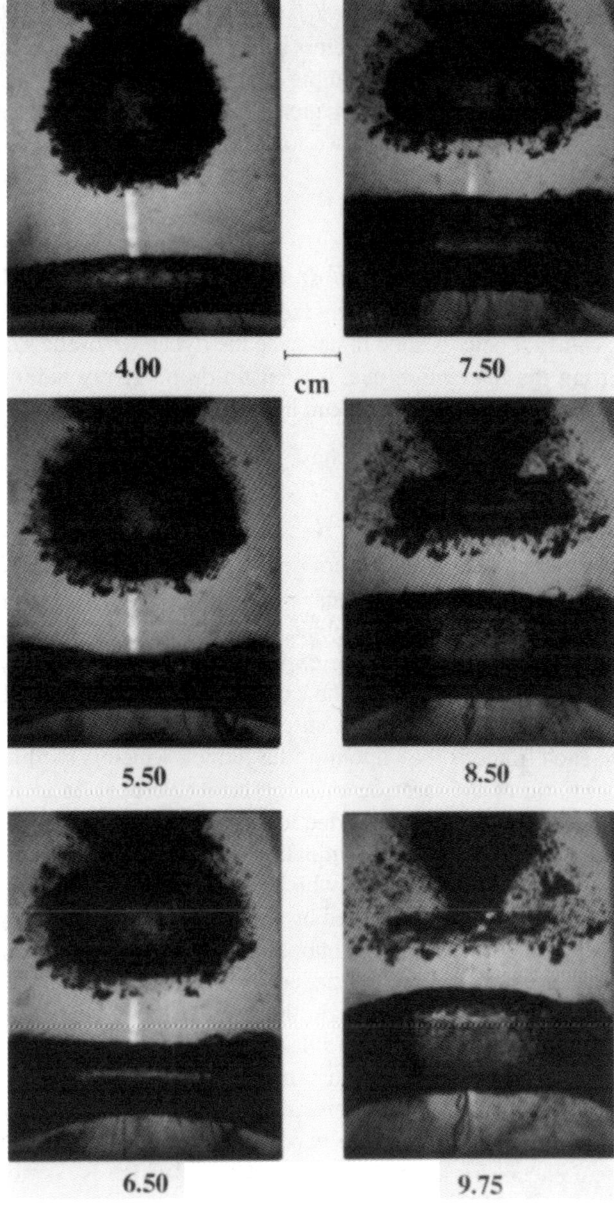

Fig. 5 (continued) High-speed photographs of the fragmenation of a 0.5 g drop of tin (at 700°C). Dark rectangle at top of each frame is a funnel immersed within the water to guide the drop. Bubble that comes into view at bottom is generated by exploding wire trigger.

then the nondimensional breakup time T is two (here $T = (\rho_f/\rho_d)^{1/2}Ut/d$ where the subscript d refers to the drop, f refers to the surrounding fluid, and U is the relative velocity between the drop and the fluid). This value is consistent with that obtained by earlier investigators (typical values measured for the nondimensional total breakup time range from two to five, Pilch and Erdman[13]).

B. Hot Drop Experiments

To illustrate the steam explosion of a single hot melt drop, both regular high-speed photographs and flash X-ray radiographs will be presented. Regular photographs will first be presented to describe the dynamics of the steam bubble generated during the interaction of a molten tin drop. X-ray radiographs will then be used to illustrate the drop breakup behavior.

1. Vapor Bubble Dynamics

During the steam explosion of a hot drop, a steam bubble is formed shortly after the triggering shock wave initiates collapse of the vapor film. An overall view of the dynamics of the steam bubble is shown in Fig. 5 which includes a series of Hycam photographs from the explosion of a 0.5-g tin drop initially at 700°C immersed in water at 65°C. The exit of the vertical tube containing the triggering system is visible at the bottom of each frame. The dark rectangle at the top of the each frame is the bottom of the funnel, which is used to guide the drop to the position of the laser beam. The bright vertical line which can be seen in the center of all the frames is due to a crack in the glass diffuser. At t = 0, the exploding wire is triggered, illuminating the funnel from below and generating cavitation bubbles in the water which are caused by the sharp expansion behind the initial shock wave generated by the exploding wire. The typical lifetime of the majority of the cavitation bubbles is about 300 μs. The bubble generated by the trigger system first appears at the bottom of the frame at a time of about 0.5 ms, moving upwards. The initial drop shape appears to be slightly elongated, indicating the presence of a small amount of air that is trapped around the drop when it first contacts the water surface. The shock arrives at the drop 135 μs later, and the steam bubble generated during the interaction grows to a maximum diameter at 0.75 ms. The vapor bubble at this time is almost perfectly spherical except for a small bulge on the top, which is due to the initial entrained air. The bubble overshoots its equilibrium radius and collapses to a minimum at about 1.5 ms. After the collapse of the first bubble a second much larger bubble is produced as the result of a second interaction between the water and the drop. This second bubble collapses asymmetrically at 9.75 ms due to the ambient coolant flow with no significant third bubble generated. Apparently only two bubble cycles are required to release the thermal energy of the drop. At the time of the second bubble collapse the drop is completely broken up and the

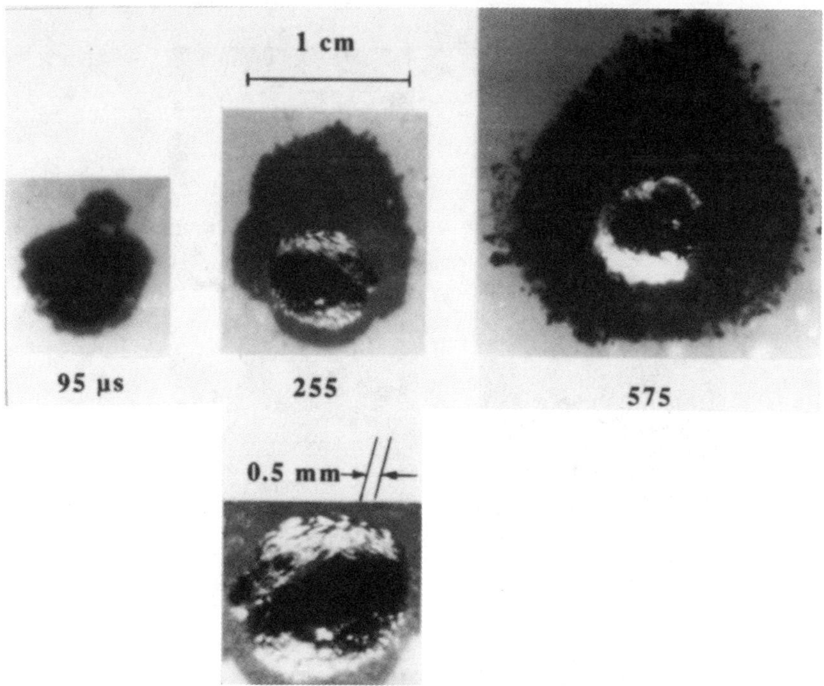

Fig. 6 Double exposures of explosion of 0.5 g tin drops at 700°C. Initial drop shape illuminated by exploding wire. Enlargement of drop in center shows surface ripples on initial vapor film.

fragments are uniformly dispersed over a volume roughly equal to that occupied by the second bubble at its maximum.

The initial growth of the steam bubble is illustrated in Fig. 6, which shows several *double*-exposure open-shutter photographs of exploding drops. Initiation of the exploding wire-trigger system generates a flash that illuminates the drop from below. A preset time later, a second flash is fired, showing the profile of the steam explosion bubble at that time. At time zero on the photographs, the shock arrives at the drop location. At 95 μs the expansion of the vapor bubble is clearly evident. An enlargement of the drop at 255 μs is shown, illustrating the ripples (with a typical scale of about 0.5 mm) that exist on the vapor film surrounding the drop. At 575 μs the vapor bubble is near its maximum diameter and drop fragments are visible in the water surrounding the bubble. These fragments are torn from the drop surface during the first interaction stage following the initial melt/water contact and accelerated during the expansion of the vapor bubble. The inertia of the fragments causes them to penetrate the bubble surface when the bubble begins to decelerate near its maximum diameter.

To obtain a more accurate measurement of the bubble growth, drop interactions were recorded with the Cordin camera running at 25,000 frames/s for a

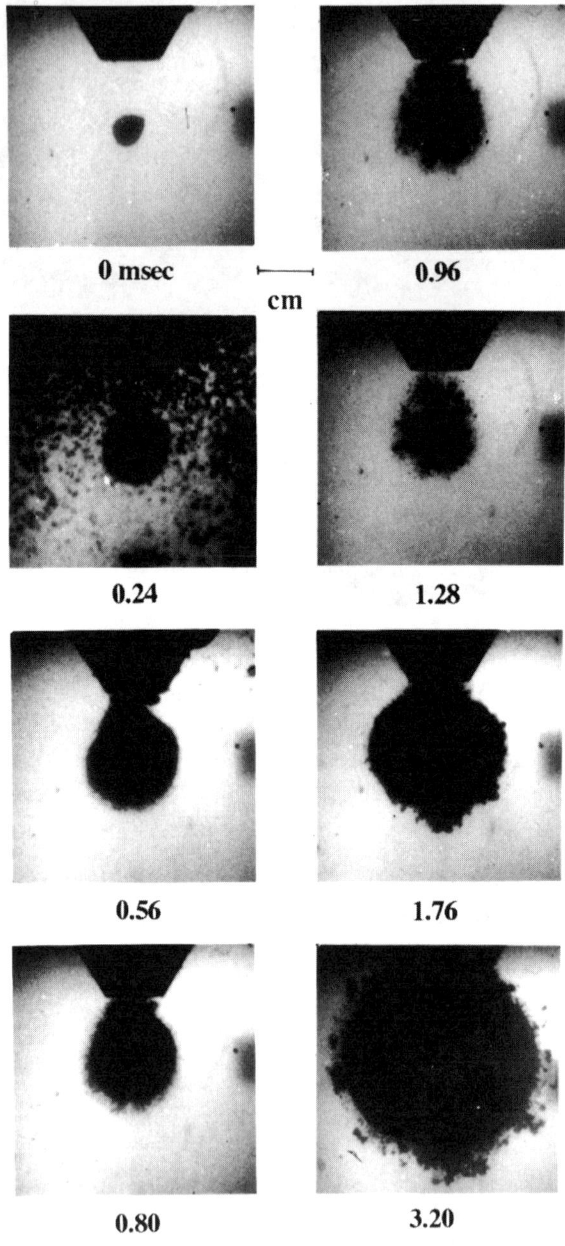

Fig. 7 High-speed photographs of the explosion of a 0.5 g drop of tin at 700°C immersed in water at 65°C.

temporal resolution of 40 μs. An example is shown in Fig. 7 for a drop with the same conditions as the trials described above. Shown in Fig. 8 is a plot of the horizontal bubble diameter versus time taken from every second frame from the high-speed film. The bubble diameter is normalized with the effective initial drop diameter, $d_{eff} = 5.1$ mm. The solid circles correspond to the frames shown in Fig. 7. Time $t = 0$ identifies the frame before the shock wave arrives at the drop location. The initial drop shape appears to be almost spherical except for a small bulge on the top right indicating the presence of a small amount of entrapped air. By 240 μs the vapor bubble has expanded to more than double the initial drop diameter. The bubble surface appears very smooth at this point in the expansion phase. Also visible in this frame are cavitation bubbles in the bulk of the water. At 560 μs the vapor bubble reaches its maximum horizontal diameter of about three times the initial drop diameter and melt fragments are visible in the water surrounding the bubble. Due to the upward water flow generated by the trigger bubble the steam explosion bubble is displaced upwards and the fragments first appear on the upstream side of the bubble.

At 800 and 960 μs the bubble is collapsing and leaves behind a trail of fine fragments. The vapor bubble collapses until a time of about 1.2 ms at which point the bubble is obscured by the fragments within the interaction region. The

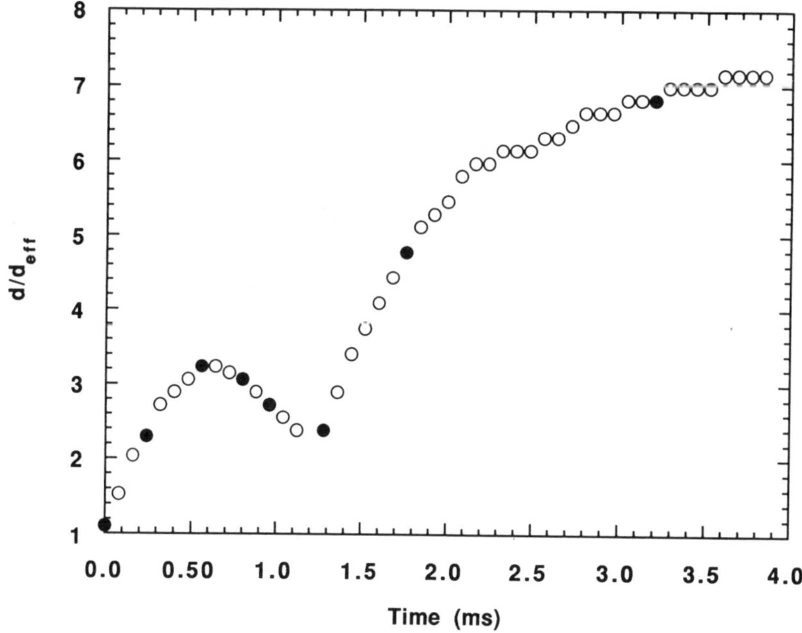

Fig. 8 Radial bubble growth time history for drop shown in Fig. 7. Diameter normalized with effective diameter of drop. Solid circles indicate frames shown in Fig. 7.

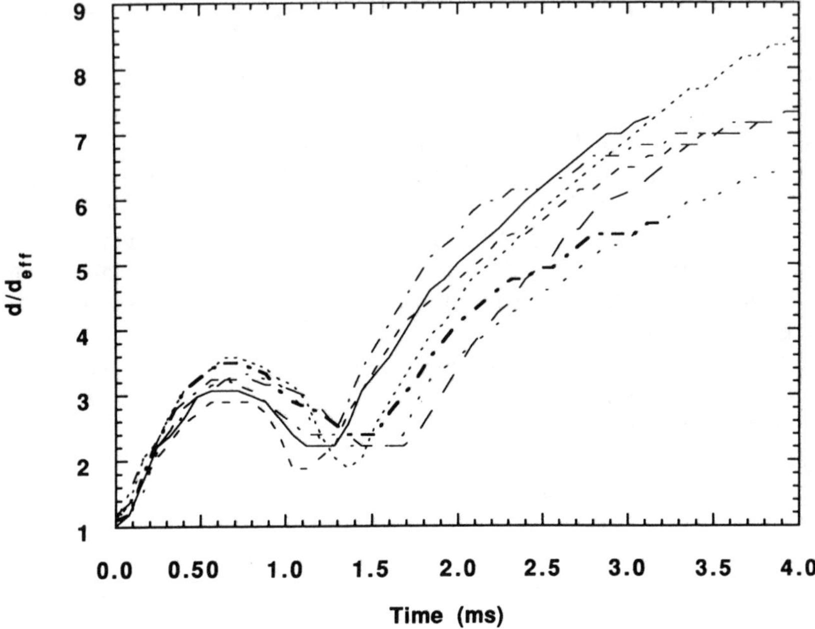

Fig. 9 Bubble growth curves for seven different trials with 0.5 g tin drops at 700°C.

rebound and expansion of a second vapor bubble is first visible at a time of 1.28 ms. No vapor production is evident around the small fragments in the water, indicating that the small fragments have solidified. The second vapor bubble expands, reaching a diameter of four times the initial drop diameter at 1.760 ms and reaches a maximum of over six times the initial diameter at a time of about 4 ms. Again, as in the first bubble maximum, fragments emerge from the vapor bubble as it approaches its maximum. The bubble then collapses to a minimum and no further bubble oscillations are observed.

Although there are small variations in the drop shape and vapor film thickness from one trial to the next, the overall shape of the bubble diameter–time profile is quite repeatable. Figure 9 shows a composite of seven trials illustrating the variation in the details of the normalized growth of the bubble diameter for the same initial drop conditions. The first vapor bubble maximum is typically about three times the initial drop diameter. Collapse of the first bubble occurs between 1.1 and 1.6 ms. There is more scatter in the growth of the second bubble, which depends on the complex heat transfer and fragmentation processes that occur during the first collapse.

To investigate the influence of melt fragmentation on the initial explosion and size of the first bubble maximum, experiments were carried out in a system in which no fragmentation was possible. In particular, a solid steel sphere with a similar diameter (5 mm) and initial temperature (700°C) as the hot tin drops

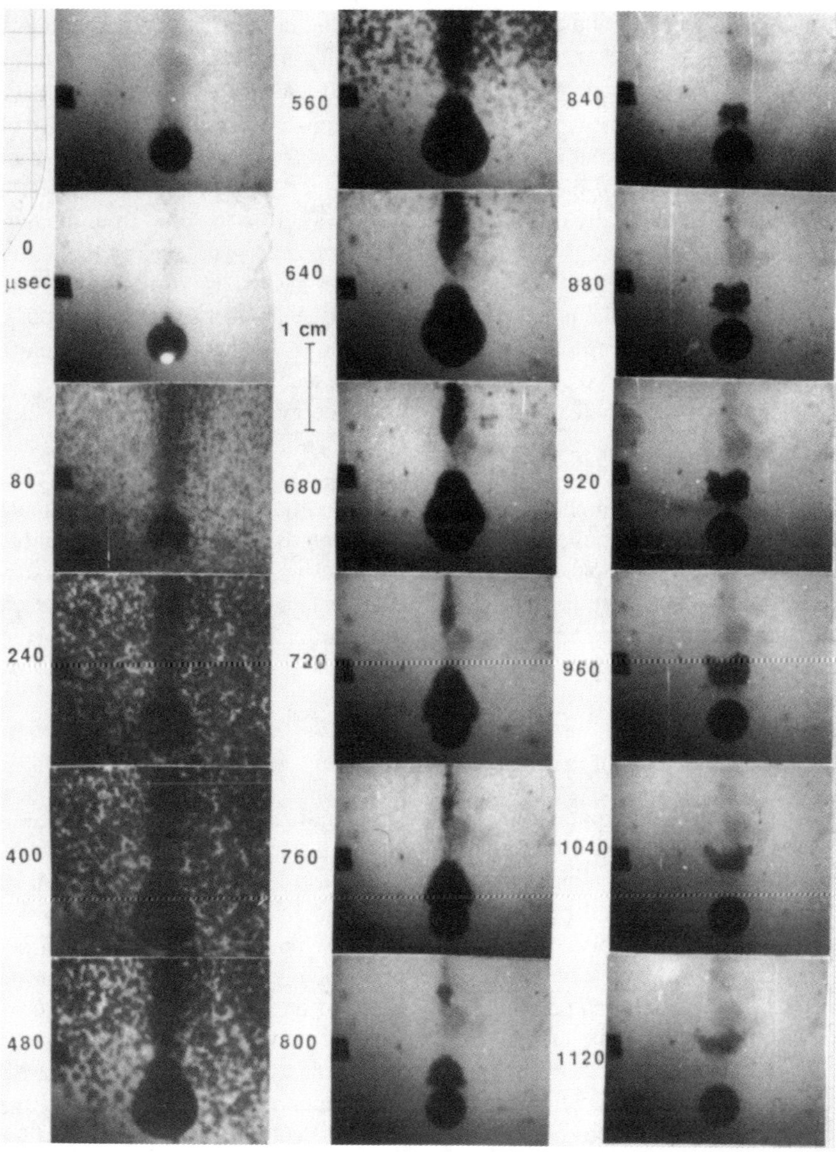

Fig. 10 High-speed photographs of the interaction of solid- steel sphere at 750°C with a shock.

were dropped into water and subjected to the same triggering shock conditions. Figure 10 shows an example of a solid hot sphere interacting with a shock and the growth of the vapor bubble around the drop. After the shock passes the drop, the vapor film surrounding the sphere collapses and a steam bubble is generated that grows to a maximum diameter about twice the initial sphere diameter about 500 μs later. The presence of a small bubble of gas or vapor on the top of the sphere (visible in the photograph at time $t = 0$) leads to an asymmetry in the bubble shape. The cavitation of the water in the tank generated by the passage of the triggering blast wave is most intense in the wake of the sphere because of the hot wake region of the sphere itself. When the steam bubble collapses, it collapses asymmetrically and appears to *peel* away from the drop from the bottom. This effect is due to the vertical convective flow generated by the steam bubble associated with the exploding wire trigger at the base of the tank. Note that after collapsing, the bubble rebounds slightly, then disappears after about 1 ms. In the absence of fragmentation, the heat due to conduction from the center of the drop to the surface is insufficient to form a second steam bubble.

For a hot solid sphere, most of the thermal energy contained near the surface of the sphere is transferred following the initial collapse of the vapor film surrounding the sphere. The steam bubble that is generated grows to a maximum size in a similar time as that for a molten drop, although the maximum bubble diameter attained is smaller. The similarity in the dynamics of the first bubble generated between a molten drop and solid sphere suggests that the energy transfer is limited to a thin surface layer of the drop and that significant fragmentation of the molten drop does *not* take place during the short time that the water is in contact with the drop prior to the generation of vapor and expansion of the first bubble.

2. Fragmentation of a Molten Tin Drop

With regular photography, very little information can be extracted concerning the dynamics of the drop fragmentation that occurs within the vapor bubble. In order to observe the fragmentation process, flash X-ray radiography (with simultaneous high-speed photography) was used in the present study to see through the vapor bubble. Since only one X-ray photograph can be taken per experiment, it is necessary to take a number of X-rays at different times during different experiments to obtain a time history of the drop breakup. Figure 11 shows such a composite of trials where the X-ray photograph is shown on the right and the regular photograph (taken simultaneously but in a perpendicular direction) is shown on the left. The times shown are normalized relative to t_c, the time for the first bubble collapse esti nated from the high-speed film record for each trial. In this way, the fragmentation behavior can be compared at different stages in the bubble growth cycle. Figure 12 shows a representative trial with the solid circles representing the times shown in Fig. 11.

At a time of $t/t_c = 0.22$ the vapor bubble has grown to twice its initial size. From the X-ray photograph, the surface of the drop appears highly perturbed

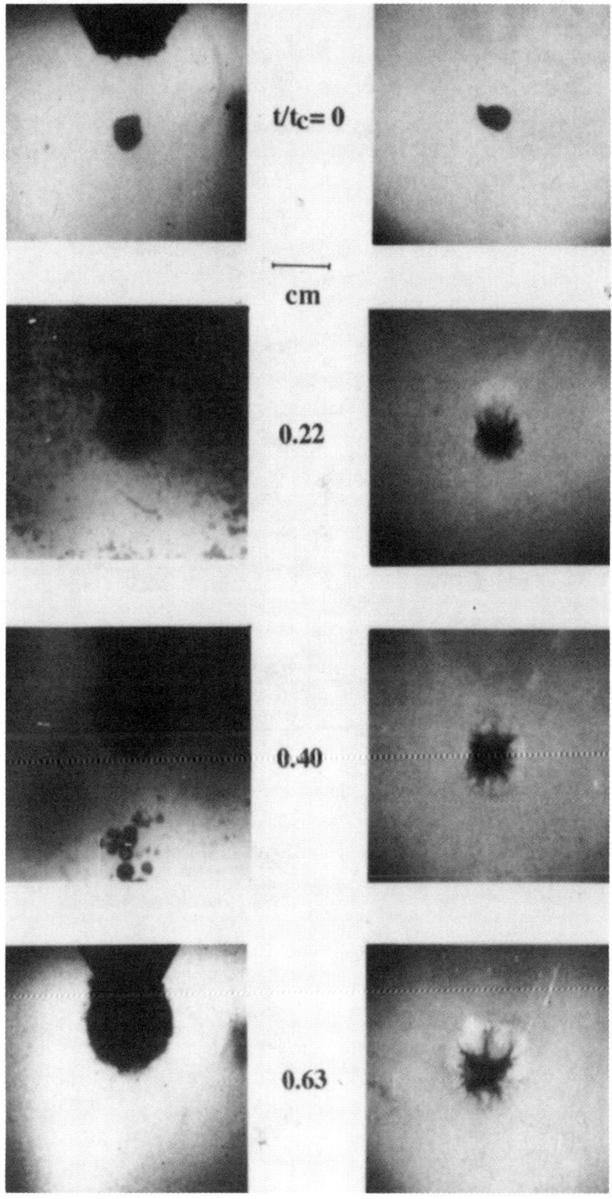

Fig. 11 Simultaneous X-ray radiographs (on the right) and regular photographs (on the left) of the vapor explosion of 0.5 g tin drops initially at 700°C. Each set of photographs corresponds to a different experiment. The times given are normalized with t_c, the time required for the collapse of the first bubble.

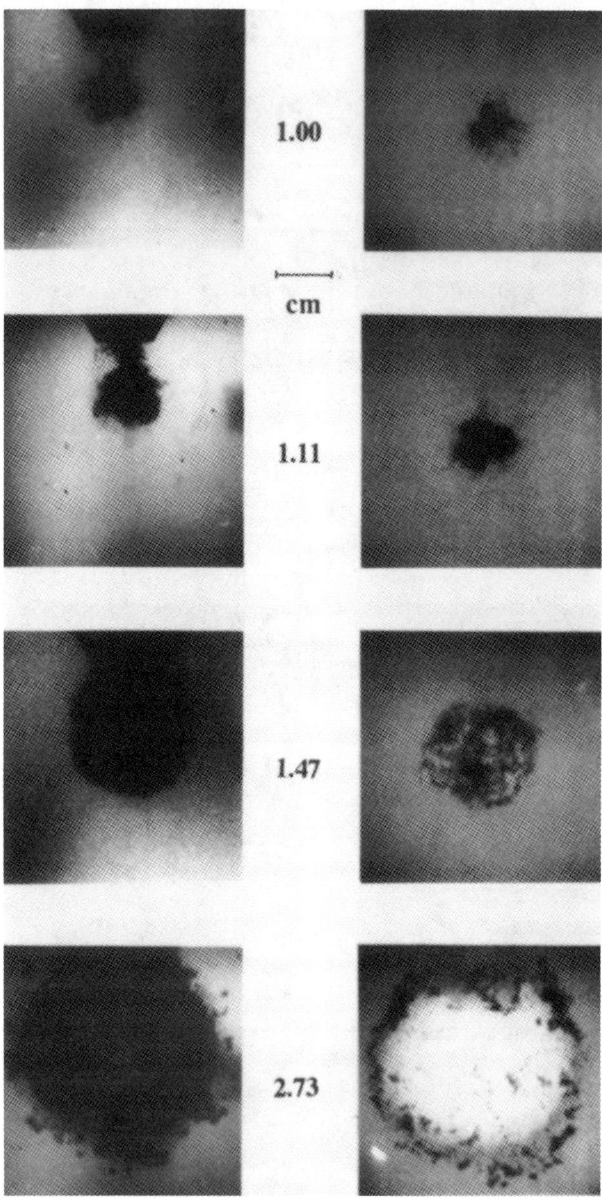

Fig. 11 (continued) Simultaneous X-ray radiographs (on the right) and regular photographs (on the left) of the vapor explosion of 0.5 g tin drops initially at 700°C. Each set of photographs corresponds to a different experiment. The times given are normalized with t_c, the time required for the collapse of the first bubble.

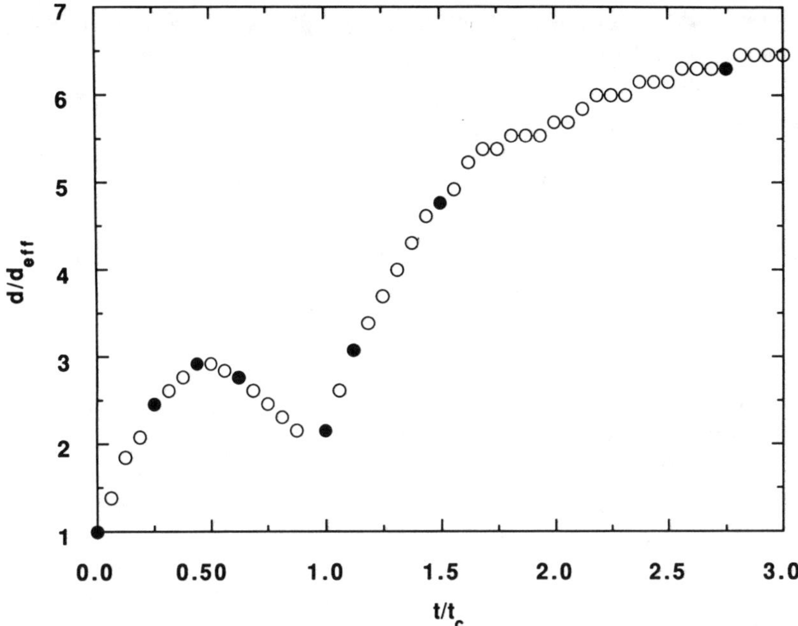

Fig. 12 Radial bubble growth time history for tin drops shown in Fig. 11. Solid circles indicate photographs shown in Fig. 11.

with thin filaments of metal extending radially out from the drop symmetrically around the drop. At $t/t_c = 0.4$ the bubble has almost reached its maximum diameter. At this time, growth of the protruding spikes of metal is apparent and close inspection of the X-ray photograph shows that a thin "shell" of fragments is present inside the bubble near the bubble surface. At $t/t_c = 0.63$ the drop surface is highly convoluted and the vertical displacement of the vapor bubble is evident. Fine fragments are visible near the spikes probably due to freezing and shattering of the smallest filaments of metal. Note that the distortion of the drop occurs *inside* the vapor bubble where there is little drag to impede the breakup. This can be compared with the breakup of a cold drop where in the absence of vapor it takes considerably more energy (and time) to disperse the drop fragments in a *liquid* medium. At bubble collapse the regular and X-ray photographs look similar. The impact of the incoming water with the drop shatters the fine spikes, and the subsequent heat transfer and turbulent motion of the vapor generated leads to total fragmentation of the drop. At a nondimensional time of 1.11, the second high-pressure steam bubble is expanding and, from the X-ray photograph, the orderly melt fingers evident at early times have been completely destroyed. What remains is an expanding compact region containing fragments, water droplets, and vapor. As the bubble grows the fragments are carried out with the expanding vapor. At a time of $t/t_c = 1.47$, the fragments are dispersed throughout the bubble and continue to move radially outwards as the bubble decelerates. As the bubble approaches the second maxima ($t/t_c = 2.73$) the inside of the bubble is largely void of particles.

3. Fragmentation of a Molten Cerrolow Alloy Drop

To study the effect of drop properties on the drop fragmentation behavior, a second series of experiments was carried out with 0.5 g drops of cerrolow alloy with the same trigger conditions an: initial conditions (i.e., T_{drop} = 700°C, T_{water} = 65°C) as for the molten tin drops. The properties of the alloy differ from tin in that it is heavier (density = 9.2 g/cm^3 vs. 7.0 g/cm^3 for tin), has a lower heat capacity (c_p = 154 J/kgK vs. 223 J/kgK for tin), and remains liquid throughout the interaction.

Figure 13 shows the horizontal bubble diameter time history for five trials with cerrolow alloy drops (cf. Fig. 9 for tin drops). The overall shape of the curves is similar to that obtained for tin drops. The first bubble maximum is a similar size as for the tin drops, although the second bubble reaches a smaller maximum size (d/d_{eff} ~ 4-5) as compared to tin (d/d_{eff} ~ 6-9). This is probably due to the fact that at the same temperature, the tin drops contain about 50% more thermal energy than the alloy drops and the majority of the thermal energy is released during the collapse of the first bubble which determines the eventual size of the second bubble.

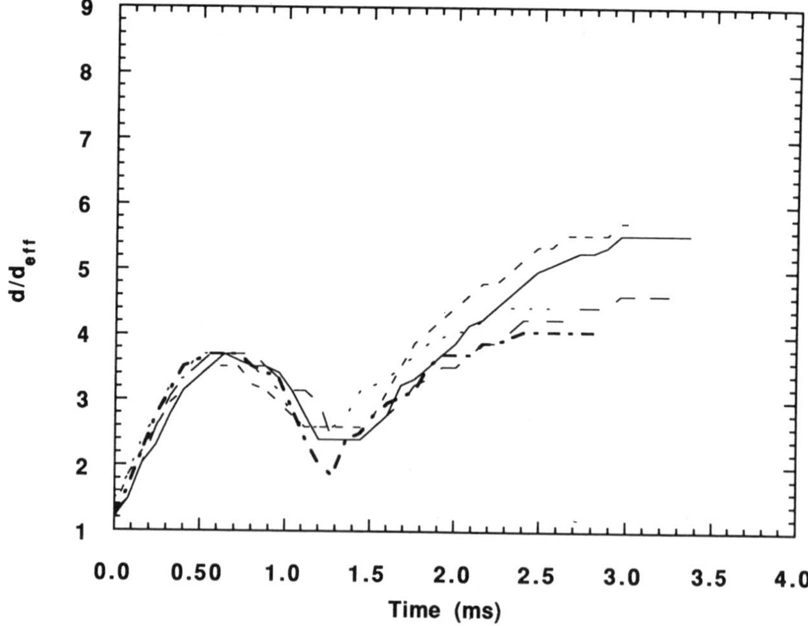

Fig. 13 Bubble growth curves for five different trials with 0.5 g cerrolow alloy drops at 700°C.

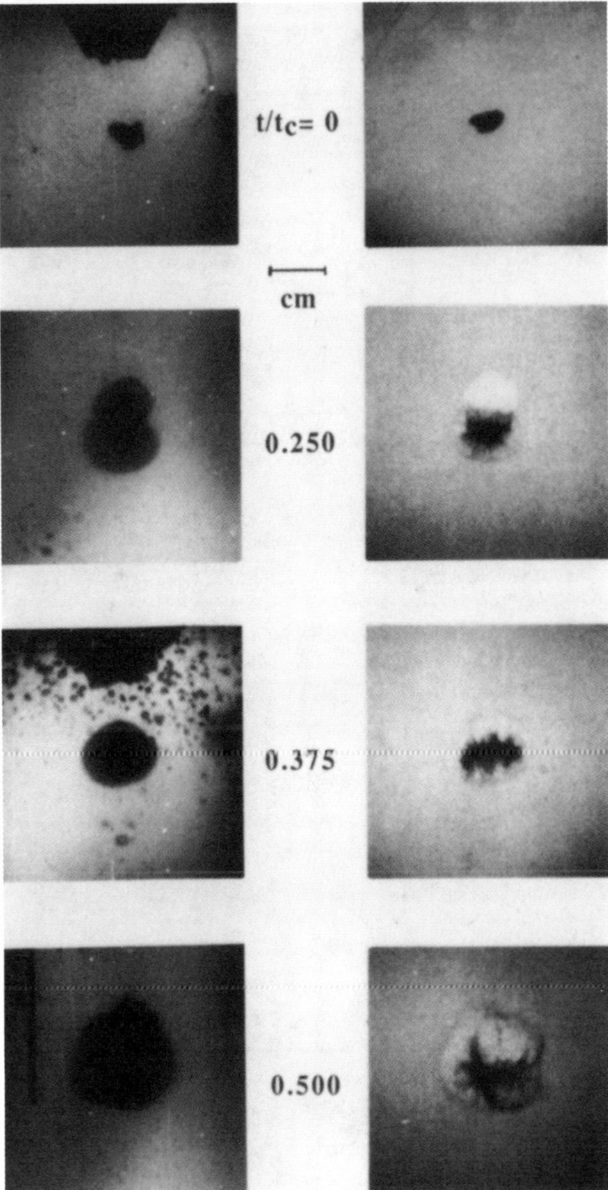

Fig. 14 Simultaneous X-ray radiographs (on the right) and regular photographs (on the left) of the vapor explosion of 0.5 g alloy drops initially at 700°C. The times given are normalized with t_c, the time required for the collapse of the first bubble.

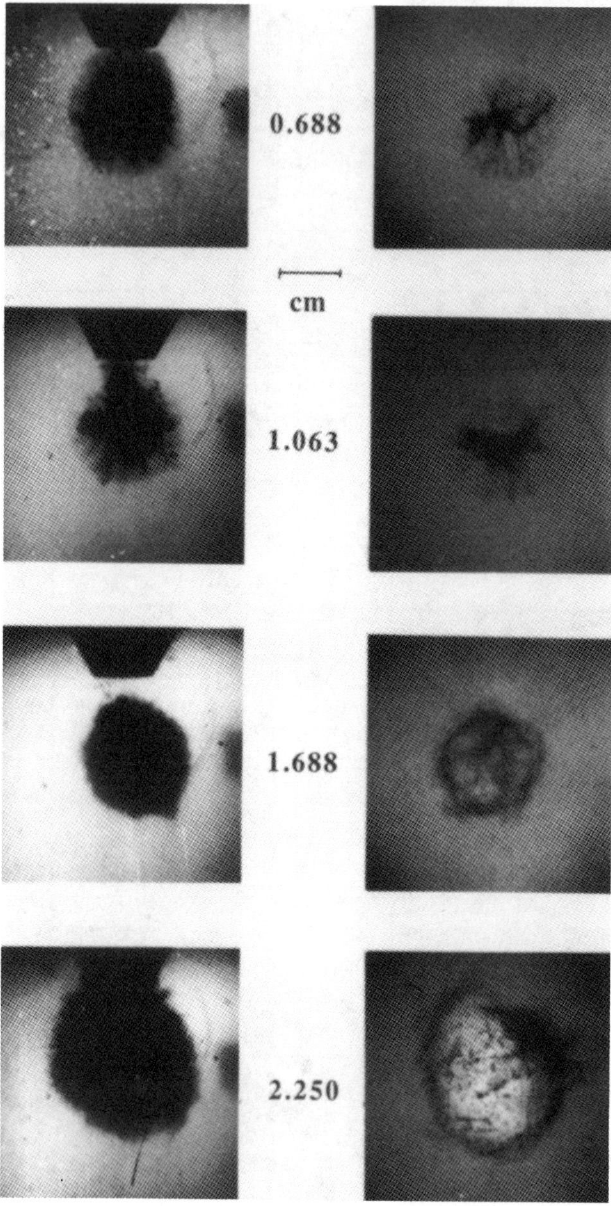

Fig. 14 (continued) Simultaneous X-ray radiographs (on the right) and regular photographs (on the left) of the vapor explosion of 0.5 g alloy drops initially at 700°C. Each set of photographs corresponds to a different experiment. The times given are normalized with t_C, the time required for the collapse of the first bubble.

Figure 14 shows a series of trials for alloy drops showing simultaneous regular (on the left) and X-ray (on the right) photographs (cf. Fig. 11 for tin drops), and Fig. 15 shows the characteristic bubble growth history with the solid circles again representing the trials shown in Fig. 14. At the time of the arrival of the shock ($t/t_c = 0$) the initial drop has an ellipsoidal shape as compared with the more spherical tin drops (due to the higher surface tension of tin). At $t/t_c = 0.25$, the drop surface appears highly distorted, with filaments of melt again extending radially from the surface of the drop. At $t/t_c = 0.375$ the drop appears elongated in the horizontal direction relative to the initial drop diameter. The X-ray photograph taken with the bubble at its maximum diameter ($t/t_c = 0.5$) shows that the drop is highly deformed. The highly turbulent motion of the vapor within the bubble plays a role in the large deformation of the melt fingers and parent drop. The absence of any freezing effects is in contrast with molten tin drops where freezing of the surface of the melt filaments probably limits the extent of deformation. As the bubble collapses ($t/t_c = 0.688$) the drop appears very elongated forming thin strands. Shortly after the collapse of the bubble, the hydrodynamic impact appears to have shattered the distorted drop into fine fragments and the surface no longer appears continuous. As the bubble expands the fragments are dispersed and eventually penetrate the water as the bubble decelerates.

IV. Discussion

A. Vapor Bubble Dynamics

Insight into the explosion process can be obtained by analyzing the growth of the vapor bubble generated. If the energy deposited in the bubble following film collapse occurs in a time that is short relative to the time for bubble expansion and if the mass flux across the bubble surface due to evaporation or condensation is negligible then a good approximation to the bubble growth will be given by the solution to the classical Rayleigh-Plesset equation for bubble dynamics, i.e.,

$$R\ddot{R} + \frac{3}{2}\dot{R}^2 = \frac{1}{\rho}\{p_B - p_\infty - 2\sigma/R - 4\mu\dot{R}/R\} \tag{1}$$

where R, $\dot{R}$, $\ddot{R}$, ρ, p_B, p_∞, σ, and μ are the bubble radius, radial velocity, radial acceleration, liquid density, pressure in the bubble, pressure at infinity, liquid surface tension, and viscosity, respectively. Conservation of energy for the bubble, assuming a closed system, can be written,

$$\frac{dU}{dt} = \dot{Q} + p\frac{dV}{dt} \tag{2}$$

where U and V are the internal energy and volume of the bubble and $\dot{Q}$ is heat transfer to the bubble from the hot melt drop. If an estimate for the heat transfer

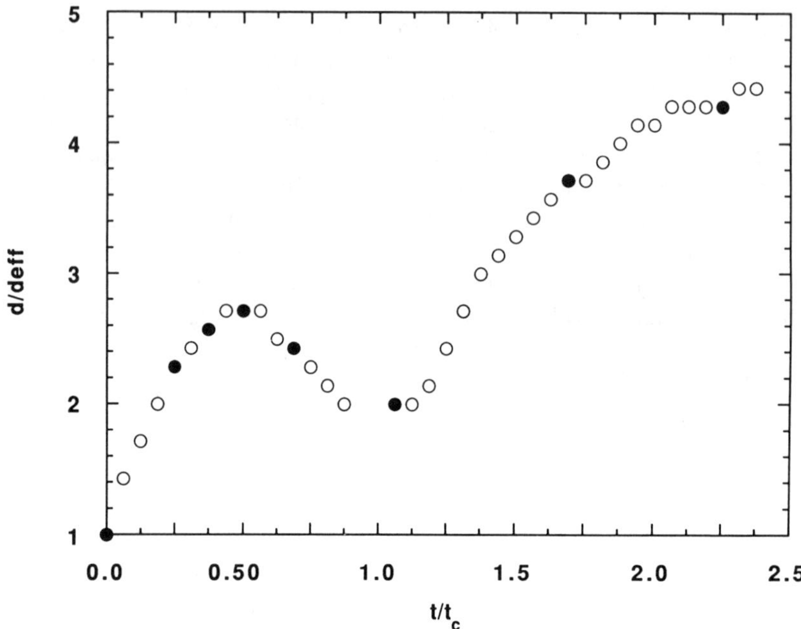

Fig. 15 Radial bubble growth time history for alloy drops shown in Fig. 14. Solid circles indicate photographs shown in Fig. 14.

$\dot{Q}$ is made, then Eqs. (1) and (2) together with an appropriate equation of state $U = U(p,V)$ and initial conditions can be solved for the time history of the bubble growth.

To estimate the initial conditions (i.e., initial bubble pressure and volume) it is necessary to consider the early stage of the interaction. When the triggering shock first arrives at the drop location, film collapse occurs within the duration of one frame (~ 40 μs). After about two frames (i.e., ~ 80 μs) the growth of the vapor bubble is first visible. So the time during which there is liquid-liquid contact and rapid heat transfer is on the order of one frame duration, or about 40 μs. When the melt first contacts the water, the interface temperature T_i can be estimated from

$$T_i = \frac{T_m\beta_m + T_w\beta_w}{\beta_m + \beta_w} \tag{3}$$

where $\beta = k/(\alpha)^{1/2}$ where k is the thermal conductivity and α is the thermal diffusivity, and the subscripts m and w refer to the melt and water, respectively. For melt and water temperatures of 600°C and 65°C, respectively, the interface temperature is estimated from Eq. (3) to be 524°C. If we assume that diffusive heat transfer occurs between the melt and water for about 40 μs, then the mass of water that is heated above the superheat limit of water is about 0.3 μg. If this

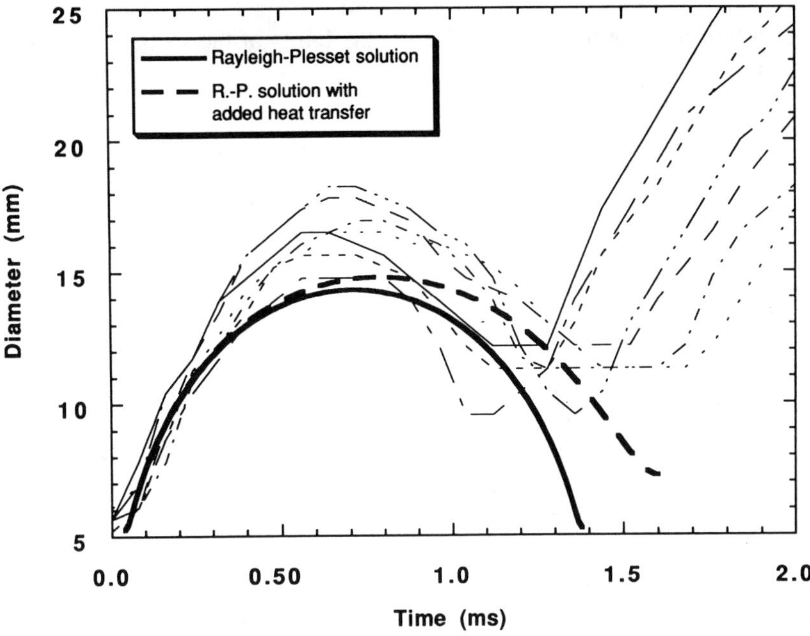

Fig. 16 Comparison of the predictions of the Rayleigh-Plesset equation (with and without added heat transfer from the drop to the vapor bubble) with experimental bubble growth curves.

mass of water is assumed to vaporize at the superheat limit, then the initial pressure in the bubble will be about 90 bar. To estimate the rate of heat transfer from the drop to the bubble, a heat transfer coefficient of 10^3 was used based on previous investigators (e.g., Fletcher and Thyagaraja[14]), and it was assumed that the area available for heat transfer increases by an order of magnitude during the bubble growth due to the distortion of the drop surface. This yields maximum heat transfer rates on the order of 250 W.

Equations (1) and (2) were solved with the assumption of ideal gas behavior and the initial conditions described above, and the results are shown in Fig. 16 for the case of no heat transfer as well as with the assumption of additional heat transfer from the drop to the bubble during expansion. Also shown in Fig. 16 are the experimental curves taken from Fig. 9. The Rayleigh-Plesset solution reproduces the shape of the bubble growth curves as well as the period quite well considering the assumptions made in estimating the initial conditions. However, the maximum bubble diameter is underpredicted. Heat transfer to the bubble during the bubble expansion increases the peak diameter and the period of the bubble oscillation a small amount. To obtain a more accurate model for the bubble dynamics it is necessary to provide a model for the nonequilibrium evaporation process as well as a better estimate of the fragmentation (and hence heat transfer) rate for the drop during the bubble growth. The added complexity of such a model is not justified considering the limits of the spatial and temporal resolution of the experimental results.

B. Phenomenological Model for Fragmentation

Models that have been proposed in the past for the fragmentation mechanism for a hot drop interaction were based on indirect evidence from regular high-speed photography. Flash X-ray radiography allows the fragmentation of the melt drop to be observed directly. The fragmentation mechanisms described earlier in the introduction will now be reconsidered in light of the experimental data and a new phenomenological model for the fragmentation process will then be described based on the flash X-ray results.

From a direct comparison between the fragmentation of cold and hot cerrolow alloy drops (cf. Figs. 3 and 14), it is clear that direct hydrodynamic fragmentation of the drop by the relative flow of the ambient water cannot account for the breakup of a hot drop. From the X-ray photographs, there is also no evidence that the impact of the water with the drop following film collapse directly causes fragmentation of the drop. Therefore the energy required to break up the drop must be derived from the thermal energy of the drop itself, i.e., heat transfer during liquid-liquid contact leads to the rapid generation of high-pressure vapor that expands doing mechanical work on the drop surface. The X-ray photographs do not show any evidence that water is injected *into* the drop by the formation of water jets formed during asymmetric film collapse. If the formation of coolant jets is the primary mechanism for drop fragmentation, then the explosion process should be quite sensitive to the initial shape of the vapor film (i.e., thickness and asymmetry) as well as the strength of the triggering shock wave which will determine the amount of coolant injected into the drop. However, recent results by Frost and Ciccarelli[15] indicate a very weak dependence on the triggering strength for shock pressures up to 20 MPa and that the triggering process serves only to collapse the vapor film and initiate the subsequent explosion. They also found that the explosion of a drop with a large attached air bag was qualitatively similar to the explosion of a drop with only a thin vapor film surrounding the drop.

The fragmentation of a hot drop can be divided conceptually into three stages: 1) collapse of the initial vapor film and formation of the melt spikes (t = 0–100 μs), 2) growth of perturbations on drop surface and distortion of drop by turbulent vapor flow inside bubble (t = 100 μs – 1 ms), and 3) collapse of bubble, impact of the water with the drop, and fine fragmentation of the bulk of the drop (t > 1 ms). Figure 17 shows a schematic of a scenario for the events that occur following film collapse. Prior to film collapse, small-scale (~ 0.5 mm) ripples are present on the vapor film surrounding the drop (see Fig. 6) that are caused by Kelvin-Helmholtz instability as the drop descends in the water. During shock-induced film collapse (picture b of Fig. 17) of the vapor film, these perturbations will grow on the film surface due to Rayleigh-Taylor instability. After the passage of the triggering shock, the collapsing film will first

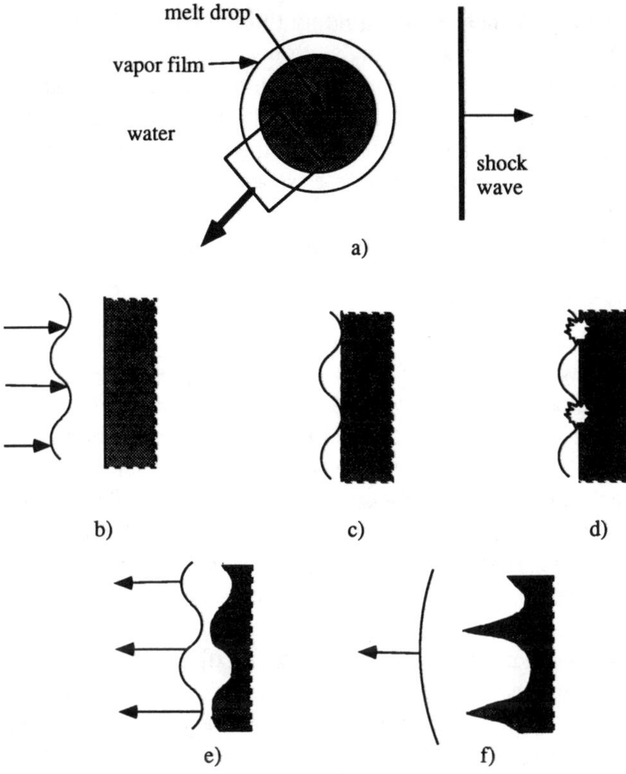

Fig. 17 Schematic of scenario for fragmentation of melt drop following film collapse.

contact the drop at a series of local points (picture c). Rapid heat transfer and the formation of high-pressure vapor bubbles at these points will lead to a nonuniform pressure distribution on the surface of the drop. The local generation of high-pressure vapor at the drop surface will cause the formation of a surface wave and a small crater or indentation on the drop surface (pictures d and e). The metal squeezed between the indentations form thin filaments of metal that are ejected radially from the surface at high speed (picture f in Fig. 17). The filaments then break up into small fragments. The shape of the drop surface at early times (see Figs. 11 and 14) suggests that the drop surface is distorted by strong localized forces much as a ball of putty is deformed when squeezed sharply by hand; extruding the putty between the fingers. The mechanism of melt jet formation by rapid bubble growth at the melt surface has been explored further in experiments in an analogous system which will be described later.

The filaments of melt that extend from the drop surface continue to grow as the vapor bubble expands. The highly turbulent flow of vapor within the bubble further distorts the surface of the drop. This is particularly pronounced for the cerrolow alloy drops which remain a liquid throughout the interaction. During the expansion of the bubble, although the drop surface appears severely distorted and some fine fragments are present within the bubble and near the bubble surface, the parent drop appears to be largely intact. When the bubble collapses and contact between the water and the drop is re-established, a second melt/water interaction occurs. The hydrodynamic force associated with the inrush of the water and the subsequent rapid boiling effectively shatter the remainder of the drop. The melt fragments are then dispersed by the expansion of the vapor.

B. Formation of Melt Jets

To gain insight into the mechanism of the formation of the jets of melt that are visible at early times in the interaction, further experiments were carried out in a stratified water/liquid metal system. The experiments were used to qualitatively investigate the behavior of a liquid metal surface following the generation of local high pressure vapor bubbles on the surface. The apparatus consists of a narrow channel (5 cm wide and 1.2 cm thick) filled to a height of 5 cm of liquid metal with an equal height of water above the metal. A low melting point alloy was used (Wood's metal) and the apparatus was heated above the melting point of the alloy with a surface heater. Two exploding wires were placed less than a millimeter above the metal surface in the water and located 2.5 cm apart. The exploding wires were connected in parallel to a 8 μF capacitor charged to 4 kV and discharged simultaneously, generating two line sources of high-pressure vapor just above the liquid metal surface. The subsequent vapor bubble growth and growth of the surface perturbations were recorded with regular and X-ray photography.

Figure 18 shows four X-ray radiographs (from different experiments) at different times, illustrating the growth of waves on the liquid metal surface. At $t = 0$, prior to triggering the exploding wires, the tips of the electrodes holding the wires are just visible above the metal surface. At $t = 1$ ms, the metal surface has become highly perturbed with cavities forming at the locations of the electrodes. By 2 ms a definite structure has emerged. Four distinct surface waves have formed on the metal surface, two between the electrodes moving towards each other and one on either end moving towards the side walls. There are also three very fine filaments of metal which form at the exploding wire locations and one at the midpoint between the exploding wires. After 5 ms the two surface waves between the exploding wires have merged forming one large spike at the midpoint. The surface waves on the sides have collided with the walls and have also formed spikes. The three large spikes are remarkably symmetric. The two fine filaments at the exploding wire locations are still present and appear to have grown vertically and spread laterally slightly. At later times the spikes break up into fine particles.

The initial waves formed on the liquid metal surface by the impulsive disturbances are reminiscent of the waves observed by Naugolnykh et al.[16] during the local generation of high-pressure vapor on a liquid surface with a laser beam. The formation of liquid metal jets has also been observed by Greene et al.[17] in

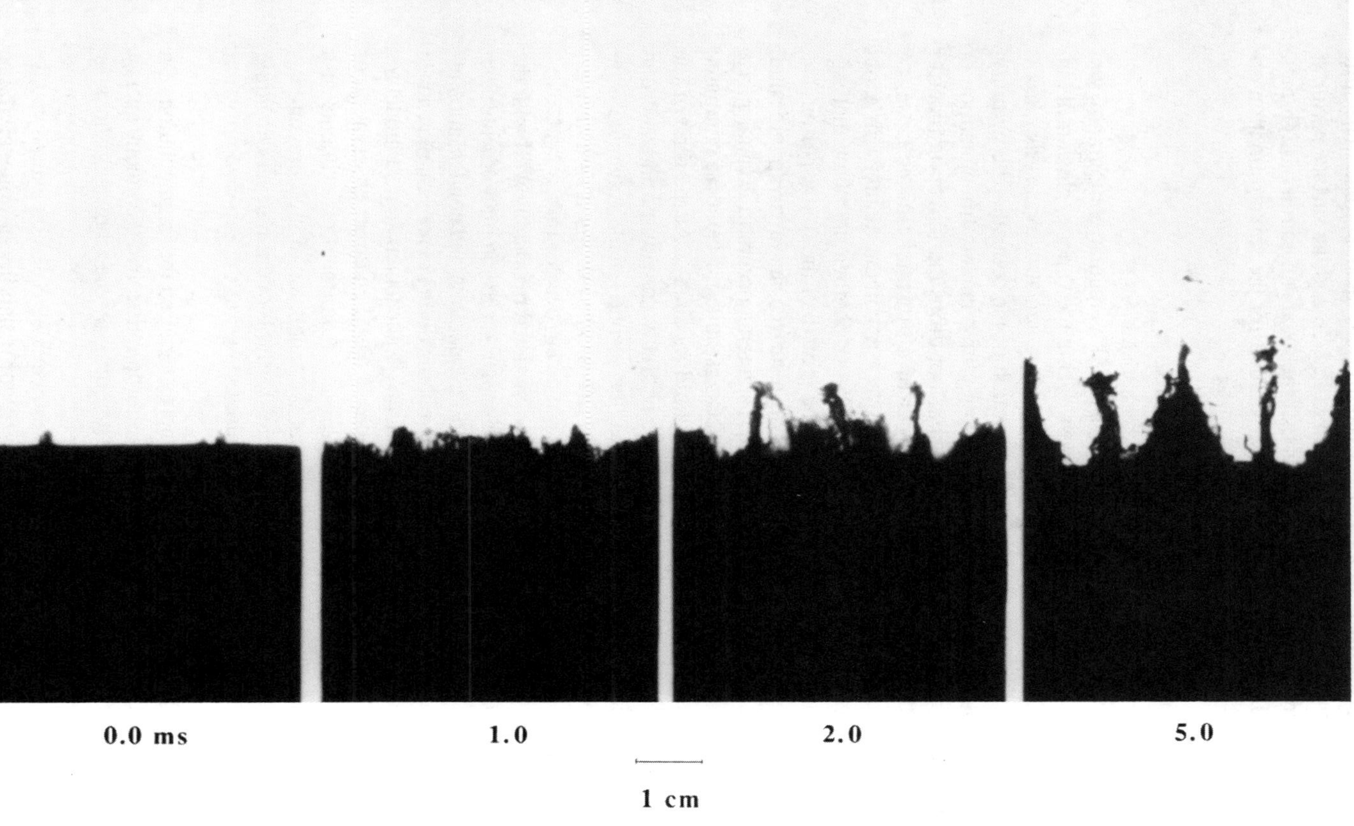

Fig. 18 X-ray radiographs of the growth of disturbances at a liquid metal/water interface subject to rapid bubble growth at the interface at two locations. Two exploding wires, located 2.5 cm apart, are visible just above the surface of the metal in the first radiograph.

film boiling experiments with a stratified water/liquid metal system. The similarity in appearance between the jets formed in Fig. 18 and the surface perturbations in the hot drop experiments suggests that the local generation of high-pressure vapor at an interface is a plausible explanation for the spike formation observed in Figs. 11 and 14.

V. Conclusions

In the present study X-ray radiography as well as regular photography was used to investigate the fragmentation process during the vapor explosion of single drops of melt. For cold liquid metal drops fragmentation occurs as mass is stripped off the surface due to the relative motion of the coolant. The hydrodynamic fragmentation process is relatively slow with complete fragmentation of the drop evident only after about 7.5 ms. For hot drops the induction time for the initiation of the explosion is on the order of 40 μs after film collapse. A first vapor bubble is generated, which grows to a maximum diameter in about 600 μs (for 0.5 g drops) and then collapses. X-ray photographs show that significant distortion of the drop occurs during the growth of the first vapor bubble. Fine filaments of metal are ejected from the drop surface and break up into small fragments. Since the dispersion of the fine fragments occurs in a vapor-phase medium with little drag, the dispersion process is quite efficient (as compared with the dispersion of fine fragments in a liquid medium). At the time of the collapse of the first bubble, the drop surface is highly convoluted and a second interaction is initiated. For molten tin drops, the second bubble generated is always larger than the first bubble. For hot cerrolow alloy drops, a similar interaction occurs although the second bubble is not as large as for tin.

The dynamics of the vapor bubble can be modeled reasonably well with the Rayleigh-Plesset equation together with an estimate for the amount of water that participates in the interaction. Nonequilibrium evaporation and condensation effects and heat transfer from the drop to vapor bubble during expansion play a role in the difference between theory and experiment. A phenomenological model for the fragmentation mechanism has been proposed based on direct observation of the drop breakup with flash X-ray visualization. Additional experiments in a stratified liquid metal/water system show that the generation of high-pressure vapor bubbles at the metal/water interface leads to the formation of jets of liquid metal similar to that observed during the initial stages of the explosion of a hot drop.

The present experiments show that the energy required for fragmentation of a hot drop is derived from the thermal energy of drop rather than the kinetic energy of the ambient flow or direct shattering of the drop by the triggering shock or film collapse. The majority of current models for the escalation and propagation stages of a vapor explosion include fragmentation models based on hydrodynamic fragmentation mechanisms (e.g., Rayleigh-Taylor instability or boundary layer stripping). It is likely in the initial escalation stage of a coherent vapor explosion, when system pressures are still relatively modest (i.e., subcritical), that thermal fragmentation effects will dominate, particularly for low-energy melts. Only at higher pressures and flow velocities will hydrodynamic fragmentation play a dominant role. More information is required on the fragmentation, heat transfer, and momentum exchange mechanisms before significant improvement in the state-of-the-art models can be made.

Acknowledgments

The authors would like to thank J. H. S. Lee of McGill University for many stimulating discussions and J. Shepherd for his comments on the analysis of bubble dynamics. Assistance with the experiments involving X-ray photography was provided by R. Katofsky. Financial support was provided by the Natural Sciences and Engineering Research Council of Canada, the Atomic Energy Control Board of Canada, and Sandia National Laboratories.

References

[1]Reid, R. C., "Rapid Phase Transitions from Liquid to Vapor," *Advances in Chemical Engineering* , Vol. 12, 1983, pp. 105-208.

[2]Cronenberg, A. W., and Benz, R., "Vapor Explosion Phenomena with Respect to Nuclear Reactor Safety Assessment," *Advances in Nuclear Science and Technology,* Vol. 12, 1979, pp. 247-334.

[3]Buchanan, D. J., "A Model for Fuel-Coolant Interactions," *Journal of Physics D., Applied Physics*, Vol. 7, 1974, pp. 1441-1457.

[4]Kim, B. J., "Heat Transfer and Fluid Flow Aspects of a Small-Scale Single Droplet Fuel-Coolant Interaction," PhD Thesis, University of Wisconsin, Madison, WI, USA, 1985.

[5]Drumheller, D. S., "The Initiation of Melt Fragmentation in Fuel-Coolant Interactions, "*Nuclear Science and Engineering* , Vol. 72, 1979, pp. 347-356.

[6]Ochiai, M., and Bankoff, S. G., "Liquid-Liquid Contact in Vapor Explosions," *Proceedings of the International Conference on Fast Reactor Safety,* American Nuclear Society, Chicago, IL, October 5-8, 1976.

[7]Corradini, M. L., "Analysis and Modelling of Steam Explosion Experiments," Sandia National Laboratory Report, SAND80-2131, 1981.

[8]Fletcher, D. F., and Anderson, R. P., "A Review of Pressure-Induced Propagation Models of the Vapour Explosion Process," *Progress in Nuclear Energy,* Vol. 23, 1990, pp. 137-179.

[9]Oh, M. D., and Corradini, M. L., "A Propagation/Expansion Model for Large Scale Vapour Explosions," *Nuclear Science and Engineering,* Vol. 95, 1987, pp. 225-240.

[10]Burger, M., Miller, K., Buck, M., Cho, S. H., Schatz, A., Schins, H., Zeyens, R., and Hohmann, H., "Analysis of Thermal Detonation Experiments by Means of a Transient Multiphase Detonation Code," *Proceedings of the 4th Topical Meeting on Nuclear Reactor and Thermal Hydraulics*, Karlsruhe, K.F.G. 10-13 October 1989, Vol. 1.

[11]Corradini, M. L., Kim, B. J., and Oh, M. D., "Vapour Explosions in Light Water Reactors: a Review of Theory and Modelling," *Progress in Nuclear Energy*, Vol. 22, 1988, pp. 1-117.

[12]Nelson, L. S., Duda, P. M., Fröhlich, G., and Anderle, M., "Photographic Evidence for the Mechanism of Fragmentation of a Single Drop of Melt in Triggered Steam Explosion Experiments," *Journal of Non-Equilibrium Thermodynamics*, Vol. 13, 1988, pp. 27-55.

[13]Pilch, M., and Erdman, C. A., "Use of Break Up Time Data and Velocity History Data to Predict the Maximum Size of Stable Fragment for Acceleration-Induced Break Up of a Liquid Drop," *International Journal of Multiphase Flow*, Vol. 13, No. 6, 1987, pp. 741-757.

[14]Fletcher, D. F. and Thyagaraja, A., "A Mathematical Model of Melt/Water Detonations," *Applied Mathematical Modelling*, Vol. 13, 1989, pp. 339-347.

[15]Frost, D. L., and Ciccarelli, G. "Propagation Mechanisms of Molten/Fuel Moderator Interactions," *Atomic Energy Control Board of Canada Report*, INFO-0382, June, 1991.

[16]Naugolnykh, K. A., Rybak, S. A., P'ichenkov, O. V., and Zosimov, V. V., "Laser Stimulated Instabilities of Waves in a Subsurface Layer of a Liquid," *Nonlinear Waves in Active Media,* edited by J. Engelbrecht, Springer-Verlag, Berlin, 1989.

[17]Greene, G. A., Park, N. A., Burson, S. B., Klages, J., Klein, J., Sanborn, Y., and Schwarz, C. E., "Some Observations on Simulated Molten Debris-Coolant Layer Dynamics," *Proceedings of the International Meeting on Light Water Reactors,. Severe Accident Evaluation*, Vol. 2, Cambridge, Mass., Aug. 28-Sept. 1, 1983.

Onset of Boiling Liquid Expanding Vapor Explosion

C. K. Chan* and K. N. Tennankore*
Whiteshell Laboratories, Pinawa, Manitoba, Canada
and
C. A. McDevitt† and F. R. Steward‡
University of New Brunswick, Fredericton, New Brunswick, Canada

Abstract

Two sets of experiments were performed to examine the onset of boiling liquid expanding vapor explosions (BLEVE). Commercially available 1-L containers were used in the outdoor BLEVE experiments (where the containers usually disintegrated completely). Indoor experiments were performed in a steel channel equipped with glass windows to allow the boiling process to be observed photographically. Either Freon 12 or Freon 22 was used in these two sets of experiments. Results from this study show that a BLEVE follows the initial depressurization due to the venting of the pressure-liquified gas through a break that leaves a part of the remaining fluid in the container in a superheated state. A local explosion caused by a rapid evaporation resulting from the boiling of the superheated fluid occurs in the vicinity of the break. The blast wave from such an explosion is the cause for the catastrophic failure of the container observed in many accidents.

*Scientist, Containment Analysis Branch, AECL Research.
†Research Assistant, Mechanical Engineering Department.
‡Professor, Mechanical Engineering Department.

Introduction

According to Ref. 1, the occurrence of a boiling liquid expanding vapor explosion (BLEVE) is as follows: "When a pressure-liquified gas tank is heated by an external fire which heats the metal wall at the vapor space level, the vessel will rupture into a number of large pieces. Because of the rapid evaporation of the depressurized liquid, these pieces of the container will rocket considerable distances. This is accompanied by a large fireball and some explosive pressure effects produced from the liquid rapidly expanding during the propagation of fracture as the vessel ruptures." Such a phenomenological description of BLEVEs does not reveal the nature of the explosion. For example, the mechanisms causing the tank to rupture are not clarified. As Walls[2] pointed out, a BLEVE is a physical process and is independent of the chemical content of the pressurized liquid. Therefore, the fireball observed in some accidents is merely a consequence of a BLEVE. Presently, the mechanism which causes the catastrophic failure of the container is not understood. In a recent book on explosion hazards, Baker et al.[3] described the onset of a BLEVE as "for some reason, the ductile tank tears." Pressure-time histories obtained in recent BLEVE experiments[4-6] show a pressure increase after the initial decrease in pressure suggesting that the phenomenon consists of two steps: venting of superheated fluid followed by a vapor explosion within the container. This idea was also proposed by Reid.[7] However, for highly transient phenomena, the pressure development within the containment is not uniform. As a result, a definitive conclusion regarding the nature of BLEVEs cannot be drawn from the pressure trace alone. We believe that the initial rupturing of the tank is due to a combination of a local weakening of the material caused by high temperature (heated by an external fire) and the pressurization resulting from the increased vapor pressure of the liquid. However, this rupture does not cause the whole tank to tear apart. An explosion resulting from the rapid boiling of the superheated liquid, which occurs after some fluid has vented out, is responsible for the tearing of the tank and its disintegration.

In a recent study on BLEVEs with propane and Freon 12, McDevitt et al.[8] observed that a BLEVE could be initiated by heating the container to a desired temperature and rupturing the container with a bullet shot from a rifle. They also established a set of criteria for a BLEVE to occur in terms of the volume of fluid and its initial temperature in those specific tanks. Their results suggested that the bursting of the container into pieces depended upon the physical properties of the fluid and occurred after the container had been partially depressurized. That is to say, the catastrophic failure of the container depended on the rate of depressurization and the subsequent boiling of the fluid. However, in the absence of pressure measurements inside the container, this could not be substantiated.

As a continuation of the earlier work[8], we have recently completed a study of the initial part of the phenomenon, the onset of a BLEVE. The objective of this study was to understand the processes that cause the tank to tear apart. To this end, additional diagnostics were employed. Rapid boiling (flash-evaporation) and generation of pressure waves inside the container were examined using piezoelectric pressure transducers and spark schlieren photography.

Experimental Apparatus

Two sets of experiments were performed in this study. The first set was similar to McDevitt's earlier work[8] in which commercially available 1-L containers filled with Freon 12 or Freon 22 were used. The container was heated externally using propane burners. At the desired temperature, the container was ruptured by a bullet shot from a rifle. If appropriate conditions were present, a BLEVE occurred. Besides using thermocouples to monitor the initial temperature, a piezoelectric pressure transducer was also mounted on the container to monitor the pressure variation during a BLEVE.

The second set of experiments was performed in a 2.5 cm x 3.8 cm, 91-cm-long steel channel (see Fig. 1). This channel, filled with Freon 12 or Freon

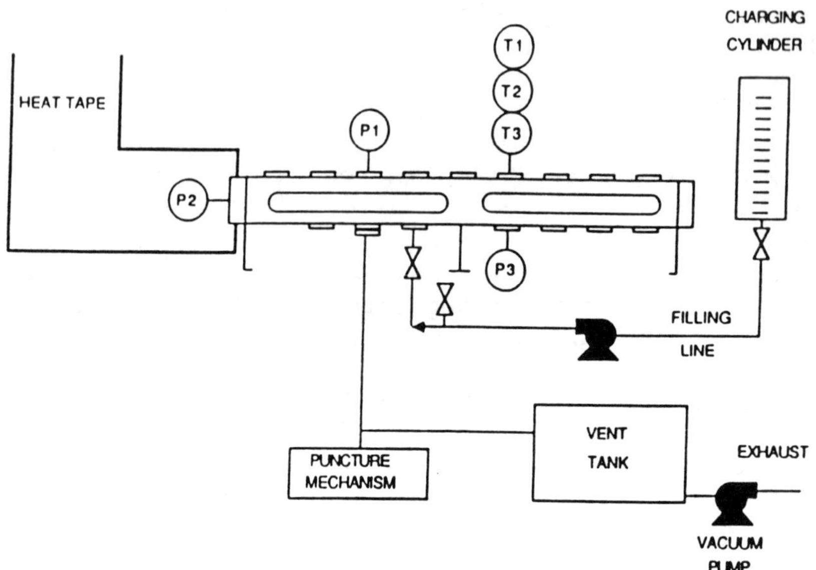

Fig. 1 Schematic of the experimental apparatus.

22, was separated from a discharge tank via a metal diaphragm. The channel, charged with Freon using a positive displacement pump, could be heated electrically to 100°C. The temperature of the fluid and the channel was monitored using thermocouples (T1, T2, and T3). The experiment was initiated by breaking the metal diaphragm (5.0 cm^2 in area) with a plunger. Piezoelectric pressure transducers mounted at various locations (P1, P2, and P3) recorded the pressure development. The channel was also equipped with glass windows to record the boiling process using spark schlieren photography.[9]

Results and Discussion

The outdoor BLEVE experiments with 1-L containers yielded results similar to those of earlier experiments.[4] If the volume of the fluid and its temperature exceeded certain critical values, rupturing of the container led to a catastrophic failure (tearing of the container into several pieces). If these critical conditions were not exceeded, the subsequent evaporation of the fluid did not tear the container into pieces. In the former case, a BLEVE is considered to have occurred as a result of the initial rupture.

Figure 2 shows the overpressure-time history at one location inside the container, recorded during an outdoor BLEVE experiment. The fluid inside the container was Freon 22 heated to 65°C. The initial pressure inside the container was about 2700 kPa (400 psi). The trace shows that the pressure dropped slightly, then rose to a maximum of 800 kPa above the initial pressure. This gives a maximum pressure of 3500 kPa. This result suggests that a BLEVE occurs after the initial partial depressurization caused by venting of the fluid and is the result of rapid vapor generation caused by the boiling subsequent to the depressurization. The fluctuations in the pressure trace reflect the violent boiling that occurs after the initial depressurization. Frost[10] demonstrated that unstable boiling of a superheated liquid can create huge pressure waves. If the container cannot withstand the blast wave from this subsequent explosion, it will be torn into pieces. However, these pressure traces do not reveal the amount of fluid involved in the explosion nor do they answer the question of why some depressurizations do not lead to explosions.

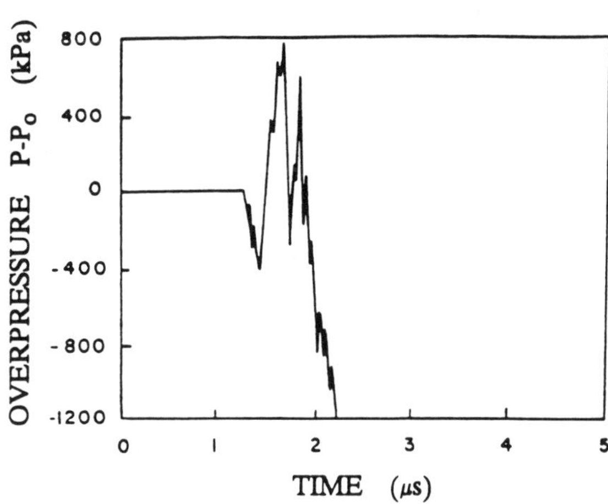

Fig. 2 Overpressure-time history during a BLEVE of Freon 22 in a 1-L Container (initial pressure P_o = 2700 kPa).

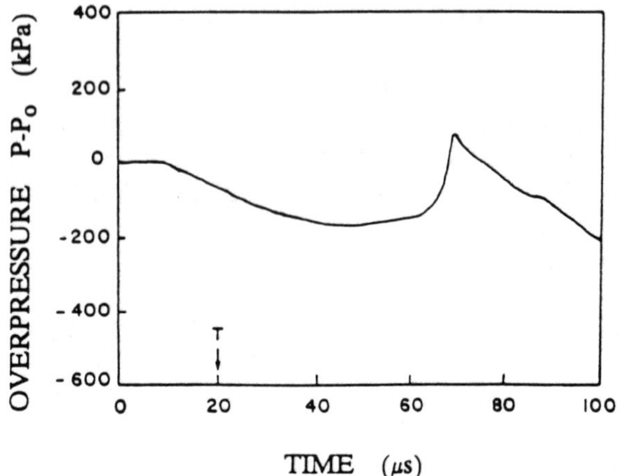

Fig. 3 Overpressure-time history (P1) of a BLEVE experiment with Freon 12 at 90°C (initial pressure P_o = 2770 kPa, T indicates the instant of arrival of the expansion waves).

Experiments performed in the steel channel with Freon 12 provided a better understanding of the phenomenon. In these experiments, besides using pressure transducers, the vapor explosion was examined by spark schlieren photography through the glass windows. Figure 3 shows the overpressure-time history at location P1 (see Fig. 1). To prevent overpressurization, the channel was not filled completely with liquid Freon. Liquid Freon heated to 90°C occupied roughly 94% of the total volume (the remainder being Freon vapor). The initial pressure of the fluid was 2770 kPa. The pressure transducer mounted on the top wall of the channel is located in the vapor phase; as a result, Fig. 3 can only qualitatively represent the pressure in the liquid. Since the oscilloscope recording the pressure was triggered by the pressure signal itself, the zero time on the trace does not represent the instant of breaking of the diaphragm. This instant was estimated to be 270 μs before the instant of arrival of the expansion waves as shown in Fig. 3 (based on the estimated sonic velocity of Freon 12 at 90°C, 270 m/s.) The trace shows a slight drop in pressure followed by a sharp increase. Even though the

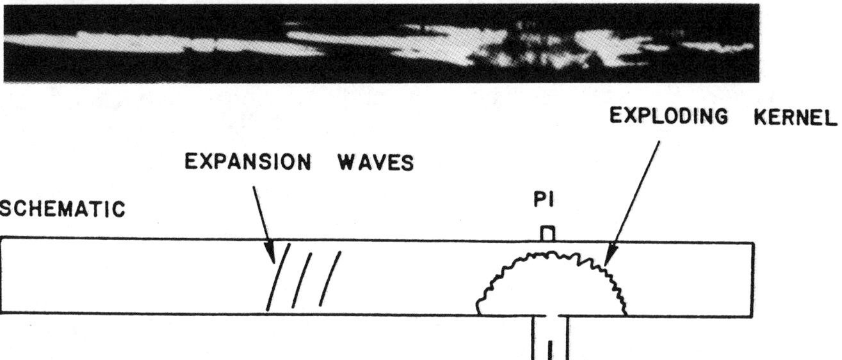

Fig. 4 Schlieren photograph showing flash-evaporation occurring in the vicinity of the break.

pressure peak in Fig. 3 is not as large as that observed in Fig. 2, the similarity of the pressure traces suggests that a rapid (flash) evaporation caused by boiling has occurred in the channel shortly after the depressurization of the fluid.

A spark schlieren photograph taken 50 μs after the arrival of the expansion waves is shown in Fig. 4. Since Freon 12 is very corrosive and it attacks the gasket material used in our apparatus, we could not achieve a leak-proof system. To prevent excessive loss of fluid we performed the experiment before the fluid had reached thermal equilibrium with the channel. Since the liquid was heated only along the bottom of the channel (glass window on both sides and Freon vapor on top) and the leakage of Freon promoted boiling along the vapor/liquid interface, there remained a temperature variation in the liquid at the time of the experiment. Although thermocouples (T1, T2, and T3) indicated that the temperatures measured at various heights were within 5°C, the temperature gradient within the liquid was sufficient to create horizontal dark bands (interference patterns) in the schlieren photograph. As expansion waves propagated through the liquid, the pattern got further distorted. The discontinuity in the pattern can therefore be used to indicate the location of the expansion waves. This location as indicated by the schematic in Fig. 4 agreed with the calculation based on sonic propagation of the waves. A hemispherical explosion kernel originating near the break is clearly visible in Fig. 4. Comparing Figs. 3 and 4, the arrival of the explosion kernel at P1 is found to correspond to the instant the pressure suddenly increases. It should be noted that the expansion waves have propagated part of the length of the channel and hence the liquid behind is in a superheated state. This photograph supports the suggestion in the pressure trace that an explosion has occurred after the initial depressurization and is caused by the explosive boiling of the superheated liquid. We estimate the timing of the explosion to be about 100 μs after the diaphragm is ruptured. This delay time should depend on the type of fluid and its temperature and pressure after the depressurization. It should be noted that once an explosion occurs, a blast wave is formed, and it will eventually overtake the expansion waves. (A pressure trace recorded at the end of the channel established this to be the case.) That is to say that only the fluid in the vicinity of the break will experience depressurization and boiling. Therefore, the volume of the exploding fluid or the size of the explosion kernel (which affects the magnitude and duration of the pressure peak) is expected to depend on the size of the break or the rate of depressurization. The geometry of the channel should also play a role. In general, a larger break will produce a stronger explosion.

Experimental results also show that the vapor explosion can trigger strong pressure oscillation within the channel. Fig. 5 shows the pressure-time history at P3. The magnitude of the oscillation in the channel is much larger than the blast wave from the initial explosion. The pressure trace shows amplification of the oscillation as the fluid vents out. The frequency of the oscillation (5000 Hz) corresponds roughly to the frequency of the transverse waves assuming the channel is filled with Freon. The amplitude of these transverse waves increases to about 400 kPa peak-to-peak before it decays gradually. The amplification of the transverse waves suggests that there exists a coupling mechanism between the wave and the boiling fluid. Such an amplification can be

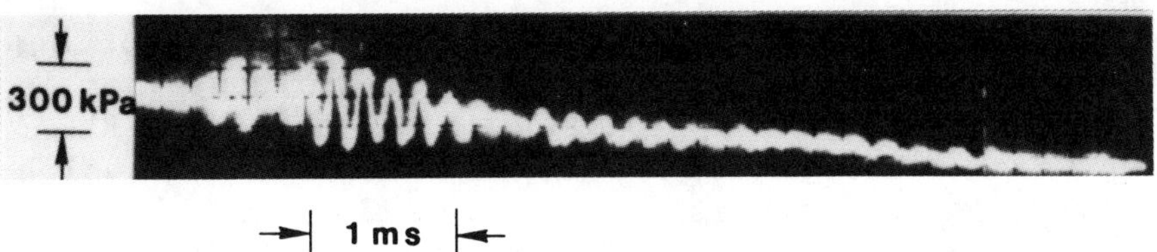

Fig. 5 Overpressure-time history (P3) of a BLEVE experiment.

explained by the Rayleigh criterion[11]: "If heat is periodically added to and taken from a mass of oscillating fluid, the effect produced will depend upon the phase of the oscillation at which the transfer of heat takes place. If heat is given to the fluid at the moment of highest pressure, or taken from it at the moment of lowest pressure, the oscillation is encouraged." Essentially, if heat (energy) is added to the fluid at the instant corresponding to the top of the pressure wave and removed at the instant corresponding to the bottom, the wave will amplify.

In general, a wave can amplify itself in a superheated fluid near the saturation line. As the pressure decreases, nucleation of bubbles and evaporation remove energy from the fluid. As the pressure increases, condensation and collapsing of bubbles release energy back to the fluid. Consequently, an oscillating wave will amplify itself. Since the channel is filled with Freon initially, the rate of amplification is very rapid. As the amount of Freon in the channel decreases, the amplification mechanism vanishes and the wave decays gradually. Unstable oscillating waves are known to be capable of causing severe damage to combustors and rocket motors.[11] In accidents involving pressure-liquefied gas, whether these oscillating transverse waves can cause damage to the container leading to a "delayed BLEVE" is uncertain. Nevertheless, it has been observed[3] that a BLEVE can still occur after a pressure relief valve has opened.

Since the channel used in the present experiment was designed to withstand the blast waves produced by the vapor explosion, a BLEVE (catastrophic failure of the vessel) did not actually occur. However, it is reasonable to believe that a vapor explosion resulting from the initial depressurization of the container is a necessary step in the onset of a BLEVE. For a BLEVE to occur, the strength of the blast waves, which depends on the initial state of the fluid and the rate of depressurization, has to be strong enough to tear the container apart.

So far, we have only examined the vapor explosion resulting from the venting of superheated liquid. Our present experimental assembly does not permit us to perform experiments in which venting occurs in the vapor phase. Although in most observed BLEVE incidents the initial break occurred in the wall surrounding the vapor phase, we believe that the overall phenomenon is similar to that observed in our experiment in which superheated fluid was vented.

Summary

This study shows that a BLEVE following the initial depressurization from the venting of pressure-liquefied gas through a break leaves a part of the remaining fluid in the tank in a superheated state. The onset of a BLEVE is a local explosion caused by a rapid (flash) evaporation resulting from the boiling of the superheated fluid. Results indicate that the local explosion originates in the vicinity of the break before the expansion waves have propagated a long distance from the break. The blast wave from the explosion will eventually overtake the expansion waves and stop any further boiling. This indicates that the volume of the fluid involved in the explosion, and hence the magnitude of the blast wave, depends on the size of the initial break. This blast wave is the cause of the catastrophic failure of the container observed in many accidents.

References

[1]Lewis, D. J., "Unconfined Vapor-cloud Explosions - Historical Perspective and Predictive Method Based on Incident Records," *Prog. Energy Combustion Science,* Vol. 6, 1980 p. 151.

[2]Walls, W., "The BLEVE - Part 1", Fire Command, May 1979, p. 24.

[3]Baker, W. E., Cox, P. A., Westine, P. S., Kulesz, J. J. and Strehlow, R. A., Explosion Hazards and Evaluation, Elsevier Scientific Publishing Company, 1983, Chapter 2.

[4]Nolan, P. F., Pettitt, G. N., Hardy, N. R. and Bettis, R.J. "Release Conditions Following Loss of Containment," *J. Loss Prev. Process Ind.*, Vol. 3, 1990, pp. 97-103.

[5]Bettis, R. J., Nolan, P. F. and Moodie, K., "Two Phase Flashing Release Following Rapid Depressurization due to Vessel Failure," *Inst. Chemical Engineering Symposium* Ser. 102, 1987, pp.247-263.

[6]Friedel, L., Molter, E. and Purps, S., "Delay Boiling in Refrigerant on Sudden Pressure Release from the Equilibrium State," *Chem. Ing. Tech.*, Vol. 57, No. 2, 1985, pp. 154, 155.

[7]Reid, R. C., "Possible Mechanism for Pressurized-Liquid Tank Explosion or BLEVE," Science, Vol. 203, March 1979, p. 23.

[8]McDevitt, C. A., Steward, F. R. and Venart, J. E. S., "Boiling Expanding Vapor Explosion An Update," Proceedings of the *5th Technical Seminar on Chemical Spills*, Pergamon Press, New York, 1989.

[9]Chan, C. K., and Greig, D. R., "The Structures of Fast Deflagrations and Quasi-detonations," *22nd Symposium (International) on Combustion, The Combustion Institute*, 1988, p. 1733.

[10]Frost, D. L., "Dynamics of Explosive Boiling of a Droplet," *Phys. Fluid*, Vol. 31, No. 9, 1988, p. 2554.

[11]Markstein, G. H. (ed.), Nonsteady Flame Propagation, a Pergamon Press, MacMillan, New York, 1964.

Models of Rapid Evaporation in Nonequilibrium Mixtures of Tin and Water

S. McCahan* and J. E. Shepherd†
Rensselaer Polytechnic Institute, Troy, New York 12180

Abstract

Rapid mixing of hot metals and water can result in an explosive interaction, producing high pressures and damaging the surroundings. It has been proposed by Board et al.[1] that these explosions may propagate as a steady wave analogous to a combustion wave such as a detonation. We have determined the families of steady waves that are possible for selected cases in tin-water mixtures. We construct shock adiabats by numerical solution of the Rankine-Hugoniot equation using realistic equations of state for both tin and water. Both detonation and deflagration branches are obtained. The lower (deflagration branch) Chapman-Jouguet point is always observed and can be either a pure water vapor or a liquid-vapor mixture state. An upper Chapman-Jouguet point is found in most cases. We propose that a deflagration-type solution has probably been observed in experiments conducted to date. A simple model for a deflagration-type wave in a tube with one end closed is formulated and solved for one set of initial conditions.

Introduction

Destructive explosions are possible when hot liquids, such as molten tin or aluminum, are mixed with cold liquids such as water. Although chemical reaction is possible with some liquid combinations, the principal phenomenon appears to be rapid production of vapor due to fragmentation and heat transfer between the hot and cold liquids.[2] Models and experiments[3] pertaining to these explosions indicate that a wide range of events occur. An extreme possibility is phase change through a wave propagation process similar to a chemical detonation.

---

* Graduate Student, Department of Mechanical Engineering

† Associate Professor, Department of Mechanical Engineering

The concept of such a "thermal detonation" was first proposed by Board et al.[1] They presented a simple analysis of explosions in tin-water systems and predicted that a Chapman-Jouguet detonation state could exist. In later work,[4] the structure of such thermal detonation waves was examined from a theoretical point of view. Since that time, a number of transient simulations (reviewed by Corradini et al.[3]) of multiphase flow have been carried out to examine this hypothesis. However, because of the inherent complexity of these problems many gross assumptions must be made to complete a model and, as a consequence, the simulation results are highly suspect. In any case, simulations of thermal detonation waves all predict peak pressures and propagation velocities far in excess of what is observed in experiments. It is the purpose of the present research to examine all possible steady wave solutions to a simple thermal explosion problem. Our analysis is based on our previous research on rapid evaporation waves.[5,6] This analysis was developed independently of, but is similar to, the earlier work of Labuntsov and Avdeev.[7]

In the present study we apply this model to mixtures of tin and water. We treat the tin-water mixture as a homogeneous substance with mass-weighted average properties. Initially the mixture of liquid tin, liquid water, and water vapor is not in thermal equilibrium. Following some transient initiation process, we suppose that a steady evaporation wave that propagates through the mixture is established. This wave transforms the metastable mixture into an equilibrium mixture. We do not consider the actual structure or mechanism of propagation in the present study. In our model this rapid evaporation wave is analyzed as a thermodynamic discontinuity. The structure of the wave is included inside a control volume. Thus, although the wave may be unstable and nonplanar, it can be analyzed as a fixed adiabatic control volume. The metastable liquid moves at a constant average velocity into the control volume and an equilibrium mixture of tin and water exits the control volume. By applying equations of mass, momentum and energy conservation, and the second law of thermodynamics to the control volume, a family of solutions is calculated for the equilibrium state.

Computation Procedure

Thermodynamic States

To begin the calculation process the thermodynamic properties of each of the components in the metastable upstream state are needed along with information on the composition of the mixture. A $P(\rho,T)$ equation of state (EOS) is used for the properties of the water,[8] however, no such equation of state was available for tin. Instead a two-fold constant β - κ construction was used, where β is the volumetric thermal expansion coefficient, $\beta = (1/v)(\partial v/\partial T)_P$, and κ is the isothermal coefficient of compressibility, $\kappa = -(1/v)(\partial v/\partial P)_T$. In this construction one set of β - κ data is used for liquid tin and another set of β - κ data is used for solid tin. We assume that the tin melts at a constant 505.0 K without regard to pressure, and no attempt is made to fit the liquid tin properties to the solid tin properties at the melting boundary. The liquid tin values used for β and κ, taken from Frost et al.,[9] are $\beta = 8.68\times10^{-5}$ K^{-1} and $\kappa = 4.0\times10^{-11}$ Pa^{-1}. Similarly the values used for solid tin are $\beta = 7.6\times10^{-5}$ K^{-1} and $\kappa = 1.81\times10^{-11}$ Pa^{-1}, where β is taken from Cavaleri et al.[10] and Deshpande and Sirdeshmukh,[11] and κ is taken as an average value from Refs. 10-12. Note that the isothermal coefficient of compressibility κ is the reciprocal of the bulk modulus of a substance. As stated previously we assume that the liquid tin β and κ values remain constant from 505.0 to 2543.0 K, the boiling point of tin and, similarly, that the solid tin β and κ values remain constant up to 505.0 K. The upstream mixture, state 1, is assumed to be in mechanical equilibrium, and the initial temperatures of the water and the tin are specified. The liquid water and water vapor states are considered to be at the same temperature, 373.15 K, for all cases, and each is at saturation conditions. From this information the $P(\rho, T)$ equation of state for water is used to find all the other state 1 properties for liquid and vapor water and the pressure for the system. For the tin we use the constant β-κ equation of state to find v_1,

$$v_1 = v_o \exp\left[\beta(T_1 - T_o) - \kappa(P_1 - P_o)\right] \tag{1}$$

where $v_o = 1.368\times10^{-4}$ m^3/kg, $T_o = 293.15$ K, and $P_o = 101325.0$ Pa as a reference state. The other two essential properties needed for the calculation are the initial enthalpy and entropy of the tin. Formulas for the enthalpy and entropy can be constructed by starting with the definitions for these properties using the constant β-κ assumption. We find

$$dh = c_p dT + v(1 - \beta T)dP \tag{2}$$

$$ds = \left(c_p/T\right)dT - (\beta v)dP \tag{3}$$

By integrating these expressions first from the reference state (T_o, P_o) to the state (T_o, P) along a line of constant temperature, and then integrating from the (T_o, P) intermediate state to any final state of interest (T, P) along a line of constant pressure, general expressions for the enthalpy and entropy for the state of interest can be developed. We find

$$h(T,P) = h_o + c_{po}(T - T_o) + \int_{T_o}^{T}\int_{T_o}^{T} \frac{\partial c_p(P_o)}{\partial T} dT dT$$

$$- \frac{v_o}{\kappa}(1 - \beta T)\left\{\exp\left[-\kappa(P - P_o)\right] - 1\right\}\exp\left[\beta(T - T_o)\right] \tag{4}$$

$$s(T,P)=s_o+\int_{T_o}^{T}\frac{c_p(P_o)}{T}dT$$

$$+\frac{\beta v_o}{k}\{\exp[-\kappa(P-P_o)]-1\}\exp[\beta(T-T_o)] \tag{5}$$

A correlation, $c_p=C(a+bT)$, for the specific heat is given in Smithell's metals.[12] For solid tin

$$c_{ps} = 35.2691\ (5.16 + 4.34\times10^{-3}\ \mathrm{T})\ \mathrm{J/kg\cdot K} \tag{6}$$

and for liquid tin

$$c_{pl} = 35.2691\ (8.29 - 2.2\times10^{-3}\ \mathrm{T})\ \mathrm{J/kg\cdot K} \tag{7}$$

Although these correlations do not take into account any variation of the specific heat with pressure, that is not necessary for our purposes since we only use the specific heat evaluated at the reference state c_{po} and the expression for the specific heat at the constant reference pressure $c_p(P_o)$, as it varies with temperature. So, finally, the full expressions for the enthalpy and entropy at state 1 are as follows:

$$h_1(T,P)=h_o+C\{(a+bT_o)(T_1-T_o)+(b/2)(T_1-T_o)^2\}$$

$$-\frac{v_o}{\kappa}(1-\beta T_1)\{\exp[-\kappa(P_1-P_o)]-1\}\exp[\beta(T_1-T_o)] \tag{8}$$

$$s_1(T,P)=s_o+C\{a\ln(T_1/T_o)+b(T_1-T_o)\}$$

$$+\frac{\beta v_o}{k}\{\exp[-\kappa(P_1-P_o)]-1\}\exp[\beta(T_1-T_o)] \tag{9}$$

where, for solid tin, $a=$ 5.16 and $b=$ 4.34×10^{-3} and, for liquid tin, $a=$ 8.29 and $b=$ -2.2×10^{-3}. In addition, for liquid tin an extra term representing a fixed latent heat of melting, $h_{ftin}=$ 58000.0 $\mathrm{J\cdot kg^{-1}}$, is added to the enthalpy expression, and an extra term representing the entropy of melting is added to the entropy expression, $h_{ftin}/T_m=$ 114.85 $\mathrm{J\cdot(kg\cdot K)^{-1}}$.

The composition at state 1 is specified by volume fractions and these are converted to mass fractions using the set of equations

$$v_1=1/\left(Z_t/v_{1t}+Z_f/v_{1f}+Z_g/v_{1g}\right) \tag{10}$$

$$Y_t = (Z_t v_1)/v_{1t} \tag{11}$$

$$Y_{1f} = \left(Z_f v_1\right)/v_{1f} \tag{12}$$

$$Y_{1g} = \left(Z_g v_1\right)/v_{1g} \tag{13}$$

where Z is the volume fraction, Y is the mass fraction, and the subscripts t, f, and g refer to the tin, liquid water, and water vapor components, respectively. v_1 is the mass weighted average specific volume at state 1 and v_{1t}, v_{1f}, and v_{1g} are the component specific volumes. The mass weighted average enthalpy and entropy are defined as follows:

$$h_1 = Y_t h_{1t} + Y_{1f} h_{1f} + Y_{1g} h_{1g} \tag{14}$$

$$s_1 = Y_t s_{1t} + Y_{1f} s_{1f} + Y_{1g} s_{1g} \tag{15}$$

The mass fraction of the tin is referred to simply as Y_t because it does not change during the rapid evaporation process.

Solution of Jump Conditions

Once the metastable upstream state has been completely defined, we choose a range of downstream temperatures and apply the conservation equations for mass, momentum and energy, and the second law of thermodynamics to obtain a family of solutions for the downstream equilibrium state. The conservation equations and statement of the second law are the same as those used for shock discontinuities and can be stated in the form of jump conditions

$$[\rho w] = 0 \tag{16}$$

$$\left[h + \left(w^2/2\right)\right] = 0 \tag{17}$$

$$\left[P + \rho w^2\right] = 0 \tag{18}$$

$$[s] \geq 0 \tag{19}$$

where w is the velocity of the fluid in a reference frame where the wave is held stationary. The first three of these expressions can be combined into the familiar Rankine-Hugoniot relation

$$h_2 - h_1 = (P_2 - P_1)(v_2 + v_1)/2 \tag{20}$$

This relation is applied differently depending on whether the downstream state lies inside or outside the transformed water saturation boundary. The transformed saturation boundary is defined as

$$v_{sat} = Y_t v_t + (1 - Y_t) v_{h_2o} \tag{21}$$

where, for a given temperature, v_{h_2o} is the corresponding water liquid or vapor saturation volume; v_t is the tin volume, as given by Eq. (1), at the water saturation pressure; and Y_t is the mass fraction of tin in the mixture of interest.

We begin by assuming that the downstream state 2 lies inside the transformed saturation dome. For this case there is a direct solution for state 2. We have already chosen a downstream temperature; if we assume this is a saturation temperature, then we can get all of the other saturation properties for the water using the subroutines from Reynolds' compilation.[8] Using the saturation pressure obtained for the water and the chosen temperature, we use Eqs. (1), (8), and (9) to calculate the downstream tin properties. We express h_2 and v_2 in terms of the component properties

$$h_2 = Y_t h_{2t} + Y_{2f} h_{2f} + Y_{2g} h_{2g} \tag{22}$$

$$v_2 = Y_t v_{2t} + Y_{2f} v_{2f} + Y_{2g} v_{2g} \tag{23}$$

and note that

$$Y_{2f} = 1 - Y_t - Y_{2g} \tag{24}$$

By substituting expressions (22-24) into the Rankine-Hugoniot relation (20), and rearranging, an explicit expression for Y_{2g} can be found

$$Y_{2g} = \left\{ h_1 - Y_t h_{2t} - h_{2f}(1 - Y_t) + \frac{1}{2}(P_2 - P_1)\left[v_1 + Y_t v_{2t} + v_{2f}(1 - Y_t)\right] \right\} \cdot \left\{ h_{2g} - h_{2f} + \frac{1}{2}(P_2 - P_1)(v_{2f} - v_{2g}) \right\}^{-1} \tag{25}$$

From the mass fraction of the water vapor at state 2, Y_{2g}, the quality of the water which is the fraction of water that is vapor, can easily be assessed

$$X = Y_{2g} / (1 - Y_t) \tag{26}$$

If X is between 0 and 1 then the downstream equilibrium state does indeed lie inside the saturation boundary and all the state 2 properties have been solved for.

However, if X is less than 0 or greater than 1, then the downstream state lies outside the transformed saturation dome and the saturation properties are no longer applicable. In this case an iterative method must be used to find a solution. A volume for the water is guessed, and all of the state 2 water properties are found using the $P(\rho,T)$ EOS. The pressure obtained from the EOS and the chosen temperature are substituted into Eqs. (1), (8), and (9) to find the volume, enthalpy, and entropy for the tin. Then the mass averaged mixture properties are calculated as follows:

$$v_2 = Y_t v_{2t} + (1 - Y_t) v_{2w} \tag{27}$$

$$s_2 = Y_t s_{2t} + (1 - Y_t) s_{2w} \tag{28}$$

$$h_2 = Y_t h_{2t} + (1 - Y_t) h_{2w} \tag{29}$$

The subscript w denotes the single phase water property. The enthalpy at state 2 found in this manner is compared to the enthalpy at state 2 calculated using the Rankine-Hugoniot relation (20), the pressure returned from the water EOS, and the state 2 volume calculated in Eq. (27). A root-finder routine is used to iterate on the downstream water volume until the two enthalpy calculations agree. This method is also used when the chosen downstream temperature is supercritical. The entropy production, that is, $s_2 - s_1$, is computed and any points violating the second law statement are discarded.

When the downstream temperature is 505.0 K the tin changes phase from solid to liquid. To account for this transition we define the "quality" of the tin to be the mass fraction which is liquid. Then we increase the quality from 0 to 1 while the downstream temperature is kept constant. At each point the tin properties are considered to be the mass weighted average properties of the liquid and solid tin. In this way we "step" through the transition.

Results

This computational method was applied to a variety of tin-water mixtures. The results show that the shock adiabats for tin-water mixtures are well behaved and exhibit a range of configurations depending on the starting composition of the mixture and the initial temperature of the tin. Analogous to the situation for shock waves in reacting gases,[13] the shock adiabats usually have a deflagration and a detonation branch with identifiable Chapman-Jouguet points. Some of the shock adiabats have deflagration branches which stay inside the transformed saturation dome, while other shock adiabats lie partially or completely in the vapor region.

Beginning with relatively low-temperature tin at state 1 and a considerable volume fraction of water vapor produces a shock adiabat like the one shown in Fig. 1. The deflagration branch for this adiabat falls entirely inside the transformed saturation dome. The deflagration Chapman-Jouguet (CJ) point is the point of maximum velocity and maximum entropy production on the deflagration branch. The detonation branch begins just inside the two-phase region and, in some cases, has a true detonation CJ point, a point of minimum

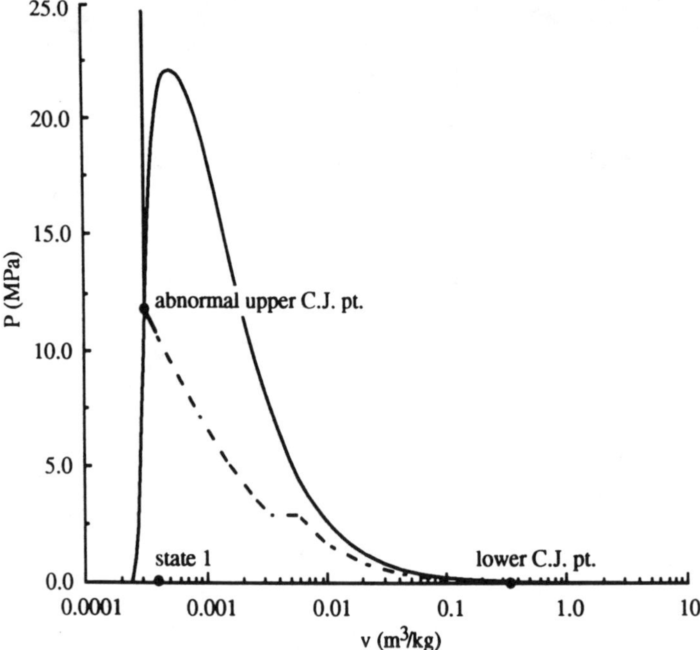

Fig. 1 Shock adiabat with the usual deflagration Chapman-Jouguet point and an abnormal detonation Chapman-Jouguet point: tin initially at 1000 °C; composition at state 1 is 33% tin, 33% liquid water, and 33% water vapor by volume.

velocity on the detonation branch, and minimum local entropy production on the portion of the solution with supersonic wave speeds and $v_1 < v_2$. In many cases, the minimum wave speed occurs where the shock adiabat crosses the liquid saturation boundary. The phase change results in a "kink" in the adiabat and derivatives fail to exist at this point, so this is not a CJ point in the usual sense. This behavior is analogous to that observed for computed shock adiabats for superheated liquids.[6,14]

As the initial fraction of water vapor is decreased, a true detonation CJ point appears as shown in Fig. 2. The deflagration branch stills lies completely within the two phase region and contains a regular CJ point. However, the detonation branch has been pushed up the adiabat. The detonation branch begins outside the saturation dome, usually in the supercritical region.

As the initial temperature of the tin or the vapor fraction in state 1 is increased the detonation branch remains unchanged, however, the deflagration branch begins to move outside the transformed saturation dome into the vapor region. In Figs. 3 and 4, the rightmost portion of the deflagration branch lies in the vapor region, and in both cases the lower CJ point lies in the vapor region. If the initial temperature of the tin or the vapor fraction continues to increase, the deflagration branch will move completely outside the saturation dome. Figure 5 shows this behavior for three vapor fractions. In all three cases shown

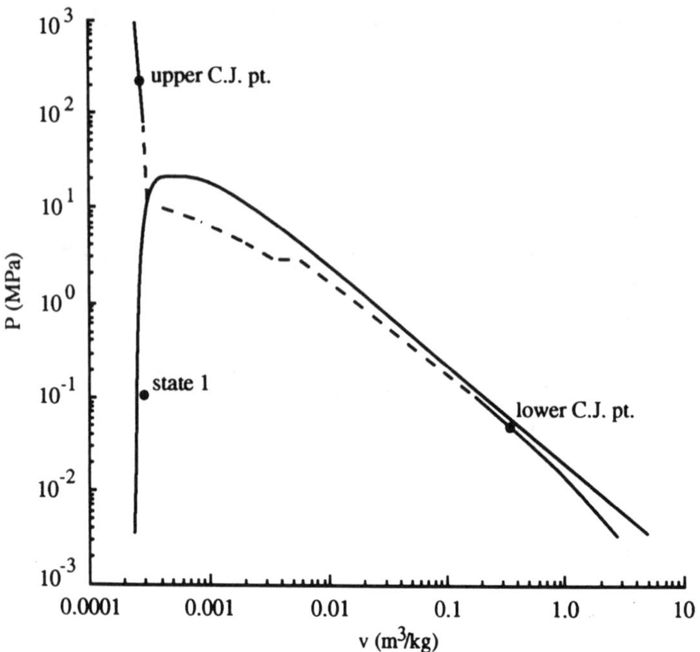

Fig. 2 Shock adiabat shown in logarithmic coordinates: tin initially at 1000 °C; composition at state 1 is 45% tin, 45% liquid water, and 10% water vapor by volume.

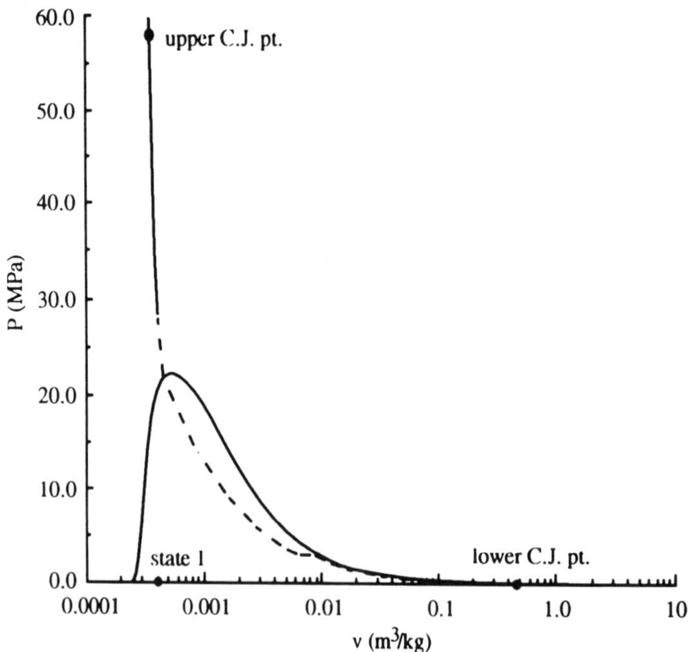

Fig. 3 Shock adiabat exhibiting ordinary Chapman-Jouguet points for both the detonation and deflagration branches: tin initially at 1500 °C; composition at state 1 is 33% tin, 33% liquid water, and 33% water vapor by volume.

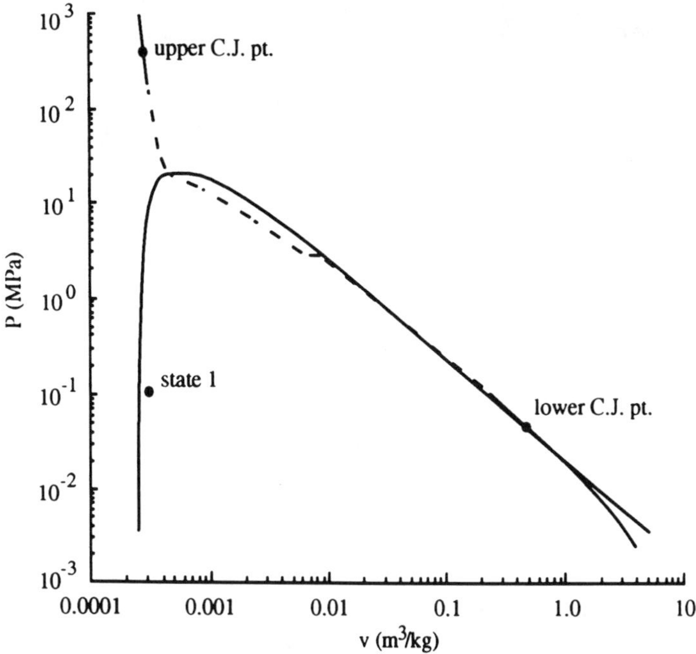

Fig. 4 Shock adiabat shown in logarithmic coordinates: tin initially at 1500 °C; composition at state 1 is 45% tin, 45% liquid water, and 10% water vapor by volume.

in Fig. 5, state 1 contains 87% tin by mass at 1500 °C. However, the percentage of water vapor in the water liquid-vapor mixture varies from 0 to 100%. Table 1 shows pressure, temperature, volume, and wave velocity data for the CJ points shown in Figs. 1-5.

In the limit when the temperature of the tin at state 1 is increased to 2543 K, that is, the boiling point of tin, and the fraction of tin is extremely high, above 96% by volume, the detonation branch is out of the range of states that can be handled by the water EOS. A detonation branch probably exists, however, a different equation of state for water will have to be employed to define such an adiabat.

Discussion

This method for computing the downstream properties and wave velocities of a rapid evaporation interface has previously been applied to water, a variety of hydrocarbons, and refrigerants 12 and 114 (refs. 6 and 15). In some of these cases experimental data were available for comparison. However, to date there are limited quantitative data[9,16] on the behavior of rapid evaporation waves in metal-water mixtures with which to compare our results. It is possible to compare the present results with the earlier computations of Board et al.[1] They predict that for an initial mixture of equal parts tin, liquid water and water vapor by volume, where the tin is initially at 1000 °C and the water is saturated at

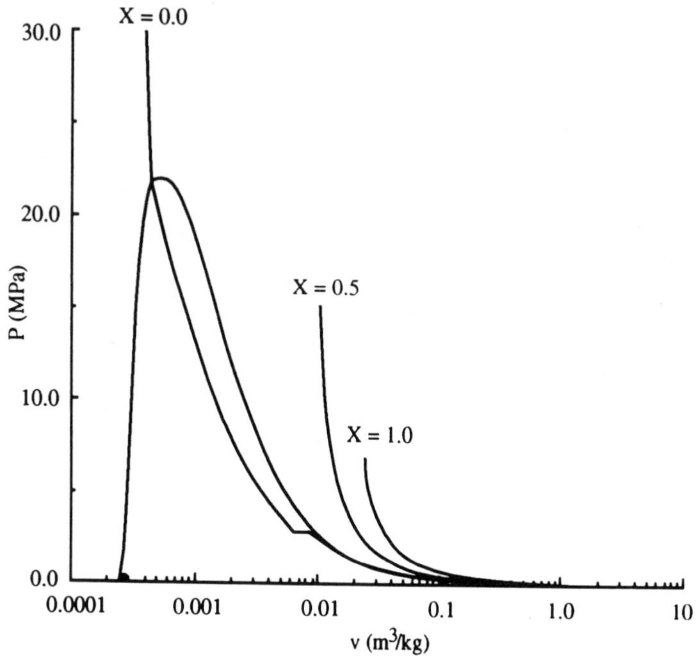

Fig. 5 Three shock adiabats for various fractions X of water vapor: tin initially at 1500 °C and mass fraction of tin is 0.870; for the X=0.0 case the initial mixture contains only liquid water; for the X=0.5 case the initial mixture contains half liquid and half vapor water by mass; and for the X=1.0 case the initial mixture contains only water vapor.

100 °C, the upper Chapman-Jouguet point will fall at about 1 kbar with a wave velocity of 300 m/s. Both the pressure and velocity predicted by Board et al.[1] are significantly larger than our values for this case, shown in Table 1; also we predict an abnormal or discontinuous detonation Chapman-Jouguet point. However, further up our shock adiabat we do find a point which lies very close to these values at 1 kbar with a state 1 velocity of 386 m/s. This is the first time we have applied this procedure to a mixture rather than a pure fluid. Although we had to assume that the mixture is homogeneous at both the upstream and downstream states, we are confident that this does not effect the results substantially. It is interesting to note that in the case of refrigerants 12 and 114, where we were able to compare our results to experimental data,[17] we found that the actual downstream state did not necessarily lie at the Chapman-Jouguet point. This may also turn out to be the case for tin-water mixtures.

Multiphase flows have the interesting property that the sound speed for the mixture can be significantly lower than that for any of the constituents individually. Because of this characteristic, it is interesting to consider the case where a mixture of hot tin and water is in a closed-ended tube and the triggering and subsequent propagation of the rapid evaporation wave begins at the closed

Table 1 Chapman-Jouguet properties for various initial conditions

State 1					State 2					
T_{1t}, °C	Z_t	Z_f	Z_g	v_1, m^3/kg	CJ	T, K	P, MPa	v, m^3/kg	W. vel.[a], m/s	Phase
1000.	0.33	0.33	0.33	3.910e-4	Lower	353.7	0.0485	0.3449	0.153	Liq./vap.[b]
					Upper	596.7	11.81	3.128e-4	151.3	Sat. bd.[c]
1000.	0.45	0.45	0.10	2.896e-4	Lower	353.7	0.0485	0.3447	0.113	Liq./vap.
					Upper	644.3	222.2	2.718e-4	1022.	Supercrit.[d]
1500.	0.33	0.33	0.33	4.061e-4	Lower	375.4	0.0471	0.4740	0.137	Vapor
					Upper	691.4	57.92	3.548e-4	431.0	Supercrit.
1500.	0.45	0.45	0.10	3.008e-4	Lower	375.5	0.0472	0.4734	0.102	Vapor
					Upper	771.0	407.2	2.786e-4	1286.	Supercrit.
1500.	0.50	0.50	0.0	2.707e-4	Lower	375.5	0.0471	0.4734	0.092	Liq./vap.
					Upper	845.2	403.2	2.570e-4	2738.	Supercrit.
1500.	1.24e-3	6.2e-4	0.99814	0.1091	Lower	559.2	0.0520	0.6444	33.14	Vapor
					Upper	702.7	0.6691	6.264e-2	381.4	Vapor
1500.	6.2e-4	0.0	0.99938	0.2182	Lower	845.9	0.0540	0.9427	55.79	Vapor
					Upper	1041.2	0.4860	0.1289	452.8	Vapor

[a]wave velocity
[b]liquid and vapor mixture
[c]on the saturation boundary
[d]supercritical

end, see Fig. 6. A situation similar to this is examined experimentally by Frost et al.[9] In this case the downstream state will be restrained and can be modeled as stationary. Transforming our frame of reference we find

$$W_w = (v_2/v_1)W_{10} \tag{30}$$

$$W_1 = W_{10}[(v_2/v_1) - 1] \tag{31}$$

where W_w is the transformed wave velocity, W_1 the transformed upstream velocity, and W_{10} the upstream velocity with respect to the wave front. The results of applying this transformation to the deflagration Chapman-Jouguet points for the first four cases in Table 1 are shown in Table 2. Notice that the wave appears to be "pushing" the upstream state ahead of it and that the rate of flow into the wave remains slow. Now consider the speed of sound in the upstream fluid. Using a homogeneous flow approximation and assuming that the vapor volume fraction is not close to either zero or one

$$c^2 = P/[\rho_f \beta(1-\beta)] \tag{32}$$

from Van Wijngaarden,[18] where c is the sound speed, P the pressure, ρ_f the liquid density, and β the volume fraction of the vapor. For our three-component

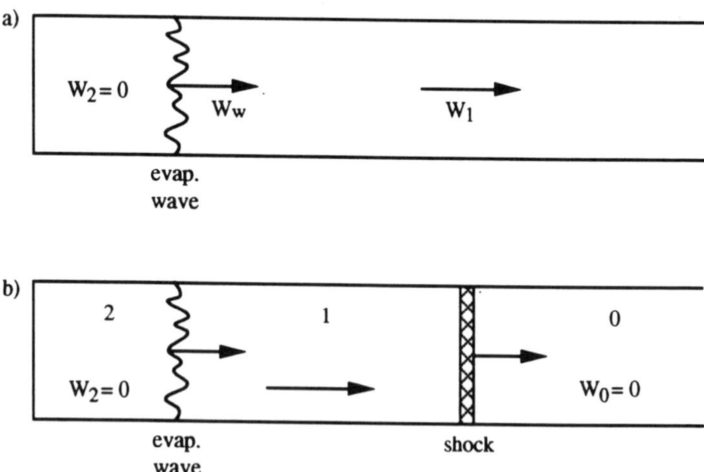

Fig. 6 Evaporation waves within a tube with one end closed: a) the result of restraining state 2 such that $W_2 = 0$, which implies the situation shown in b) where the evaporation wave is preceded by a shockwave which runs out into the undisturbed mixture.

mixture, we will let ρ_{fa} be an averaged density for the tin and liquid water

$$\rho_{fa} = \left(\frac{Z_t}{Z_t + Z_f} \right) \rho_t + \left(\frac{Z_f}{Z_t + Z_f} \right) \rho_f \tag{33}$$

Using this value in the Van Wijngaarden equation for the liquid density, and taking $\beta = Z_g$ and $P = 1$ bar, we find upstream sound speeds between 10 and 20 m/s for the four cases shown in Table 2. By comparison, the transformation given by Eqs. (30) and (31) applied to the deflagration Chapman-Jouguet points yields wave and flow velocities six to eight times larger.

This result suggests that the triggering and subsequent propagation of the evaporation wave acts as a piston initiating a strong shock wave that travels out in front of the evaporation wave. The shock wave accelerates and compresses the flow. This configuration is shown in Fig. 6. If we require that the initial tin-water mixture and the final evaporated state next to the closed end of the tube be stationary in the laboratory reference frame, then we can construct a matching condition for the state where the mixture has been shocked but not evaporated

$$w_{0s} - w_{1s} = w_{2w} - w_{1w} \tag{34}$$

where state 0 is the undisturbed mixture, state 1 the state downstream of the shock wave, and state 2 the state downstream of the evaporation wave. Subscripts s and w refer to the frames of reference in which the shock and the evaporation wave are held stationary, respectively. We investigated solutions to this problem for the case of the initial conditions considered by Board et al.[1] and discussed previously. We assume that the shock is isentropic, and that the tin and water in the downstream state are in mechanical but not thermal equilibrium, and that the water vapor and liquid are in thermal equilibrium. This is an ad hoc assumption which gives physically reasonable solutions but leads to a discrepancy in the conservation of energy across the wave by as much as 25%. By choosing a point from this isentrope and constructing an equilibrium shock adiabat in the same manner as described previously we can determine the point which satisfies the matching condition, Eq. (34).

The results of this procedure are shown in Fig. 7. The isentrope, which represents possible locations for state 1, lies outside the transformed saturation boundary in the liquid region for most states 1. The corresponding "closed-end

Table 2 Transformed velocities at the deflagration Chapman-Jouguet point

T_{1t}, °C	Z_{1t}	Z_{1f}	Z_{1g}	W_w, m/s	W_1, m/s
1000	0.33	0.33	0.33	135.0	134.8
1000	0.45	0.45	0.10	134.5	134.4
1500	0.33	0.33	0.33	159.9	159.8
1500	0.45	0.44	0.10	160.5	160.4

tube" state 2 points are shown as a curve, which remains fully inside the saturation boundary. Three specific state 1 points (a, b, c) are marked on the isentrope and there corresponding closed-end tube state 2 solutions (a', b', c') are shown on the downstream curve. Point a is the zero wave velocity point. At this point the shock does not exist, states 0 and 1 are the same, and the evaporation wave has zero velocity because the entire process is constant pressure. Point c is the last point on the isentrope used for the calculation of state 2. At this point the shock wave has a velocity of 161 m/s in the laboratory reference frame. However, the temperature rise across the shock is only about 2° for both the tin and the water. The fluid at state 1 is accelerated to 56 m/s, and the evaporation wave is moving at 116 m/s.

Point b is shown because it represents the unique case where the closed-end tube state 2 solution happens to correspond to the Chapman-Jouguet point on the deflagration branch of the shock adiabat. In this case the shock is moving at 110 m/s, the fluid at state 1 is moving at 37 m/s, and the evaporation wave is moving at 57 m/s. The curve representing possible state 2 solutions has a peculiar "wiggle" in the middle of it, this is due to the change of phase of the tin. Some of the state 2 solutions lie at the melting temperature of the tin. Whereas exact experimental duplication of the configuration shown in Fig. 6 has

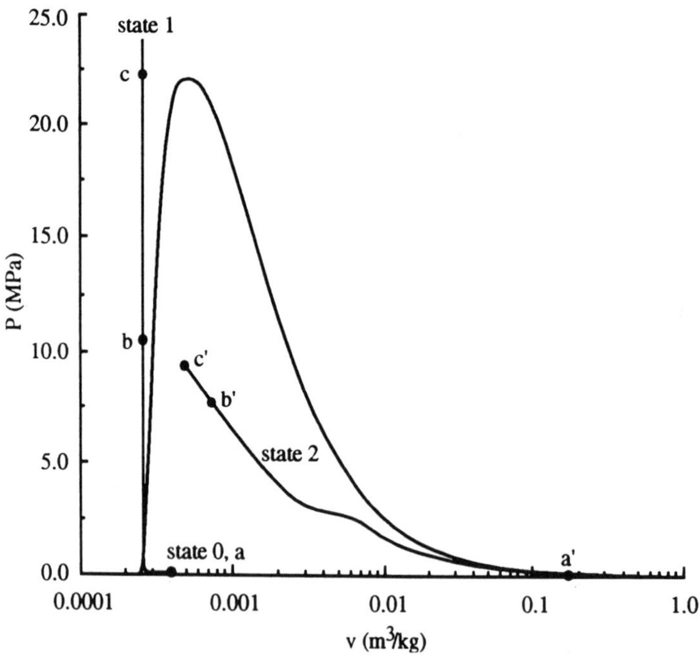

Fig. 7 Solutions for states 1 and 2 for the half-closed tube case shown in Fig. 6: state 1 loci form an isentrope computed as described in the text, the state 2 loci are the deflagration branch solutions that satisfy the boundary condition of zero velocity at the closed end of the tube; tin initially at 1000 °C; composition at state 0 is 33% tin, 33% liquid water, and 33% water vapor by volume.

never been obtained, the wave velocities and pressures observed experimentally[9] are much closer to these values than the CJ detonation solution proposed by Board et al.[1] On this basis, we propose that a reasonable explanation of metal-water vapor explosions is that they are the thermal analog of deflagrations rather than detonations.

Conclusion

A simple model for steady wave solutions to the rapid phase change of nonequilibrium metal-water mixtures has been proposed. A solution procedure and equation of states for molten and solid tin have been developed. Numerical solutions of the model have been given for several sets of initial conditions. We find that although the solutions are analogous to ordinary gaseous combustion shock adiabats, a regular detonation CJ point does not always exist. In particular, re-examination of the case previously considered by Board et al.[1] demonstrates that a true CJ detonation solution does not exist for that case. We propose an alternate explanation of vapor explosions based on the concept of a deflagration wave displacing the nonequilibrium mixture and proceeded by a strong adiabatic compression or shockwave.

The wave velocities (~150 m/s) and peak pressures (<100 bars) computed for this configuration are comparable to those observed in experiments.[9] These values are much lower than the computed detonation CJ solutions proposed by Board et al.[1] While it may be possible to experimentally produce the thermal detonations in large-scale experiments with high-energy triggers, the existing evidence suggests that a deflagration mode is a more plausible explanation for most laboratory results.

Acknowledgment

Partial support was provided by Lawrence Livermore National Laboratories under Contract B055778. We thank M. M. Abbott of RPI, and D. Frost and G. Ciccarelli of McGill Univ. for useful discussions.

References

[1]Board, S.J., Hall, R.W., and Hall, R.S., "Detonation of Fuel Coolant Explosions," Nature, Vol. 254, No. 5498, 1975, pp. 319-321.

[2]Reid, R.C., "Rapid Phase Transitions from Liquid to Vapor," Advances in Chemical Engineering, Vol. 12, edited by J. Wei, Academic Press, New York, 1983, pp. 105-208.

[3]Corradini, M.L., Kim, B.J., and Oh, M.D., "Vapor Explosions in Light Water Reactors: A Review of Theory and Modeling," Progress in Nuclear Energy, Vol. 22, No. 1, edited by T.D. Beynon and B.R. Sehgal, Pergamon Press, Oxford, England, 1988, pp. 1-117.

[4]Hall, R.W. and Board, S.J. "The Propagation of Large Scale Thermal Explosions," International Journal of Heat and Mass Transfer, Vol. 22, No. 7, 1979, pp. 1083-1092.

[5]Cho, J.H., Evaporation Waves in Superheated Liquids. MS Thesis, Dept. of Mechanical Engineering, Rensselaer Polytechnic Instit., 1988.

[6]Shepherd, J.E., McCahan, S., and Cho, J.H., "Evaporation Wave Model for Superheated Liquids," Adiabatic Waves in Liquid-Vapor Systems, edited by G.E.A. Meier and P.A. Thompson, Springer-Verlag, New York, 1990, pp. 3-12.

[7]Labuntsov, D.A., and Avdeev, A.A., "Theory of Boiling Discontinuity," Teplofizika Vysokikh Temperatur, Vol. 19, No. 3, 1981, pp. 552-556.

[8]Reynolds, W.C., Thermodynamic Properties in SI. Dept. of Mechanical Engineering Stanford Univ., Stanford, CA, 1979.

[9]Frost, D.L., Ciccarelli, G., and Zarafonitis, C., "Propagation of a Vapor Explosion in a Confined Geometry," Adiabatic Waves in Liquid-Vapor Systems, edited by G.E.A. Meier and P.A. Thompson, Springer-Verlag, New York, 1990, pp. 417-426.

[10]Cavaleri, M.E., Plymate, T.G., and Stout, J.H., "A Pressure-Volume-Temperature Equation of State for Sn(b) by Energy Dispersive X-Ray Diffraction in a Heated Diamond-Anvil Cell." Journal of Physics and Chemistry of Solids, Vol. 49, No. 8, 1988, pp. 945-956.

[11]Deshpande, V.T., and Sirdeshmukh, D.B., "Thermal Expansion of Tetragonal Tin," Acta. Crystallographica, Vol. 14, No. 4, 1961, pp. 355-356.

[12]Smithell's Metals Reference Book, Butterworth, Boston, MA, 1983.

[13]Thompson, P.A., Compressible-Fluid Dynamics, RPI Bookstore, Troy, NY, 1988, pp. 347-358.

[14]Fowles, G.R., "Vapor Detonations in Superheated Fluids," Adiabatic Waves in Liquid-Vapor Systems, edited by G.E.A. Meier and P.A. Thompson, Springer-Verlag, New York, 1990, pp.407-416.

[15]McCahan, S., and Shepherd, J.E., "Computational Analysis of Evaporation Waves in Superheated Refrigerants," Bulletin of American Physical Society, Vol. 35, No. 10, 1990, p. 2279.

[16]Lee, J.H., and Frost, D.L., "Steam Explosions: Major Problems and Current Status," Progress in Astronautics and Aeronautics, edited by A.L. Kuhl, J.R. Bowen, J.-C. Leyer, and A. Borisov, AIAA Washington D.C., Vol. 114, 1988, pp. 436-450.

[17]Hill, L.G., and Sturtevant, B., "An Experimental Study of Evaporation Waves in a Superheated Liquid," Adiabatic Waves in Liquid-Vapor Systems, edited by G.E.A. Meier and P.A. Thompson, Springer-Verlag, New York, 1990, pp. 25-38.

[18]Van Wijngaarden, L., "One-dimensional Flow of Liquids Containing Small Gas Bubbles," Annual Review of Fluid Mechanics, Vol. 4, 1972, pp. 369-394.

Shock Waves by Sudden Expansion of Hot Liquid

S. P. Medvedev,* A. N. Polenov,* B. E. Gelfand,† and S. A. Tsyganov‡
Russian Academy of Sciences, Moscow, Russia

Abstract

Depressurization of hot liquid during industrial accidents may cause emergence of shock waves in the environment. The aim of the present work is to investigate parameters of planar shock waves generated by sudden expansion of a high temperature pressurized liqid-saturated vapor system. Experiments were carried out in a properly designed, conventional vertical shock tube 3 m long and 50 mm in diameter. The setup consists of a heating chamber, high pressure, and test sections. The heating chamber was equipped with electrical heating bands and filled with water. The high pressure chamber was made from aluminum and was immersed into the hot water. The liquid under investigation was inserted into the high pressure chamber before the experiment. The high pressure chamber was separated from the test section by a diaphragm. During heating the water in the heating chamber, the pressure in the high pressure section increases. After diaphragm rupture, the hot liquid-saturated vapor system suddenly expanded and a shock wave formed in the test section. Parameters of the wave were measured by piezo-electric pressure gauges. Water, ethanol, and freon-113 were studied within a temperature range of 360 - 550 K and a pressure range of 0.1 - 5 MPa. The experiments showed that the value of shock impulse is determined mainly by the evaporation of liquid but not the expansion of vapor. The impulse (compression-phase time duration) rises with an increase in the mass of liquid. It was found that the shock amplitude depends on the initial pressure ratio at the diaphragm but not on the liquid properties. Shock impulse with pressure ratio increase and liquid molecular weight decreases. A simple approximate method is developed to predict the shock intensity.

Introduction

The study of shock waves which arise in the environment during the sudden expansion of a hot liquid is interesting for simulation of

---

* Senior Researcher, Institute of Chemical Physics.
† Professor, Institute of Chemical Physics.
‡ Head of Laboratory, Institute of Chemical Physics.

industrial accidents. The depressurization of a hot liquid accompanies the intensive evaporation. The fast expansion of a vapor-droplet cloud may lead to the appearance of shock waves. Usually the consideration of hot liquid expansion deals with the dynamics of evaporation and pressure-fall inside the volume. There is a gap of systematical study of shock parameters in literature, although the industrial accident consequences reveal the great mechanical effect by explosive expansion of boiling liquid.[1]

Gelfand et al.[2] offered to study shock parameters of the phenomenon under consideration in a shock tube that is equipped with the heating high pressure chamber. The intensity and structure of the shock and rarefaction waves due to expanding of a water-saturated vapor system were studied early.[2,3] It was found that the positive shock duration is rather large and therefore the measurements of shock impulse is difficult. This paper presents the description of shock tube which differ from our previous works.[2,3] New modification is suitable for investigation of planar finite duration shock waves. The aim of the present work is to obtain the dependencies of shock amplitude and impulse on parameters of expanded volume. It studied three kinds of liquids.

Experimental

Experiments have been carried out in a properly designed conventional vertical shock tube 3 m long and 50 mm in diameter. The scheme of the experimental setup is presented in Fig. 1. The high pressure chamber of the tube consists of a heating chamber (HC) and high pressure volume (HPV). HC (1) is equipped by electrical heating bands (2) and filled with water. The water temperature was measured by the thermocouple (3). HPV (4) was made from aluminum and immersed into hot water. The liquid under investigation (LI) was inserted into the HPV before the experiment. The HPV is separated from the low pressure chamber (LPC) (5) by a bursting diaphragm (6). Initially LPC is filled with air at normal conditions in all runs (temperature T_1= 300 K, pressure p_1= 0.1 MPa). During the heating process the pressure and the temperature in HPV rise in agreement with the LI saturated curve. When the pressure achieves appropriate valuep_4 (at temperatureT_4) the diaphragm bursts and liquid seems to be superheated in comparison with the LPC conditions. Outflow of liquid and vapor leads to the shock wave formation. The parameters of shock are measured by piezo-electric pressure gauges 7 - 10. In addition, photogauge 11 is used for registration of two-phase flow in LPC. Gauge 7 is positioned at a distance 0.1-0.2 m from the bursting diaphragm. The distance between gauges is 0.24 m. The maximum pressure and temperature values in HPV are, respectively, 6 MPa and 550 K. The scheme, described above differs from our previous one.[2,3] The present high pressure volume is rather small. So the short-duration shock waves

Table 1 The properties of liquids under investigation

Liquid	μ	T_S,K	ρ_s, kg/m^3	T_C,K	ρ_C, kg/m^3	A, kg/m^{3}K	B, kg/m^3
Water	18	373	958	647	318	1.175	1396
Ethanol	46	351	757	521	267	1.312	1217
Freon-113	187	321	1510	487	576	2.157	2202

are formed, and it is easy to measure the positive shock impulse. For the sake of convenience the signal of gauge 9 was integrated by means of a special electronic scheme.

The pressure of diaphragm rupture p_4, the HPV volume V, the mass m, and the properties of liquids were varied in the experiments. The available HPV volumes were $V = 26$, 41, and 70 sm^3 . The main parameters of liquids under investigation are presented in Table 1. Here is molecular weight; T_S, normal boiling temperature;ρ_S, density of liquid fraction at the temperature T_S; T_C, critical temperature; and ρ_C, critical density. Coefficients A,B are defined later.

Results

It is a matter of fact that liquid occupies the HPV only partially. The space over the liquid is filled with saturated vapor-air mixture. There are two main cases in a liquified gas tank accident.[1] The first case deals with the tank crush over the liquid surface, and the second case deals with the tank crush below the liquid surface. It is interesting to compare shock parameters in both cases. The first case was realized when the HPV was mounted at the bottom part of the shock tube (i.e., as shown in Fig. 1). In the second case HPV is mounted at the top part of the shock tube. Fig. 2a and Fig. 2b show pressure histories behind shock waves, registered by the use of a gauge mounted at the distance 0.83 m from the diaphragm. Fig. 2a and Fig. 2b illustrate the cases when the HPV is located at the bottom part of the shock tube and at the top part, respectively. It is necessary to note that the parameters p_4, V, and m are similar in both cases. As seen from Fig. 2b, shock amplitude and impulse (positive time duration) are equal to each other in both cases, but the wave structure is different. As seen from Fig. 2a, the pressure profile consists of a leading shock of a triangular shape and a pressure wave without a shock front. Such a structure is analogous to that in our previous work.[2,3] A leading shock wave deals with the expansion of the saturated vapor-air system. Further, one can see rather slow increasing of pressure up to $p_2 = p_1 + \Delta p_2$, which is defined by the process of expansion in a LPC vapor-droplets cloud, which is formed due to liquid-phase boiling. The expansion of superheated liquid (Fig. 2b) cause the formation of the pressure wave without sharp shock front. Note that

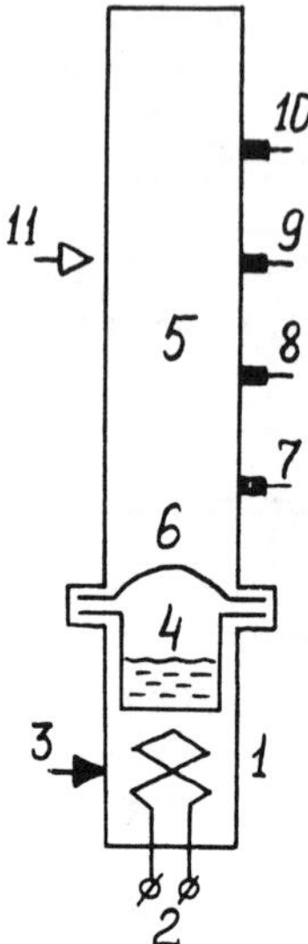

Fig. 1 The scheme of the experimental setup.

according to Ref. 1 the situation of envelope rupture over the level of liquid is the most probable. The base scheme of the experiments is the setup with HPC, which is located in the bottom part of LPC. The following parameters were obtained from the records: $\beta = \Delta p_2 / p_1$, where Δp_2 is the overpressure relative to the p_1 at the "plato" behind the shock of a triangular shape, and I is the shock wave compression-stage impulse.

The influence of liquid mass on the shock wave parameters is illustrated by Fig. 2c and Fig 2d. In these figures and below, the data from the gauge placed at the distance 0.68 m from membrane (gauge 9 in Fig. 1) are drawn. It can be seen that with the increase of m the duration (impulse) of a shock wave rises significantly, while the amplitude of shock wave remains constant. The experimental dependencies of the relative impulse and intensity of shock wave on water mass are presented in Fig. 3 (p_4= 1.4-1.6 MPa, T_4= 460-470 K). The value of I_0 for the fixed volume of HPV was obtained by the use of extrapolation method and is in accordance with the value of shock wave impulse from the expansion of a gas volume filled by a saturated vapor in the absence of liquid phase at the moment of diaphragm rupture. As it follows from Fig. 3a, the contribution of a mass addition process from evaporated liquid in the shock wave impulse formation becomes dominant with low values of m. With liquid- volume fracture about 50%, the shock wave impulse exceeds the value of I_0 by seven-ten times. The values of β (Fig. 3b) are practically independent upon the mass of liquid, so the significant increase of shock wave impulse is conditioned by the compression-stage duration rise.

For practical purposes, it is very useful to know the dependence of shock wave parameters on pressure ratio at the diaphragm. Taking

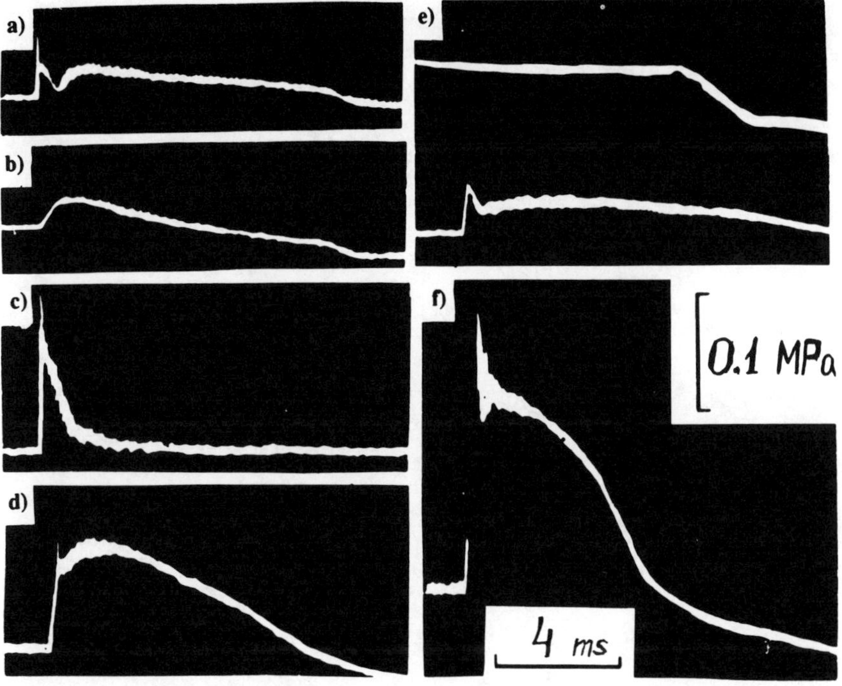

Fig. 2 Pressure histories:
a, b) F-113, $m=12$ g, $V=26$ sm^3, $p_4=0.5$ MPa; c) water, 5 g, 41 sm^3, 15 MPa; d) water, 22 g, 41 sm^3, 1.5 MPa; e) bottom trace, ethanol, 22g, 76 sm^3, 0.5 MPa; e) top trace, photogauge signal; f) ethanol, 22 g, 76sm^3, 3.1 MPa.

ethanol as an example, Fig. 2e and Fig. 2f represent the obtained specific feature - with fixed mass of liquid and volume of HPV and the increase of the ratio p_4/p_1 (T_4/T_1) the intensity of shock wave increases but the duration decreases. Quite sharp increase of the parameter β causes the total increase shock wave impulse. The dependencies of shock wave impulse on pressure ratio at diaphragm are presented in Fig. 4. The parameters of substances used are very different. For practical purposes and the comparison, it is convenient to present the experimental results with constant m. The analysis of the experimental results shows that the compression-stage impulse of a shock wave rises with the increase of the HPV volume and the decrease of the molecular weight of liquid. The dependence of wave amplitude obtained on p_4/p_1 ratio presented in Fig. 5 is of another type. The intensity of shock wave grows with the increase of pressure in HPC and is almost independent of LI and HPV volume. So the difference between the impulses for expanding water and freon-113 (or water and ethanol) is conditioned in general by the compression-stage duration values. Decreasing molecular weight of LI causes the increase of shock wave duration.

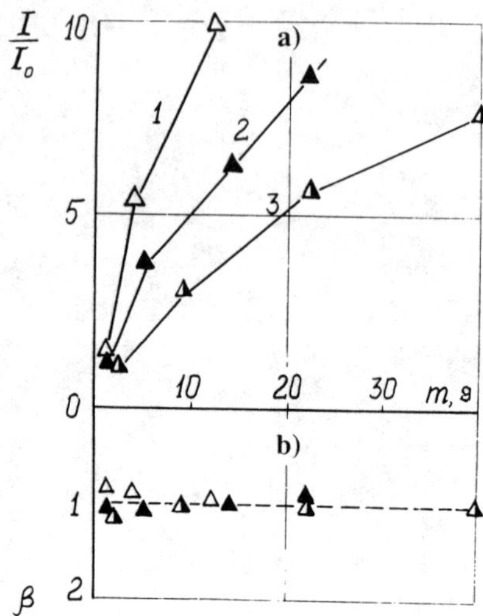

Fig. 3 Impulse (a) and shock wave intensity (b) vs. mass of water: 1) $V = 26$ sm^3, $I_0 = 35$ Pa s; 2) 41 sm^3, 60 Pa s; 3) 70 sm^3, 100 Pa s.

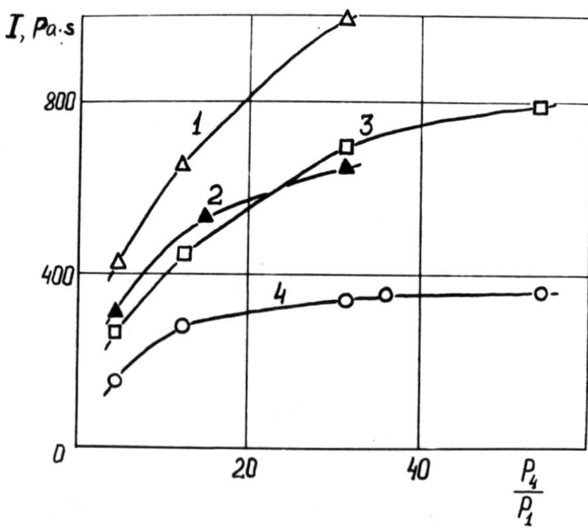

Fig.4 Shock wave impulse vs. diaphragm pressure ratio at $m = 22$ g: 1) water, $V = 70$ sm^3; 2) water, 41 sm^3; 3) ethanol, 70 sm^3; 4) F-113, 41 sm^3

Discussion

On the basis of the experimental data analysis (Figs. 2-5) and also the data,[2-4] one can propose the following scheme of the transient boiling liquid expansion process. The first stage of the process is characterized by the "quick" rarefaction wave with acoustic velocity about 10^3 m/s propagating on motionless single-phase liquid. At the same time a weak shock wave is formed in LPC. Parameters of this wave are defined by the liquid-phase compressibility. At the second stage, due to the pressure decrease below the saturated limit, the formation and growing of the vapor bubbles begins. The outflow of two-phase (bubble) mixture causes the propagation of the "slow" rarefaction wave with the velocity about 10 m/s. A compression wave is formed in LPC. Moreover, due to the finite time of bubble growth and delayed boiling the wave may have no sharp shock front, at least in the vicinity of expanding volume (compare Fig. 2b and Refs. 5 and 6). Note, that the existence of the two-phase flow in LPC causes the sharp decrease of photo gauge signal (Fig.2c, upper trace) at the time of shock wave contact surface arrival.

To calculate the parameters of shock wave, which is formed at the first stage one can use the results,[5] where the expression for the evaluation of shock wave intensity in a shock tube with HPC filled by liquid was obtained. In Ref. 5 no phase transition in the process of transient expansion is supposed. Using present paper nomenclature one can write:

$$\frac{p_4}{p_1}+\frac{G}{p_1}=$$

$$\left[1+\beta+\frac{G}{p_1}\right]\left[1-\frac{n-1}{2\gamma_1}\frac{a_1}{a_4}\beta\left(1+\frac{\gamma_1+1}{2\gamma_1}\beta\right)^{-0.5}\right]^{-\frac{2n}{n-1}} \quad (1)$$

Here a_1, γ_1 is sound velocity and specific heat ratio of the gas contained in LPC. Parameters G, n are taken from the empirical equation of state for liquid: $p = G\,[(\rho/\rho_0)^n - 1]$ (ρ_0 is the initial density of liquid; p, ρ, pressure and density). For water[5] $n = 7.15$, $G = 301$ MPa, and $a_4 = [n(p_4+G)/\rho]^{0.5}$. For example, if $p_4/p_1 = 50$ then $\beta \approx 0.001$, i.e., two to three powers lower than measured values (Fig. 3). Therefore, the initial stage of boiling liquid expansion cause the formation of the weak acoustic precursor, which propagation velocity close to the sound velocity in the environment. It is obvious that for $G = 0$, $n = \gamma_4$, Eq. (1) transforms into a well-known shock tube formula, where a_4, γ_4 are, respectively, the sound speed and the specific heat ratio of the HPC gas.

To describe the dominant second stage (i.e., two-phase mixture expansion from HPC) one can use the hypothesis of flow velocity and temperature equilibrium. Such a scheme gives satisfactory correspondence to the experimental data for the hot-liquid outflow from pipe and therefore can be applied for the evaluation of the

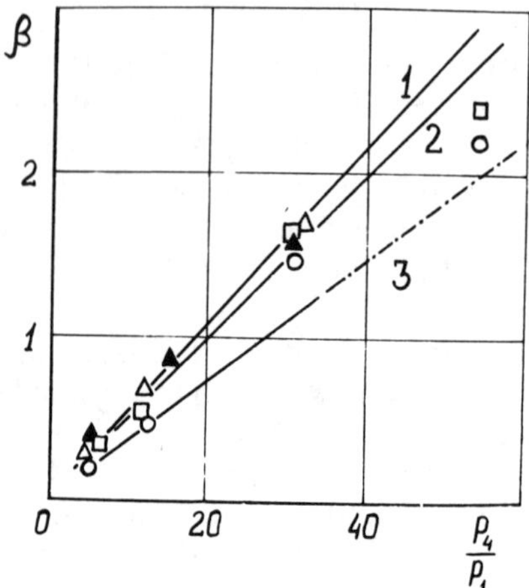

Fig. 5 Shock wave intensity vs diaphragm pressure ratio m = 22 g. (Experimental data points are the same in Fig. 4.) 1-3 — calculations by use of (2),(6),(7) for ethanol (curve 1), water (2) and F-113 (3).

shock wave intensity in the connected volume. In this case one should substitute to Eq. (1) the "equilibrium" values of "effective" specific heat ratio $\gamma_4 = \gamma(1+\eta\delta)(1+\gamma\eta\delta)^{-1}$ and sound velocity $a_4{}^2 = \gamma_4 a^2 [\gamma(1+\eta)]^{-1}$ (η,δ is the mass and heat capacity ratios of condensed and vapor phases; a,γ the parameters of vapor phase). It is supposed that the Claperon-Clasius low is valid for the mixture, the liquid phase is incompressible, and there is no phase transitions. For the liquids under consideration $\gamma \approx 1.08$-$1.3, \delta = 1$. So for $\eta > 1$ one has $0 < (\gamma_4 - 1) < 0.04$-$0.15$. Expanding the right of Eq. (1) (for $G=0$) into series in small parameter $(\gamma_4 - 1)$, and taking into account only the first member one has:

$$\frac{p_4}{p_1} = [1+\beta]\exp\left[\frac{a_1\beta}{a_4\gamma_1}\left(1+\frac{\gamma_1+1}{2\gamma_1}\beta\right)^{-0.5}\right] \qquad (2)$$

Such a simplified equation is more applicable for the practical evaluation because a set of HPC gas parameters are eliminated from the consideration. The expression for "equilibrium" sound velocity a_4 can be rewritten as follows: $a_4 = [\gamma_4 p_4 \rho_V^{-1}(1+\eta)^{-1}]^{0.5}$. The absolute value of "effective" γ_4 slightly depends on η, so for upper evaluation of a_4 we may set $\gamma_4 = \gamma =$ const and thus:

$$a_4 = \left[\frac{\gamma\, p_4}{\rho_V(1+\eta)}\right]^{0.5} \qquad (3)$$

It is convenient to express the parameter through the densities of vapor and liquid phases ρ_V and ρ_L and volume vapor fraction φ in HPC:

$\eta = \rho_L(1-\varphi)\rho_V^{-1}\varphi^{-1}$. According to Ref. 4 far from the critical region (at $T_4 < 0.9T_c$) $\varphi = 0.5$- 0.9. For approximate evaluation we set $\varphi = 0.66$ and thus: $\eta \approx \rho_L/2\rho_V$ ($\rho_L >> \rho_V$). In the vicinity of the critical region ($0.9T_c < T_4 < T_c$), the intensity of vapor generation sharply increase due to the mechanism of homogeneous nucleation. This leads to $\varphi \approx 1$. Besides, for $T_4 > T_c$ the difference between phases disappears, i.e., $\rho_L = \rho_V$. Therefore, for $T_4 > 0.9T_c$ we have $\eta \approx 0$. A simple approximate expression for parameter , which is valid for $T_4 > T_s$, can be written as follows:

$$\eta = \frac{\rho_L - \rho_V}{2\rho_V} \tag{4}$$

Substituting Eq. (4) in Eq. (3) one can obtain:

$$a_4 = \left[\frac{2\gamma p_4}{\rho_V + \rho_L}\right]^{0.5} \tag{5}$$

Equation (5) is not suitable for practice because usually the temperature dependence of ρ_L and ρ_V for chosen liquid is unknown beforehand. The well-known "rectilinear diameter" rule can be used in the analysis. In accordance with this rule, the sum of densities of liquid and vapor phases at saturation state is a linear function of temperature: $\rho_L + \rho_V = B - AT_4$. Thus, for a_4 one has:

$$a_4 = \left[\frac{2\gamma p_4}{B - AT_4}\right]^{0.5} \tag{6}$$

Coefficients A and B are defined by the tabulated values of ρ_s, T_s, ρ_c, and T_c and can be expressed as follows:

$$A = \frac{\rho_s - 2\rho_c}{T_c - T_s}\ ; \qquad B = 2\rho_c + AT_c \tag{7}$$

The values of A and B are presented in Table 1. The results of calculations based on Eqs. (2), (6), and (7) are presented in Fig. 5. The comparison with the experimental results shows that the proposed methodic can be used for evaluation of the shock wave intensity. Note that the dependence of Eq. (2) is in a good acquaintance with the experimental data at supercritical temperature region (broken part of curve 3 for Freon-113).

Conclusion

The impulse-amplitude characteristics of planar shock waves formed by a sudden expansion of hot liquid were investigated. It was shown that the main contribution to impulse of the shock wave-compression stage is conditioned by the process of evaporation of liquid phase. When initial pressure is fixed, then the amplitude of shock wave remains the same, but impulse (duration) increases with the increase of liquid mass. It was found that the wave amplitude slightly depends upon the properties of the liquid and in general is defined by the pressure ratio between expanding volume and surrounded media. Impulse (duration) of a shock wave grows with the increase of pressure ratio and the decrease of the molecular weight of liquid. The method of approximate evaluation of the intensity of the shock wave is proposed.

References

[1]Marshall, V. C., *Major Chemical Hazards*, Ellis Horwood Ltd., New York, 1987.

[2]Gelfand, B. E., Medvedev, S. P., and Frolov, S. M., "Shock Waves by Expanding Superheated Liquid," *Doklady Akademii Nauk SSSR*, Vol.301, No. 6, 1988, pp. 1413-1416.

[3]Gelfand, B. E., Medvedev, S. P., Polenov, A. N., and Frolov, S. M., "Shock Waves by Expanding Saturated Vapor-Liquid Systems," *Teplofizika Vysokih Temperatur*, Vol. 27, No. 6, 1989, pp. 1159-1166.

[4]Edwards, A. R. and O'Brien, T. P., "Studies of Phenomena Connected with the Depressurization of Water Reactors," *Journal of British Nuclear Energy Society*, Vol. 9, 1970, pp. 125-135.

[5]Nayfeh, H. and Hassan, S. D., "On Liquid Driver Shock Tube," *Journale de Mecanique,* Vol. 8, No. 2, 1969, pp.193-206.

[6]Nigmatulin, R. I., *Dynamics of Multiphase Media*, Pt.2, Nauka, Moscow, 1987.

Thermal Detonation in Molten Sn-Water Suspension

B. E. Gelfand,* A. M. Bartenev,† S. M. Frolov,‡ and S. A. Tsyganov§
Russian Academy of Sciences, Moscow, Russia

Abstract

A violent large-scale interaction between molten tin and water is considered in the frame of a steady thermal detonation model. On the basis of experimental data, a modified fragmentation mechanism is introduced into the model.This mechanism takes into account some time delay between lead shock arrival and the incipience of drop breakup in liquid-liquid systems. A set of equations of a separated one-dimensional, two-phase flow has been studied qualitatively and solved numerically. It is found that momentum losses due to friction at tube walls result in decreasing the detonation velocity and maximum overpressure in the system. It is also shown that within a certain range of wall surface roughness steady thermal detonation does not exist.The influence of the initial vapor content on detonation parameters is analyzed.

Particular analysis is undertaken to determine singularity conditions at the C-J plane. It appears that, contrary to the ideal case, certain differences of phase velocities take place when choking condition is satisfied. This phenomenon is analogous to incomplete fuel burnout in the reaction zone of nonideal chemical detonations.

---

*Chief Researcher, Institute of Chemical Physics.
†Researcher, Institute of Chemical Physics.
‡Senior Researcher, Institute of Chemical Physics.
§Vice Director, Institute of Chemical Physics.

Introduction

Safety requirements in the metallurgy plants and nuclear energy gave rise to investigations of explosion processes due to the contact interaction between molted metal and coolant (water). As a result of such an interaction, both a calm vaporization regime with small pressure rise and an intensive regime with shock formation can be realized. The former regime can be suppressed by conventional methods, although the latter may cause significant damage to plant equipment.

To explain large-scale interactions in molted metal- coolant systems a model of thermal detonation was suggested in Ref.1. The model is based on the ZDN-model of chemical detonation . It was assumed that the interaction proceeds as follows: large- scale mixing of molted metal drops with water leads to the formation of a stable vapor film boiling on the drop surface. As a result of shock wave arrival from an arbitrary source, the vapor film collapses. In the water flow behind the shock front metal drops involve into the movement. Hydrodynamics interaction results in fragmentation, and the formation of numerous small (10-15 m) debris of large interfacial surface. The temperature of the coolant increases due to intensive heat transfer from debris. When saturation conditions are attained, an intensive vapor generation begins. Heat flux from debris and unfragmented metal drops to the coolant play the role of energy source, which is able to support propagation of the shock wave at nearly constant velocity (compare with the energy of chemical reaction).

Recently numerous theoretical models were developed aimed to describe the process of thermal detonation propagation. A review of the models is given in Ref.3. Considering multiphase flow with detailed fragmentation kinetics allowed to treat such particular features of propagation as incomplete fragmentation [4,5], detailed structure of the reaction zone, etc. Transient computer codes[6-9] revealed the effects of parameter distribution in initial mixture; attempts were made to derive criteria of detonation initiation [10]. Advanced mathematical modeling of the phenomenon is accompanied with numerous difficulties, e.g., definition of sound speed in two-phase media, heat transfer and drag coefficients in bubbly media, fragmentation kinetics. Complex models include a whole number of governing parameters. Since experimentalists often report incomplete sets of relevant parameters, comparison of predicted and measured results appears often to be inconsistent. This seems to be a reason of the trend to simplify mathematical description with emphasis to physical analysis of the phenomenon [11-13]. The use of Hugoniot analysis [12,13] showed the physical background of thermal detonations.

When considering the problem of thermal detonation a set of principal questions arise. These are the range of permitted velocity of steady thermal detonations, the influence of the mechanism of energy release, and energy losses on the process. In the present work the method commonly used for studying chemical detonations is applied to solve some problems of thermal detonation.

Theory

When solving a problem of thermal detonation several phases should be considered. One can involve four phases: molted metal drops, coolant, coolant vapor, and metal debris. By applying some physical assumptions the total number of phases can be deduced[4]. Thus we assume that the debris attain instantaneously thermodynamic and kinematics equilibrium with coolant. It is valid when the debris are very small and the interfacial surface is very large. The appearance of a vapor phase cause saturated state so that the pressure and temperature of the coolant become interrelated. Therefore we may consider only two phases, i.e., the molten metal drops and the fluid consisting of metal debris, the coolant, and the vapor (if the latter exist).

The fact that the slow processes of interface heat and mass transfer have a dominant role in the thermal detonation propagation, energy and momentum losses at confining walls become the question of principle. It is conditioned by the lengthy zones of heat transfer from the drops to the coolant.

A set of governing equations of the steady one-dimensional, two- phase flow with irreversible interfacial exchange and losses at confining walls can be written as follows:

$$
\begin{aligned}
\frac{d(\alpha_1\rho_1 u_1)}{dx} &= -M \\
\frac{d(\alpha_2\rho_2 u_2)}{dx} &= M \\
\frac{d(\alpha_1\rho_1 u_1^2)}{dx} &= -\alpha_1\frac{dP}{dx} - F - u_1 M \\
\frac{d(\alpha_2\rho_2 u_2^2)}{dx} &= -\alpha_2\frac{dP}{dx} + F + u_1 M + H_f \qquad (1)\\
\frac{d(\alpha_1\rho_1 u_1(h_1 + u_1^2/2))}{dx} &= -u_1 F - \varphi - M(h_1 + u_1^2/2) \\
\frac{d(\alpha_2\rho_2 u_2(h_2 + u_2^2/2))}{dx} &= u_1 F + \varphi + M(h_1 + u_1^2/2) + DH_f
\end{aligned}
$$

where 1 and 2 are molten metal drops and fluid respectively, ρ,u,α,h is the density, the velocity, the void fraction, and the specific enthalpy of a given phase; P is the pressure, M is the mass flow term; F,H_f are the resultant drag force due to phase- walls interaction; and ,φ is the interfacial heat flux per unit volume. Heat flux into the wall is not taken into account here.

Equations (1) should be coupled with the phase void fraction relation:

$$\alpha_1 + \alpha_2 = 0 \tag{2}$$

and thermal and caloric equations of state for every phase:

$$h_i = h_i(T_i, \rho_i, \varepsilon, \gamma); \rho_i = \rho_i(T_i, P, \varepsilon, \gamma); i = 1, 2 \tag{3}$$

where ε is the fraction of mass stripped from a single drop, T_i is the phase temperature, and γ is the volume void fraction of coolant vapor. Note that initial volume ratio of the mixture components (metal/coolant liquid/coolant vapor) is: $\alpha_1 / \alpha_2 (1\text{-}\gamma)/\alpha_2\gamma$

To complete the set of equations appropriate conditions at the shock front should be defined on the basis of the parameters in the undisturbed three-phase coarse mixture. Assuming the shock front to have a zero width and neglecting energy changes in comparison with density changes, one can write the following equations for shock discontinuity:

$$\begin{aligned} \alpha_{si}\rho_{si}u_{si} &= \alpha_{0i}\rho_{0i}D \\ \alpha_{si}\rho_{si}u_{si}^2 + \alpha_{si}P_s &= \alpha_{0i}\rho_{0i}D^2 + \alpha_{0i}P_0 \end{aligned} \qquad i=1,2 \tag{4}$$

where D is the shock front velocity and s and o denote the parameters at and in front of the shock, respectively.

Thus, Equation.(4) calculates parameters at the beginning of the reaction (fragmentation) zone on the basis of undisturbed coarse mixture properties. Equations(2),(3), and (1) can be integrated up to the Chapman-Jouguet plane (C-J).

Singularity conditions

An attempt to obtain the singularity conditions for the thermal detonation velocity on the basis of analysis and numerical modeling was made in Ref.4. It was shown that when the processes of heat transfer from molted drops to coolant were being neglected, the C-J plane is placed at the point where phases attain kinematics equilibrium and the choking conditions are satisfied simultaneously. Analogous calculations performed[5] aimed to interpret experimental observations on detonation wave propagation in the molted tin-water and molted aluminum-water mixtures. To define the detonation velocity it was assumed[4] that at the C-J plane phase velocity equilibrium is attained simultaneously with the incipience of vapor generation. Further modeling[6,10] of transient thermal detonation shows a possibility of more than one stable propagation regime with vapor production in the fragmentation zone, as well as the stable regime without vapor. Nevertheless, without any fragmentation kinetics data it was found in[14] a wide spectrum of kinetics-independent permissible C-J states.

In the present work the analysis of sewing the time-dependent expanding zone with the steady reaction zone at the C-J plane is performed following the approach in chemical detonation theory[2,15,16]. In Refs. 2 and 16 the permissible range of steady detonation velocities was found on the basis of the singularity analysis of the derivative dP/dv (v-is the specific volume).

Rewriting Eqs. (1) like it was done in Ref. 16 gives the following expression for dP/dv_2 :

$$\frac{dP}{dv_2} = \frac{\zeta \dfrac{u_2^2}{v_2^2} \dfrac{1}{1 + \dfrac{\alpha_1 \rho_2 u_2^2}{\alpha_2 \rho_1 u_1^2}} - \dfrac{c_2^2}{v_2^2}}{1 - \zeta} \tag{5}$$

$$\zeta = \frac{\beta_2 c_2^2}{C_{p2}} \frac{M(h_1 - h_2 + (u_1 - u_2)^2/2 - m\alpha_2 u_2 C_{p2}/\beta_2) + \varphi + F(u_1 - 2u_2) + H_f(D - u_2)}{M((u_1 - 2u_2)/u_2 + \rho_2 u_2/\rho_1 u_1) + F(1/u_2 - \rho_2 u_2/\rho_1 u_1^2) + H_f/u_2} \left(1 + \frac{\alpha_2 \rho_2 u_2^2}{\alpha_2 \rho_1 u_1^2}\right)$$

Here c_2, C_p, β_2 are local sound speed, heat capacity, and thermal expansion coefficient of the fluid, respectively.

When deriving Eq. (5) it was assumed the reaction zone contains no vapor, energy changes are negligible in comparison with the fluid energy, and metal density is constant. Equations of state for the fluid were taken in the form[14]:

$$d\rho_2 = \frac{dP}{c_2^2} - \frac{T_2 \beta_2 \rho_2}{C_{p2}} dS_2 + mMdx$$

$$m = \frac{k\rho_2}{(1 - \varepsilon)(j - \rho_1 \alpha_1 u_1)} (1 - \varepsilon j/(j - \rho_1 \alpha_1 u_1))$$

$$k = (\rho_{H_2O}(1/\rho_{H_2O} - \rho_{FR}) + \frac{\beta_2}{C_{p2}}(h_{FR} - h_{H_2O}))$$

Equation (5) is analogous to the corresponding equation in Refs. 2 and 16. Therefore one can conclude that the self-sustaining thermal detonation exists only when condition :

$$1\text{-}\zeta=0$$

and condition:

$$\frac{\alpha_1}{\rho_1 u_1^2} - \frac{\alpha_2}{\rho_2 u_2^2}(1 - \frac{u_2^2}{c_2^2}) = \Phi = 0$$

are simultaneously satisfied.

The condition $\Phi = 0$ is equivalent to the condition of equality of the local sound velocity in the fluid and the detonation velocity with respect to the products in two-phase mixture. Two cases are possible: there exists no vapor phase (then the situation is analogous to the choking conditions for chemical detonations); and there exists the saturation state (then the change in Φ-sign is conditioned by a sharp decrease of sound velocity caused by vapor production).
Assuming holding the condition $\Phi=0$,the condition $1\text{-}\zeta=0$ can be rewritten as follows:

$$\begin{aligned}
&-Mk_m + Fk_f + H_f k_h - \varphi k_\varphi = 0 \\
&k_m = \frac{\beta_2}{C_{p2}}(h_1 - h_2 + \frac{(u_1 - u_2)^2}{2}) - m\alpha_2 u_2 - \frac{(u_1 - 2u_2)}{u_2} - \frac{\rho_2 u_2}{\rho_1 u_1} \\
&k_f = \frac{1}{u_2} - \frac{\rho_2 u_2}{\rho_1 u_1^2} + \frac{\beta_2}{C_{p2}}(2u_2 - u_1) \\
&k_h = \frac{1}{u_2} + \frac{\beta_2}{C_{p2}}(u_2 - D) \\
&k_\varphi = \frac{\beta_2}{C_{p2}}
\end{aligned} \tag{6}$$

To analyze Eq. (6) let us write expressions for the losses. Investigations done show that in liquid-liquid systems the breakup mechanism with drop deformation and boundary layer stripping is realized[17]. It was also pointed out that drop breakup starts after a certain time delay relative to the beginning of the interaction between the flow and the drop. This fact wasn't taken into account previously. For this case it is convenient to use the modified Reinecke-Waldman approach[18].

$$M = 0; 0 < t < t_i$$

$$M = u_1 \alpha_1 \rho_1 \frac{r_{d0}}{r_d} \frac{\pi}{4 r_{d0} (t_b - t_i)} (\rho_2 / \rho_1)^{1/2} \sin\left(\frac{\frac{z}{2 r_{d0}} (\rho_2 / \rho_1)^{1/2} - t_i}{t_b - t_i} \pi\right); t > t_i \qquad (7)$$

where r_d is the radius of a metal drop, z is the relative displacement of the drop[4], t_b - is the dimensionless breakup time, and t_i - is the time delay.

The term F is given by Ref.4

$$F = 3 / 8 \rho_2 C_w (u_2 - u_1)^2 \alpha_1 / r_d$$

where C_w is the drag coefficient, which is related to the Reynolds number.

The analogous expression can be written for H_f[19]:

$$H_f = \rho_2 \frac{C_f}{R} (D - u_2)^2 \alpha_1$$

where R is the tube radius and C_f is the wall drag coefficient.

Heat flux between molten drops and the fluid can be expressed as follows[4] :

$$\varphi = 3h \frac{(T_1 - T_2)}{r_d} \alpha_2$$

where h is the heat transfer coefficient.

When analyzing the singularity in the absence of losses at walls and neglecting φ it was found[4] that the second critical condition (together with the sound condition) causes zero values of the M and F if $u_1 = u_2$. It is clear that Eq.(6) being satisfied when H_f and φ equal zero at the point of phase velocities equilibrium. It should note that the strict mathematical solution of particles acceleration problem in the flow shows that the kinematics equilibrium may be attained only at infinite distance from the shock. So $u_1 = u_2$ means $u_1 - u_2 < 1$, where l is a small value given a priori. From Eq. (6) one can see, that the losses tender F and M of nonzero values and therefore there can be no real velocity equilibrium at the critical point.

It is analogous to the effect of incomplete fuel burnout at the C-J plane of chemical detonation with heat and momentum losses[16] Moreover, in the problem under consideration chemical reactions are substituted by the process of heat transfer from debris to the coolant. If the rate of heat transfer is very high then the process of drops fragmentation becomes governing. Incomplete fuel burnout in this case means the nonzero fragmentation rate, which is conditioned by phase kinematics nonequilibrium.

Another consequence of Eq. (6) should be discussed. In transient calculations[10] the fragmentation was terminated when the vapor generation began. In the absence of losses under such an assumption the expression $1-\zeta$ will always change signs with vapor appearance. In fact, neglecting φ at zone of intensive fragmentation we have: $Fk_f - Mk_m = 1-\zeta < 0$. Setting $M=0$ (fragmentation interrupt) and noting that $Fk_f > 0$ we have :$1-\zeta > 0$. Thus one can interpret the possibility of several regimes of thermal detonation with vapor generation in the zone of fragmentation obtained in Ref.10.

Nevertheless, the experiments on the breakup of liquid nitrogen drops[20] showed that the rate of fragmentation is independent of vapor phase existence. Moreover, one can suppose, that the process of hydrodynamics fragmentation is followed by the vapor jet penetration mechanism[21], or they proceed together.

The pressure profile of detonation wave is analogous to that in nonideal chemical detonation[15]. Using Eq. (5) for dP/dx one can write:

$$\frac{dP}{dx} = \frac{-Mk_m + Fk_f + H_f k_h - \varphi k_\varphi}{-1/c_2^2 + \left(1 + \alpha_1\rho_2 u_2^2 / \alpha_2\rho_1 u_1^2\right) / u_2^2} \tag{8}$$

Equation (8) is analogous to Eq.(52) in Ref. 4, but in the denominator we have infinite sound speed in the metal and in the numerator the term with H_f is added. If $\zeta=0$,then $M=0$((7)), φk_φ is small, Fk_f, $H_f k_h > 0$ so $dP/dx > 0$. Afterwards the term with M dominates, so the maximum arises the pressure dependence .

It has to be mentioned that the numerator of Eq. (8) represents the condition seen in Eq. (6), and the denominator is the sound condition. Thus at the C-J plane, the derivative dP/dx becomes indefinite.

Calculation procedure

The calculations were performed for the molted tin-water mixture with the following initial conditions:
P_0=0.1MPa, C_w =2, T_2=360 K , T_1 =1070 K, T_b =3, r_{d0} =0.5 cm, $\rho_1\alpha_1/\rho_2\alpha_2$=4, h=0.6·10^5Wm^{-2}K^{-1}

The equations of state[22,23] for water and tin were used.

Temperature dependence of tin heat capacity was defined in Ref.24.

The density of liquid metal was assumed constant. During calculations we iterated the velocity of a lead shock conditions 1-ζ=0 and Φ=0 hold simultaneously.

In order to avoid difficulties concerning determination of the sound velocity in a two - phase medium[4] the following numerical procedure was used.

1. At first, a certain value of the detonation velocity D is been chosen .
2. Parameters at the shock front are then calculated by the use of Eqs.(3,4).
3. Equations (1) are then solved by iterations with above mentioned empirical equations of state for tin, water and steam.
4. Approaching C-J state the pressure profile changes dramatically. As is seen from Eq.(8),$dP/dx \to -\infty$if condition 1-ζ=0 holds before condition Φ=0; and $dP/dx \to \infty$ if condition Φ=0 holds before condition1-ζ=0.Thus,the variation D and the analysis of pressure profile allow to obtain an approximate solution. The error in the determined values of D is less than 0.1 m/c.

Results of calculations

The initial conditions for the calculations were chosen to fit the experimental results of Fry-Robinson taken from Refs.5,6 on tin -water interaction.

In Fig.1 the profiles of pressure and phase velocities in the reaction zone are presented. Fig.1a corresponds to the ideal case, while Fig. 1b represents the detonation with momentum losses at the tube walls. The star sign denotes the C-J plane position. Near before this point an intensive evaporation begins, which causes a sharp decrease in fluid sound velocity, so that Φ=0. In Fig.1a, in spite of the absence of losses, a certain phase velocities difference exists at the C-J plane. Besides the above mentioned mathematical reasons related with

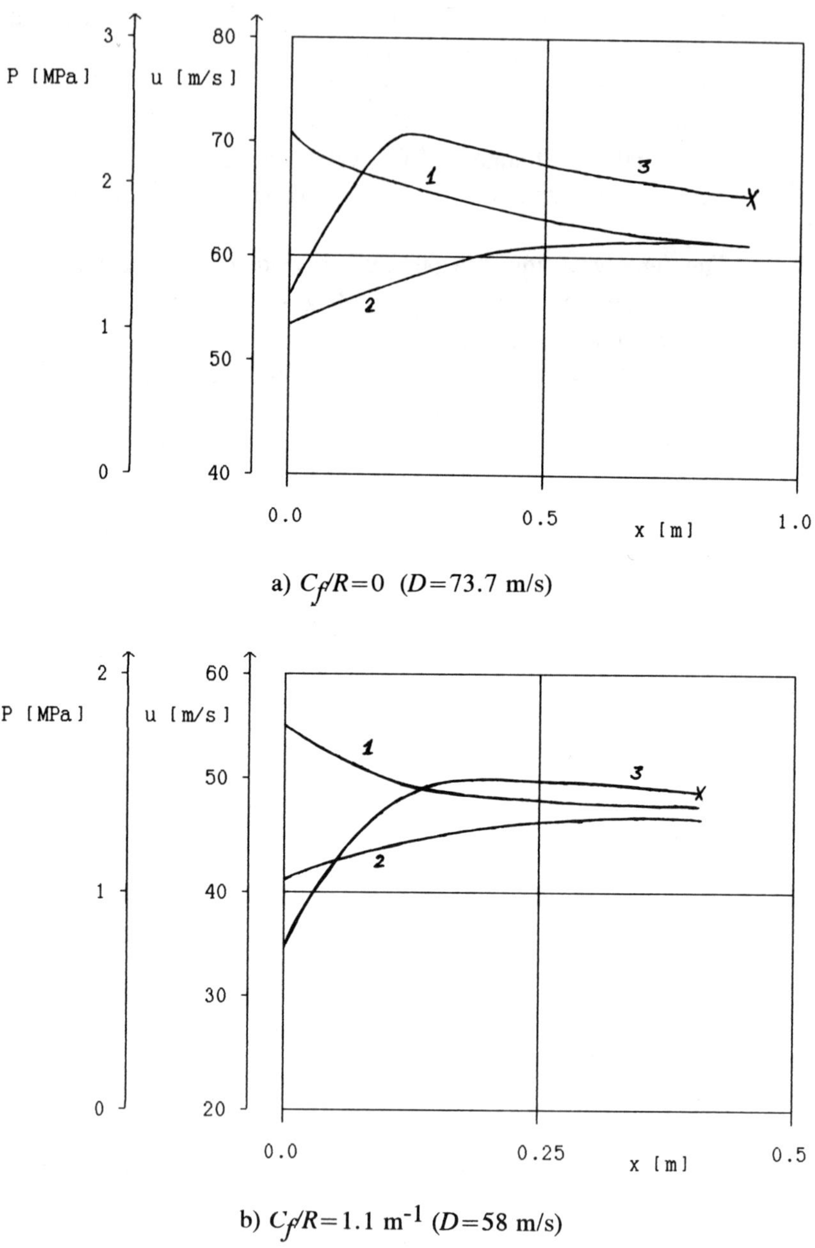

a) $C_f/R=0$ ($D=73.7$ m/s)

b) $C_f/R=1.1$ m^{-1} ($D=58$ m/s)

Fig. 1 Distribution of 1) liquid metal drops velocity, 2) suspension velocity, and 3) pressure in the reaction zone of detonation

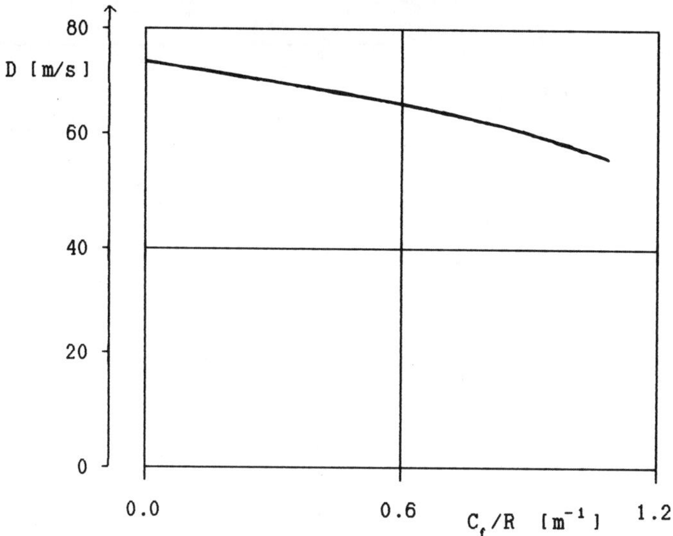

Fig.2 Influence of wall friction on detonation velocity.

asymptotic, this effect arises due to the heat transfer term in Eq. (6). Nevertheless this difference is only 0.1%, so we may assume the velocity equilibrium take place. Introducing the momentum losses (case b), one observes the marked difference in phase velocities up to 6%, and in this case the criteria of singularity of the steady detonation velocity, suggested in Ref.4 become unacceptable.

The presence of momentum losses due to friction at walls leads to a decrease in the thermal detonation velocity and finally in detonation decay. It is illustrated in Fig.2, where the dependence of steady detonation wave velocity on the parameter C_f/R is presented. A limit value of C_f/R characterizes the limiting diameter of the tube and the tube roughness, where the steady thermal detonation is possible. For evaluation of the limiting diameter it is convenient to use the formulas for fluid flow in tubes with rough walls[19]:

$$C_f = (2\lg(Rk_s) + 1.74)^{-2}$$

where k_s -is the equivalent sand roughness. For $k_s = 1$ mm and the value of $C_f/R = 1.1\ m^{-1}$ found in calculations we find that the critical diameter is about 7.6 cm.

The dependence of the steady detonation velocity on the initial void fraction γ is presented in Fig.3. This dependence is in a

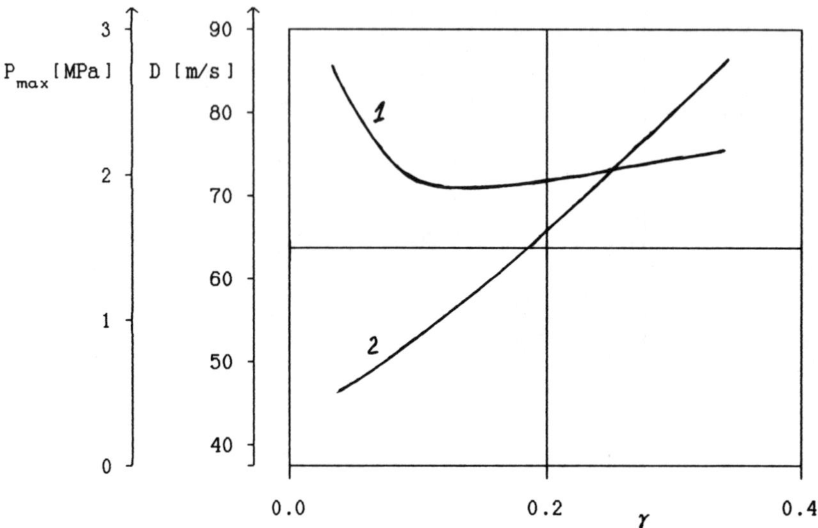

Fig.3 Influence of initial vapor content on the 1) velocity and 2) maximum pressure of detonation wave .

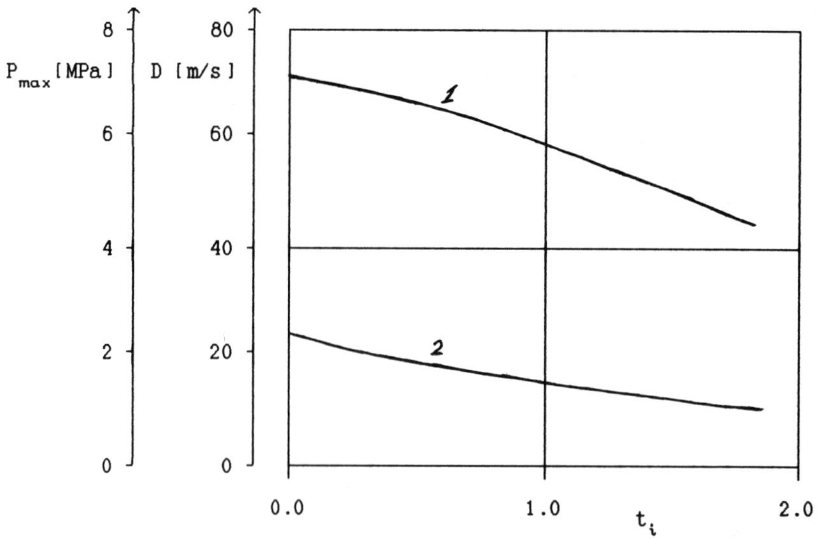

Fig.4 1) Velocity and 2) maximum pressure of steady thermal detonation wave vs. fragmentation time delay.

qualitative agreement with the dependence of the speed of small perturbations on the vapor content with the characteristic minimum at the region of small γ[25]. The maximum pressure in the wave is also plotted in Fig.3. It is a quiet difficult experimental problem to obtain the value of a vapor content. This value will be influenced by the current states of the molted metal and the coolant, the drop residence time in the coolant, etc. So during experiments in vertical channels, gradients of g form during the drop falling. Nevertheless, it follows from Fig.3, that in a wide range of γ= 0.07 - 0.35 velocity changes are very small (4%), while the maximum pressure varies significantly (up to 6.5 times). Therefore, the model under study calculates the velocity of propagation of pressure waves in molted metal- coolant system, but not for the maximum pressure.

The influence of the fragmentation process on the steady thermal detonation velocity is illustrated in Fig.4. The delayed fragmentation results in a decrease in the thermal detonation velocity and maximum pressure. The presence of t_i leads to the lower phase velocity difference at the beginning of the fragmentation zone, so the lower intensity of energy flows into the fluid realized. When $t_i > 2/3t_b$ the process of fragmentation is so weak that the steady regime doesn't exist.

Special calculations performed aimed to detect the regime of propagation without vapor generation[4,10]. Nevertheless, in spite of the numerous calculations in a wide range of the initial parameters, such a stable regime was not found.

It was found in [12,13] that classical CJ detonation solutions only exist if the initial mixture has a low initial void fraction (less than 26% void for tin at 1000C). In the present model no steady solution was obtained for γ>0.35. In terms accepted in Ref. 12 it means that the critical volume fraction of steam in initial mixture is approximately 24.5% (for tin at 800C). It seems that the disagreement is due to incomplete fragmentation in present model.

References

[1]Board, S.J., Hall, R.W., and Hall, R.S., "Detonation of Fuel Coolant Explosions," *Nature*, No.254, 1975, p.319.

[2]Zeldovich, Ya.B., and Kompaneets, A.S., "*Theory of Detonation,*" Gostechizdat, Moscow, 1955.

[3]Fletcher, D.F., and Anderson R.P., "A Review of Pressure-Induced Propagation Models of the Vapor Explosion Process," *Progress in Nuclear Energy*, Vol.23, No.2, 1990, pp.137-179.

[4]Sharon, A., and Bankoff, S.G., "On the Existence of Steady Supercritical Plane Thermal Detonations," *International Journal of Heat and Mass Transfer*, Vol.24, No.10, 1981, p.1561.

[5]Schwalbe, W., Burger, M., and Unger, H., "Investigation of Fry-Robinson Experiments by Means of a Thermal Detonation Model", CSNI/OESD Joint Interpretation Exercise on Fuel- Coolant Interaction, Karlsruhe, 1980.

[6]Carachalios, C., Burger, M., and Unger, H., "A Transient Two-Phase Model to Describe Thermal Detonations Based on Hydrodynamic Fragmentation",. International Meeting on Light-Water Reactor Severe Accident Evaluation, Cambridge, 1983.

[7]Fletcher D.F., "An Improved Mathematical Model of Melt/Water Detonations", parts 1,2, *International Journal of Heat and Mass Transfer*, Vol.34, No.10, 1991, pp.2435-2459.

[8]Burger, M., Muller, K., Buck, M., et.al., "Analysis of Thermal Detonation Experiments by Means of a Transient Multiphase Detonation Code," Proceedings of NURETH-4, Karlsruhe, 10-13 October, Vol.1, 1989, pp.304-311.

[9]Medhekar, S., Amarasooriya, W.H., and Theofanous, T.G., "Integrated Analysis of Steam Explosions", Proceedings of NURETH-4, Karlsruhe, 10-13 October,Vol.1, 1989, pp.319-326.

[10]Carachalios, C., Burger, M., and Unger, H., "Triggering and Escalation Behavior of Thermal Detonations," 23rd ASME/AIChe/ANS Nat. Heat Transfer Conference, Denver, 1985.

[11]Lee, J.H.S., and Frost, D.L., "Steam Explosions: Major Problems and Current Status", Dynamics of Explosions, *Progress in Astronauts and Aeronautics*, Vol.114, 1988, pp.436-450.

[12]Frost, D.L., and Ciccareli, G., "Implications for the Existence of Thermal Detonations from Equilibrium Hugoniot Analysis," Proceedings of 13th ICDERS, Nagoya, Japan, 1991.

[13]Frost, D.L., Lee, J.H.S., and Ciccarelli, G., "The Use of Hugoniot Analysis for the Propagation of Vapor Explosion Waves," *Shock Waves*, Vol.1, No.2, 1991, pp.99-110.

[14]Duane, W.C., "Contributions Concerning Quasi-Steady Propagation of Thermal Detonations Through Dispersions of Hot Liquid Fuel in Cooler Volatile Liquid Coolants", International Journal of Heat and Mass Transfer, Vol.25, No.1, 1982, p.87.

[15]Zeldovich,Ya.B.,Gelfand, B.E.,Borisov A.A.,et al.,"The Reaction Zone of Low-Speed Detonation of Gases in Rough Tubes",*Khimicheskaja Fizika*,Vol.4,No.2,1985, p.279.

[16]Frolov, S.M., Polenov, A.N., Gelfand, B.E., et al., "Detonation Features in Systems with Losses of Arbitrary Type, *Khimicheskaja Fizika*, Vol.5, No.7, 1986, p.978.

[17]Burger, M., Kim, D.S., Schwalbe, W., et al., "Two-phase Description of Hydrodynamic Fragmentation Processes within Thermal

Detonation Waves," Transactions of the ASME, *Journal of Heat Transfer,* Vol.106, 1984 , p.728.

[18]Reinecke, W.G., and Mckay, W.L.,"Free Flight Measurement of Catastrophic Water Drop Breakup", *AIAA Journal*,. Vol.14, No.11, 1976, p.1635.

[19]Schlichting, H., *Theory of Boundary Layer*, Inostrannaya Literatura, Moscow, 1956.

[20]Gelfand, B.E., Gubin, S.A., Kogarko, S.M., et al., "The Breakup of Cryogenic Liquid Droplets by Shock Waves," *Doklady AN SSSR*, Vol.20, No.6, 1972, p.1113.

[21]Frost, D.L., "Dynamics of Explosive Boiling of a Droplet," *Physics of Fluids*, Vol.31, No.9, 1988, p.2554.

[22]Nigmatullin, R.I.,*Multiphase Fluid Dynamics*, Nauka, Moscow, Vol.1,2, 1987.

[23]Stanukovich, K.P.,and Baum, F.A., *Physics of Explosion*, Fizmatgiz, Moscow, 1959.

[24]*Thermal Constants of Non-Organic Substances*, edited by A.V. Briske, and A.F. Kapustinski, USSR Academy of Science Publishing, Moscow, 1949.

[25]Radovski, I.S., "*Speed of Sound in Two-Phase Steam-Water Mixture*," Gosstandart, VNIC GSSSD, MIFI, 1982.

Chapter IV. Nonsteady Flows

Analysis of Combustion Processes in a Mobile Granular Propellant Bed

Tony W. H. Sheu* and Shi-Min Lee†
National Taiwan University, Taiwan, Republic of China
Ming-Yih Chen‡
Tamkang University, Republic of China
and
Vigor Yang§
Pennsylvania State University, University Park, Pennsylvania 16802

Abstract

This paper presents the preliminary results using a first-order Lax finite difference scheme to study the gas-solid combustion physics in a gun tube with a mobile projectile. One-dimensional analysis is conducted for the time being that space-sharing interspersed continua assumption is made for the investigated two-phase flowfield. Noble-Abel gas equation of state, molecular viscosity, and thermal conductivity, interphase drag, heat transfer relation, and burning rate correlation for propellants form a set of constitutive equations. The developing computer code is used to predict the pressure wave buildup and flame front acceleration in a gun barrel full of granulated solid propellants.

Introduction

Both physical and chemical processes taking place in internal ballistics are complex. The investigation of such phenomena by experimental approach is difficult to perform and subject to a high degree

---

* Associate Professor, Institute of Naval Architecture and Ocean Engineering.
† Graduate Student, Institute of Naval Architecture and Ocean Engineering.
‡ Graduate Student, Institute of Mechanical Engineering.
§ Associate Professor, Department of Mechanical Engineering.

of uncertainity and inaccuracy. The advances of computer hardware as well as state of the art computing capacity make such two-phase flow study possible and playing an increasingly important role.

The major physics involved in the gun interior during the short time interval were well described by Kuo.[1] The chemical processes following the ignition of solid propellant are completed only within a few milliseconds in the gun barrel. The penetration of hot gases, produced by the black powder in the center-core igniter near the front end of the barrel, into the voids leads to the ignition of solid granules by convective heating. The granule compaction and rapid pressurization are then set up during the afterward ballistic cycle. The ignition front accelerates from a moderate to a more rapid pace due to the resulting pressure buildup. Such ignited propellants give off more hot gases, which are pushed forward by the developed pressure gradient and thus can ignite more propellants. Such steep pressure gradient; like shock-line front, is thus set up inside the combustion chamber and the pressure behind the shock-line is very high. The resulting accelerated gaseous products finally cause the projectile to move. The flame spreading and possible shock-front formation in porous propellant charges are, without doubt, important in the design and analysis of loading appropriate solid propellants for achieving the propulsion purpose.

A lead-foil decoppering agent is embedded to reduce the rate of copper buildup on the surface between the gun tube and propelling charge. Granular suppressant (K_2SO_4) is installed between the former axial ullage and projectile base for reducing the secondary flash when the fuel-riched combustion gases are vented into the atmosphere following the discharge of projectile. The above dual functionalities are not considered here for simplicity.

Numerical studies of combustion physics in a gun interior have been attempted in the past. Most of the computer codes were developed in the context of one-dimensional analysis. The major contributors are as follows: 1) MGBC code by K. K. Kuo and his colleagues;[2] 2) CALSPAN code by E. B. Fisher et al.;[3] 3) NOVA code by P. S. Gough et al.;[4] and 4) PHOENICS code by N. C. Markatos et al..[5] The extensions to multidimensional analysis for such problems follow: 1) TDNOVA code by P. S. Gough;[6] 2) ALPHA[7] and ALPHA2 codes by H. J. Gibeling et al.;[8] and 3) PHOENICS code by N. C. Markatos et al..[9]

The present analysis is restricted to one-dimensional content. The separated approach, or continuum-mixture approach, was employed as a two-phase model. The mass, momentum, and energy fluxes are balanced over the control volume occupied by space-sharing interspersed continua. Gas and solid phases are present within the same space. Their shares are measured by the value of volume fraction.

Basic Formulation

The prediction of transient combustion processes in a granular propellant bed is made numerically by solving the following two-phase basic equations. This set of equations is formulated by assuming the gas and solid phases are interdispersed and coupled with appropriate interaction relations between them. Each phase itself is a continuum. The change of mass, momentum, and energy for the gas and solid phases can be represented by the following expressions for dependent variables ϕ :

Gas phase

$$\frac{\partial}{\partial t}(\rho_1 R_1 \phi) + \frac{\partial}{\partial x}(\rho_1 R_1 u_1 \phi) = S_1 \tag{1}$$

Solid phase

$$\frac{\partial}{\partial t}(\rho_2 R_2 \phi) + \frac{\partial}{\partial x}(\rho_2 R_2 u_2 \phi) = S_2 \tag{2}$$

where $R_1 + R_2 = 1$ and

	Gas phase		Solid phase	
	ϕ	S_1	ϕ	S_2
Mass	1	$\dot{m}_{comb}$	1	$-\dot{m}_{comb}$
Momentum	u_1	$-R_1 \frac{\partial P}{\partial x} + \frac{4}{3}\frac{\partial}{\partial x}(\mu R_1 \frac{\partial u_1}{\partial x})$ $-D + \dot{m}_{comb} u_2$	u_2	$-R_2 \frac{\partial P}{\partial x} - \dot{m}_{comb} u_2$ $+D + \nabla \cdot (R_c \tau)$
Energy	E_1	$-\frac{Q}{A} + \dot{m}_{comb}(E_{ch} + \frac{u_2^2}{2gJ})$ $-\frac{Du_2}{J} - \frac{1}{J}\frac{\partial}{\partial x}(R_1 p u_1)$	E_2	$\frac{Q}{A\rho_2}$

The intergranular stress in momentum equation of solid phase vanishes as R_1 is greater than the specified critical volume fraction. The Noble-Abel equation,[1]

$$P = (\frac{1}{\rho} - b)^{-1} RT$$

is used as the equation of state for the gas phase where R is the specific gas constant and b is the co-volume. The employed constitutive equations for the closure of Eqs. (1) and (2) are described below.

Intergranular stress

The intergranular stress arises in a partially packed barrel to keep the particles apart. This isotropic normal stress τ in solid phase affects the particles only and is represented by the following form:

$$\begin{cases} \tau = 0 & \text{for } 0 \leq R_c \leq R_1 \\ \tau = a^2 \rho_2 R_c \frac{R_1 - R_c}{R_1 R_2} & \text{for } R_1 < R_c \end{cases}$$

where a is the speed of sound in the solid phase. There is no direct contact between particles as porosity exceeds the critical porosity R_c.

Interphase drag

The average steady-state interphase drag D appearing in momentum equations is represented by the following correlation for a packed bed:

$$D = R_2 \frac{S}{V} F$$

where $S = 4\pi r_p^2$, $V = \frac{4}{3}\pi r_p^3$ are the surface area and volume of a spherical particle, respectively. The interphase drag per unit area of solid phase F can be modeled by Ergun's correlation:[10]

$$F = \frac{\rho\, |u - u_p|\, (u - u_p) \hat{f}}{6}$$

$$\hat{f} = \begin{cases} 1.75, & \text{for } R_1 \leq R_c \\ 1.75[\frac{R_2 R_c}{R_1(1-R_c)}], & \text{for } R_c < R_1 \leq R_l \\ 0.3, & \text{for } R_l < R_1 < 1 \end{cases}$$

$$R_l = \left[1 + 0.01986 \left(\frac{1 - R_c}{R_c}\right)\right]^{-1}$$

We used $\hat{f}$ =1.4 here for simplicity.

Particle burning rate

An empirical pressure-dependent burning law is employed here for simplicity to model the particle surface regression rate under steady-state condition:

$$\dot{Z} = BP^n$$

where the proportionality constant B and propellant burning rate index are specified for the investigated propellant.

The rate of burned propellant is given by

$$\dot{m}_{comb} = R_2 \ \rho_2 \ \frac{\dot{V}}{V}$$

where $\frac{\dot{V}}{V}$ is the fractional rate of change in propellant grain volume. As far as the number of burning grains in a cell N and the number of perforations in a grain f are concerned, the propellant burned in a cell is given by

$$\Delta\dot{m}_{comb} = R_2(t)\rho_2(t)\Delta V(t)$$

where

$$\Delta V(t) = \frac{\pi}{4}(D_o{}^2 - fD_i{}^2)L - \frac{\pi}{4}(L - \beta(t))\left[(D_o - \beta(t))^2 - f(D_i + \beta(t))^2\right]$$

$$\beta(t) = 2\dot{Z}\Delta t$$

Gas-solid interphase heat transfer

The interphase heat transfer Q, contributed by the propellant heating, is given by

$$Q = Sh_t\Delta T$$

where $h_t = Nu\frac{k}{d}$ is total heat transfer coefficient and $d = \frac{6V}{S}$ is equivalent propellant grain diameter. The relationship for propellant grain heating is given by

$$Nu = \max\begin{cases} 1.8 * 10^{-4}\frac{d}{k}P^{0.56} \\ 0.3R_e{}^{0.62} \end{cases}$$

where the pressure is measured by MPa.

Molecular transport properties

We simply used Sutherland's law for the time being to compute molecular viscosity and thermal conductivity. They are assumed to be the functions of temperature given by

$$\mu = \frac{C_1 T^{\frac{3}{2}}}{C_2 + T}$$

$$k = \frac{C_3 T^{\frac{3}{2}}}{C_4 + T} = \frac{C_p}{P_r}\mu$$

where

$C_1 = 0.7535*10^{-6} lbm/ft - s - \sqrt{R}$
$C_2 = 262.5\ R$
$C_3 = 0.291*10^{-6} Btu/ft - s - R^{\frac{3}{2}}$
$C_4 = 170.1\ R$
$P_r = \frac{4\gamma}{9\gamma - 5}$
$\gamma = 1.27$

Discretization Scheme

The solutions ϕ for the set of governing equations are solved by a finite difference method. The solution domain between breech and projectile is first divided into 100 nonuniform cells. During the course of pressurization, the physical domain can be expanded as the force acting on the projectile exceeds the spindle force and finally results in the movement of projectile.

The prediction of possible development of shock formation during the ballistics cycle may encounter oscillatory wiggles near the high gradient region of a few mean free path thickness. The suppression of such numerical oscillations can be made by employing first-order upwind scheme, central schemes with explicit or implicit artificial viscosity, or high resolution schemes such as TVD[11] and ENO.[12] This paper presents a typical first-order scheme, namely Lax scheme, to solve the investigated nonlinear system. The dissipative mechanism will be incorporated into the formulation during the discretization phase such that the high-gradient profile will be smeared a bit. P. Lax[13] in 1954 proposed a discretization scheme for the following partial differential equation in conservative form:

$$\frac{\partial \underline{u}}{\partial t} + \frac{\partial \underline{F}}{\partial x} = 0$$

The corresponding discretization equation is given by

$$\underline{u}_j{}^{n+1} = \frac{1}{2}(\underline{u}_{j+1}{}^n + \underline{u}_{j-1}{}^n) - \frac{\Delta t}{2\Delta x}(\underline{F}_{j+1}{}^n - \underline{F}_{j-1}{}^n)$$

The basic Eqs. in (1) and (2) are formulated in conservative forms and will be discretized by Lax method. The quality of the computed nonoscillating solutions, in terms of accuracy, will be polluted due to the addition of artificial dissipation. The improvement of such drawback by increasing the solution accuracy can be made by Fromm scheme,[14] which will be attempted later.

Computed Results

The physical domain on which the internal ballistic cycle to be simulated is illustrated in Fig. 1. The major components are 1) basepad, 2) centercore igniter, 3) solid propellant, and 4) projectile. Several modifications will be made for simplicity.

The basepad as well as centercore igniter normally consist of black powder. A jet of hot gases is discharged from a primer, taken to be 16.67% of the length of chamber, in the breechblock. This jet is intended to ignite the basepad and then, in turn, ignites the centercore. The representation of ignition train is modeling by specifying the given flow rate and energy. When the burning through the centercore starts, a predominantly radial convection is established in the voids of propelling charges. Here a bag of M6 propellant is considered, which is uniformly distributed.

The initial loading of the solid propellant in internal ballistics is a fundamental issue in achieving optimal burning of propellants to impart the propulsion force on the base of projectile. A low initial loading density may not deliver strong enough force to propel the projectile to a desired distance. A high-loading density for a tightly packed bed, on the other hand, may result in an incomplete combustion process due to the lack of inferior surface area to volume ratio. The initial density of solid propellant is 98.5 $\frac{lbm}{ft^3}$ in the present study. The propellants initiate combustion as the particle surface temperature exceeds the ignition temperature, Tign = 450k

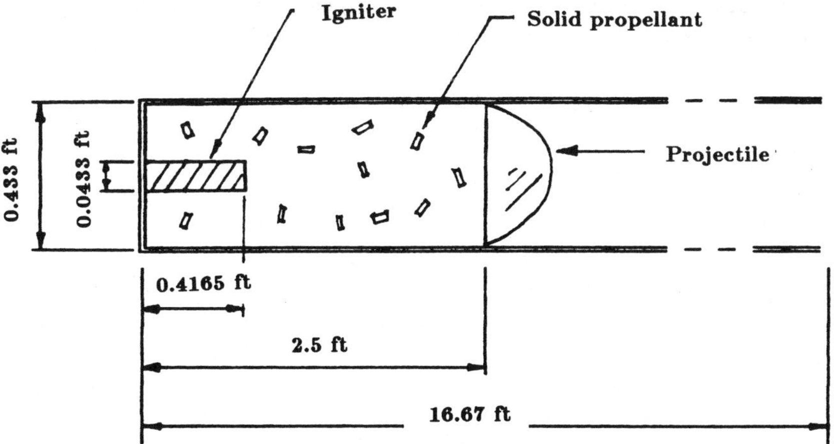

Fig. 1 Illustration of computed physical domain.

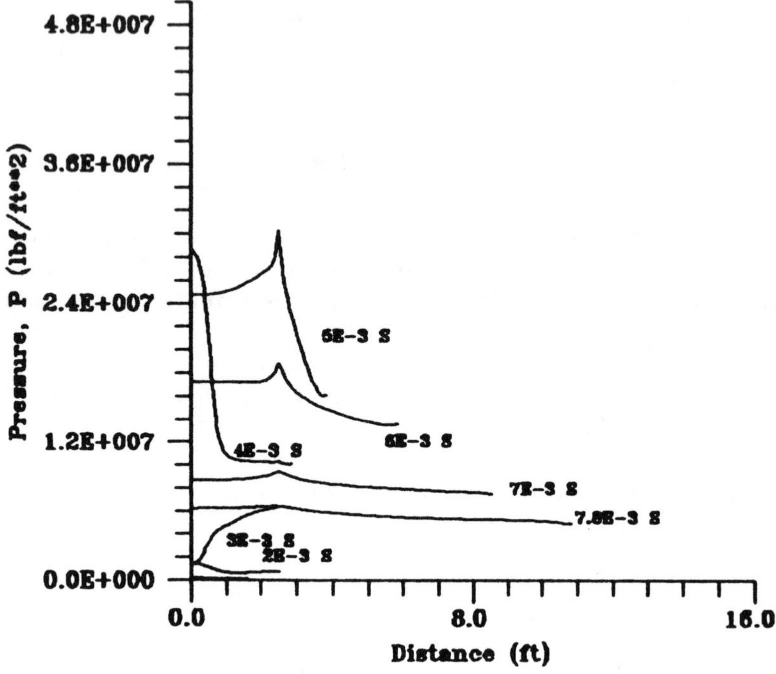

Fig. 2 Distributions of computed pressure.

($810R$). For the sake of clarity, the input data are prepared in a tabulated Table 1.

The computed peak pressure in Fig. 2 remains near the ignited end during the initial stage of ballistic cycle and then moves rapidly toward the ignition front in the bed. The pressurization process at the downstream portion lags that at the upstream. The computed pressure front and flame front are illustrated in Fig. 3 and provide the information about wave propagations within the gun tube.

The peak of gas velocity profile in Fig. 4 at each time is observed near the location of steep pressure front. The solid particle velocity profile in Fig. 5 looks like that of gas phase but with lower value due to the greater inertia of solid particle.

High gas temperature is formed due to the combustion, and it will decay as shown in Fig. 6 since gas loses heat to the solid particle by convection. The gas temperature may be compressed and exceeds that near the front end due to the presence of steep pressure front. The steady increase of particle temperature in the beginning period can be expected due to the convective heat transfer from gas. The computed gas densities at different times are illustrated in Fig. 7.

The computed gas volume fraction in Fig. 8 near the entrance end increases fairly rapidly due to a reduction of particle volume by burning

Table 1 Input data

Gun tube:	Diameter	=	0.433 ft, constant
	Breech face	=	0.0 cm
	Propellant bed	=	2.5 ft
	Muzzle	=	19.17 ft (230 in)
	Surface temp.	=	535 R
Projectile:	Mass	=	100 lbm
	Density	=	500 lbm/ft^3
	Elastic stress wave	=	16000 ft/sec
Propellant:	Total mass	=	21 lbs
	Density	=	98.5 lbm/ft^3
	Grain length	=	0.0833 ft
	Grain diameter	=	0.0375 ft
	Perforation	=	0.00375 ft
	Burn rate	=	Bp^n cm/sec
	B	=	0.2218
	n	=	0.7864
	Ignition temperature	=	810 R
	Thermal conductivity	=	0.356×10^{-4} Btu/ ft-sec-R
	Thermal diffusivity	=	$0.935\times10^{-6} ft^2$/sec
	Emissivity	=	0
	Chemical energy	=	1607 Btu/lbm
	Molecular weight	=	21.36
	Co-volume	=	0.0173611 ft^3/lbm
	Plastic deformation stress wave	=	1900 ft/sce
	Elastic stress wave	=	3800 ft/sec
	Porosity	=	0.4217
	Ultimate failure stress	=	1.584×10^6 lbf/ft^2
	Yield stress	=	8.21×10^5 lbf/ft^2
	Initial temperature	=	535 R
Igniter: (black powder)	Chemical energy	=	675 Btu/lbm
	Molecular weight	=	36.13
	Diameter	=	0.00433 ft
	Co-volume	=	0.016493055 ft^3/lbm
	Thermal conductivity	=	0.2×10^{-4} Btu/ft-sce-R
	Thermal diffusivity	=	1.0×10^{-6} ft^2/sce
Initial conditions:	Ambient air temperature	=	535 R
	Ambient air pressure	=	2116.8 lbf/ft^2
	Propellant bed temperature	=	535 R
	Velocity	=	0.0 ft/sec

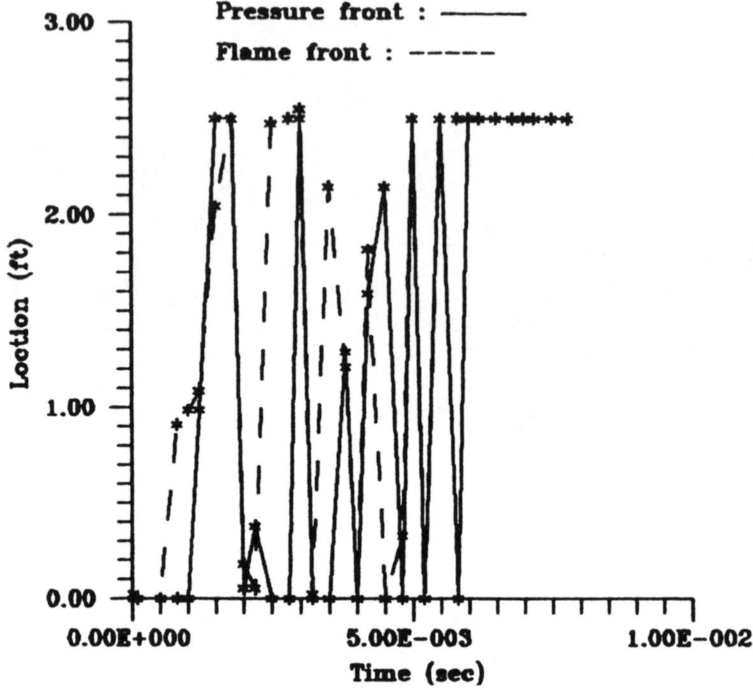

Fig. 3 Computed pressure and flame front location.

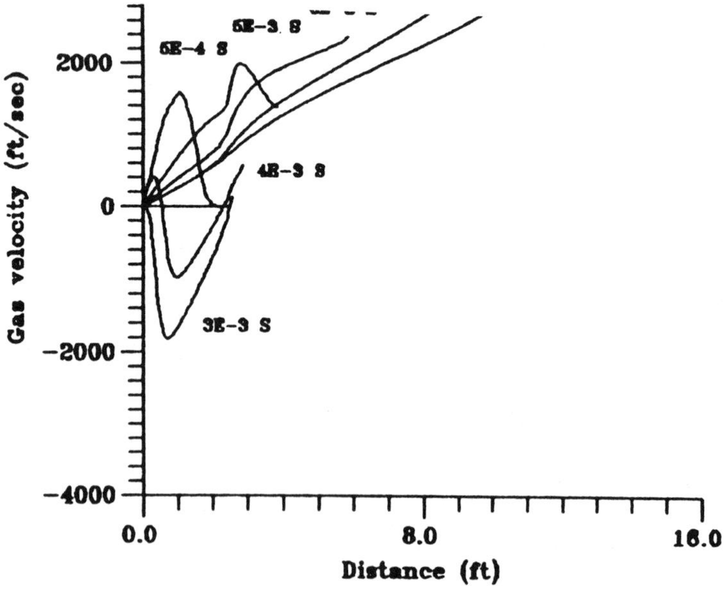

Fig. 4 Distributions of computed gas velocity.

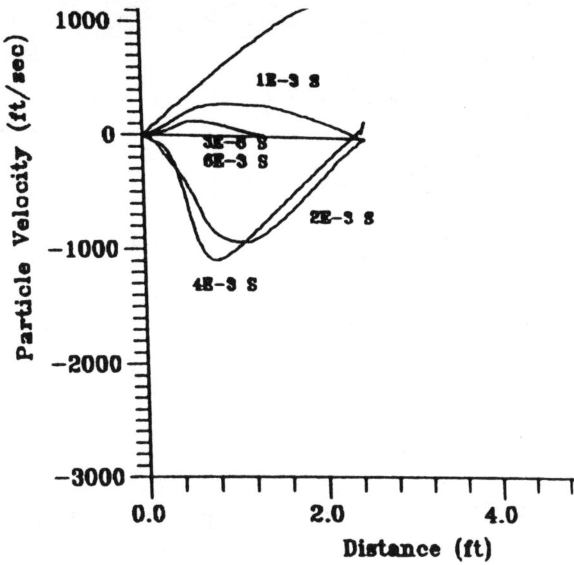

Fig. 5 Distributions of computed solid velocity.

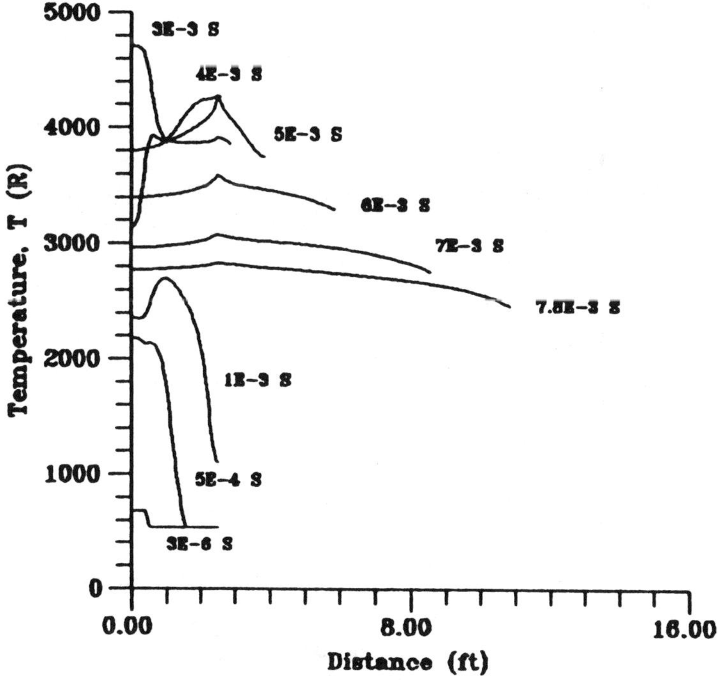

Fig. 6 Distributions of computed gas temperature.

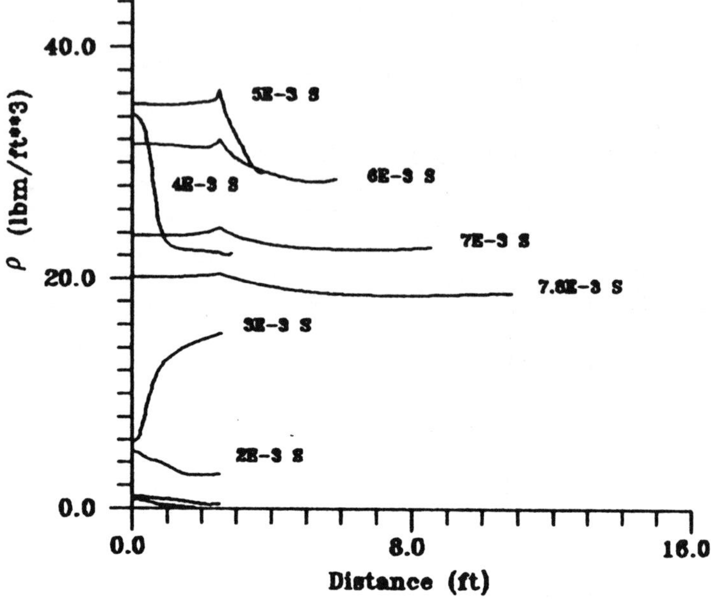

Fig. 7 Distributions of computed gas density.

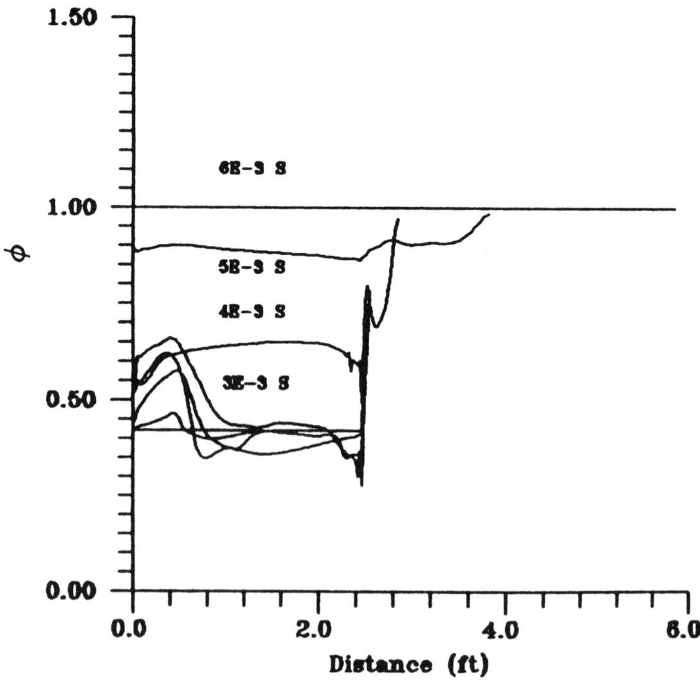

Fig. 8 Distributions of computed gas volume fraction.

and the forward motion of these particles carried along by the gas. The compaction of particles at the further downstream subsequently results in the exceeding of original solid-volume fraction due to the forward drag acting on the particles.

The computed kinks in Figs. 2, 6, and 7 near the original location of projectile base may be reduced by inserting a control volume attached to the projectile base.[15] The lumped parameter analysis along with MOC treatment was emloyed to compute the flow properties at that location. This boundary treatment has not been implemented in the present study.

Conclusions

The preliminary computed results of gas-solid flowfield in a gun tube are summarized in this paper. The global trend of combustion physics can be well-predicted by the present computer code. Higher-order solution scheme, implementation of various enhanced constitutive equations, and incorporation of characteristic analysis on the control volumes attached to the mobile projectile base and breech end are required for further study in the context of one-dimensional analysis.

Acknowledgments

This paper was conducted under a contract from the National Science Council. The author acknowledges with thanks the valuable comments and criticisms offered by K. K. Kuo during the course of work. Thanks are also expressed to H-C Cheng, Director of the Ballistic Research Center Ordnance Production Service Combined Service Force, and his colleagues who made this project possible.

References

[1]Kuo, K. K., Principles of Combustion, John Wiley & Sons, New York, 1986.

[2]Kuo, K. K., Koo, J. H., Davis, T. R., and Coates, G. R., "Transient Combustion in Mobile Gas-Permeable Propellants", Acta Astronautica, Vol. 3, 1976, pp. 573-591.

[3]Fisher, E. B., and Graves, K. W., "Mathematical Model of Double Base Propellant Ignition and Combustion in the 81mm Mortar", Calspan Report No. GD-3029-D-1, Aug. 1972.

[4]Gough, P. S., "Numerical Analysis of a Two-Phase Flow with Explicit Internal Boundaries", IHCR 77-5, Naval Ordnance Station, Indian Head, MD, April 1977.

[5]Markatos, N. C., "Modelling of Two-Phase Transient Flow and Combustion of Granular Propellants", *Int. J. Multiphase Flow*, Vol. 12, No. 6, 1986, pp. 913-933.

[6]Gough, P. S., "Two-Dimensional Convective Flame Spreading in Packed Beds of Granular Propellant", ARBRL-CR-00404, July 1979.

[7]Gibeling, H. J., Buggeln, R. C., and McDonald, H., "Development of a Two-Dimensional Implicit Interior Ballistics Code", ARBRL-CR-00411, January 1980.

[8]Gibeling, H. J., and McDonald, H., "Development of a Two-Dimensional Implicit Interior Ballistics Code", ARBRL-CR-00451, March 1981.

[9]Markatos, N. C., and Kirkcaldy, D., "Analysis and Computation of Three-Dimensional Transient Flow and Combustion Through Granulated Propellants", *Int. J. Heat Mass Transfer*, Vol. 26, No. 7, 1983, pp. 1037-1053.

[10]Ergun, S., "Fluid Flow Through Packed Columns", *Chem. Eng. Progr.*, Vol. 48, 1952, pp. 89.

[11]Harten, A., "A High Resolution Schemes for Hyperbolic Conservation Laws", *J. Comput. Phys.*, Vol. 49, 1983, pp. 357-393.

[12]Harten, A., and Osher, S., "Uniformly High-Order Accurate Non-Oscillatory Schemes", I, *SIAM J. Numer. Anal.*, Vol. 24, No. 2, 1987, pp. 279-309.

[13]Lax, P. D., "Weak Solutions of Nonlinear Hyperbolic Equations and Their Numerical Computation", *Comm. Pure Appl. Math.*, Vol. 7, 1954, pp. 159-193.

[14]Fromm, J. E., "A Method for Reducing Dispersion in Convective Difference Schemes", *J. Comp. Phys.*, Vol. 3, 1968, pp. 176-189.

[15]Chen, D. Y., Yang, V., and Kuo, K. K., "Boundary Condition Specification for Mobile Granular Propellant Bed Combustion Processes", *AIAA Journal*, Vol. 19, No. 11, 1981, pp. 1429-1437.

Unstable Wall Layers Created by Shock Reflections

A. L. Kuhl*
Lawrence Livermore National Laboratory, El Segundo, California 90245
and
R. E. Ferguson,† K.-Y. Chien,* and P. Collins†
Naval Surface Warfare Center, Silver Spring, Maryland 20903

Abstract

This paper describes numerical simulations of the unstable wall layer created by the reflection of planar shock waves from wedges. Four cases were considered: a normal shock case, a regular-reflection case, a single-Mach-reflection case, and a double-Mach-reflection case. It was assumed that the wedge contained an initial dense layer, similar to those formed in dusty boundary layers. Shock interactions with the dense layer created vorticity near the wall by the baroclinic mechanism. The wall shear layer was unstable and rolled up into vortical structures that entrained mass from the fluidized bed. This led to a chaotically striated mixing layer. The flowfield was time-averaged in similarity coordinates (i.e., lines of constant r/t and z/t) to establish the mean and fluctuating-flow profiles of the wall layer. Analysis of the results showed that the wall-layer thickness δ grew as a power-law function of the distance behind the shock ($\delta \sim \xi^{3/5}$) and the mass entrainment rate decayed with distance behind the shock ($\dot{m}_o \sim \xi^{-2/5}$) for the specific cases studied.

Introduction

Boundary layers are an inherent feature of explosions occurring over real surfaces. Knowledge of boundary layer features is important to explosion effects analysis. For example, in dust explosions in mines and industrial processes, the dust is initially deposited in a loose layer along the wall but is subsequently entrained into the turbulent boundary layer induced by the shock wave. One is typically interested in how much dust is scoured from the wall and how it is transported within the turbulent boundary

---

*Aerospace Engineer.

†Mathematician.

layer. Another example comes from explosion safety analysis. To calculate the drag loads on structures, one must know the boundary layer flowfield impinging on the structure.

Initial investigations of such nonsteady boundary layers utilized analytical methods, such as the momentum integral equation, and were limited to clean, viscous flows. Typical examples are H. Mirels classic solutions of the turbulent boundary layer behind a normal shock,[1,2] and the boundary layer induced by a self-similar hemispherical explosion.[3,4] More recently, Mirels and others have used similar analytical techniques to calculate the boundary layer induced by a thermal precursor [5] and to estimate the dust scouring behind shocks.[6,7,8] However, much of our fundamental knowledge of turbulent, dusty boundary layers comes from the pioneering shock tube experiments performed by R. Batt[9] and others.[10,11]

Considerable information exists on shock reflections from wedges. For example, parametric numerical simulations [12] of shock reflections from wedges have revealed details on the flowfields associated with a variety of shock structures. However, those studies were focused on the inviscid laminar flow solutions, so turbulent dusty boundary layer effects were not considered.

Recently we have pursued a new approach — the direct calculation of the mixing occurring in unstable wall layers by following the dynamic evolution of the rotational structures on the computational grid. This approach was used in numerical simulations of turbulent, dusty boundary layers behind a normal shock[13] and behind a double-Mach-reflection (DMR) shock structure on a wedge.[14] In the latter case, however, interpretation of the boundary layer growth was quite complex due to the presence of a turbulent wall jet embedded in the DMR structure. In order to learn more about the boundary layer growth, one must consider more simplified shock structures.

This paper presents a comparative study of unstable wall layers created by shock reflections from wedges which had an initial loose-dust layer on the wedge surface. Numerical simulations of regular reflection (RR), single-Mach-reflection (SMR) and double-Mach-reflection (DMR) cases are considered. The calculated flowfield was stored along similarity lines (i.e., lines of r/t = constant and z/t = constant), and the solution was time-averaged to produce the mean-flow and R.M.S. fluctuating-flow profiles in the wall layer, and to evaluate the boundary layer thickness along the wedge. The next section presents the Formulation of the calculations. The Results section describes the rollup and mixing in the layer by flow visualization techniques, similarity scaling equations and empirical relations for the growth of the mixing layers, and time-averaged results of the mean and R.M.S. profiles of the wall layers. The Discussion section utilizes the Mass and Momentum Integral Equations to intrepret the results in the context of boundary layer theory. This is followed by a Summary and Conclusions.

Formulation

Figure 1 shows a schematic of the problem considered: the reflection of a square wave shock of Mach number M_I from a planar wedge which contained an initial loose layer of dust. The analysis was based on the following idealizations: 1) the air behaves as a perfect gas; 2) the dust particles have a very small diameter, so the dust and the air are in thermal and mechanical equilibrium; 3) the dust-air mixture behaves like a continuum fluid whose equation of state can be approximated by a dense-gas model; 4) the loose dust bed fluidizes immediately behind the shock; 5) the flow is two-dimensional (2-D); and 6) fluid viscosity is zero (i.e., dust density effects dominate the dynamics of the flow near the wall).

According to the preceding assumptions, the dynamics of the flow is governed by the 2-D inviscid conservation laws of gasdynamics:

$$\frac{\partial}{\partial t}\rho + \nabla \cdot (\rho \boldsymbol{u}) = 0 \tag{1}$$

$$\frac{\partial}{\partial t}\rho \boldsymbol{u} + \nabla(\rho \boldsymbol{u}\boldsymbol{u}) = -\nabla p \tag{2}$$

$$\frac{\partial}{\partial t}\rho E + \nabla \cdot (\rho E \boldsymbol{u}) = -\nabla \cdot (p\boldsymbol{u}) \tag{3}$$

where $\boldsymbol{u}$ denotes the velocity and E represents the total energy: $E = e + 0.5\boldsymbol{u}\cdot\boldsymbol{u}$. The pressure p was related to the density ρ and internal energy e by the perfect gas equation of state:

$$p = (\gamma - 1)\rho e \tag{4}$$

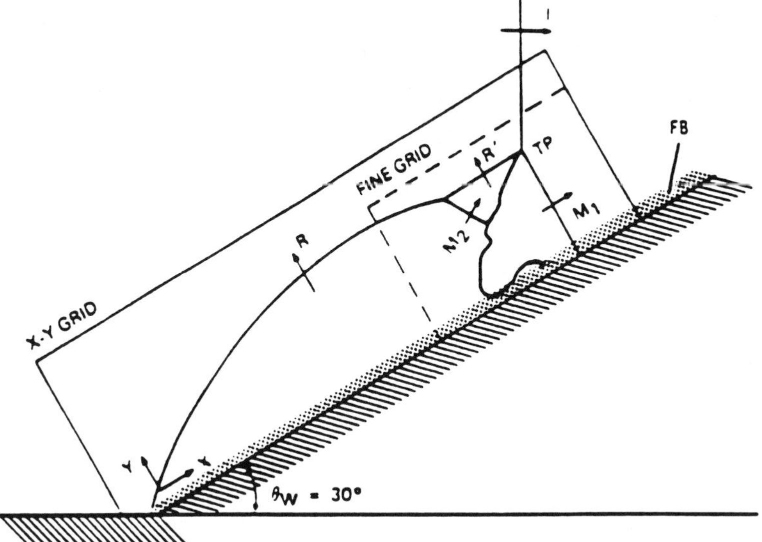

Fig. 1 Schematic of the computational grid.

where γ represents the effective γ of the mixture. For simplicity a value of $\gamma = 1.4$ was used in the calculations. This value is an adequate model for diatomic gases such as air. For the dust-air mixture, it also takes into account the kinetic pressure created particle-particle interactions of the dust which occur in the fluidized bed. If one employs a multifluid formulation, more complete equations of state are possible, however, this was considered to be beyond the scope of the present investigations.

In the above equations, ρ actually represents the mixture density. The dust density ρ_d may be calculated from the relation:

$$\rho_d = \rho C \tag{5}$$

This requires an additional transport equation for the dust concentration C, namely:

$$\frac{\partial}{\partial t} C + (\boldsymbol{u} \cdot \nabla) C = 0 \tag{6}$$

The above equations were integrated numerically by means of a high-order Godunov scheme.[15]

A two-dimensional Cartesian grid, aligned with the wedge surface, was used for the computational mesh. It consisted of a fine-mesh region ($100 \leq i \leq 600$ with initial $\Delta r = 0.025$m; $1 \leq j \leq 322$ with initial $\Delta z = 0.025$m) that followed the shock reflection point; and a stretched-mesh region ($1 \leq i < 100$ with Δr variable) to capture the flow well behind the reflection point. The mesh was initialized with ambient air conditions (state 1):

$$p_1 = 1.01325 \times 10^6 \text{ dy/cm}^2;\ \rho_1 = 1.293 \times 10^{-3} \text{ g/cm}^3;\ \rho_d = 0;$$
$$C = 0;\ e_1 = 1.96 \times 10^9 \text{ erg/g};\ u_1 = 0; a_1 = 3.31 \times 10^4 \text{ cm/s}.$$

and a three-cell-thick fluidized dust bed (subscript FB):

$$p_{FB}/p_1 = 1;\ \rho_{FB}/\rho_1 = 38.67;\ \rho_d = 50 \times 10^{-3} \text{ g/cm}^3;$$
$$C = 50/51.293 = 0.9748;\ e_{FB}/e_1 = 0.0258;\ \boldsymbol{u}_{FB} = 0$$

along the wedge surface ($10\text{m} \leq r \leq \infty$; $0 \leq j \leq 3$). We found that a three-cell-thick layer was adequate (i.e., did not run out of dust mass) in the present calculations. The left boundary of the mesh $(0, z, t)$ was driven by the conditions (state 2) behind a M_I shock, which are given in Table 1. Wall drag was neglected at the bottom of the fluidized bed, hence an inviscid, slip boundary condition ($v = 0$, $\partial u/\partial z = 0$ $\partial p/\partial z = 0$) was used at the bottom boundary. A sliding in-flow boundary condition (corresponding to state 2 behind the incident shock) was used at the top boundary of the mesh to drive the reflection process.

Calculations were run for 5000 computational cycles for each case, in order to accumulate enough data for a good statistical analysis of the fluctuating flow. This required about 10 hours on the Cray XMP computer. The results are described in the next section.

Table 1 Incident Shock Front Conditions

Case		M_I	θ_W	p_2/p_1	ρ_2/ρ_1	e_2/e_1	u_2/a_1	v_2/a_1
1	Normal	1.7	0	3.24	2.22	1.46	0.939	0
2	RR	2	60	4.54	2.70	1.68	0.630	-1.09
3	SMR	2	27	4.54	2.70	1.68	1.12	-0.572
4	DMR	10	30	116.5	5.71	20.39	7.145	-4.125

Results

Four calculations were performed, according to the conditions listed in Table 1. Case 1 consisted of a $M_I = 1.7$ normal shock propagating along a loose dust layer. Previous calculations[13] of this case were performed in shock-fixed coordinates, while this calculation was performed in stationary coordinates with a leading edge to the dust layer; hence, it served to verify the previous solution for a propagating shock. Case 2 corresponded to the regular reflection of a $M_I = 2$ shock wave from a 60° wedge. By changing the wedge angle to $\theta_W = 27°$ in Case 3, the shock structure changed to a single Mach reflection. By increasing the shock Mach number to $M_I = 10$ in Case 4, the shock structure changed to a double Mach reflection.

Flow Visualization

Figure 2 shows a flow visualization of a normal shock wave propagating along a loose dust bed (Case 1: $M_I = 1.7$ and $\theta_W = 0$). When the shock-

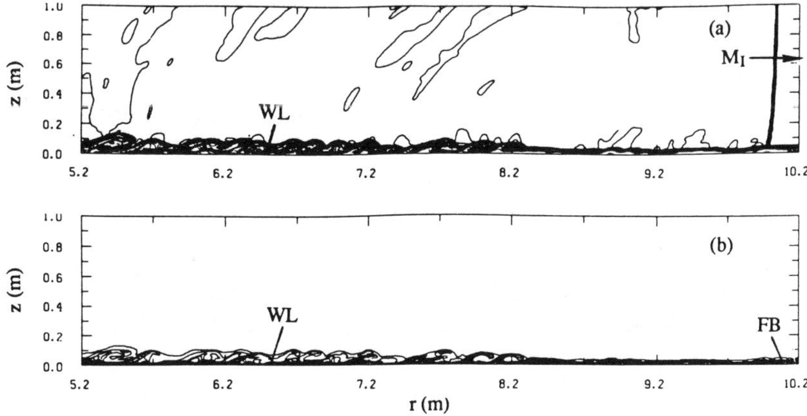

Fig. 2 Contour plots showing the rollup of the wall layer (WL) inducted by a normal shock wave ($M_I = 1.7$, $\theta_W = 0°$): (a) internal energy; (b) density.

front interacts with the top of the dense fluidized bed, it creates vorticity by the baroclinic mechanism: $\nabla p \times \nabla \rho$. The flow is unstable and rolls up into a chaotic mixing layer. Dense material from the fluidized bed is entrained up and around the rotational structures, leading to density striations. The wall layer grows by entrainment and merging of vortices.

We call this chaotic mixing process an unstable wall layer (which is dominated by baroclinically-generated vorticity and density effects) to distinguish it from turbulent boundary layers (which are dominated by viscous wall drag and three-dimensional effects and contain a spectrum of length scales).

Figure 3 presents a visualization of the flowfield induced by the regular-reflection shock structure on the wedge (Case 2: $M_I = 2$, $\theta_W = 60°$). The incident shock I compresses the fluidized bed FB. It reflects off the top of the layer generating shock R′ and off the wall generating shock R″; these shocks coalesce to form the main reflected shock R and the slipline SL. This modifies the shock structure near the reflection region. Shock interactions with the fluidized bed create vorticity, and the unstable wall layer rolls up into chaotic mixing layer. This generates acoustic noise which radiates away from the reflection region. Near the tip of the wedge ($r = 0$), all the material from the fluidized bed is scoured from the wall. This forms a leading-edge LE to the wall layer.

Figure 4 presents a visualization of the flowfield induced by a single-Mach-reflection shock structure on the wedge (Case 3: $M_I = 2$, $\theta_W = 27°$). Figure 5 presents the details of the flow near the SMR structure. The

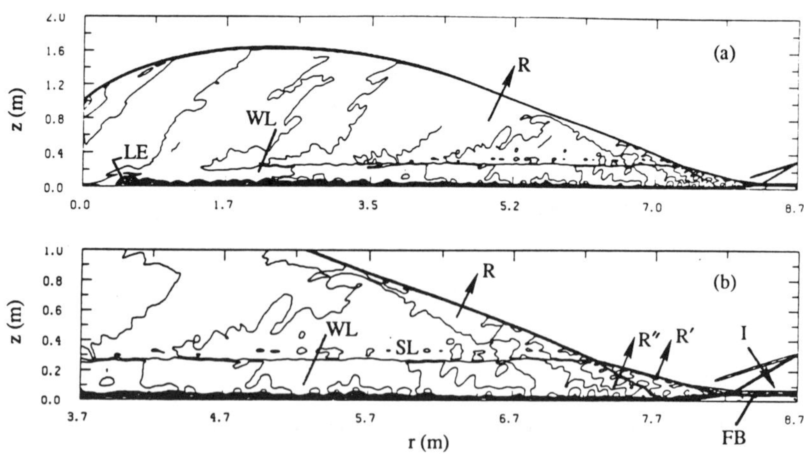

Fig. 3 Contour plots showing the rollup of the wall layer (WL) inducted by a regular-reflection shock structure on a wedge ($M_I = 2$, $\theta_W = 60°$): (a) internal energy contours; (b) internal energy contours near the front.

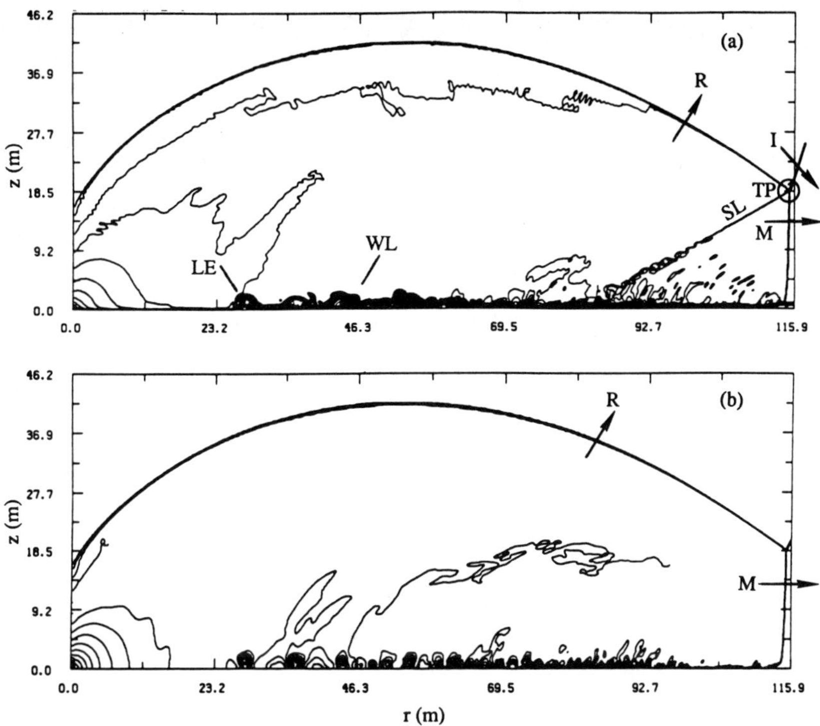

Fig. 4 Contour plots showing the rollup of the wall layer (WL) inducted by single-Mach-reflection shock structure on a wedge ($M_I = 2$, $\theta_W = 27°$): (a) internal energy contours; (b) pressure contours.

slipline SL is unstable and rolls up. The vorticity contours show that the Mach stem M generates vorticity at the top of the fluidized bed. When the vorticity from the slipline SL approaches the wall, it interacts with the opposite-sign vorticity of the wall layer WL, merging of these vortex structures of opposite signs changes growth of the wall layer. (This effect will be more evident in Fig. 7, to be presented in the next subsection.) As in the previous case, the flow near the tip of wedge sweeps the fluidized bed along the wall to form the leading edge rollup shown in Figure 4.

Figure 6 presents a visualization of the flowfield created by a double-Mach-reflection shock structure on the wedge (Case 4: $M_I = 10$, $\theta_W = 30°$). The internal energy contours show the main Mach stem M_1 and secondary Mach stem M_2 and the reflected shocks R′ and R. The slipline SL_1 from the main triple point TP is unstable and rolls up into large-scale rotational structures. These are entrained into the supersonic wall jet WJ and actually form the free shear layer FSL portion of the wall jet. The vortex structures of the free shear layer perturbs the wall layer and causes

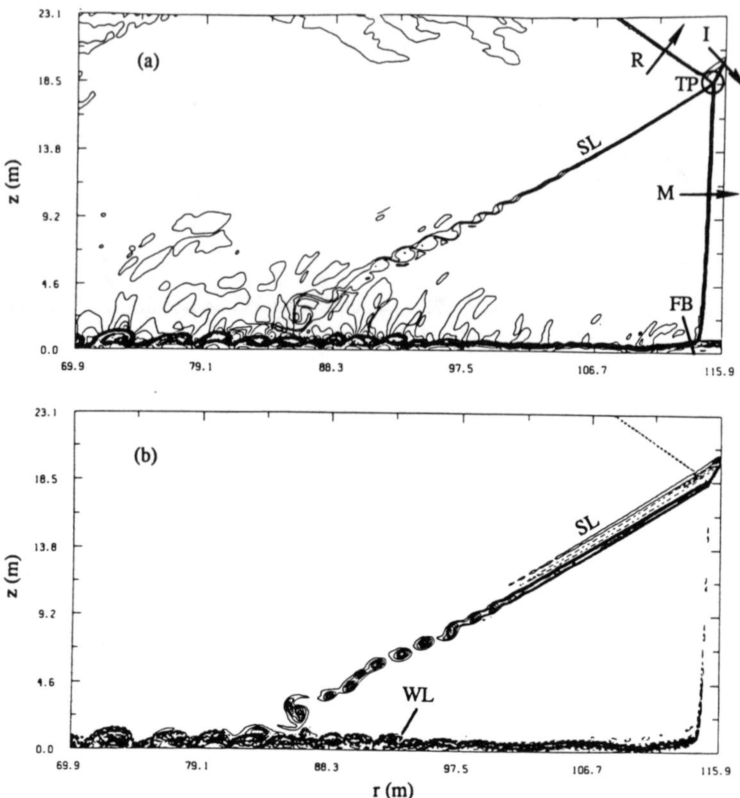

Fig. 5 Contour plots showing the details of the flowfield of Fig. 4 near the front ($M_I = 2$, $\theta_W = 27°$): (a) internal energy; (b) vorticity.

it to separate at many locations along the wall (e.g., three in this particular figure). Flow separation causes dense material to explode off the wall and, of course, modify the growth of the wall layer.

Similarity Scaling

Because the incident shock is a square wave (i.e., with uniform properties behind the front), there are only two characteristic length scales in the problem: 1) the shock front position R_s along the wedge:

$$R_s = W_s t \tag{7}$$

where W_s denotes the velocity of the reflection point or the Mach stem velocity along the wedge and $t = 0$ when the shock arrives at the wedge tip; and 2) the thickness of the fluidized bed z_{FB}. If we neglect the initial

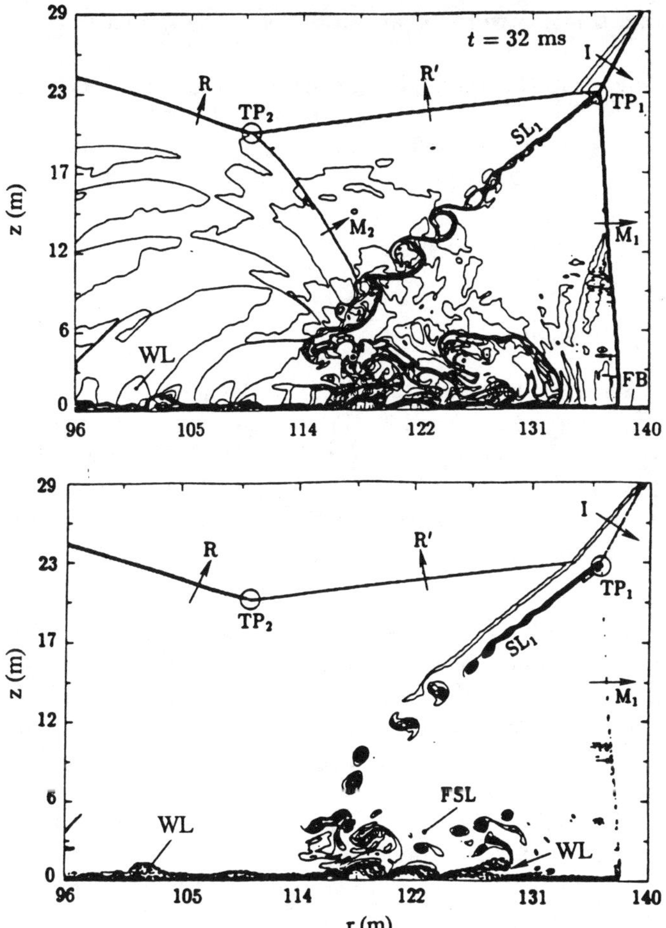

Fig. 6 Contour plots showing the rollup of the wall layer (WL) induced by a double-Mach-reflection shock structure on a wedge ($M_I = 10$, $\theta_W = 30°$): (a) internal energy contours; (b) vorticity contours.

interactions near the beginning of the fluidized bed, i.e., we limit ourselves to the case where $R_s/z_{FB} \rightarrow \infty$ or times $t \gg z_{FB}/W_s$, then only a single characteristic length scale remains, namely R_s, the distance the shock front has propagated along the wedge. Under such circumstances, the flowfield is self-similar,[16] and the number of independent variables may be reduced from three (r, z, t) to two $(r/t,\ z/t)$ namely:

$$x = r/R_s \tag{8}$$

$$y = z/R_s \tag{9}$$

The laminar solution will remain stationary in the similarity coordinates x and y.

Each time step, the flowfield was sampled along similarity lines:

$$x = 0.1, 0.2, 0.3, 0.4, 0.5, 0.6, 0.7, 0.8 \textit{ and } 0.9$$
$$0 < y \leq 0.04$$

and stored for statistical analysis.

Wall Layer Growth

Let us define the top of the wall layer as the height y_{BL} where the mean streamwise velocity $\overline{u}$ reaches 0.99 of the freestream value U_∞ (i.e., $y_{\mathrm{BL}} = y$ where $\overline{u}/U_\infty$ 0.99). And let us define the bottom of the wall layer as the mean top of the fluidized bed y_{FB} (the height where the Reynolds stresses go to zero and where the density profiles converge). Then the wall layer thickness δ becomes:

$$\delta/R_s = y_{\mathrm{BL}} - y_{\mathrm{FB}} \tag{10}$$

The calculated wall layer thickness δ/R_s is presented in Figure 7 as a function of nondimensional position along the wedge, $x = r/R_s$. The regular reflection case (RR denoted by the circles) is the simplest. The wall layer thickness is zero at the reflection point and grows with distance behind the shock, attaining a value of $\delta = 0.012\,R_s$ at $x = 0.4$. The wall layer thickness also grows from the leading edge of the fluidized bed ($x_{\mathrm{LE}} = 0.05$). The two types of boundary layers merge at $x \simeq 0.3$ where the thickness reaches a maximum value of $\delta = 0.0146\,R_s$.

The single-Mach-reflection case (SMR denoted by the squares) is somewhat more complicated. The wall layer thickness increases with distance behind the Mach stem, similar to the *RR* case, until it encounters the effects of the slipline at $x \simeq 0.7$, where it balloons up due to merging of countersign vortices from the slipline and the wall layer. It reaches a peak value of $\delta = 0.0205\,R_s$ at $x = 0.5$, and then decays to zero at the leading edge of the fluidized bed ($x_{\mathrm{LE}} \simeq 0.22$). The double-Mach-reflection case (DMR denoted by the diamonds) is even more complicated. The wall layer thickness is zero at the foot of the Mach stem but rapidly approaches infinity at the leading edge of the wall jet ($x \simeq 0.95$) due to flow separation at the tip of the jet. The layer thickness rapidly decreases and then slowly increases with distance behind the Mach stem. It reaches a peak value of about $\delta = 0.009\,R_s$ at $x = 0.65$ and then decays to zero at the leading edge of the fluidized bed ($x_{\mathrm{LE}} \simeq 0.3$).

For comparison purposes, Figure 7 also presents the results of a numerical simulation of a self-similar precursor shock propagating over a loose dust bed.[17] These results were scaled with the precursor wall jet position R_P (i.e., $x = r/R_P$). The wall layer balloons up at the tip of the jet ($x = 1$)

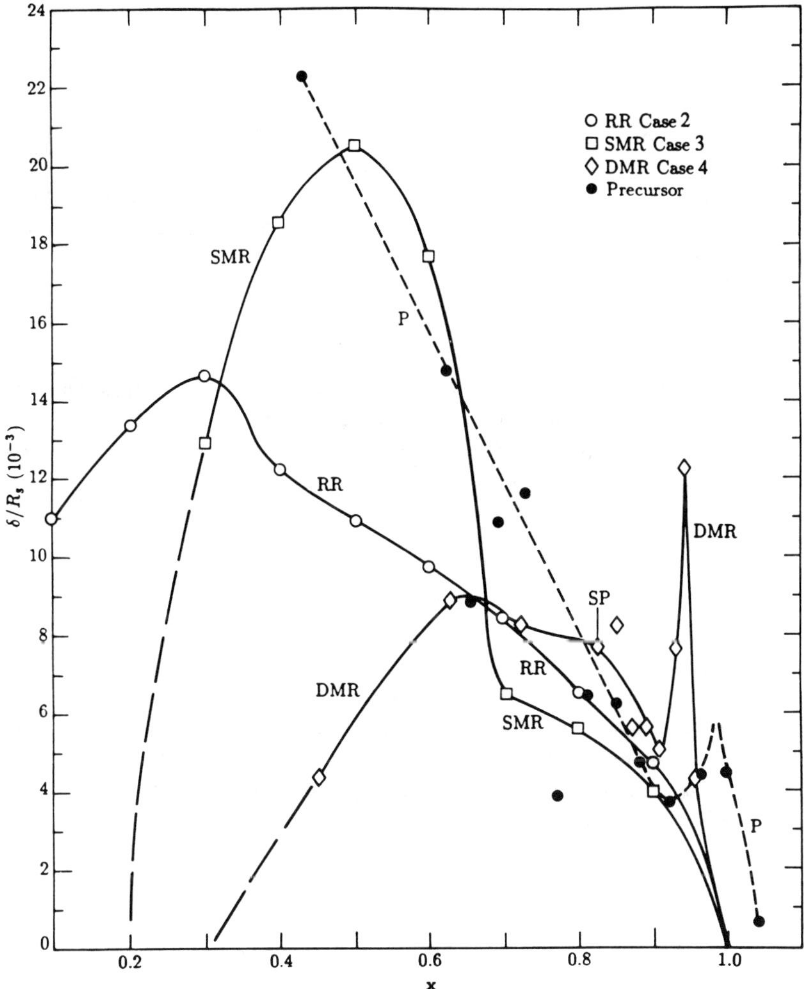

Fig. 7 Dusty wall layer thickness δ/R_S versus distance, $x = r/R_S$, for regular reflection (RR), single-Mach-reflection (SMR), and double-Mach-reflection (DMR) from a dusty wedge and for a precursor (P).

similar to the DMR case, but then continues to grow with distance behind the front. The growth rate for the precursor case seems to be quite a bit larger than the shock reflection cases.

In order to analyze the boundary layer growth, it is convenient to define a nondimensional distance behind the shock:

$$\xi = 1 - x \tag{11}$$

and plot the wall layer thickness versus ξ in log-log coordinates. Figure 8 shows that the calculated thicknesses for the normal shock (Case 1) do indeed fall on a straight line, thus indicating a power-law growth as a function of ξ over the domain $0 < \xi \leq 0.7$ (i.e., neglecting the leading edge boundary layer). This may be expressed in the form:

$$\delta/R_s = a\,\xi^{\beta} \tag{12}$$

where $\beta = 3/5$. The RR, SMR, and DMR data points follow the same power-law trend but have slightly different constants, as shown in Table 2. Note that this is the same functional form as that found for a turbulent boundary layer on a clean flat plate:[18]

$$\begin{aligned} \delta/\chi &= 0.37 Re_{\chi}^{-1/5} \\ \delta &\sim \chi^{4/5} \end{aligned} \tag{13}$$

where χ denotes the distance from the leading edge.

Examination of Table 2 shows that the wall layer growth with distance behind the shock was somewhat slower for the wedge cases ($\delta \sim \xi^{3/5}$) than for our previous calculation[13] performed in shock-fixed coordinates, where $\delta \sim \xi$. That calculation had an infinitely-long fluidized bed, without any length scale other than the distance behind the shock, while Cases 1–4 contained a characteristic lengthscale of distance up the wedge. Perhaps this caused the linear growth rate ($\delta \sim \xi$) in the previous case. The precursor case had a faster growth ($\delta \sim \xi^{5/6}$) than any of the wedge cases. This was a result of the enhanced mass entrainment from the fluidized bed due to the intense mixing in the precursor wall jet.

Mean-Flow Profiles

The flowfield variables ϕ were time-averaged along similarity lines (i.e., along lines of x = constant and y = constant) to establish the mean-flow profiles:

$$\overline{\phi}(x,y) = \int \phi(x,y,t)dt/\tau \tag{14}$$

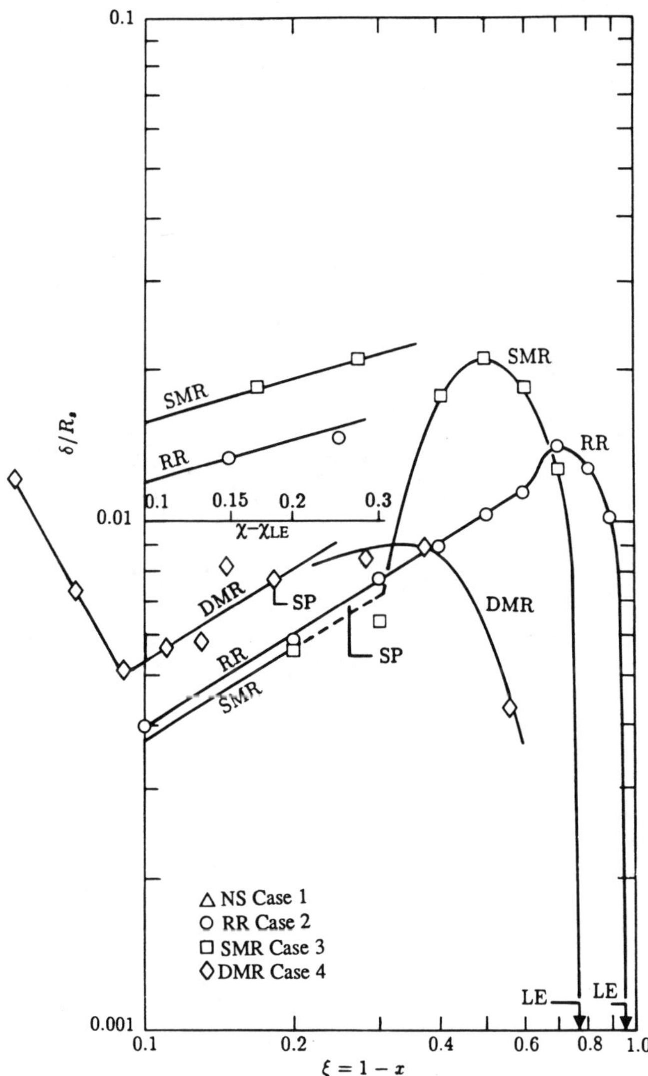

Fig. 8 Dusty wall layer thickness δ/R_S versus distance behind the shock, ξ for a normal shock (NS), regular reflection (RR), single-Mach-reflection (SMR), and double-Mach-reflection (DMR) from a dusty wedge.

Table 2 Boundary Layer Growth of Unstable Wall Layers

Case	Boundary layer growth	β
Clean flat plate[18]	$\delta/\chi = 0.37 Re_{\chi}^{-1/5}$ $\delta \sim \chi^{4/5}$	4/5
Calculations		
Normal shock - Case 1 $(M_I = 1.7,\ \theta_w = 0°)$	$\delta/R_s = 0.037\,\xi^{3/5}\,(0.1 < \xi < 0.7)$	3/5
RR - Case 2 $(M_I = 2,\ \theta_w = 60°)$	$\delta/R_s = 0.0157\xi^{3/5}\,(0.1 < \xi < 0.6)$	3/5
SMR - Case 3 $(M_I = 2,\ \theta_w = 27°)$	$\delta/R_s = 0.0147\xi^{3/5}\,(0.1 < \xi < 0.3)$	3/5
DMR - Case 4 $(M_I = 10,\ \theta_w = 30°)$	$\delta/R_s = f_1(\xi)$ $\simeq 0.0213\,\xi^{3/5}\,(0.1 < \xi < 0.25)$	3/5
Precursor Case[17] $(M_I = 1.7,\ \rho_{\text{TL}}/\rho_1 = 0.1,\ \rho_{\text{FB}}/\rho_1 = 50)$	$\delta/R_J = f_2(\xi)$ $\simeq 0.0325\,\xi^{5/6}$	5/6
Normal shock, infinitely-long fluidized bed[13] $(M_I = 1.7,\ \rho_{\text{FB}}/\rho_1 = 50)$	$\delta/R_s = 0.024\xi$	1
Shock tube experiments		
Normal shock over loose soil bed[9] $(M_I = 1.7)$	$\delta_t = 0.0325(\Delta\chi)^{5/6}$ $[\chi] = \text{cm}$ δ_t = tangent slope thickness	5/6
Normal shock along a clean wall[9] $(M_I = 1.7)$	$\delta_t = 0.00983(\Delta\chi)^{0.93}$	0.93

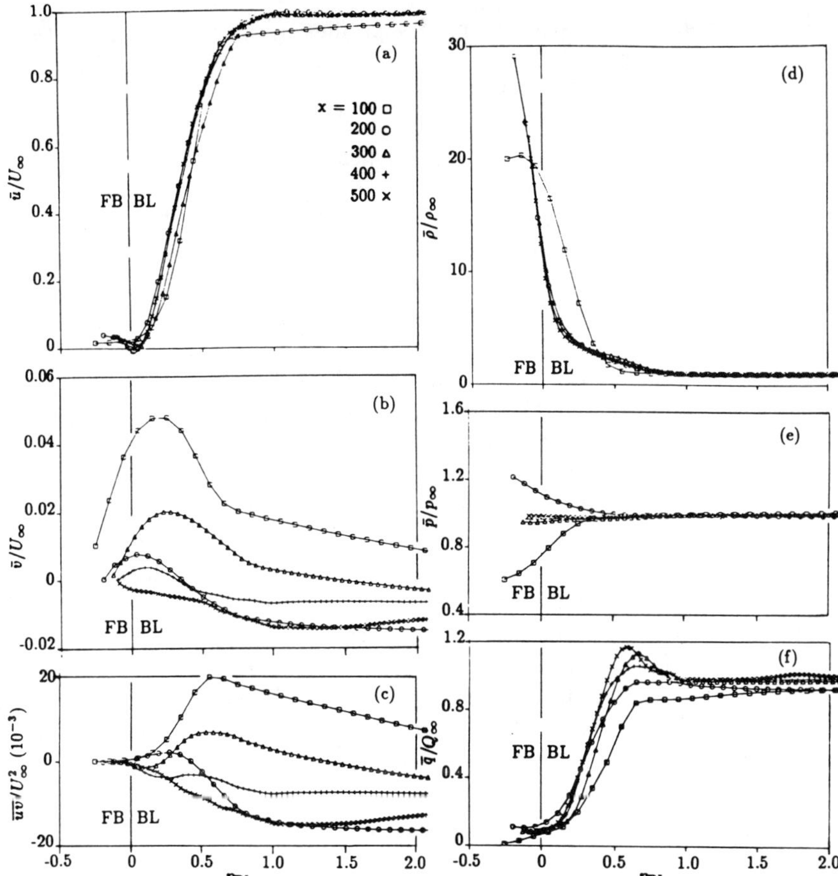

Fig. 9 Mean-flow profiles of the wall layer for the normal shock (Case 1): (a) stream-wise velocity; (b) transverse velocity; (c) shear stress; (d) density; (e) pressure; (f) dynamic pressure.

where the integration duration τ was taken as the last 3000 cycles of the calculation. The profiles were then scaled with the boundary layer thickness, i.e.:

$$\eta = \frac{y - y_{\mathrm{FB}}}{(\delta/R_s)} \tag{15}$$

Note that the boundary layer region of the flow corresponds to the domain $0 < \eta \leq 1$, while the region of $\eta < 0$ corresponds to the flowfield inside the fluidized bed.

The mean-flow profiles of the wall layer induced by a normal shock (Case 1) are shown in Fig. 9. The flow variables have been nondimensionalized by the local freestream conditions, denoted by the subscript ∞. Using the boundary layer scaling Eq. (15), the mean streamwise velocity and density profiles collapse to similarity profiles $\overline{u}/U_\infty = f(\eta)$ and $\overline{\rho}/\rho_\infty = g(\eta)$ that are independent of the x distance behind the shock. The profiles are essentially identical to the profiles from our previous calculation of this problem,[13] which was performed in shock-fixed coordinates. They also agree with the velocity and density profiles measured by Batt[9] in his shock tube experiments of normal shocks interacting with a loose dust bed. The mean density reached a peak value of about $10\,\rho_\infty$ at $\eta = 0$; this caused the mean streamwise velocities to be very small near the wall. The transverse velocities were small ($\overline{v} \simeq 0.01\,U_\infty$) but positive near the wall because of the net entrainment of material from the fluidized bed, and then decayed to negative values away from the wall due to the negative-displacement-thickness effect of shock-induced layers. The mean static pressure profiles were essentially constant throughout the layer. The mean dynamic pressure profiles $\overline{q}/Q_\infty$ were somewhat steeper than the velocity profiles, and they seemed to overshoot by about ten percent (i.e., $\overline{q}/Q_\infty \simeq 1.1$) in the middle of the layer. Thus, the wall-layer thickness, as inferred from the dynamic pressure profiles was only about half the thickness of the mean velocity profile. This is a density effect that is characteristic of dusty boundary layers. Such effects must be carefully taken into account when trying to infer the boundary layer thickness from stagnation pressure measurements.

The mean-flow profiles of the wall layer induced by regular-reflection (Case 2) and single-Mach-reflection (Case 3) shock structures are shown in Figs. 10 and 11, respectively. The profiles are quite similar to the normal-shock case of Fig. 9. Neglecting the stations near the leading edge (i.e., $0 < x \leq 0.4$ for the RR case, and $0 < x \leq 0.6$ for the SMR case), the profiles are approximately independent of x, however, they are not as well converged as in Case 1.

The profiles for the double-Mach-reflection shock structure (Case 4) were published previously,[14] and will not be repeated here. Those calculations showed that the wall layer profiles for the DMR case depended on x, due to the more-complex mixing occurring in the unstable wall jet.

Fluctuating-Flow Profiles

The R.M.S. fluctuating-flow profiles were calculated from the following relation:

$$\phi'(x,y) = \left[\int \{\phi(x,y,t) - \overline{\phi}(x,y)\}^2 \, dt/\tau\right]^{1/2} \tag{16}$$

where the integration duration τ was taken over the last 3000 cycles of the calculation.

The fluctuating-flow profiles for the wall layer induced by a normal shock (Case 1) are shown in Fig. 12. Again, the profiles are essentially identical to the R.M.S. profiles from our previous calculation of this problem[13] that was performed in shock-fixed coordinates. Streamwise velocity fluctuations reached a peak value of $u' \simeq 0.2U_\infty$ at $\eta \simeq 0.25$, while the transverse

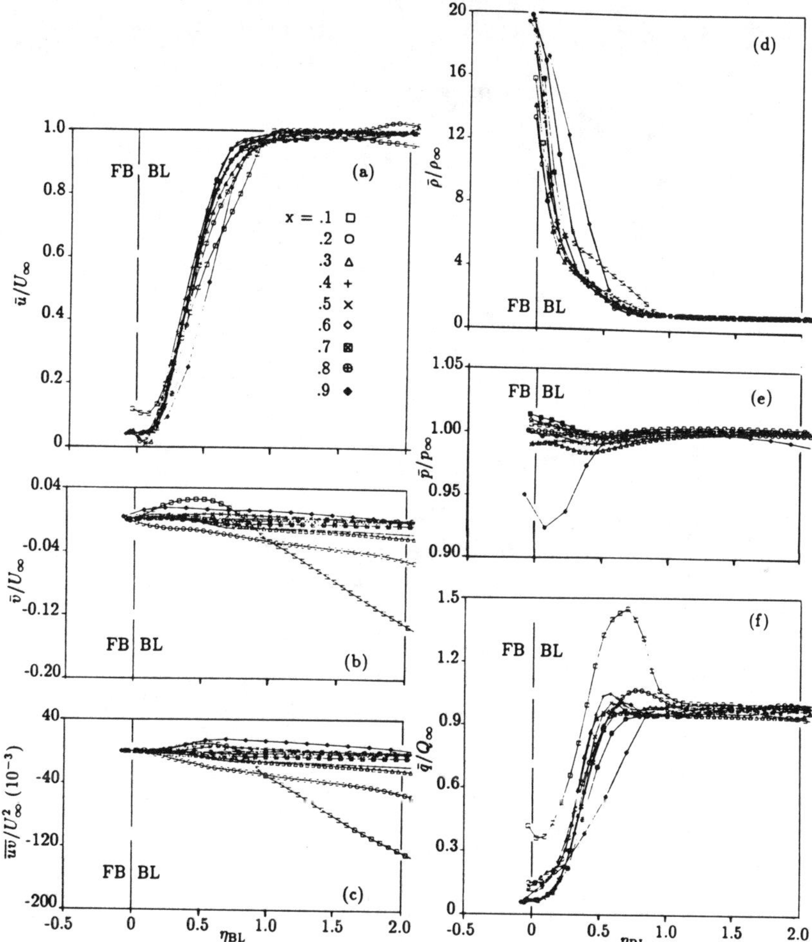

Fig. 10 Mean-flow profiles of the wall layer for the regular-reflection (Case 2): (a) stream-wise velocity; (b) transverse velocity; (c) shear stress; (d) density; (e) pressure; (f) dynamic pressure.

velocity fluctuations reached a peak value of about half that value ($v = 0.12\,U_\infty$ at $\eta = 0.6$), similar to clean turbulent boundary layers. Velocity fluctuations within the fluidized bed were less than ten percent of U_∞. The Reynolds stress was zero inside the fluidized bed and reached a peak value of about $\overline{u'v'} = 0.003\,U_\infty^2$, which is also characteristic of clean turbulent boundary layers. The density fluctuations reached a peak of about six times the freestream value because of turbulent entrainment of dense material from the fluidized bed. Static pressure fluctuations in the wall layer reached a peak value of $p' = 0.15\,p_\infty$ at the bottom of the layer. Dynamic pressure

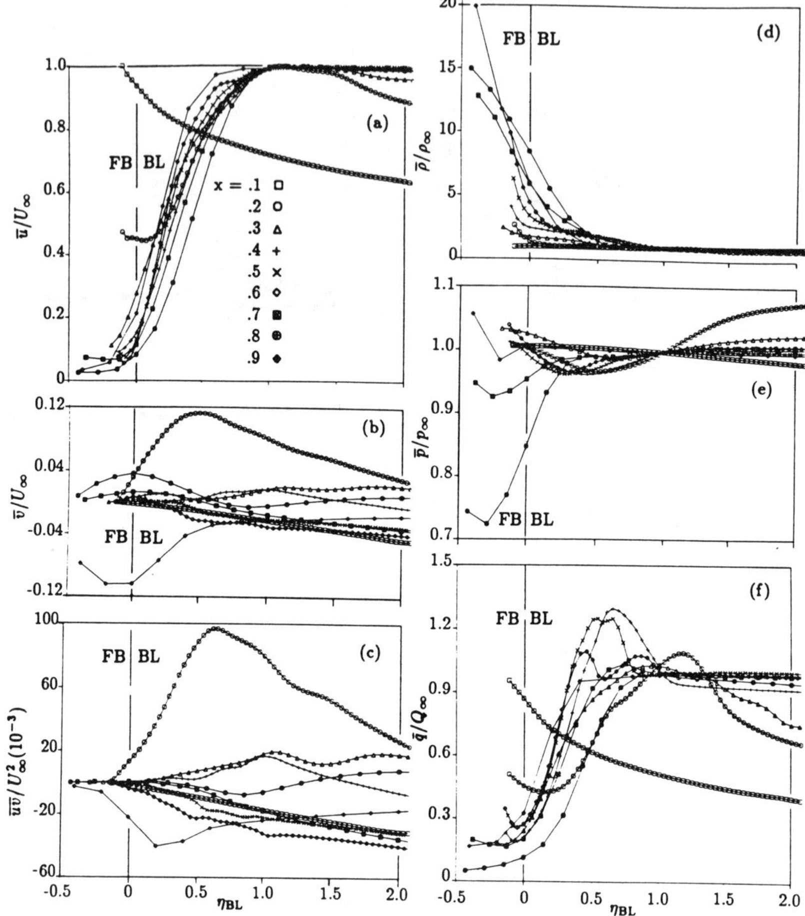

Fig. 11 Mean-flow profiles of the wall layer for the single-Mach-reflection (Case 3): (a) stream-wise velocity; (b) transverse velocity; (c) shear stress; (d) density; (e) pressure; (f) dynamic pressure.

fluctuations were considerably larger; they reached a peak value of almost $q' = 0.5\,Q_\infty$ because of density effects.

The fluctuating-flow profiles for the regular-reflection (Case 2) and single-Mach-reflection (Case 3) shock structures are presented in Figs. 13 and 14, respectively. The profiles are quite similar to the normal-shock case presented in Fig. 12. Neglecting the stations near the leading edge (i.e., $0 \leq x \leq 0.4$ for the RR case, and $0 < x \leq 0.6$ for the SMR case), the profiles again collapse to similarity profiles that are essentially independent of x. However, the profiles are not as well converged as in Case 1; no doubt, a longer statistical averaging is required for complete convergence of these second moments.

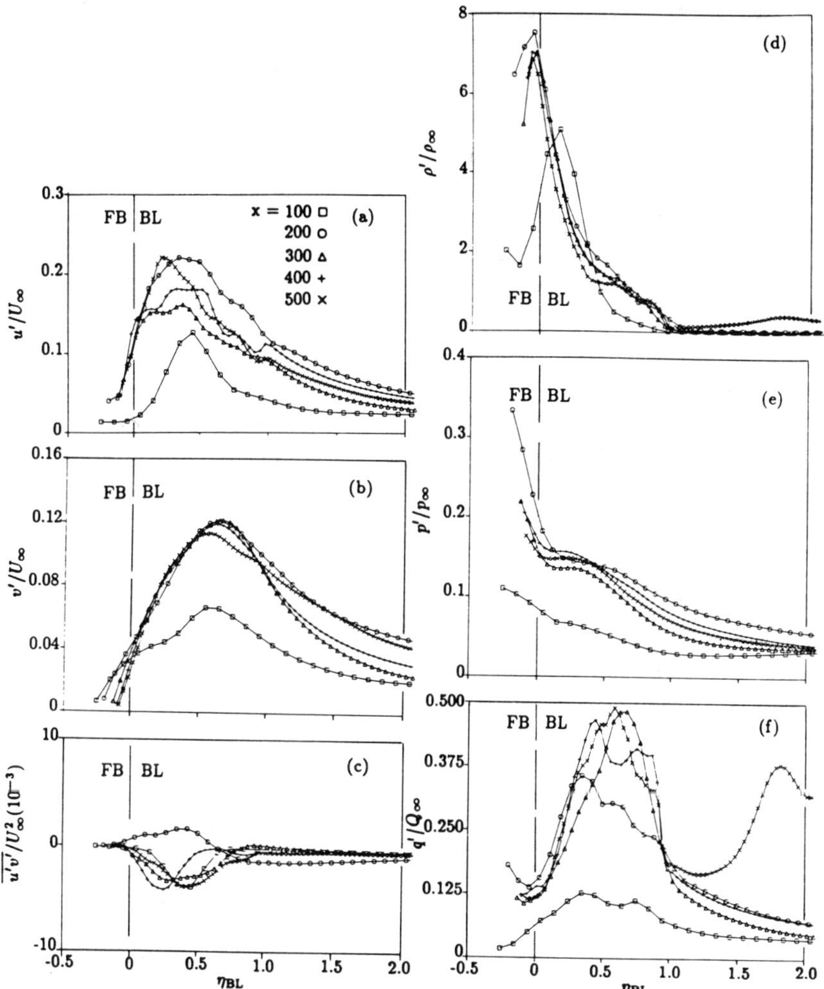

Fig. 12 RMS fluctuating-flow profiles of the wall layer for the normal shock (Case 1): (a) stream-wise velocity; (b) transverse velocity; (c) shear stress; (d) density; (e) pressure; (f) dynamic pressure.

The profiles for the DMR shock structure (Case 4) were published previously[14] and will not be repeated here. Those calculations showed that the wall layer profiles for the DMR case were considerably more complex and depended on x, due to the chaotic mixing in the wall jet.

Discussion

This section explores the mechanisms of the growth of the wall layer in the context of boundary layer theory. (A complete derivation of the equations can be found in Ref. 19). We start by defining the mass thickness

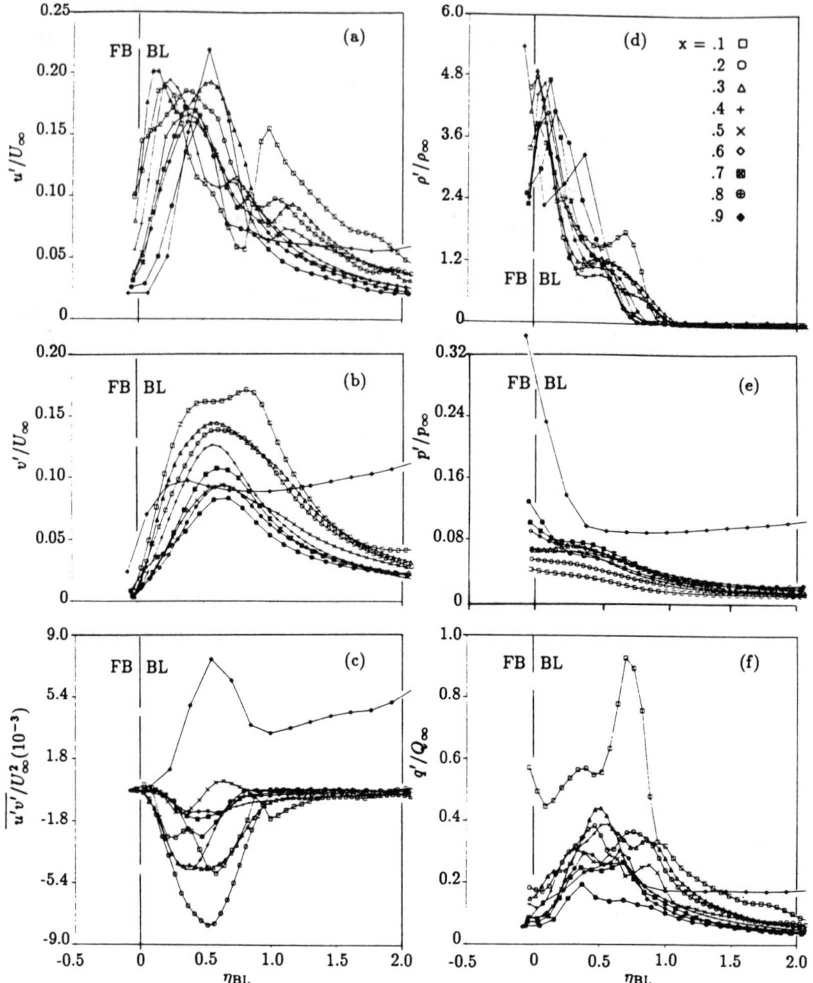

Fig. 13 RMS fluctuating-flow profiles of the wall layer for the regular-reflection (Case 2): (a) stream-wise velocity; (b) transverse velocity; (c) shear stress; (d) density; (e) pressure; (f) dynamic pressure.

δ_m, which is related to the boundary layer thickness δ according to:

$$\delta_m = I_m \delta \tag{17}$$

Here, I_m represents the integral of the mass and mass-flux profiles taken over the boundary layer:

$$I_m = \frac{W_s}{u_2} \int_0^1 (h-1)d\eta + \int_0^1 (1-hf)d\eta \tag{18}$$

where u_2 denotes the gas velocity behind the shock. If the density and velocity profiles in the wall layer are self-similar (i.e., if they are independent

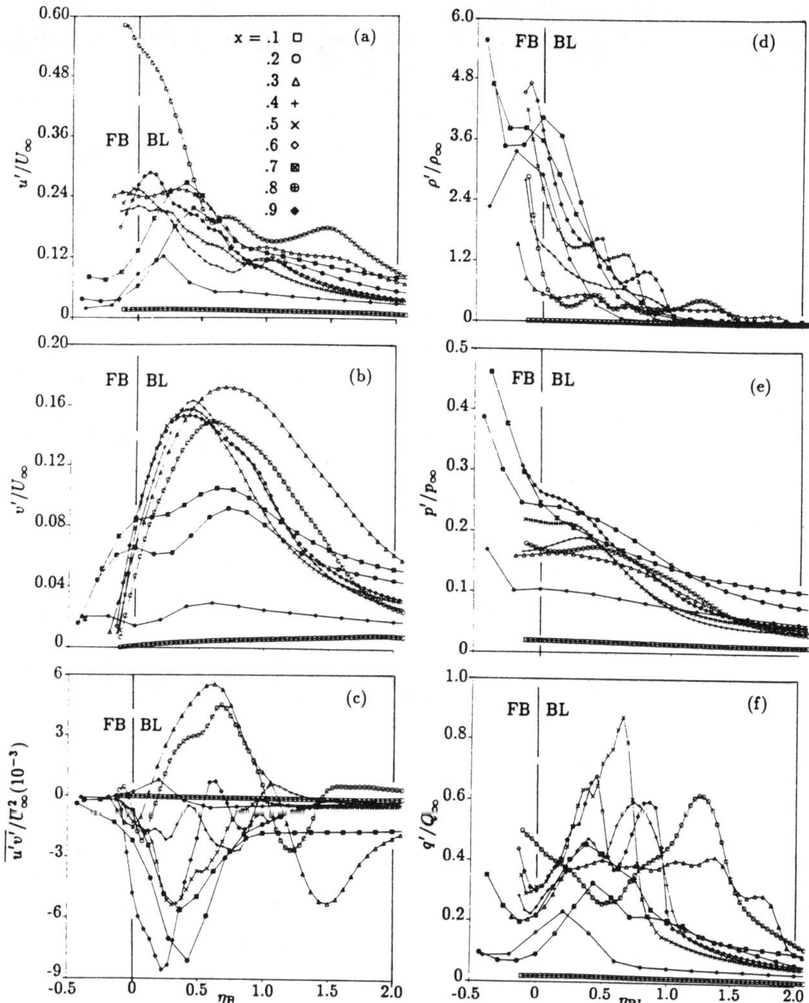

Fig. 14 RMS fluctuating-flow profiles of the wall layer for the single-Mach-reflection (Case 3): (a) stream-wise velocity; (b) transverse velocity; (c) shear stress; (d) density; (e) pressure; (f) dynamic pressure.

of x so that $h \equiv \bar{\rho}/\rho_\infty = h(\eta)$ and $f \equiv \bar{u}/U_\infty = f(\eta)$, respectively), then the mass integral I_m equals a constant; such was the case for the normal shock calculation (see Fig. 9) where $I_m \simeq 1.85$.

Next, consider the Mass Integral Equation

$$\delta_m' = \frac{d}{d\xi}\left[I_m\, \delta/R_s\right] = \dot{m}_o - v_\infty/u_2 \tag{19}$$

which may be derived from a control volume analysis of the mass flux in the boundary layer. In the above, $\dot{m}_o$ represents the rate that mass is being entrained into the bottom of the boundary layer at $\eta = 0$ due to turbulent mixing:

$$\dot{m}_o = \overline{\rho_o\, v_o}/\rho_2 u_2 \tag{20}$$

Assuming that the velocity and density profiles are self-similar, then $\delta'_m = I_m \delta'$ and Equation (19) becomes:

$$\delta' = (\dot{m}_o - v_\infty/u_2)/I_m \tag{21}$$

Thus the Mass Integral Equation states that the fundamental reason that the dusty boundary grows is because of turbulent mass entrainment from the fluidized bed (i.e., because of $\dot{m}_o$). This is true regardless of momentum considerations.

In a similar manner, one can define a momentum thickness δ_θ of the boundary layer:

$$\delta_\theta = I_\theta\, \delta \tag{22}$$

where

$$I_\theta = \frac{W_s}{u_2} \int_0^1 h(1-f)d\eta - \int_0^1 hf(1-f)d\eta \tag{23}$$

Again, using a control volume analysis of the momentum flux in the boundary layer, one can derive the Momentum Integral Equation (for Case 1):

$$\delta'_\theta = \frac{d}{d\xi}\left[I_\theta\, \delta/R_s\right] = \dot{m}_o + c_f/2 \tag{24}$$

where c_f denotes the local wall drag. This states that the momentum thickness grows because of mass entrainment and wall drag. Typically, $\dot{m}_o \gg c_f/2$ (e.g., $\dot{m}_o \simeq 0.04$ and $c_f/2 < 0.001$), so again one finds that the boundary layer growth is caused by mass entrainment.

Solving Eq.(21) for the mass entrainment rate, one finds

$$\dot{m}_o = I_m\, \delta' + v_\infty/u_2 \tag{25}$$

The boundary layer slope δ' may be evaluated from the empirical relation Eq. (12):

$$\begin{aligned}\delta' &= \beta a\, \xi^{\beta-1} \\ &= 0.022\, \xi^{-2/5} \text{ for Case 1}\end{aligned} \tag{26}$$

and eliminated from Eq. (25), yielding:

$$\dot{m}_o = I_m\, \beta a\, \xi^{\beta-1} + v_\infty/u_2 \tag{27}$$

Evaluating the constants in the above for the normal-shock calculation (Case 1, for $0.1 < \xi < 0.7$) gives:

$$\dot{m}_o = 0.041\, \xi^{-2/5} - 0.01 \tag{28}$$

This relation shows that the mass entrainment rate decays as $\xi^{-2/5}$ of the distance behind the shock front.

Summary and Conclusions

Shock interactions with the dense fluidized bed generated vorticity near the wall by the baroclinic mechanism: $\nabla p \times \nabla(1/\rho)$. The wall shear layer was unstable and rolled up into large-scale vortical structures that entrained dense material from the fluidized bed. This led to a chaotically striated mixing layer.

The wall layers grew due to merging of large vortex structures and due to entrainment of dense material from the fluidized bed. For the cases studied, the wall layers grew as a power function of the distance behind the shock:

$$\delta/R_s = a\,\xi^{3/5} \tag{29}$$

where $0.015 \leq a \leq 0.037$ depending on the specific case considered. This is qualitatively similar to the growth of a turbulent boundary layer on a clean flat plate $\delta \sim \chi^{4/5}$. The exponent for the wall layer calculations is somewhat smaller than the flat plate boundary layer (i.e., 3/5 versus 4/5, respectively). This value of the exponent may depend on the problem (e.g., shock reflections from dusty wedges versus a clean flat plate case), or it may be a consequence of the two-dimensional flow assumption. Boundary layers are actually three-dimensional, and the extra degree of freedom may allow the boundary layer to grow more rapidly. Three-dimensional calculations should be performed to investigate this effect.

Using the Mass Integral Equation of boundary layer theory, it was shown that the wall layer grew because of turbulent mass entrainment from the fluidized bed. The mass entrainment rate decayed as a power-law function of distance behind the shock: $\dot{m}_o \sim \xi^{-2/5}$.

The mean-flow velocity and density profiles in the wall layer calculations were qualitatively similar to the dusty boundary layer profiles measured in shock tube experiments[9] using a normal shock propagating along a loose dust bed. The peak values of the R.M.S. fluctuations in the wall layer are qualitatively similar to those found in turbulent boundary layers. Nevertheless, experimental data on dusty boundary layers behind reflected shocks are needed to quantitatively check the accuracy of these calculations.

The numerical simulations described here provide a useful tool for studying mixing layers that are dominated by the evolution of baroclinically-generated vorticity (e.g., dusty boundary layers). This method should be used to calculate turbulent mixing in unstable wall layers induced by nonsteady blast waves.

Acknowledgments

This work was sponsored by the Defense Nuclear Agency (DNA) with a contract to the Naval Surface Warfare Center (MIPR-91-670 and MIPR-

91-558) and a contract to the Lawrence Livermore National Laboratory (IACRO number 91-853, Work Unit 00354). Part of the work was also performed under the auspices of the U.S. Department of Energy (LLNL contract W-7405-ENG-48). Their support is gratefully acknowledged.

References

[1]Mirels, H., "Boundary Layer Behind a Shock or Thin Expansion Wave Moving into a Stationary Fluid," National Advisory Committee for Aeronautics, Wash., D.C., TN-3712, 1956.

[2]Mirels, H., "The Wall Boundary Layer behind a Moving Shock Wave," in *Boundary Layer Research, Proceedings of the International Union of Theoretical and Applied Mechanics*, edited by H. Görtler, Springer-Verlag, Berlin, 1958, pp. 283–293.

[3]Mirels, H., and Hamman, J., "Laminar Boundary Layer behind a Strong Shock Moving with Nonuniform Velocity," *Physics of Fluids*, Vol. 5, No. 1, 1962, pp. 91–96.

[4]Crawford, D. R., Quan, V., and Ohrenberger, J. T., "Blast Wave Turbulent Boundary Layers," DNA-2768F, 1972.

[5]Mirels, H., "Interaction of a Moving Shock with Thin Stationary Thermal Layer," Report SD-TR-86-08, Aerospace Corp., El Segundo, CA, Vol. 33, 1986.

[6]Mirels, H., "Blowing Model for Turbulent Boundary-Layer Dust Ingestion," *AIAA Journal*, Vol. 22, No. 11, 1984, pp. 1582–1589.

[7]Mirels, H., "Turbulent Boundary Layer Behind Constant-Velocity Shock Including Wall Blowing Effects," *AIAA Journal*, Vol. 22, No. 8, 1984, pp. 1042–1047.

[8]Denison, M. R. and Baum, E., "Dusty Boundary Layer Modeling," DNA-001-84-C-0107, 1984.

[9]Batt, R. G., Kulkarny, V. A., Behrens, H. W., and Rungaldier, H., "Shock-Induced Boundary Layer Dust Lofting," *Shock Tubes and Waves*, edited by H. Grönig, VCH, Weinheim, Germany, 1988, pp. 209–215.

[10]Hartenbaum, B., "Lofting of Particles by a High-Speed Wind," DNA-2737, 1974.

[11]Ausherman, D., "Initial Dust Lofting: Shock Tube Experiments," DNA-31-62F, Wash., D.C., 1973.

[12]Glaz, H. M., Colella, P., Glass, I. I., and Deschambault, R. L., "A Numerical Study of Oblique Shock Wave Reflections with Experimental Comparisons," *Proceedings of the Royal Society of London*, Serial A, Vol. 398, 1985, pp. 117-140.

[13]Kuhl, A. L., Chien, K.-Y., Ferguson, R. E., Collins, J. P., Glaz, H. M. and Colella, P., "Simulation of a Turbulent Dusty Boundary Layer Behind a Shock," *Current Topics in Shock Waves*, edited by Y. W. Kim, American Institute of Physics Press, New York, 1990, pp. 762–769.

[14]Kuhl, A. L., Ferguson, R. E., Chien, K.-Y., Glowacki, W., Collins, J. P., Glaz, H., and Colella, P., "Turbulent Wall Jet in a Mach Reflection Flow,"

Dynamics of Detonations and Explosions: Explosion Phenomena, edited by A.L. Kuhl, J.-C. Leyer, A.A. Borisov, W.A. Siriginano, Vol. 134, Progress in Astronautics and Aeronautics, New York, 1991, pp. 201–232.

[15]Colella, P., and Glaz, H. M., "Efficient Solution Algorithms for the Riemann Problem for Real Gases," *Journal of Computational Physics*, Vol. 59, No. 2, 1985, pp. 264–289.

[16]Sedov, L. I., *Similarity and Dimensional Methods in Mechanics*, Academic Press, New York, 1959.

[17]Kuhl, A. L., Glowacki, W., Chien, K.-Y., Ferguson, R. E., Collins, J. P., Glaz, H. M., and Colella, P., "Simulation of a Turbulent Wall Jet in a Precursor Flow," *Proceedings of the Eleventh International Symposium on Military Applications of Blast Simulation*, edited by A. Mark of Ballistics Research Laboratories, 1989.

[18]Schlichting, H., *Boundary Layer Theory*, McGraw–Hill, New York, 1955.

[19]Kuhl, A. L., Ferguson, R. E., Chien, K.-Y, and J. P. Collins, "Parametric Studies of Turbulent Dusty Boundary Layers Behind Shocks," Report RDA-TR-2-1263-2201-003, Logicon RDA, Los Angeles, CA, 1992.

Numerical Prediction of Mechanism on Oscillatory Instabilities in Shock-Induced Combustion

Akiko Matsuo* and Toshi Fujiwara†
Nagoya University, Nagoya, Japan

Abstract

The characteristics of the oscillatory instabilities in shock-induced combustion around a spherical projectile flying at Mach number 4.8 is studied on the basis of numerical simulation using a two-step α–β chemical reaction model: The unknown mechanism on the generation of oscillatory instabilities is investigated by a series of simulations with varied tip radii. The oscillatory instabilities are clearly observed when the tip radius is about 10 times the induction length. To explain this phenomenon, a new feedback mechanism on the wave interaction between the bow shock wave and the stagnation point of the body surface along the stagnation streamline is proposed. It is clarified that the oscillatory instability is closely related to the relation between a shock stand-off distance and an induction length; i.e., the instabilities are caused not only by the wave interaction between the bow shock and the reaction front but also by the compression wave reflected from the stagnation point of the projectile surface toward the bow shock. In this work, a fine grid distribution such as 451 × 451 is used since the complicated interaction among the low-amplitude weak waves in front of the tip is important.

Introduction

One of the famous phenomena in shock-induced combustion is a periodic instability around a hypersonic flying body. There are many reports about such a

* Graduate Student, Department of Aeronautical Engineering.
† Professor, Department of Aeronautical Engineering.

periodic instability observed in the experiments in the 1970s. Figure 1 has been well known for a long time as a typical case, done by Lehr[1] in 1971. In this experiment, the projectile has a 15-mm diameter with a hemispherical tip and cylindrical afterbody. The projectile is flying at Mach number 5.04, which is nearly equal to the C-J detonation velocity of the gas mixture, into a stoichiometric hydrogen/air-mixture at an initial pressure 0.421 atm. There are several notable features in the flowfield in Fig. 1. The first is the separation between the bow shock and the reaction front. The second is the corrugated reaction front and the striations in the reaction region that are connected with the corrugated patterns. The third is the existence of many waves between the bow shock wave and the reaction front that are also connected with the corrugated reaction front. In this flowfield, high frequency oscillations are observed in front of a spherical projectile, and the frequency of oscillation is reported to be 1.04 MHz. However, these periodic phenomena appears only under certain conditions. These phenomena are the interesting features, but their mechanism is not well understood.

The shock-induced combustion around a blunt body has been studied for a long time. During the last few years, in particular, many papers were reported under the motivation to investigate hypersonic air-breathing engines using shock-induced combustion. With the advance of recent computational fluid dynamics, some of them[2-4] were studied numerically, where the chemically reacting flowfields were computationally simulated and the results were compared with the previous

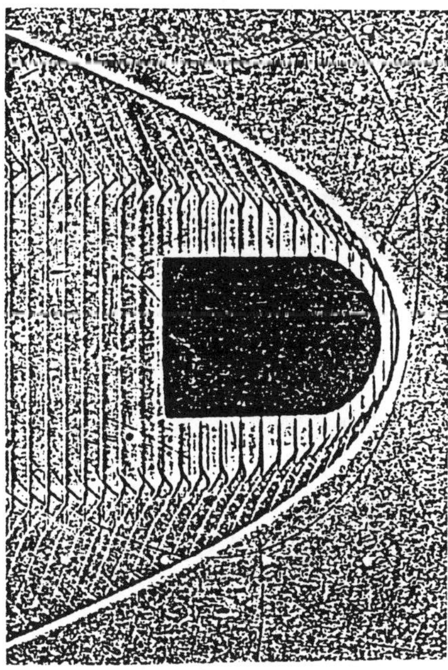

Fig. 1 Shadow graph of a spherical-nose projectile moving at Mach 5.04 into a premixed stoichiometric hydrogen-air mixture, taken by Lehr (from Ref. 1).

experiments with emphasis on the accuracy and reliability of the developed code. Note, however, that in these studies the projectile velocity was much higher than the C-J detonation velocity of the test gas mixture and only the steady-state solutions with no instability were discussed.

Previous Work

Basically the one-dimensional instability is hardly observed in detonable gases whereas the three-dimensional mode often appears. However, Ref. 5 discussed the generation mechanism of the cyclic combustion by interpreting the observed instability as one-dimensional longitudinal movement. In particular, the periodic mechanism was discussed using the history of physical variables along the stagnation streamline, which can be regarded as the one-dimensional mode. Their works give the feedback mechanism on the longitudinal instability; as seen in Fig. 2, the wave interaction model proposed by McVey[5] and Toong is given in the x-t diagram of the waves existing between the bow shock wave and the reaction front along the stagnation streamline (they call the model "regular regime").

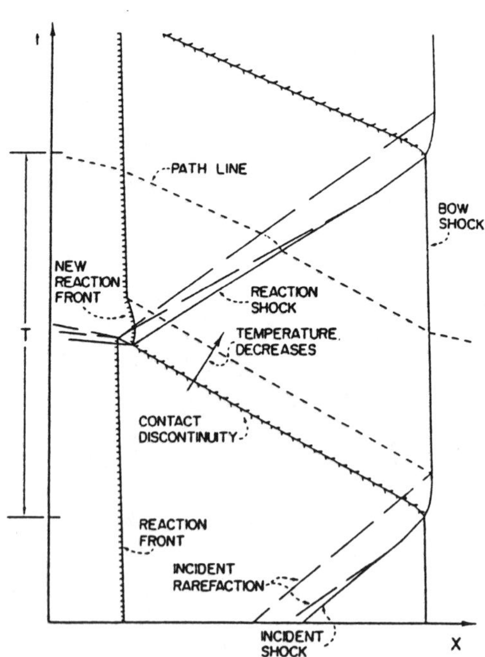

Fig. 2 X-t diagram of the waves along the stagnation streamline proposed by McVey and Toong (from Ref. 5).

Numerical Approach

The Zel'dovich-von Neumann-Doring (ZND) wave structure model is useful to understand the steady 1-D Chapman-Jouguet (C-J) detonation. However, this one-dimensional detonation wave model has shown that it is dynamically unstable. In fact, careful observations confirmed that all the detonations propagated with a complex unsteady shock structures. This shock system consists of a number of triple shock waves traveling in the direction transverse to the main flow. Fujiwara et al. were successful in numerically simulating these shocks in two- and three-dimensional detonations.[6-9]

A problem of shock-induced combustion containing periodic instability can be numerically formulated by using a simplified chemical model based on the ZND structure. The model consists of two-step chemical reactions, i.e., the induction and exothermic chemical reactions, instead of all the elementary reactions. Two reaction progress variables, α and β, are introduced when the reacting flowfields around an axisymmetric blunt body is numerically simulated. The specific reaction rate of the induction reaction is given by ($0 \leq \alpha \leq 1$)

$$\omega_\alpha \equiv \frac{d\alpha}{dt} = -\frac{1}{\tau_{ind}} = -K_1 \rho \exp\left(-\frac{E_1}{RT}\right) \tag{1}$$

and the one for the exothermic reaction is given by ($\beta_{eq} \leq \beta \leq 1$)

$$\omega_\beta \equiv \frac{d\beta}{dt} \begin{cases} = 0, & \alpha > 0 \\ = -K_2 P^2 \left[\beta^2 \exp\left(-\frac{E_2}{RT}\right) - (1-\beta)^2 \exp\left(-\frac{E_2+Q}{RT}\right)\right], & \alpha \leq 0 \end{cases} \tag{2}$$

The above model is substituted into the following equations:

$$\frac{D(\rho\alpha)}{Dt} = \rho\omega_\alpha, \qquad \frac{D(\rho\beta)}{Dt} = \rho\omega_\beta \tag{3}$$

Here the reaction parameters appearing in Eqs. (1) and (2) are selected as

$K_1 = 3.0\text{x}10^{11}$ cm^3/g/s
$K_2 = 1.875\text{x}10^{-8}$ cm^4/dyn^2/s
$E_1/R = 9800$ K
$E_2/R = 2000$ K
$Q = 4.0\text{x}10^{10}$ erg/g

in order to fit either $2H_2 + O_2 + 7Ar$ or $2H_2 + O_2 + 7He$ mixture. K_1 and K_2 are the reaction rate constants, while E_1 and E_2 are the activation energies. Q is the exothermicity per unit mass.

The governing system of equations is written below, for inviscid flows with axisymmetric geometry, under ideal gas assumptions:

$$\frac{\partial \hat{U}}{\partial \tau} + \frac{\partial \hat{E}}{\partial \xi} + \frac{\partial \hat{F}}{\partial \eta} = \hat{S} + \hat{H} \tag{4}$$

where the inviscid flux vectors in ξ and η directions are given by $\hat{E}$ and $\hat{F}$, respectively. The chemical source vector is $\hat{S}$ while the axisymmetric term is $\hat{H}$. In addition, the following equations must hold where P, T, and H are the pressure, temperature, and enthalpy, respectively:

$$\begin{aligned} P &= (\gamma\text{-}1)[e\text{-}\rho\beta Q\text{-}0.5\rho(u^2+v^2)] \\ P &= \rho RT \\ H &= (e+p) \end{aligned} \tag{5}$$

As numerical scheme, a TVD explicit method[10] and a locally adaptive grid are used in order to make the mesh finer immediately behind the bow shock wave.

Numerical Results

Although a flying projectile provides us with accurate experimental data, the data are insufficient to explain the detailed mechanism on the instability and to give the time evolution of the waves along the stagnation streamline. The present work investigates the physics of shock-induced combustion, especially the mechanism of oscillatory instability by using fine-resolution numerical simulation.

As observed in the experiment done by Lehr[1] (Fig. 1), the reaction front is gradually separated from the bow shock and the periodic instability occurs. The present computations are carried out to isolate the mechanism responsible for such periodic phenomena. Even though the two-step α–β chemical reaction model used here does not exactly describe the behavior of the gas mixture used by Lehr, the basic characteristics of the oscillatory instabilities in shock-induced combustion is considered more universal. In other words, we aim to explain the phenomenon qualitatively. Besides, our experience indicates that the grid resolution is extremely important to accurately look into wave interaction processes. One big advantage of the present chemical reaction model is the reduction of computer time.

Throughout our simulations we fix the Mach number of the incoming flow at 4.8, the C-J Mach number of the present gas mixture. It is widely known that the behavior of a detonation around the C-J velocity is unstable (Lehr experiment gives the projectile velocity 0.99 of the C-J value). The value L* is used to nondimensionalize the length scale, where L* is the induction length of the plane C-J detonation of the incoming gas mixture. We have essentially two important fluid dynamic parameters—the tip radius of the hemisphere and the incoming Mach number (fixed). The computed results are discussed for the three tip radii 5L*, 7.5L*, and 10L*.

Parametric Studies

Figure 3 shows the density contour for the tip radius 5L* case. The sudden density change along the body surface after the bow shock shows the density decrease by the exothermic reaction. The oscillation with a small amplitude and a high frequency is observed only at the reaction front near the tip. The smooth reaction front with no corrugation is observed further downstream. There is also slight interaction between the bow shock wave and the reaction front near the tip. The tip radius 5L* is understood as the case of transition from a steady solution to an unstable one.

Figure 4 is the density contour for the tip radius 7.5L* case. Many small bumps are shown over the entire reaction front where they have high frequency and low amplitude. The cyclic instabilities are observed in this case in comparison with 5L* case.

Figure 5 shows the density contour for the tip radius 10L* case. The corrugated reaction front is observed behind the bow shock and the phenomenon is clearly periodic. Also the waves exist between the bow shock and the reaction front as in Fig. 1, reproducing the experimental results obtained in the past. The corrugated reaction front is reproduced not only by the density plot (Fig. 5), but also by the contour of $\rho\omega_\beta$, the mass production rate of final product (Fig. 6). It is recognized that the exothermic chemical reaction is occurring at the corrugated reaction front only in the neighborhood of the spherical tip, and the extent is rapid especially near the stagnation streamline. This calculated contour indicates that the corrugated pattern taken in the experimental shadow graph is the trace of exothermic reaction.

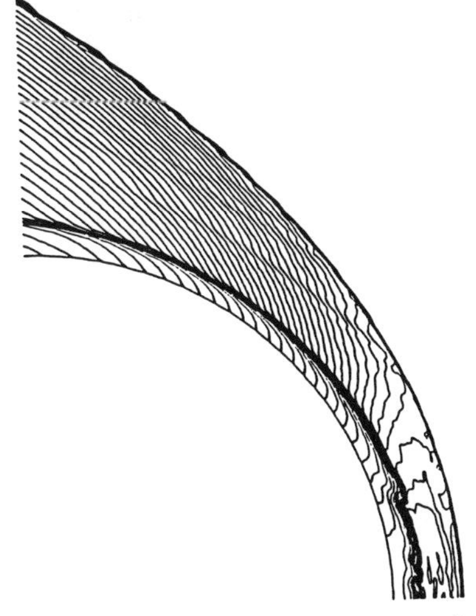

Fig. 3 Density contours for tip radius 5L*.

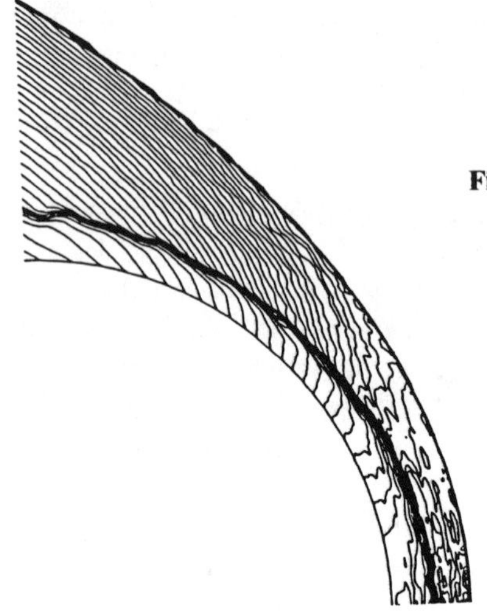

Fig. 4 Density contours for tip radius 7.5L*.

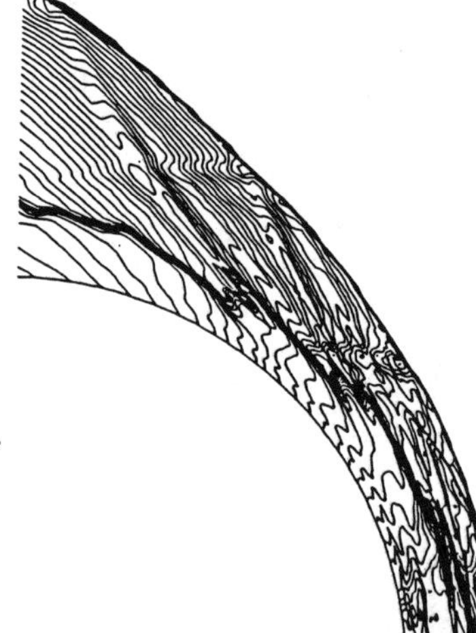

Fig. 5 Density contours for tip radius 10L*.

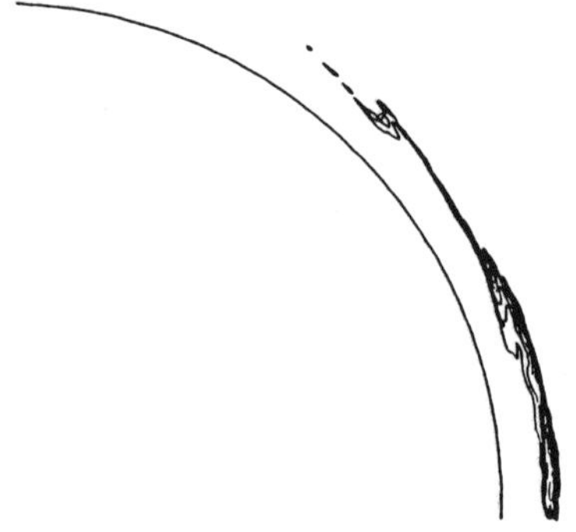

Fig. 6 $\rho\omega_\beta$ contour for tip radius 10L*.

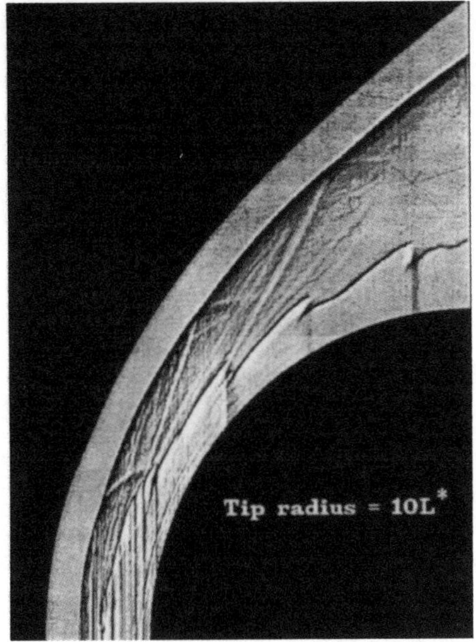

Fig. 7 Computational shadow graph picture.

The striations in the reaction region observed in Fig. 1 do not appear in Fig. 5. A post processor,[11] which generates a new 3-D picture corresponding to the experiment from the computed data, is used to give a three-dimensional shadow graph; in Fig. 7 the vertical lines or striations in the reaction region are observed as clearly as in Fig. 1. Now it is clearly understood that the striation in the reacting region in Fig. 1 comes from the corrugated reaction front.

The relation between the projectile radii and instabilities is studied for the above cases, and it has become clear that the instabilities highly depend on the relation between the shock standoff distance and induction length. In chemically reacting flowfields, more than two length scales exist—the body length and the reaction lengths. Thus, the flowfield loses similarity, and the behavior of the results becomes more complicated and phenomenological, which makes the extraction of essential features difficult.

Time Evolution of Oscillatory Cycle

Figure 8 is the closeup view of the time-evolving contours of $\rho\omega_\beta$ near the tip, corresponding to one cycle of the periodic instability. The corrugated contour is clearly moving with the fluid along the projectile body and disappears further downstream, where the exothermic chemical reaction does not occur any more. The instability cycle is described as follows: 1) at Time 2, a new reaction region is generated in front of an original reaction front near the stagnation streamline; 2) from Times 2 to 5, the new reaction region grows in front of the original reaction front; 3) at Time 6, the original reaction front is coupled with the new reaction region; 4) from Times 6 to 9, the corrugated contour is moving along the original reaction front. After Time 10, a new cycle is started and the same cycle is repeated.

A series of such time-evolving contour plots give us the unstable phenomena that are observed in the experiment. Although the instability has been reproduced in the simulation, the mechanism cannot be deduced in detail from Fig. 8. In order to understand the basic phenomena qualitatively, the physics on the stagnation streamline are discussed in the next section.

Instability Mechanism

Wave Interaction

The mechanism for generating oscillatory instability is discussed using the numerical result for the tip radius 10L* (Fig. 5). The history (x-t diagram) of the density distribution between the stagnation point of body surface and the bow shock along the stagnation streamline is shown in Fig. 9; the history of density distribution along the stagnation streamline where each line shows the instantaneous density profile. Figure 9b is the schematic representation of Fig. 9a, showing the wave interaction.

Several remarkable features are observed between the reaction front and the stagnation point in comparison with Fig. 2. In the wave interaction model in Fig. 2, proposed by McVey and Toong, the waves propagating toward the stagnation point were neglected and thus the region between the reaction front and the stagnation point did not come into play. However, our numerical result indicates the new reaction front generates the compression waves not only toward the bow shock wave but also toward the stagnation point. The shock standoff distance is approximately 0.24 × 10L* (radius) = 2.4L* in this result. Since the induction

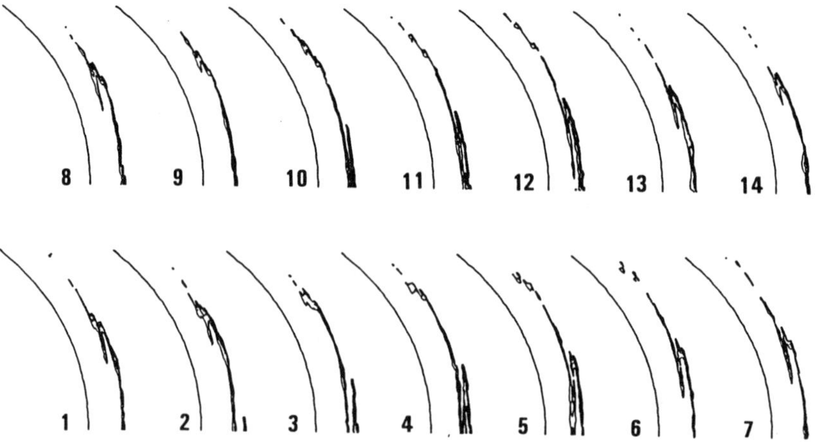

Fig. 8 The closeup time-evolving contours of $\rho\omega_\beta$.

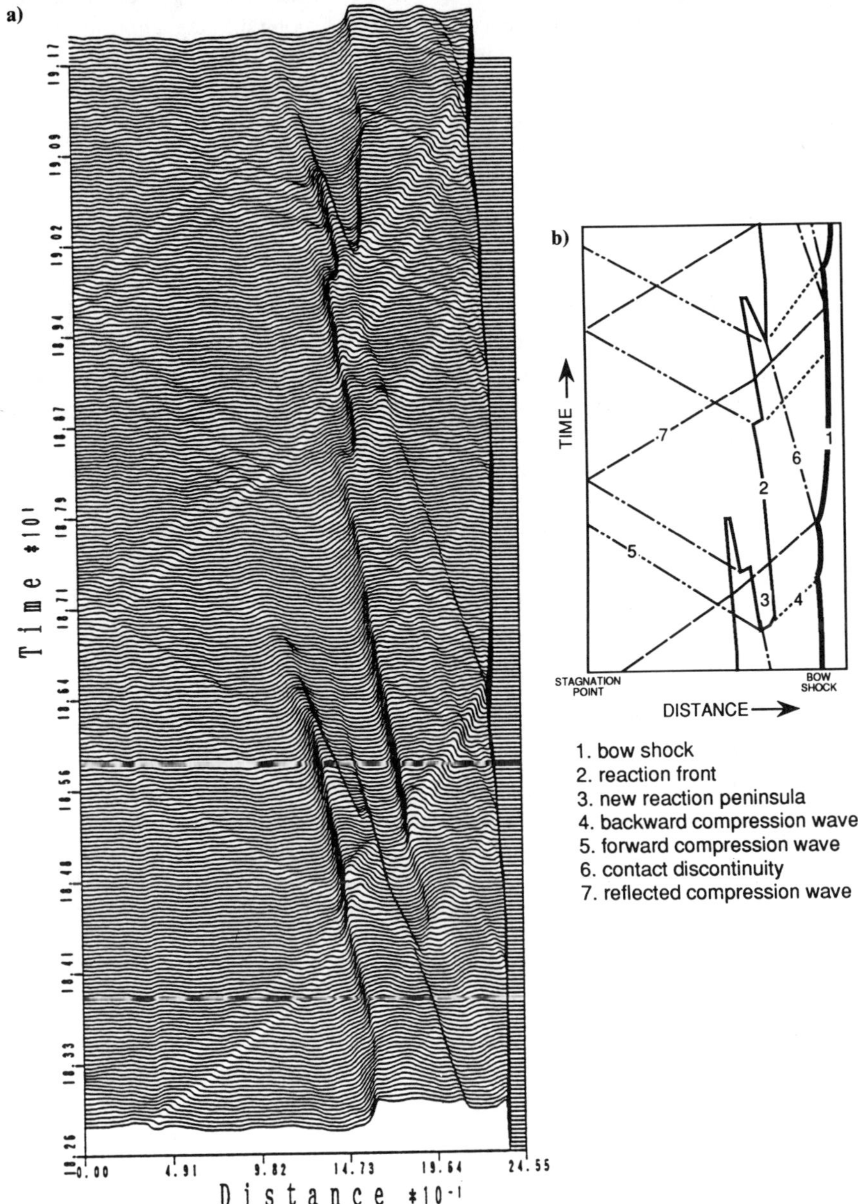

Fig. 9 History (x-t diagram) of density distribution along the stagnation streamline. (a) Density distribution, (b) schematic representation of density waves.

length L* is defined as the one at C-J velocity where the ZND structure holds, the induction zone must be theoretically close to L* along the stagnation streamline behind the normal segment of the bow shock. Therefore, about 40% of the shock standoff distance is considered as the induction zone, whereas the region between the reaction front and the stagnation point occupies about 60% of the shock standoff distance, totally non-negligible in the wave interaction. Actually, Fig. 9 shows the induction zone approximately of L*. Our subsequent proposal on the instability mechanism is based on the entire domain between the bow shock and the stagnation point along the stagnation streamline.

All kinds of waves showing the density change are observed in the density history. In Fig. 9a the history of the bow shock is given by the sudden density increase of an incoming uniform flow, and the reaction front is indicated by the sudden density decrease behind the bow shock. There are new reaction zones between the original reaction front and the bow shock: Some of them are generated as isolated areas in front of the original reaction front. After being connected with the original one, they form a peninsula. A number of other wave interactions between the bow shock wave and the stagnation point are also observed. In connection with them, there are two kinds of feedback mechanisms. Naturally, compression waves are generated at the new reaction front. One wave propagates backward (to the bow shock) and the other forward (to the stagnation point). The former interacts with the bow shock and generates a contact discontinuity. In general, the temperature on one side of a contact discontinuity is higher than that of the other side. The problem is more like a one-dimensional shock-tube. The induction length is highly dependent on the temperature determining the induction time (see Eq. (1)). The higher temperature fluid has a shorter induction time since the fluid velocity across the contact discontinuity is identical. On the other hand, the latter wave propagating toward the projectile body is reflected on the stagnation point. The reflected wave eventually hits the bow shock and generates a contact discontinuity like the former compression wave. These two feedback mechanisms are basically repeated, yielding an imperfect periodicity.

Occurrence and Extinguishment of New Reaction Front

The strength of a compression wave has an important role to determine the position and the strength of a new reaction front. The strength of the compression wave depends on that of explosive reaction. The mass production rate $\rho\omega_\beta$ is useful as a parameter to measure the strength of reaction and to know where the exothermic reaction progresses. If the mass production per unit time is high, a strong energy release occurs and strong compression waves are generated. Figure 10a shows the history of the mass production rate distribution along the stagnation streamline for the same result as Fig. 9, whereas Fig. 10b is a schematic representation of Fig. 10a, representing the behavior of the reaction front. The exothermically reacting region can be recognized in Fig. 10a. The reaction hardly progresses behind the reaction front, where the condition reaches chemical equilibrium. Also there is a notable difference between the original reaction front (1) and the regions (2, 3) where compression waves are generated. Along the original reaction

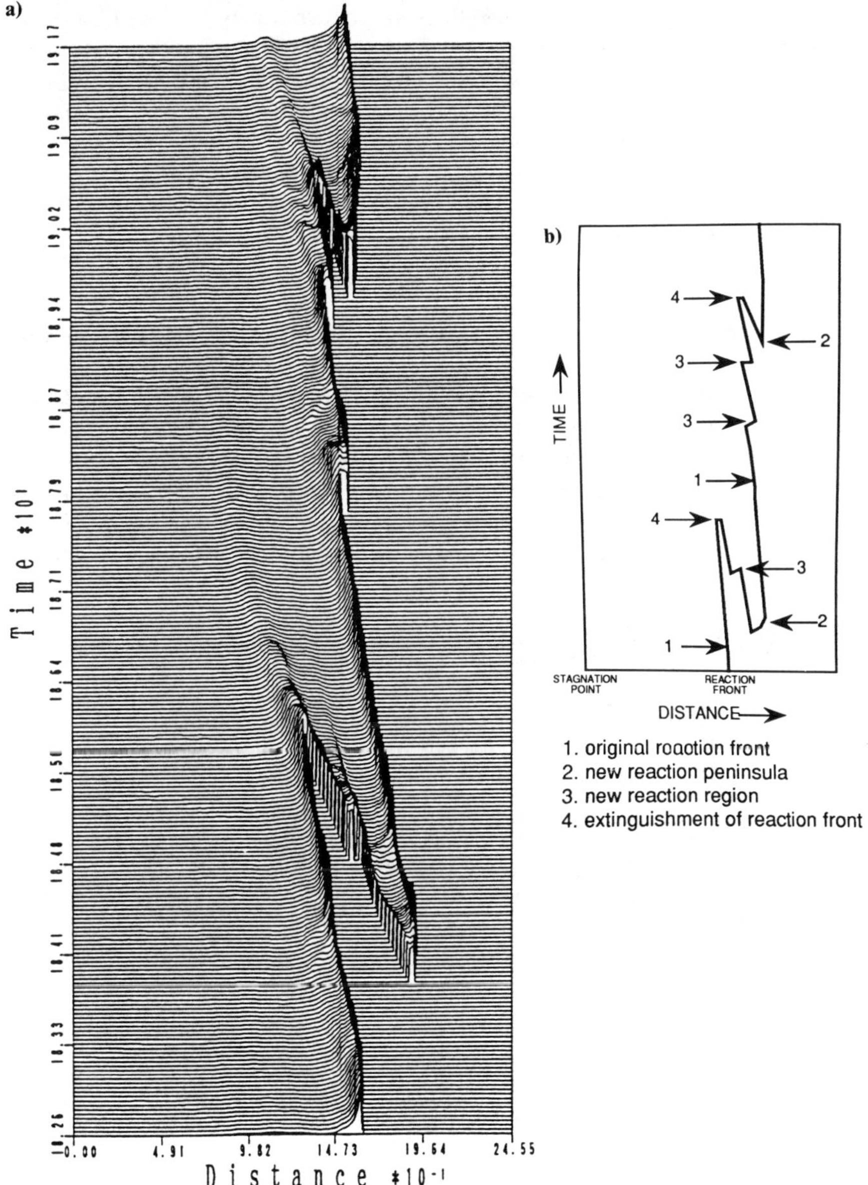

Fig. 10 History (x-t diagram) of $\rho\omega_\beta$ distribution along the stagnation streamline. (a) distribution of $\rho\omega_\beta$, (b) schematic representation of reaction front.

front, the rate distribution shows only one peak point. The peak height is an important factor to know the strength of the released energy. If we observe the new reaction peninsula (2) in Fig. 10, the highly-elevated region is found inside the peninsula. Although the new reaction region (3) is not shaped like a peninsula, there is a peak in the distribution as high as in the peninsula; the compression waves are generated there, as seen in Fig. 9.

The extinguishment of the reaction front is observed in Figs. 9 and 10, which is shown as (4) in Fig. 10b. As seen in Fig. 10a, the reaction is still progressing at (4), but not strongly enough to generate any waves.

Path Lines

As seen in Fig. 10a, weak exothermic reaction occurs along the original reaction front. In order to understand this phenomenon, the particle path lines superposed on the density profile history are used in Fig. 11, where the reaction front, contact discontinuity, and many waves are shown. As is naturally observed in Fig. 11, the particle path line is attached to the contact discontinuity. The new reaction front which generates the compression waves has a deep angle with the path lines. On the other hand, the original reaction front, which does not yield a rapid reaction progress, is nearly parallel to the path lines. Mathematically the

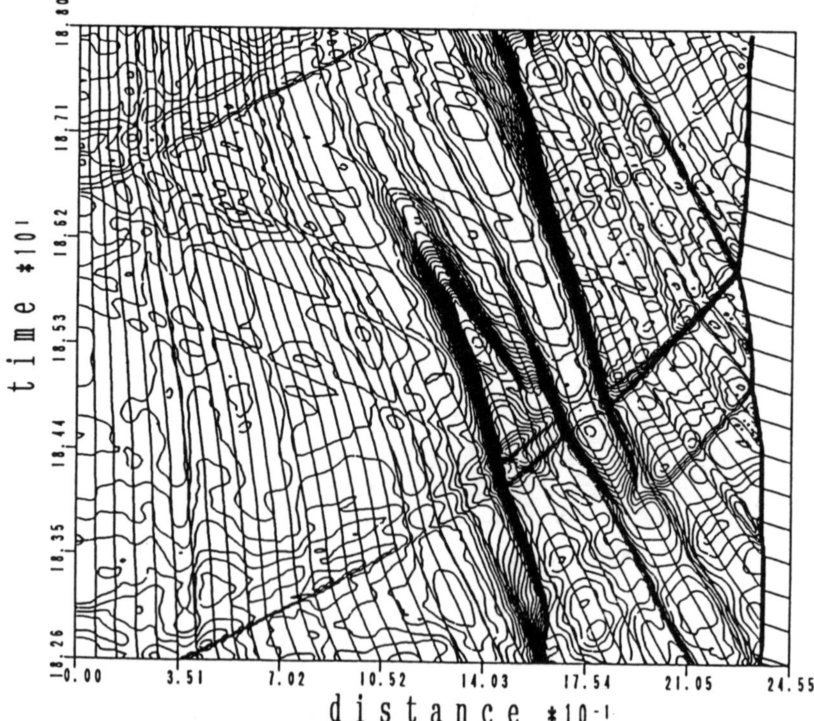

Fig. 11 History (x-t diagram) of path lines superposed by the density contours along the stagnation streamline.

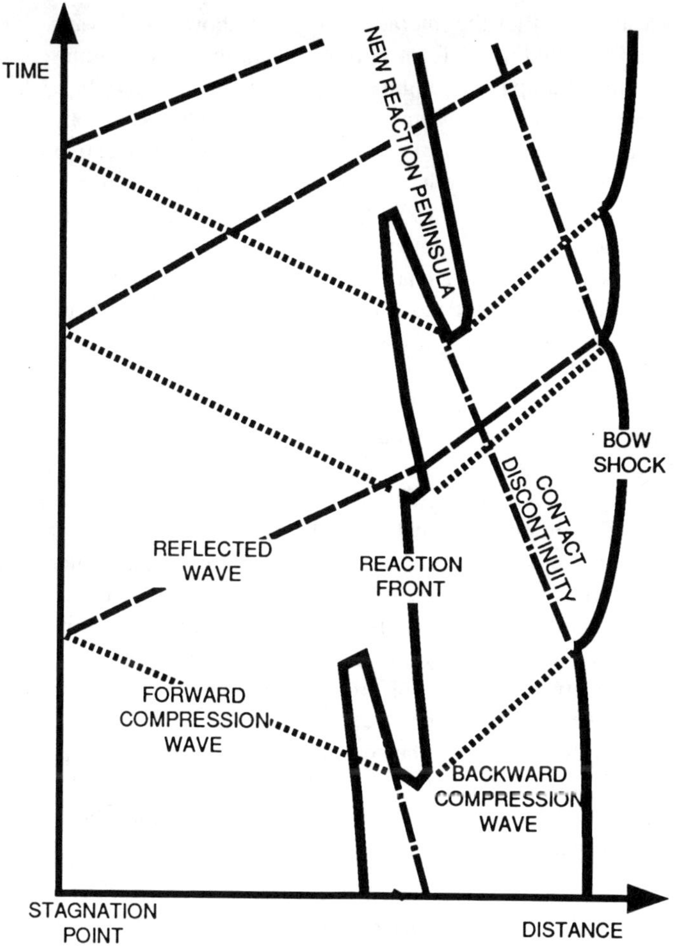

Fig. 12 X-t diagram representation of the proposed mechanism.

induction time is determined by Eq. (1) so that Fig. 11 shows "effective doubling of the reaction rate."

Proposed Mechanism

Figure 12 summarizes the x-t diagram of the instability mechanism proposed here, in comparison with Fig. 2. It shows the entire region between the stagnation point and the bow shock along the stagnation streamline. In this proposed diagram, the compression waves generated at a new reaction region have an important role to repeat the oscillatory behavior. One of the compression waves propagates backward toward the bow shock, while the other propagates forward to the stagnation point on the body surface and is reflected, finally resulting in a contact discontinuity after the interaction with the bow shock wave. The new reaction front is created in the form of a peninsula in the induction region or in the form of a small bump in the original reaction front, as seen in Fig. 12.

Although this feedback mechanism is caused by the two kinds of compression waves (forward and backward), the waves are not always effective because of their dependence on the strength. The form of the new reaction front can be arbitrary regarding its strength. In fact, their basic patterns are indicated in Fig. 12, but the detailed features are not exactly identical.

The proposed feedback mechanism is based on the balance between the shock standoff distance and the induction length; a periodic oscillation should be observed when a proper balance is satisfied.

Conclusions

The characteristics of shock-induced combustion has been qualitatively clarified using fine-resolution numerical simulation. An instability mechanism discovered in the present simulation can properly explain the phenomena found in the previous work.[5] Since the previous works are based on the experimental results and the simplified one-dimensional wave interaction theory, the exact time-evolving data on the stagnation streamline are not incorporated. A numerical simulation is confirmed as a useful probing technique for such complex and detailed phenomena.

The feedback mechanism for the generation of oscillatory instability is newly constructed; i.e., important also is the wave interaction between the stagnation point and the bow shock on the stagnation streamline and not just the compression wave from a new reaction region.

References

[1]Lehr, Hartmuth F., "Experiments on Shock-Induced Combustion," *Astronautica Acta*, Vol. 17, 1972, pp. 589-597.

[2]Lee, S. H., and Deiwert, G. S., "Flux-Vector Splitting Calculation of Nonequilibrium Hydrogen-Air Reactions," *Journal of Spacecraft and Rockets*, Vol. 27, No. 2, Mar.-Apr. 1990, pp. 167-174.

[3]Wilson, G. J., and MacCormack, R. W., "Modeling Supersonic Combustion Using a Fully-Implicit Numerical Method," AIAA Paper 90-2307, 1990.

[4]Yungster, S., Eberhardt, S., and Bruckner, A. P., "Numerical Simulation of Hypervelocity Projectiles in Detonable Gases," *AIAA Journal*, Vol. 29, No. 2, Feb. 1991, pp. 187-199.

[5]McVey, J. B., and Toong, T. Y., "Mechanism of Instabilities of Exothermic Hypersonic Blunt-Body Flow," *Combustion Science and Technology*, Vol. 3, 1971, pp. 63-76.

[6]Taki, S., and Fujiwara, T., "Numerical Analysis of Two-Dimensional Nonsteady Detonation," *AIAA Journal*. Vol. 16, No. 1, Jan. 1978.

[7]Sugimura, T., Fujiwara, T., and Lee, J. H., "Cellular Detonation - Instability and Sub-Structure," *Memoires of the Faculty of Engineering*, Nagoya University, Vol. 42, No. 2, 1990.

[8]Reddy, K. V., Fujiwara, T., and Lee, J. H., "Role of Transverse Waves in a Detonation Wave - a Study Based on Propagation in a Porous Wall Chamber," *Memoirs of the Faculty of Engineering*, Nagoya University, Vol. 40, No. 1. May, 1988.

[9]Wang, Y., and Fujiwara, T., "Three-Dimensional Standing Oblique Detonation Wave," AIAA Paper 88-0478, 1988.

[10]Yee, H. C., "Upwind and Symmetric Shock Capturing Schemes," NASA TM-89464, 1987.

[11]Tamura, Y., and Fujii, K., "Visualization for Computational Fluid Dynamics and the Comparison with Experiments," AIAA Paper 90-3031, 1990.

Influence of Nonequilibrium Processes on Gasdynamic Parameters of Nonstationary Supersonic Jets

T. V. Bazhenova,* V. V. Golub,† A. V. Emelyanov,‡ A. V. Eremin,§
A. M. Shulmeister,¶ O. D. Miloradov,** and V. T. Ziborov††
Russian Academy of Sciences, Moscow, Russia

Abstract

The influence of the internal degrees of freedom of various gases on density distribution and gas front motion in impulse supersonic jets is investigated. The gas jets were visualized by shadow and interference methods. The density distribution in the jet was obtained by means of electron beam absorption. Emission and absorption measurements were made at various wavelengths, including infrared, for the impulse nitrogen and carbon dioxide jets. Then the information about the conditions of the vibrational degrees of freedom in these jets was obtained. The influence of the internal degrees of freedom of the outflowing gas on the gas front velocity was revealed.

Introduction

The gasdynamics of supersonic jets possessing a stagnation temperature in a range of 1500-3000 K is connected to a great extent with a complicated

---

*Principal scientist. General guidance and discussion.

†Leading scientist. Shadow pictures and discussion.

‡Senior scientist. Shadow pictures, empirical equations.

§Leading scientist. General guidance and discussion. Electron beam measurements.

¶Senior scientist. Shadow pictures.

**Post-graduate student. Shadow pictures.

††Senior scientist. Spectroscopic measurements.

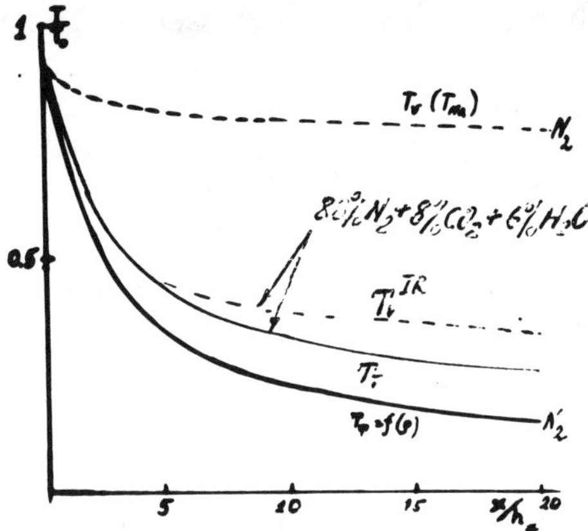

Fig. 1 The variation in the translational T_T and vibrational T_v temperatures along an axis of a two-dimensional jet of N_2 when the vibrational degrees of freedom are frozen, and of $86\% N_2 + 8\% CO_2 + 6\% H_2O$ when the gas is in equilibrium.

physicochemical phenomena that take place in the gas under such temperatures. High power-consuming qualities of internal degrees of freedom of these phenomena can have, in a number of cases, a dominant influence on shock wave patterns and on the distribution of parameters on the stage of the jet start, as well as on the regularity of the stationary outflowing stage.

Experimental Setup

In order to investigate the interaction of gasdynamic and kinetic processes, geometric characteristics, as well as density and temperatures distribution in the forming supersonic jets, have been measured.

To carry this out, a nozzle has been installed at the end of the shock tube and the shock-heated gas flows out into the gas chamber. The outflowing gas has been visualized with the help of schlieren and interference methods. The emission and absorption measurements were made at the flow of carbon dioxide and its mixtures with argon, nitrogen, and water vapor outflowing from a sonic slot nozzle (half-width 1 mm) placed at the end of the shock tube. Measurements were carried out at the spectral ranges at 4.3 μm and 2.7 μm, corresponding to intensive IR bands of CO_2, and at the wavelengths 285 and 488 nm at the region of recombination radiation of the band $(^1B - X^1\Sigma)CO_2$. Using the electron beam absorption method, the density distribution in certain sections of argon and nitrogen forming jets have been obtained.

Influence of Nonequilibrium Processes on the Translational Temperature in a Jet

At a stagnation temperature near 2000 K, approximately half of the internal energy of the three-atomic molecules belong to the vibrational degrees of freedom. At the same time, the efficiency of the vibrational energy exchange with the other degrees of freedom is very low (the possibility of a V-T exchange is 10^{-3} to 10^{-6} times less than of an R-T exchange). Since the vibrational energy in the jet is very often frozen, the vibrational degrees of freedom did not take part in the gasdynamic processes. Figure 1 presents the translational temperatures in N_2 and in the mixture $86\%N_2 + 8\%CO_2 + 6\%H_2O$. The vibrational temperature measured by the infrared spectroscopic method is frozen in the jet of pure N_2, and the translational temperature calculated from the measured density data falls very fast. When a low amount of CO_2 and H_2O is added to N_2, the efficiency of the vibrational energy exchange increases and the translational temperature is higher (near the equilibrium value).

At higher stagnation temperatures, the chemical reactions take part in the energy distribution in the jet. At temperatures near 3000-4000 K a large number of molecules are dissociated, and the endothermic reaction of recombination occurs when the flow is cooled in the supersonic jet. It can be shown that if only 2-3% of molecules are dissociated, the heat of recombination can increase the energy of the gas twice when its temperature falls to 1000 K. But the rate of the recombination processes falls rapidly with decreasing temperature and density in the expanding flow. In this case, the situation can occur when the recombination processes do not change the chemical composition of the gas and the gasdynamic properties of flow

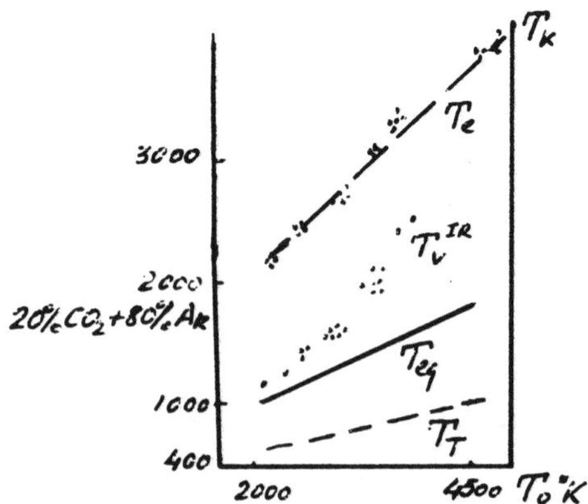

Fig. 2 The dependence of the translational T_T, vibrational T_v, and electronic T_e temperatures on the stagnation temperature To for the jet of dissociated CO_2 at a distance $X=8r_0$.

but sometimes leads to the overpopulation of highly excited molecular levels and to intensive nonequilibrium radiation. Figure 2 presents the values of the translational T_T, vibrational T_v, and electronic T_e temperatures vs the stagnation temperature T_0 of the jet of dissociated CO_2 at the distance $x=8r_0$ from the nozzle output. The solid line presents the equilibrium temperature T_{eq}. It can be seen that the measured electronic and vibrational temperatures significantly exceed the equilibrium and translational temperatures. The electronic temperature in this case is not frozen at the stagnation level but rises by means of energy addition to the electronic levels from the release of heat of the recombination processes. The experiment showed that the overpopulation decreases with decreasing pressure.

Influence of Internal Degrees of Freedom on the Front of Gas Velocity in the Impulse Jets

The formation of a shock ahead of the outflowing gas front in the nonstationary jet makes the flow structure very complex. The level of excitation of internal degrees of freedom of molecules and the rate of energy release during the quenching processes in expanding gas as well as in the shock wave influence the flow parameters.

In our experiments, we studied the flow structure of the nonstationary flat jets of three sorts of gases: monoatomic (Ar), biatomic (N_2), and three-atomic (CO_2) are compared. Figure 3 is a schlieren picture of a jet formation ($N = P_0/P_\infty = 150$, $P_0 = 42$ mm Hg), seaturing: 1,the starting shock wave; 2,the front of the outflowing gas; and 3,the secondary shock front. During the jet formation, the starting shock wave degenerates into the sonic wave and disperses, the secondary shock wave forms the Mach disc, and the side shocks are placed nearer the axis (the jet becomes narrower).

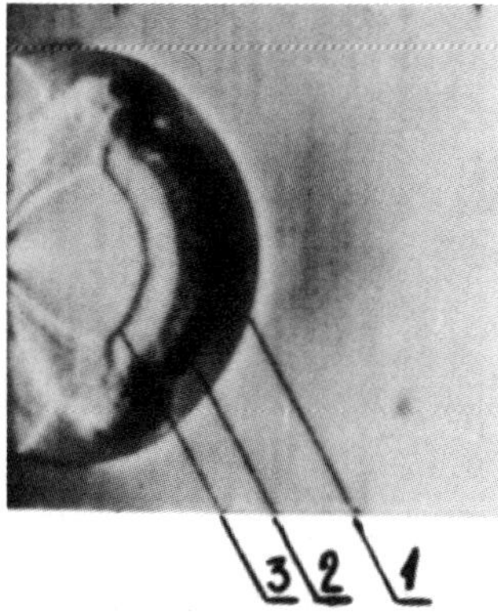

Fig. 3 Shadow picture of the impulse N_2 jet outflowing from a slit nozzle.

The law of gas front propagation can be presented by an empirical equation,[1]

$$\Theta_1 = 0.1\Psi_1^2 + 0.1\Psi_1$$

where

$$\Theta_1 = \frac{t}{h_* N} \cdot \frac{\gamma - 1}{2} C_0, \quad \Psi_1 = \frac{x}{h_* N} \cdot (T_0/T_\infty)^{\frac{\gamma-1}{2}}$$

C_0=sound velocity in the undisturbed gas
$C_0 = \sqrt{\gamma \cdot R \cdot T_0/\mu}$
$\gamma = C_p/C_v = 1.67 - Ar$
$\gamma = 1.4 - N_2$
$\gamma = 1.33 - CO_2$
h_*= the height at nozzle critical section
T_0= gas temperature in the forechamber ahead of the nozzle
T_∞= undisturbed gas temperature in the surrounding gas

Figure 4 shows the dependence $\Theta_1(\Psi 1)$. As can be seen, the approximation is good enough for the different gases. It is remarkable that the results for very low initial pressures (P_0=10 mm Hg, N=10^8), when the starting shock are absent, are described by the same curve.

As seen from the empirical equation, the speed of the front of a gas essentially depends on the adiabatic index γ. This dependence is a relative reversal of the maximum flow velocity um ($u_m = 2C_0/(\gamma - 1)$). In the gas with the frozen rotations of molecules (γ=1.67), the front speed is higher

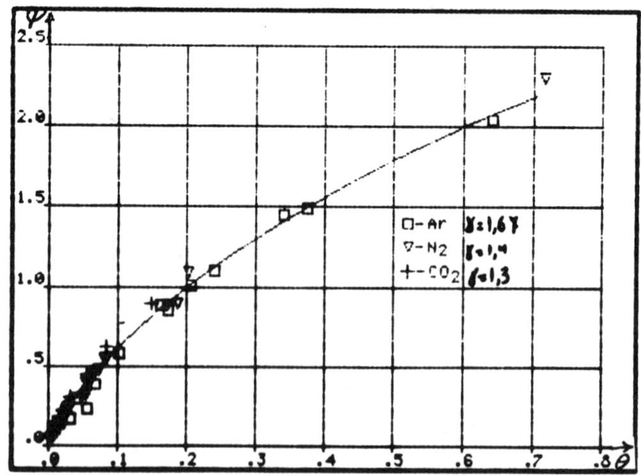

Fig. 4 The dependence of the front position of outflowing Ar, N_2, CO_2 on time in dimensionless coordinates for a wide range of pressures and temperatures.

than in a diatomic gas with the frozen vibrations of molecules (γ=1.4). In the gas with partially exited vibrations (γ=1.3), the front speed is the lowest.

Influence of Rotational Nonequilibrium on a Nonstationary Jet Structure

In our experiments Eremin, a low-density nitrogen jet transformation was observed that can be explained by the change of rotational energy input on the nitrogen expansion law.

As shown by Lukyanov, the rotational degrees of freedom can be frozen in an N_2 jet expanding in a vacuum from the sonic nozzle. The needed collision number for establishing rotational equilibrium is z=10. If the initial pressure P_0 is equal to 240 mm Hg for diameter of the nozzle critical section d_* equal to 1 mm, the rotational temperature is frozen at the distance X_f=(40-50)r_* (T_r=10 K). It can be suggesed that, at extremely low translational temperatures, the influence of the rotational energy on the gasdynamic properties can stop at some greater distance because of the energy input from the rare individual R-T collisions.

In our experiments, the outflowing of nitrogen and argon in the low-pressure chamber is compared. The critical diameter of the sonic nozzle

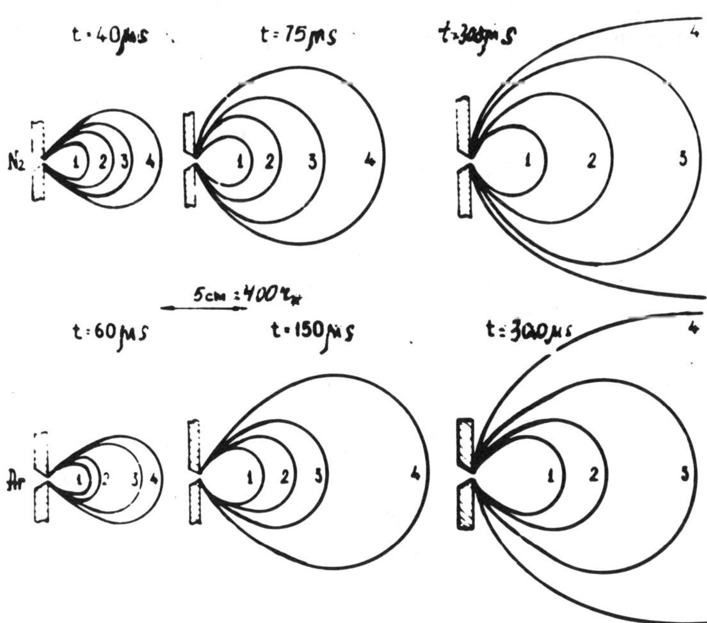

Fig. 5 Isodensity lines at different cross sections of time for nonstationary N_2 and Ar jets: 1,$\rho/\rho_* = 3.3 \cdot 10^{-6}$; 2,$\rho/\rho_* = 1.3 \cdot 10^{-6}$; 3,$\rho/\rho_* = 6.6 \cdot 10^{-7}$; 4,$\rho/\rho_* = 1.3 \cdot 10^{-7}$.

was $d_* = 0.25$ mm, the stagnation pressure P_0 was 7-8 atm, the pressure in the chamber P_∞ was $(1.5 \div 2) \cdot 10^{-5}$ mm Hg ,and the temperature was $T_0 = T_\infty = 300$ K ($P \cdot d_* = 1130$ mmHg· mm). According to Lukyanov, the results under our conditions for the distance at which the rotational exchange is frozen, X_f, must exceed the value $X_f = 250 r_*$. Figure 5 presents the isopicnics of N_2 and Ar jets.

Thus, when the characteristics of argon and nitrogen expansion are compared, it can be seen that , at a distance of up to 200-300 r_* from the nozzle exit, the nitrogen flow expands under greater angles than that of argon. At great distances, the outflowings of both gases look more alike. The differences just noted in the characteristics of the expansion of nitrogen and argon can be accounted for by the influence of the rotational energy of the molecules on nitrogen expansion. The sharp drop in collision frequencies in the expanding combustion products in jets leads to freezing of the internal degrees of freedom and, under sudden expansion in the vacuum, the process is equivalent to that in the monoatomic gas.

Conclusions

It was found that the contribution of chemical and internal energy to the kinetic energy of the jet increases with an increase in the efficiency of the energy exchanging processes.

The general equation of the outflowing jet front motion for nitrogen, carbon dioxide, and argon was obtained in a wide range of outflowing parameters.

It was found that, in the case of abrupt expansion, the trasverse dimensions of the nitrogen jet become equal to those of argon when the interal degrees of freedom become frozen.

References

[1]Eremin, A.V. and Emelyanov, A.V., "The Dynamics of the Start Shocks During the Launching of Underexpanding Jets", Proceedings of the All-Union Conference on Rarefied Gas Dynamics, Vol 3, 1991, pp. 164-171.

[2]Eremin, A.V., Kochnev, V.A., Kulikovsky, A.A. and Naboko, I.M. "Nonstationary Processes at High-expanded Jets Start", Jour. of Appl. Mech. and Techn. Phys. No.1, 1978, pp.34-40.(in Russian).

[3]Lukyanov, G.A., "Rotational Relaxation in a Free Expanding Nitrogen Jet",Jour. of Appl. Mech. and Techn. Phys. No.3, 1972, pp. 176-178.(in Russian).

Shock Waves in Self-Propagating High-Temperature Synthesis Research

Yury Gordopolov* and Alexander Merzhanov†
Russian Academy of Sciences, Moscow, Russia

Introduction

Self-propagating high-temperature synthesis (SHS) is an efficient method of obtaining a wide range of materials and represents strongly exothermic interaction of reactants in condensed medium occurring in a combustion mode.[1,2] Thermal wave arising in a mixture of reactants (powders) is spreading spontaneously over matter and transforms it into reaction products showing valuable properties. The synthesis wave velocity makes a value of 10^{-3}-10^{-2} m/s. High temperatures at the reaction zone (about 10^3K) promote diffusion and provide purification of products from contaminations.

The processes taking place at the synthesis wave may be schematically represented as shown in Fig. 1, which shows the direction of the synthesis wave propagation (arrow), temperature profile, heat release rate, and extent of conversion into reaction products. Several zones may be distinguished in the scheme. The first of them is the zone of starting substance. The second one is the zone of heating where chemical reactions have not yet been initiated. The third one is the third one is the zone of heat release in the course of chemical transformation and determines the velocity of the synthesis wave propagation. What follows are the zones of afterburning, structuring, and formation of final products, which already have no influence on the synthesis wave propagation velocity but are of great importance since they determine structure and properties of final reaction products.

The SHS products are mostly obtained in the form of powders or porous blocks, which are then ground and the articles of them are obtained by sintering using the conventional procedures of powder metallurgy. To obtain the compacted articles, the porous blanks are subject to the action of high pressures by using either presses or the well-known methods of shock wave compaction.[3,4] The latter method shows certain advantages. It needs no expensive and bulky installation, since the force-inducing part of it is replaced by an explosive that is not too expensive. Moreover, explosion provides the conditions of equal loading over the entire surface of intricated articles that could not be attained in presses. This makes it possible to obtain the long-sized rods and hollow cylinders with a high degree of

*Deputy Director, Institute of Structural Macrokinetics.
†Director, Institute of Structural Macrokinetics.

homogeneity. Such a method of explosive treatment of SHS systems (final products) is of the utmost simplicity. It will be analyzed in the second section of the present communication, mainly by obtaining the high-density, high-temperature superconductors (HTSC) from the powders synthesized by the SHS method.[5-8] Also analyzed are the data obtained in the studies on the shock wave usage for deposition of SHS powders of TiN, TiC, and TiB_2 onto steel articles,[9,10] as well as pretreatment of compacted powders of Si_3N_4 for subsequent sintering.[11]

More promising seem to be the action of high-dynamic pressures induced by detonation on the initial SHS compositions and SHS processes. These methods of shock-wave treatment of SHS systems are considered in the third and forth sections, respectively.

Especially interesting is the concomitant occurrence of SHS process and pressing. Properties of final products may be changed (improved) by the high-pressure action on hot-reaction products, since the common action of high temperatures and pressures behind the combustion wave front essentially influences the material structure. Besides, the intermediate stage of grinding is no more necessary and articles become more compacted, since at high temperatures material is in a state of ductility in the course of pressing. Pressure may have both static and dynamic character. The static pressure is reportedly[2] giving pretty good results in synthesizing hard alloys. An interest to dynamic pressure is conditioned by the fact that by using explosives, very high pressures (up to 10^5 bar) could readily be obtained. Besides, the shock wave velocities make a value of about 10^3 m/s, i.e., several orders of magnitude higher than the combustion wave velocities. This implies that the time of interaction is very small and provides a possibility of selective action of high pressure on any stage of synthesis. If the characteristic width of the post effect zone at the combustion wave (the IV-VI zones in Fig. 1) is larger than SHS charge dimension, the latter will perceive them as the consequent stages of SHS process. In such a case, by varying the detonation initiation time, one may obtain (from the same initial SHS composition) a final specimen that is

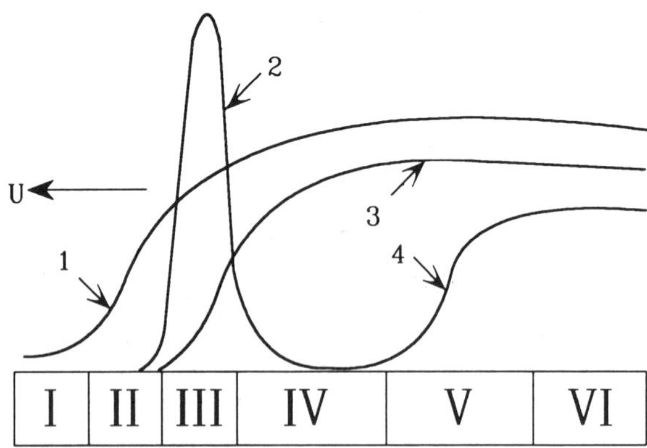

Fig. 1 Schematic representation of the synthesis wave: (1) temperature profile, (2) rate of heat release, (3) extent of conversion, (4) concentration of final product, and (U) combustion velocity.

relatively homogeneous but of different structure, and hence, properties. If the width of the post effect zone is comparable or smaller than the initial SHS charge, the shock-wave loading will result in obtaining a specimen with a gradient of structure and physico-mechanical parameters. In addition, a possibility has to be mentioned of the shock-wave initiation of combustion in SHS charges, which has a quasivolumetric nature, as well as the suppression of the locallyinitiated chemical reaction (arresting of the combustion wave). Versatile consequences of the shock-wave action on SHS systems makes them very promising in obtaining materials of modified parameters, as well as in the studies on the SHS processes.

An idea of concomitant performance of SHS and explosive compaction of hot-reaction products, born in the 1970s,[12] has only been realized recently.[13-26] Prior to the early 1970s, several cells had already been suggested for saving condensed compounds formed upon the single and multistage shock-wave loading, as well as upon the dynamic isentropic compression. Since that time, a number of physico-chemical transformations have occured, including synthesis of diamond and diamond-like modifications of boron nitride.[27-28] These cells had to be only slightly modified to provide electric ignition of a reactive mixture; remove gases formed upon combustion in a cell; and isolate explosives from hot synthesis products. Already then[12] the basic ideas had been formulated, and high density carbides and borides of IV-VI group metals had been obtained. Later, the method of combustion wave arresting by means of the shock-wave loading was suggested and realized.[15,17] The shock-wave effects on SHS in the complicated system Ti-C-Ni-Cr were studied,[13-16] as well as in the hybride system[17] Ti-N-O and in the binary systems[18-24] Ti-C, Ti-B, and Hf-C. By using the method of dynamic pressing of heated reaction products, the composites TiC-Al_2O_3 were obtained.[25] Combining combustion in the system Y_2O_3-BaO_2-Cu with dynamic compaction[26] resulted in synthesizing the high-density HTSC.

The SHS process was also suggested[12] for use as a pulsed heater (chemical furnace) for attaining favorable temperature conditions for the shock-wave loading of matter placed in the bulk of reactive mixture. This method was then used[29-31] for promoting the shock-wave compaction of poorly deformed ceramics SiC.

In the early 1970s, the feasibility of synthesizing various compounds by the shock-wave loading of initial SHS compositions was also mentioned.[12] This process was created[32-36] in the exothermic system Ni-Al to obtain intermetallide Ni_3Al, and in the system Ti-B to obtain boride TiB_2.[37] The feasibility of obtaining various solid coatings upon the shock-wave loading of powdered mixtures Ni-Al, Fe-Si, Cr-B, Cr-C, Cr-SiC, Cr-B_4C, and so forth onto a metal base was demonstrated as well.[38] The possibility of obtaining HTSC articles by the shock-wave pressing of the exothermic system of powders Y_2O_3-BaO_2-Cu with subsequent high-temperature treatment in a furnace was also studied.[39] However, it should be noted that such a shock-wave treatment of SHS systems (starting exothermic compositions) is not directly related to the SHS problem, since in all the above mentioned cases chemical reactions have no a character of self-sustaining layer-to-layer combustion. They occur either in the course of compression at a shock wave (the shock wave synthesis) or within the entire volume as a post effect (the shock wave initiation). Having no intention to present the state of the art in such an extensive field with its own history as the shock-wave synthesis of materials,[40] in the present communication we are giving only an illustrative consideration of some examples to perform more complete analysis of all the possible shock-wave treatments of exothermic compositions used for SHS.

Explosive Treatment of Final SHS Products

As a rule, final SHS products are porous ceramics. Practice, problems, and potentialities of dynamic pressing of ceramic powders (irrespectively of the method of their preparation) are widely known.[41] In view of this, the explosive treatment of various SHS materials may be expected to give pretty good results. Consider some available examples.

Impressive advances in the field of high-temperature superconductivity during previous years stimulated extensive studies in all aspects of the problem.[42-44] In the technological aspect, two branches may be distinguished: 1) development of saving processings for manufacturing HTSC powders (raw material) and 2) development of processings for manufacturing HTSC finished articles. The SHS method turned out to be productive in making HTSC powders (oxide ceramics) and producing more saving than the furnace technologies.[45-48] All the known HTSC may be obtained by the SHS method, including yttrium, bismuth, and thallium ceramics. For instance, to synthesize the yttrium ceramics (Y-Ba-Cu-O, 1-2-3), the SHS system comprising the mixture of copper, yttrium oxide, and barium peroxide powders is used. Reaction is brought about in oxygen flow; the synthesis wave velocity makes a value of 0.4-1.0 mm/s. In the absence of high pressure, the synthesis products are obtained in the form of brittle porous blocks not convenient for practical use. But they could be grinded, and thus obtained powder may be compacted by the methods of explosive pressing.[5-7] Explosive pressing combines the shock-wave compaction of synthesis products and shaping of articles (blanks). The advantages of the explosive pressing were mentioned before. In view of this, it was not surprising that the first studies on the explosive pressing of HTSC powders appeared just after discovery of HTSC phenomenon.[48-50]

HTSC ceramics obtained by the SHS method show a number of particular features different from those of ceramics obtained by other methods. In particular, it is fine-grained (grain size of about 1 μ m) and contains a large amount of slightly bound oxygen, which is probably due to the nonequilibrium character of synthesis. This feature may have an influence on the compaction process, on kinetics of phase transformations, and on chemical reactions at grain surface under the action of shock waves.

The problem, which was being solved in Refs. 5 and 6, was to obtain the high-density uniform SHS HTSC ceramics in the form of a finished article of desired shape and electrophysical parameters (depending on its designation). Uniformity (homogeneity) of HTSC is an essential condition; without its fulfillment the service parameters of articles could not usually be achieved. The problems which are to be solved here are conventional for the problems of explosive pressing of powders. For instance, the requirement of uniformity is normal in fabricating the long-sized articles. The block diagram of explosive pressing setup is presented in Fig. 2. This scheme has been used for more than two decades (with some variations) both in laboratory and industrial scales.[3,4] Starting powder is placed in a metallic cylindrical container (ampoule), which in turn is surrounded by a layer of explosive. Upon detonation onset at the upper end of a charge, a detonation wave is generated which slides alongside a cylinder element. Detonation products, pressure of which makes a value of dozens and even hundreds of kbar, compress the container at a large rate, and as a result an

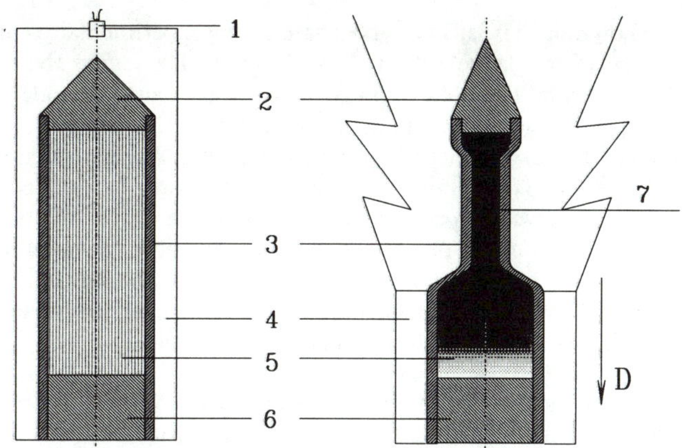

Fig. 2 Pressing of powders with cylindrical explosive piston: (1) electrodetonator, (2) upper-end-cap, (3) metallic ampoule, (4) explosive, (5) starting powder, (6) bottom end-cap, (7) high-density product, and (D) detonation velocity.

implosive conical shock wave arises which presses powder. To obtain uniform pressing, the regular reflection of shock waves should be organized within the central portion of the specimen. This is one of the central problems of the given modification of explosive treatment of SHS products.

The experiments[5] were carried out with the yttrium ceramics Y_{123} prepared by the SHS method preliminary grinded (down to size below 100 μm) and compacted to density 3.6 g/cm^3. Use was made of the cylindrical explosive press schematically presented in Fig. 2. A cell was made of copper (in some experiments, stainless steel was checked). The ratio of cell diameter to wall thickness was maintained to be constant, which enabled avoiding from a difference in the energies of shell deformation. Cell diameter and wall thickness were 15, 20, and 25 and 1.5, 2.0, and 2.5 mm, respectively. One of the most important parameters, pressure at the detonation wave front, could be varied over the range 10-100 kbar by appropriate choice of explosive: ammonite 6 ZV and its mixtures with barium saltpeter, trinitrotoluene, and RDX and their mixtures of varied composition and density. Pressure at the detonation wave front is determined by the detonation velocity: $P = 2.5\ \rho D^2$, where P is maximal pressure (in kbar), ρ is density of explosive (in g/cm^3), D is detonation velocity (in m/s). The detonation velocity was monitored with electronic pickups. Pressure in explosion products is not the only parameter determining quality of pressing. An essential role is also played by "history" of loading (i.e., pressure as a function of time); physico-mechanical parameters of powder grains and its initial porosity; and material of a cell, its wall thickness, etc. However, most important of them is maximal pressure of explosion products and time of its action, which may be characterized in an indirect way by the ratio of masses of explosive and powder to be compacted at fixed other conditions of this multiparametric process. Varying these two parameters, the authors[5] determined the curve of optimal modes of explosive compaction, shown in Fig. 3a. The curve connects the experimental points obtained under conditions of regular reflection of shock waves. The region above this curve corresponds to the conditions of overpressing, below this curve to underpressing, of specimen. Fig. 3b shows densities of homogeneously compacted specimen corresponding to the curve of optimal compaction modes. As seen, maximal relative density of a cross-sectionally uniform specimen, which may be attained within the frameworks of the given experimental setup, makes a value of about 90% of theoretically predicted density of yttrium ceramics. The obtained specimen showed the Meissner effect at liquid nitrogen temperature (77K), but the transport current was not observed without subsequent thermal treatment. To recover current superconductivity, an additional annealing is needed. Annealing was carried out in air at 905°C for 10h, then specimens were cooled down to 300°C in a furnace, kept at this temperature for 2 hours, and then cooled down to 150°C in a furnace.

Another scheme of explosive loading of the yttrium SHS HTSC ceramics was checked in Ref. 6. It differed from that shown in Fig. 2 by the presence of a metallic rod coaxially placed at the center of a cell. In this case, higher density in the compacted specimen was achieved (up to 97% of theoretically predicted density of yttrium ceramics). Homogeneity of the pressed specimen was achieved over a wider range of amplitudes and widths of pressure pulses. In experiments, the following explosives were used: RDX, mixtures of trinitrotoluene with RDX in

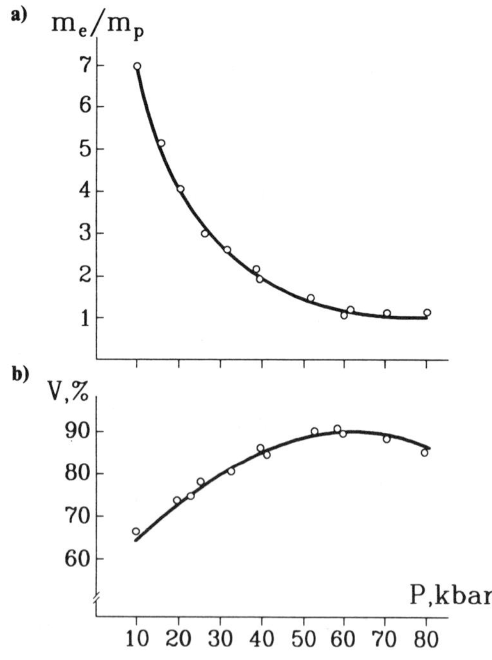

Fig. 3 Curves of optimal pressing for HTSC ceramics with cylindrical explosive press: (m_e/m_p) explosive/powder mass ratio, (V) relative density (in percent of theoretically predicted value).

ratios 50/50 and 30/70, and ammonite 6 ZV and its mixtures with barium saltpeter in ratios 50/50 and 30/70. Such parameters as pulse amplitude and its width were controlled by the explosive identity and its mass. Cell size was kept constant: 27 mm in diameter and 2 mm wall thickness. Cells were made of steel. Thickness of ceramic powder layer was 4 mm, density after preliminary pressing 3.6 g/cm^3. Maximal pressure could be varied over the large range of 10-160 kbar, but the transport current in pressed ceramics has not been obtained at any explosive treatment, and subsequent annealing was needed to recover current superconductivity. To perform this procedure, the pressed specimen had to be taken out of the metallic shell (deformed container). This operation turned out to be very difficult to perform, since brittle ceramics were usually damaged during this procedure.

This problem may be solved by using the loading scheme shown in Fig. 4. Powder of HTSC ceramics is placed in a thin-walled metallic tube, and loading is transferred via transmitting medium (water, oil, glycerol). To remove HTSC specimen from a thin-walled shell is much easier. A series of experiments was carried out within such a scheme.[6] A steel cell 25 mm in diameter and 0.5 mm wall thickness was closed with two end-caps. Density of preliminary compression of SHS HTSC ceramics was 3.6 g/cm^3. The cell was coaxially placed in a steel container 52 mm in diameter and 2 mm wall thickness containing either water or glycerol. Ammonite 6 ZV and its mixtures with barium saltpeter were used as explosives in ratios 50/50 and 30/70 of varied thickness. Pressure could be varied within the range 25-70 kbar. Specimens of high homogeneity were obtained with relative density up to 95% of the theoretically predicted density of yttrium ceramics. The experimental data on pressing conditions for this scheme are presented in Table 1. In particular, such a scheme was used in making screens

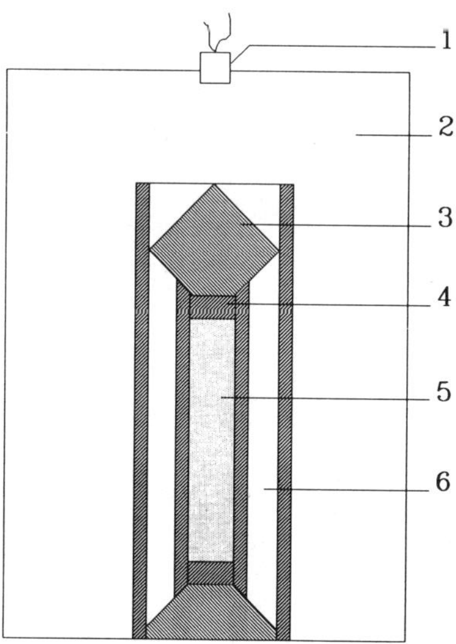

Fig. 4 HTSC pressing via transmitting medium: (1) electrodetonator, (2) explosive, (3) aligning end-cap, (4) end-cap, (5) HTSC powder, and (6) transmitting medium.

Table 1 Experimental data on explosive pressing of HTSC powders by using transmitting media

Explosive	Charge thickness, mm	Peak pressure at detonation wave, kbar	m_e/m_p,* rel. units	Experimental results
AM 6 ZV	17.0	50	2.0	Overpressing
Density of 0.8 g/cc	22.0	60	2.8	Overpressing
	27.0	70	3.7	Regular reflection mode
AM 6 ZV/$Ba(NO_3)_2$	17.0	30	2.5	Underpressing
50/50	19.5	30	3.1	Regular reflection mode
Density of 1.0 g/cc	24.5	45	4.1	Overpressing
	29.5	60	5.3	Overpressing
	24.5	45	4.1	Overpressing
AM 6 ZV/$Ba(NO_3)_2$	19.5	25	3.4	Underpressing
30/70	22.0	25	3.9	Underpressing
Density of 1.1 g/cc	27.0	40	5.2	Regular reflection mode
	32.0	50	6.5	Overpressing
	22.0	25	3.9	Underpressing

*m_e/m_p = the explosive/powder mass ratio.

against magnetic fields. Specimens obtained were annealed in air (under conditions described above). Specimens for studies on screening ability were machined from obtained material and represented hollow cylinders of 19 mm outer diameter, 2 mm wall thickness and 50 mm in length. Screening factors were measured by using induction pickups placed in the center of the specimen. The following results were obtained: In the fields 0-120 Oe at frequencies 20-150 Hz, maximal screening factors (ratio of fields outside and inside) made a value of 10^4 for longitudinal magnetic field and 116 for transverse magnetic field. Screening factors droped by as much as twice upon frequency lowering from 100 down to 20 Hz.

Other schemes were tried of explosive loading of HTSC powders which extended the range of shock-wave parameters toward higher maximal pressures. Phase composition of yttrium ceramics was found to show no visible changes over the pressure range up to 200 kbar, though sometimes the appearance of non-superconducting phase was observed in the form of films at grain interface. Pressure growth up to 250-300 kbar results in considerable increase in the amount of non-superconducting phase both at grain interface and in the bulk, which is probably due to oxygen losses during heating, to phase transformation and chemical interactions of ceramics with gases in pores. At low peak pressures 40-100 kbar, a small amount of intergrain links is formed; predominantly only mechanical contacts between grains are formed which do not conduct transport current. At pressures 200-250 kbar, the number of intergrain links was found to increase. Increase in pulse width also enhances the amount of intergrain links.

The methods of obtaining composite materials and articles of the type SHS HTSC ceramics/metal were developed by using containers of especially designed shape and different loading schemes.[7] They are based on the fact that a deformed container is transformed into metallic matrix protecting the article from damage, while HTSC layer provides the desired electrophysical parameters. In such a way, finished articles and blanks are obtained from yttrium HTSC ceramics for different purposes (Fig. 5). Typical parameters were as follows: the temperature of transition to the superconducting state 93-95 K; transition width 1.2 - 1.5 K; density of critical current reached a value of $2 \cdot 10^3 A/cm^2$ at 77 K in the absence of magnetic field (for the best specimen); and density of HTSC ceramics 90-97% (of theoretically predicted value). Cylindrical and plane blanks may be subject to turning, milling, and other types of machining with no losses in electrophysical parameters, which provides a possibility of fabricating articles of more complicated configuration.

As mentioned above, the SHS method enables synthesizing not only yttrium ceramics, but also other known HTSC. For example, SHS HTSC based on erbium (Er-Ba-Cu-O, 1-2-3) is readily pressed by explosion.[8] By using the loading schemes similar to those described above and the methods described in Refs. 48-50, the current carrying specimen of Er_{123} ceramics deposited onto metal base of copper, stainless steel, and titanium have also been obtained to date.[8]

In a similar way, the wear resistant coatings of other SHS ceramics may be cladded onto metal bases. For instance, the SHS powders of TiN, TiC, and TiB_2 in combination with metallic binder (and without) were cladded to steel articles of plane and cylindrical configuration.[9,10] Powder compaction was performed directly on a metal base by oblique shock wave induced by gliding detonation. Ammonite 6 ZV was used as an explosive.

Extensive studies are now carried out on explosive pressing of large specimen of various ceramics.[41] Ceramics Si_3N_4 seems to be one of the most promising in this respect.[51,52] SHS ceramics may also be used in this process.

Fig. 5 Articles of HTSC ceramics prepared by shock-wave compaction of SHS products.

However, to date one cannot avoid cracks in the articles obtained by this method. Sintering is another way of fabricating ceramic articles. Interesting results were obtained by combining explosive pressing of SHS ceramics Si_3N_4 by weak shock waves up to density 70-80% (of theoretically predicted) with subsequent sintering by using catalysts of sintering.[11] In a number of experiments, sintering enhancement was observed. But no ceramics activation by shock waves was found in these experiments, and in view of this the reason for sintering enhancement remains unclear.

Shock Wave Effects in Starting SHS Compositions

During the 1970s, it was stated that many SHS prepared compounds may be synthesized with no preliminary ignition of reactive mixture and no combustion.[12] This may be achieved by shock compression of reactive mixture which may be put into saving cells. Shock wave propagation in these systems results in strong shear deformations leading to disintegration and hence formation of various defects, chemical bond ruptures, and nonequilibrium temperatures at contact sites between reactants, i.e., to some highly energetic nonequilibrium state. Initiation time at shock compression is thought[12] to have a value of 10^{-6} s over the entire volume of mixture at "hot" points including gas in pores. Reaction may be completed as a post effect, or, of utmost interest, in the course of compression. It is manifested by the Hugoniot curve profile for the mixture of copper and aluminium on which there is a portion confirming heat release in the course of compression.[53] As already mentioned,[12] carbides, borides, silicides, intermetallides, solid solutions, nitrides, and hydrides may be synthesized under conditions of shock compression. A number of these compounds have indeed been obtained by this method.[27,28,54] The feasibility of a multistage chemical reaction occuring has also been outlined.

For instance, upon single shock compression of the mixtures of titanium and paraffin or polyethylene, the mixtures of titanium carbide and titanium hydride were formed.[27] The disadvantage of these syntheses is the uncompleteness of reaction and the complexity of reaction product composition, which includes new unidentified phases.

The feasibility of obtaining novel useful materials by the shock-wave treatment of exothermic powders is nowadays extensively studied.[40] For example, a modeling study for the shock-wave initiation of chemical reaction in the system Ni-Al should be mentioned,[33-36] which results in the formation of intermetallide Ni_3Al. This material may be coated by utilizing explosion energy onto a surface of parts operating at high temperature in corrosive media as a protective coating.[32] An apparatus for the shock-wave synthesis of TiB_2 by explosive loading Ti-B powders up to 295 kbar has been described.[37] The shock-wave action was shown to initiate a chemical reaction, with afterburning upon pressure release, which results in pore formation in the final product. The process was modeled, and the respective data are also presented.[37]

Of interest are the data obtained in the studies on shock-wave treatment of exothermic compositions on metallic bases,[38] since in some cases this results in the formation of solid coatings. It was found[38] that there exist regions of formation of solid and liquid coating for every powder composition, and transition from one to another occurs at certain critical pressures. In the case of solid coatings, no new compounds and phases are formed and no occurrences of chemical reaction were observed. In the case of liquid coatings, fast crystallization (quenching from liquid state), amorphization, and numerous physico-chemical transformations were found to occur. With powders Ni-Al, Fe-Si, Cr-B, Cr-C, Cr-SiC, Cr-B_4C, etc., taken as examples, feasibility of strong chemical interaction and synthesis reaction between powdered reactants in solid-liquid and liquid phases has been demonstrated.[38] As a result, continuous and layered coatings of carbides, borides, silicides, intermetallides, carboborides, etc., are formed at metallic surfaces, their metastable states being capable of stabilizing. The above mentioned coatings are characterized by high microhardness.

Attempts were undertaken[39] to obtain articles of HTSC by shock-wave pressing of exothermic mixture of powders Y_2O_3-BaO_2-Cu. Loading scheme of Fig. 2 and its modifications were used to obtain articles of complicated configuration. Superconducting phase Y_{123} is developed in the form of separate inclusions in densely pressed mass of unreacted substance. Blanks of the 90-93% density were then subject to thermal treatment. In view of high density, chemical transformation in the SHS mode did not occur. Thermal treatment in furnace[39] gave positive results. Temperature slowly increased up to ignition temperature, and chemical reaction was of a "quasivolumetric" nature. It is important that the blank shape retained its initial configuration in the course of thermal treatment. Thus obtained HTSC articles had a density of 60-70% of theoretically predicted value and standard electrophysical parameters.

Concomitant Occurrence of SHS and Explosive Pressing

As mentioned above, the direct shock-wave action on the SHS process is most interesting. The simplest experimental setup for these studies is shown in Fig. 6. In the first case (Fig. 6a), explosion products behind the gliding detonation wave compress the thin-walled cylindrical container. In the second one (Fig. 6b), the use

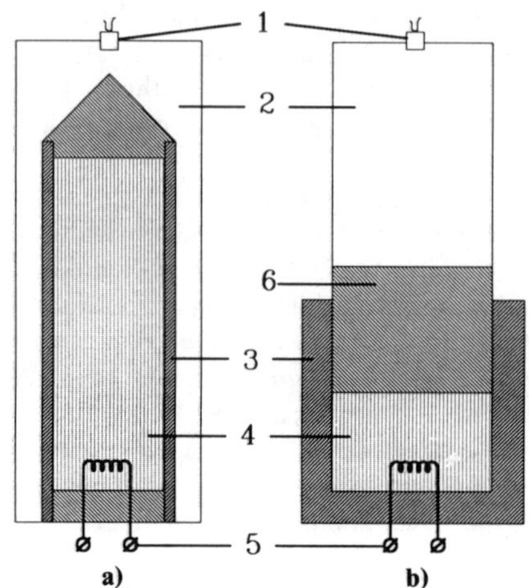

Fig. 6 Experimental setup for explosive loading of reaction products: (1) electrodetonator, (2) explosive, (3) metallic container, (4) starting mixture, (5) igniting wire, and (6) massive piston.

is made of the normally incident detonation wave to accelerate a metal piston compressing reaction products in a massive metallic container. In both cases, reaction is initiated by heated helix at the bottom of a container. At the top of the container, pickups were installed for monitoring combustion wave propagation. Detonation was initiated at some moment after wave propagation. Some difficulties arose due to the necessity of thermal insulation of the explosive, since the temperature of reaction products behind combustion wave is much greater than ignition temperature for most explosives. However, these difficulties were overcome by intrications in the structure of the container. In a similar way, the problem of removing evolved gases outside the container may also be resolved. All the experimental setups for the shock-wave loading of hot reaction products represent some modifications of the arrangements shown in Fig. 6.

It was outlined in early studies[12] that concomitant occurrence of SHS and dynamic loading is more convenient to use for the mixtures of solid combustible and solid oxidizer, combustion of which is practically gasless. Attempts were undertaken to influence the various stages of SHS of titanium carbide and boride. The mixture was placed in a cylindrical cell 10 cm long, 1 cm inside diameter. Shock compression was applied at the moment when the heating zone reached the opposite end of a cylindrical specimen. Under these conditions, the shock compression was assumed[12] to influence all the stages of SHS. Maximal pressures achieved in these experiments were in the range of 70-500 kbar. Reaction products were analyzed by the X-ray methods. Judging from the appearance of new unidentified lines in the X-ray patterns, reaction products in the zones of heating and reaction were of compound composition. The products of afterburning and structuring represented the densely packed rods of titanium carbide or boride. No new phases were found. These data may be interpreted as follows: Heat of shock compression is added to heat released in chemical reaction. As a result, the specimen may be heated up to temperatures above 2000 K. The specimen cools down at atmospheric pressure according to the heat conduction laws. This results in annealing of new phases or the pressure-induced phases (if they were formed). For this reason, the further attempts were oriented on applying dynamic pressures at the last stages of SHS to obtain densely packed articles. The experiments were carried out[12] on shock compression of IV-VI group metal carbides and borides preheated up to 1500-3000 K. Rods, tubes, and disks of 97% density were obtained. The articles showed the finely dispersed structure with grain size of 1-3 μm. Analysis showed that grains of titanium carbide were of rounded shape and surrounded with the nickel-molybdenum binder.

The schemes of the shock-wave compression were then suggested[12] for the gas-evolving systems. Densely packed specimen of TaC-Al_2O_3 and TiC-Al_2O_3 were prepared under these condition from the mixtures of tantal and titanium oxides with aluminium and soot. Porosity was of about 2%. Carbide grains were of 2-3 μm in size and uniformly distributed over oxide matrix. The articles showed enhanced resistance against heated oxygen and sulfuric acid.

These ideas received further development during further studies in the field.[13-26] The shock-wave compaction of hot SHS products was studied[18-24] for the binary systems Ti-C, Ti-B, and Hf-C. This technique was used to produce TiC and TiB_2 at greater than 98% of theoretical density and microhardness values which are equal to or greater than commercially available hot-pressed materials.[18-21] It was shown that the microstructures of the SHS materials do differ from the hot-pressed materials in significant ways leading to the possibility that new and unique structures can be fabricated. The effects of stoichiometry and the addition of third components like Cu, Fe, Mo, W, Al_2O_3, and ZrO on the product microstructures and microhardnesses were investigated. It was shown[20] that the addition of metal or oxide additives can alter not only the microstructure of the formed product but can also modify the performance characteristics. Analysis of the TiC indicated[21] that density and microhardness increase as a function of the C/Ti ratio, with maximum values at the ratio of 1.0. A dynamic, finite-difference, heat flow model that predicts SHS reaction temperatures and propagation velocities, effects of material and process conditions on SHS, and heat flow patterns during SHS compact cooling in post-densification fixtures was developed to control the processing of compacts made by simultaneous SHS and explosive consolidation.[22,23] Effects of explosive charge mass, powder compact containment design, and time delay between the SHS reaction and explosive detonation were observed to critically affect the density, hardness, and microstructure of the final product (TiC). The explosive-consolidation technique developed to fabricate combustion-synthesized titanium carbide and titanium diboride was applied to hafnium carbide and binary HfC-TiC composites.[24] This technique was also applied[25] with the aim of obtaining the high-density dispersed-phase composites of TiC-Al_2O_3. In all these cases, modifications of the shock-wave loading shown in Fig. 6b were used in the experiments.

The effects of the shock-wave loading were studied[13-16] in the system containing powdered mixtures of titanium and carbon (soot) doped with nickel and chromium. Both the schemes of shock-wave compaction of Fig. 6 were used in these experiments. Besides SHS charge composition, the delay time of detonation onset, size of explosive charge, and type of explosive varied. Microstructure and phase composition of the final product were studied by using the X-ray microanalyzers. Structure and properties of the obtained materials were found to depend markedly on the above mentioned factors and to differ from those for the materials obtained under conditions of static pressure. Fig. 7a presents a microphotograph of a specimen obtained under conditions of dynamic pressure. Fig. 7b gives the same for a specimen obtained under conditions of static pressure (press). The explosively-made materials were found to have smaller and equal-axis grains of titanium-chromium carbide (dark region) and more uniform distribution of the ductile nickel binder (bright region). Microphotograph shows the range of possible variation of grain size upon variation of loading parameters and detonation delay times. Density was 97-99% (of theoretically predicted value) hardness 90-92 HRA units. The feasibility of obtaining ultrafine-grained ceramics (0.2-0.3 μm) was demonstrated[16] for the same system (Ti-C-Ni-Cr), which opens the possibility to develop ceramic materials undergoing transition into the ductile state at relatively low temperatures.

Interesting results are obtained in the case when detonation is initiated at the moment when the combustion wave has not yet reached the specimen end. In such a case, in one experiment and at the same parameters of explosive loading, one may observe the effects of shock waves on all the stages of SHS (i.e., all the zones of the synthesis wave of Fig. 1, including the zone of initial matter).

The action of shock waves on the starting SHS compositions was shown to give results depending on the amplitude of loading, scheme of loading, and conditions of heat conduction. At other equal conditions, pressure growth initially leads to compaction of the SHS charge and deformation of grains with the formation of metallographic texture, chemical reactions being not still initiated. Upon further pressure increase, activation of reactants does occur: The large amount of defects is generated, which is accompanied by the disintegration, mixing, and heating of titanium and soot particles. The surface layers of metal undergo chemical reaction with the nearest soot particles giving titanium carbide, which is arrested at the rarefaction wave due to sharp cooling of the system. The typical

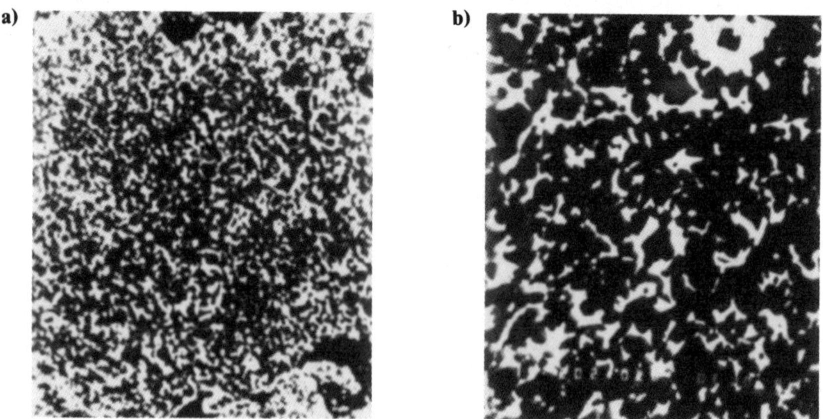

Fig. 7 Structure of specimen obtained in a combustion mode from the mixture of powdered titanium and soot doped with nickel and chromium by the action of high pressure on reaction products (section metallographic specimen COMPO×2000): (a) dynamic pressure, and (b) static pressure.

structure formed at these conditions is shown in Fig. 8. Thickness of titanium carbide film (grey region) makes a value of 2-3 μm. It separates titanium (bright region) and soot (dark region); the regions of molten titanium are absent, and the reaction probably takes place in solid state. Upon further pressure growth, the extent of conversion and temperature of reactants at shock wave become sufficient to compensate cooling at the rarefaction wave. The "quasivolumetric" initiation of exothermal mixture is brought about with subsequent afterburning in the "thermal explosion" mode.

At certain conditions of the shock wave loading of the locally initiated SHS systems, the arresting (stopping) of the synthesis wave may be observed. Arresting of chemical reaction may be explained by a sharp increase in thermal conductivity of a medium due to its compaction, leading to larger heat removal from the system. Such an analysis may help in the studies on the mechanism of SHS processes (with account for the action of high pressures).

The particular features of structuring under conditions of concomitant occurrence of SHS in the hybride system (e.g., metal-gas) and explosive loading were also reported.[17] Such an approach promises to resolve some technological problems: fabricating porousless materials in the system gas-solid, layered materials, etc. The experimental setup for these studies is presented in Fig. 9. A preliminary pressed from a metal powder specimen was placed in a specially designed ampoule of conservation. Since a free access of air had to be provided into the combustion zone, loading was performed with a thrown cylindrical striker. The striker was thrown by a charge of explosive (mixture of ammonite 6 ZV with barium saltpeter, 30/70). Combustion was initiated by tungsten helix from the

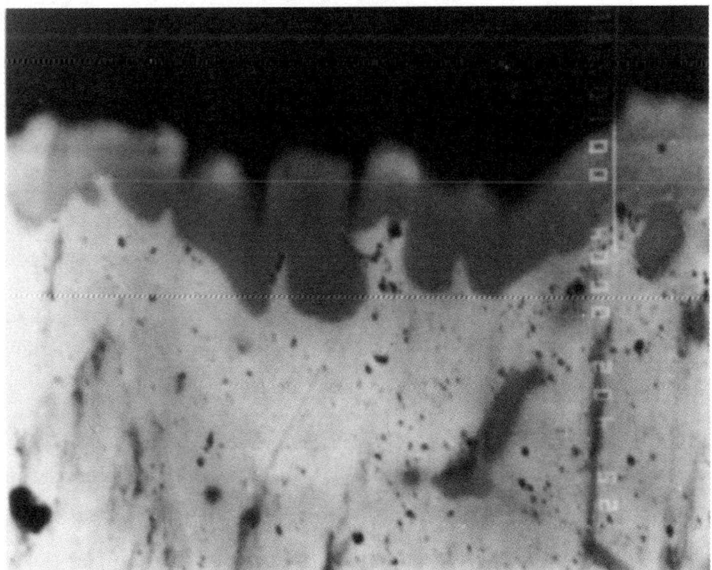

Fig. 8 Carbide film formed upon interaction of titanium grain with soot particle under conditions of shock-wave loading (section metallographic specimen COMPO×4000).

bottom side. A blasting cap was exploded at the moment the combustion wave passed over the desired portion of the specimen (monitored with thermocouples). The conditions were adjusted so that pressing should be close to the regular one. Under the action of explosion, the combustion wave stopping was observed in the system Ti-N-O. The cooling rate was found to have a value not less than 10^4grad/s (during combustion wave stopping). The following interesting results were obtained:[17] 1) titanium interaction with air in a combustion mode is occurring in two subsequent stages, the limiting stage being interaction titanium-nitrogen and oxygen enters into reaction only at the stage of afterburning (Fig. 10); 2) the "primary" structure responsible for combustion wave propagation is the formation of the 2-3 μ m film of Ti_2N in the kinetic mode, the leading temperature of process being larger than the titanium melting point; 3) at the second stage, the process is conditioned by cracking the primary nitride film and by the interaction of molten titanium with nitrogen, including dissolution of the "primary" structures giving finely dispersed (grain size of 1 μ m) phase of a composition TiN_{1-x}(x=0.1-0.2); 4) the phenomenon of microstructural irregularities of the fine-grained islet-like zones type was discovered in the formation of the TiN_{1-x} phase due to structural inhomogeneity of heterogeneous medium; and 5) material $TiN_{0.8}$ was found to have a density of 90-97% of a theoretically predicted one.

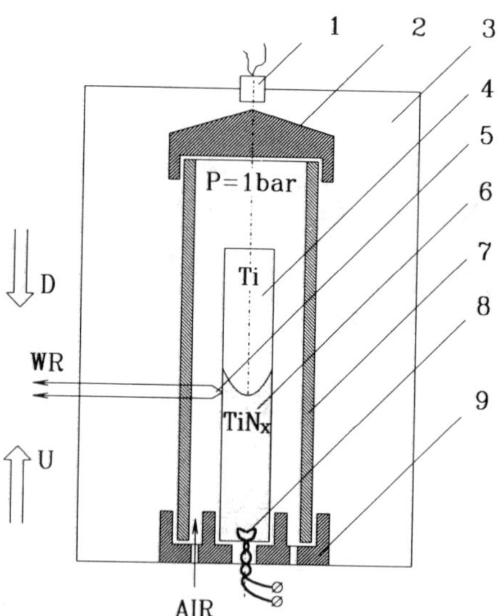

Fig. 9 Experimental setup for explosive action on the metal-gas system in the course of combustion: (1) electrodetonator, (2) upper end-cap, (3) explosive, (4) starting powder, (5) thermocouple, (6) synthesis products, (7) metallic liner, (8) tungsten helix, (9) bottom end-cap, (U) combustion velocity, and (D) detonation velocity. (From Ref. 17. Published with permission of the ISMAN.)

Attempts were undertaken[26] to combine SHS and dynamic pressing to obtain high-density HTSC. The loading scheme was as shown in Fig. 6a. A cell was designed to provide oxygen filtration through the reaction products of a composition Y_2O_3-BaO_2-Cu. The systems were also tried with internal chemical courses of oxygen. Electrophysical parameters for obtained HTSC were the same as those for the conventional systems Y_{123}. Further advances in this direction will allow to obtain HTSC with a value of critical current up to $10^4 A/cm^2$.

Another application of SHS combined with dynamic compressing was proposed[12] to be the usage of the SHS process as a pulsed heater. This may turn out to be useful in the high-pressure physics, e.g., in synthesizing diamonds, where heating under pressure is required. Various compounds and their mixtures were placed[12] (in tubes of refractory metal) inside a reactive mixture, which in turn was placed inside the ampoule. The thermocouple measurements showed that in the course of SHS the compound under investigation may be rapidly (in seconds) heated to very high temperatures. Dynamic compression of SHS products and heated compound may be done at any stage: during temperature rise, at maximal heating, or at some moment during cooling down. It was shown that the pulsed SHS heating may be used for some specific purposes, such as melting or evaporation of one or several components before compression, thermal activation of refractory component, etc., which promote the occurrence of those or other physico-chemical processes. As was stated,[12] the cubic modification of boron nitride is formed under these conditions, but not the wurtzite-like one as is the case at normal shock compression.

The method of pulsed heating was also utilized in Refs. 29-31. The SHS process in the Ti-C system was used as a heat source for enlightening the shock-wave compaction of SiC ceramics.

In conclusion it may be stated that the combination of SHS process with shock compression enables the achievement of inaccessible regions within the plane

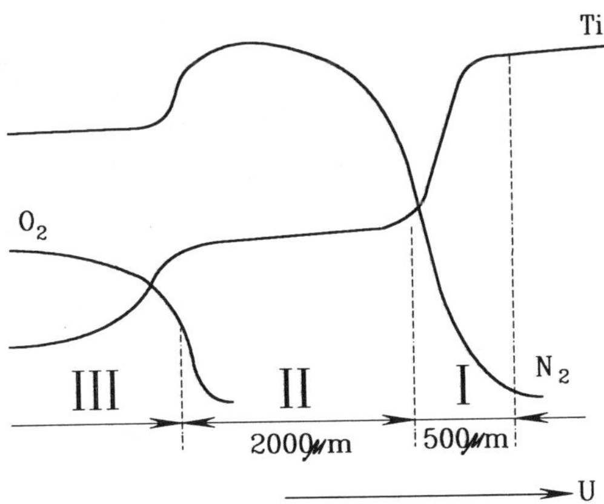

Fig. 10 Overall distribution of Ti, N, O content along the direction of combustion wave propagation: (U) combustion velocity. (From Ref. 17 Published with permission of the ISMAN.)

pressure-temperature both at high and low pressures. But there exist some limitations in the experiments with conservation of matter: very high residual temperatures upon pressure release may result in transforming new phases into normal ones. These limitations may be eliminated[12] by performing fast quenching of synthesis products, or providing conditions of their cooling down at elevated residual pressures.

Conclusion

The most important advantage of the SHS method seems to be the absence of need for external energy consumption, since the process is brought about at the expense of internal energy of the system. Since high pressures at explosive detonation may also be attained without expensive installation, the combination of these two methods of creating extreme conditions (high temperatures and high pressures) is of undoubted interest for practice. The concomitant occurrence of SHS and shock-wave treatment of synthesis products (SHS/SWT) may be considered a new, efficient method of preparing various materials which cannot be reduced to simple compaction and forming. The available data show that the action of high dynamic pressure may be used for controlling structure of materials synthesized in a combustion mode, though further studies are needed for research and development of the SHS/SWT method.

References

[1]Merzhanov, A. G., and Borovinskaya, I. P., "Self-Propagating High-Temperature Synthesis of Inorganic Compounds," *Doklady Akademy Nauk SSSR*, Vol. 204, No. 2, 1972, pp. 366-369.

[2]Merzhanov, A. G., *Self-Propagating High-Temperature Synthesis: Twenty Years of Search and Findings*, ISMAN (preprint), Chernogolovka, 1988.

[3]Rinehart, J. S., and Pearson, J., *Explosive Working of Metals*, Pergamon Press, New York, 1963.

[4]Prümmer, R., *Explosivverdichtung pulvriger Substanzen. Grundlagen, Verfahren, Ergebnisse*, Springer-Verlag, Berlin, Heidelberd, New York, London, Paris, Tokyo, 1987.

[5]Fedorov, V. M., and Gordopolov, Yu. A., *The Study of Regimes for Explosive Pressing of HTSC SHS Ceramics*, ISMAN (preprint), Chernogolovka, 1990.

[6]Fedorov, V. M., and Gordopolov, Yu. A., *Shock-Wave Compaction of HTSC SHS Ceramics*, ISMAN (preprint), Chernogolovka, 1990.

[7]Gordopolov, Yu. A., and Fedorov, V. M., *HTSC Ceramics/Metal Composite Materials Fabricated by Explosive Pressing*, VDNH SSSR (publicity), 1990.

[8]Tavadze, G. F., private communication, 1990-1991.

[9]Zotov, N. A., private communication, 1990-1991.

[10]Stertser, A. A., private communication, 1990-1991.

[11]Sharivker, Yu. S., and Fedorov, V. M., private communication, 1990-1991.

[12]Adadurov, G. A., Borovinskaya, I. P., and Merzhanov, A. G., private communications, 1972-1987.

[13]Gordopolov, Yu. A., Shikhverdiev, R. M., Molokov, I. V., Bogatov, Yu. V., Borovinskaya, I. P., and Merzhanov, A. G., *The Study of Shock Wave Loading of Heated Reaction Products at Synthesis of Refractory Alloys in Combustion Wave*, ISMAN (preprint), Chernogolovka, 1988.

[14]Gordopolov, Yu. A., Shikhverdiev, R. M., Molokov, I. V., Bogatov, Yu. V., Borovinskaya, I. P., and Merzhanov, A.G., "Effects of Shock Waves on the Formation of Structure of Refractory Alloys, Synthesized by a Combustion Process," *Proceeding of 7th International Symposium*, "Use of Explosive Energy in Manufacturing Metallic Materials of New Properties," Pardubice, Vol. 2, October 1988, pp. 324-330.

[15]Gordopolov, Yu. A., Fedorov, V. M., Molokov, I. V., Shikhverdiev, R. M., and Merzhanov, A.G., "Explosive Working of SHS Products," *Proceeding of 10th International Conference on High Energy Rate Fabrication*, Ljubljana, September 1989, pp. 144-153.

[16]Gordopolov, Yu. A., Molokov, I. V., Shikhverdiev, R. M., Pityulin, A. N., Efimov, O. Yu., Zaripov, N. G., and Petrova, L. V., "Synthesis of Hard Refractory Alloys under Shock Wave Loading," *Proceeding of 16th All-Union Conference on Powder Metallurgy*, Sverdlovsk, May 1989, p. 54.

[17]Molokov, I. V., and Mukasyan, A. S., *The Stopping of Combustion Wave by Explosive Effect on Ti-N-O System*, ISMAN (preprint), Chernogolovka, 1990.

[18]Niiler, A., Kecskes, L. J., Kottke, T., Netherwood, P. H., Jr., and Benck, R. F., "Explosive Consolidation of Combustion Synthesized Ceramics: TiC and TiB_2," *Ballistic Research Laboratory Report BRL-TR-2951*, Aberdeen Proving Cround, Dec. 1988.

[19]Niiler, A., Kecskes, L. J., and Kottke, T., "Shock Consolidation of SHS Ceramics" *Combustion and Plasma Synthesis of High-Temperature Materials*, edited by Z. A.Munir and J. B.Holt, VCH Publishers, New York, Weinheim, Basel, Cambridge, 1990, pp. 309-314.

[20]Niiler, A., Kecskes, L. J., and Kottke, T., "Consolidation of Combuston Synthesized Materials by Explosive Compacton," Paper T-5, *Proceeding 1st U.S.-Japanese Workshop on Combustion Synthesis*, Japan, Jan. 1990.

[21]Kecskes, L. J., Kottke, T., and Niiler, A., "Microstructural Properties of Combustion-Synthesized and Dynamically Consolidated Titanium Boride and Titanium Carbide," *Journal of American Ceramic Society*, Vol. 73, No. 5, 1990, pp. 1274-1282.

[22]Grebe, H. A., Advani, A., Thadhani, N. N., and Kottke, T., "Simultenеous Combustion Synthesis and Explosive Shock Consolidation of Titanium Carbide Ceramics," *Presented at the TMS Symposium on Reaction Synthesis of Materials*, New Orleans, February 1991, ***Metallurgical Transactions***, to be published.

[23]Advani, A. H., Thadhani, N. N., Grebe, H. A., Heaps, R., Coffin, C., and Kottke, T., "Dynamic Modeling of Self Propagating High Temperature Synthesis of Titanium Carbide Ceramics" *Scripta Metallurgica at Materialia* to be published.

[24]Kecskes, L. J., Benck, R. F., and Netherwood, P. H., Jr., "Dynamic Compaction of Combustion-Synthesized Hafnium Carbide," *Journal of American Ceramic Society*, Vol. 73, No. 2, 1990, pp. 383-387.

[25]Rabin, B. H., Korth, G. E., and Williamson, R. L., "Fabrication of Titanium Carbide-Alumina Composites by Combuston Synthesis and Subsequent Dynamic Consolidation," *Journal of American Ceramic Society*, Vol. 73, No. 7, 1990, pp. 2156-2157.

[26]Fedorov, V. M., and Shikhverdiev, R. M., private communication, 1990.

[27]Adadurov, G. A., and Goldansky, V. I., "Condensed Matters Transformation at Shock-Wave Compression in Thermodynamic Controlling Conditions," *Uspekhi Khimii*, Vol. 50, Issue 10, 1981, pp. 1810-1827.

[28]Adadurov, G. A., "Experimental Investigations of Chemical Processes during Dynamic Compression," *Uspekhi Khimii*, Vol. 55, Issue 4, 1986, pp. 555-578.

[29]Akashi, T., and Sawaoka, A. B., "Dynamic Compaction of SiC Powder Using Exotermic Reaction Heat," *Advanced Ceramic Materials*, Vol. 3, No. 3, 1988, pp. 288-290.

[30]Akashi, T., and Sawaoka, A. B., "Dynamic Compaction of SiC Powder Utilizing Heat of Exothermic Reaction," *Kogyo Kayaku*, Vol. 49, No. 4, 1988, pp. 278-284.

[31]Akashi, T., and Sawaoka, A. B., U.S. Patent, No. 4, 655, 830, April 7, 1987.

[32]Mazein, S. A., and Shmakov, A. M., private communication, 1990.

[33]Taylor, P. A., Boslough, M., and Horie, Y., "Modeling of Shock-Induced Chemistry in Nickel-Aluminum Systems," *Shock Waves in Condensed Matter*, 1987, North-Holland, Amsterdam, Oxford, New-York, Tokyo, 1988, pp. 395-398.

[34]Horie, Y., and Kipp, M. E., "Modeling of Chemical Reactions in the Mixture of Al-Ni Powders under Shock-Wave Compression", *Shock Waves in Condensed Matter*, 1987, North-Holland, Amsterdam, Oxford, New-York, Tokyo, 1988, pp. 387-390.

[35]Horie, Y., and Kipp, M. E., "Modeling of Shock-Induced Chemical Reactions in Powder Mixtures," *Journal of Applied Physics*, Vol. 63, No. 12, June 1988, pp. 5718-5727.

[36]Boslough, M. R., "Shock-Induced Chemical Reactions in Nickel-Aluminum Powder Mixtures: Radiation Pyrometer Measurements," *Chemical Physics Letters*, Vol. 160, No. 5,6, Aug. 1989, pp. 618-622.

[37]Maiden, D. E., Bianchini, G., Holt, B., Hornig, H., and Kingman, D., "Chemical Shock Synthesis of TiB_2," *DARPA/ARMY (Defence Advanced Research Project Agency) SHS Symposium Proceedings*, Daytona Beach, Florida, Oct. 1985, pp. 359-377.

[38]Kaunov, A. M., private communication, 1990.

[39]Molokov, I. V., private communication, 1990.

[40]Graham, R. A., Morosin, B., Venturini, E. L., and Carr, M.J., "Materials Modification and Synthesis under High Pressure Shock Compression," *Annual Review Material Science*, Vol. 16, 1986, pp. 315-341.

[41]Gourdin, W. H., "Dynamic Compaction of Ceramic Powders: Practise, Problems and Future," *Material Research Society Symposium Proceedings*, Vol. 24, 1984, pp. 307-318.

[42]Bednorz, J. G., and Muller, K.A., "Possible High Tc Superconductivity in the Ba-La-Cu-O System," *Zeitschrift fur Physik B-Condensed Matter*, Vol. 64, 1986, pp. 189-193.

[43]Chu, C. W., et al., "Evidence for Superconductivety above 40K in the La-Ba-Cu-O Compound System," *Physical Review Letters*, Vol. 58, No. 4, 1987, pp. 405-407.

[44]Wu, M. K., et al., "Superconductivity at 93K in a New-Mixed-Phase Y-Ba-Cu-O Compounds System at Ambient Pressure," *Physical Review Letters*, Vol. 58, No. 9, 1987, pp. 908-910.

[45]Nersesyan, M. D., and Merzhanov, A. G., "SHS in the High-Temperature Superconductivity Problem," *Analytical Review 1969-1989*, No. 5111, Moscow, 1990.

[46]Merzhanov, A. G., Peresada, A. G., Nersesyan, M. D., Borovinskaya, I. P., et al., "High-Temperature Superconductivity (T_c=115 K) in Ti-Ba_2-Ca-Cu-O," *Pismo v ZETF*, Vol. 47, Issue 11, 1988, pp. 604-605.

[47]Merzhanov, A. G., Lisikov, S. V., Nersesyan, M. D., Borovinskaya, I. P., et al., "Multiphase Ceramic High-Temperature Superconducto Bi-Ca-Sr-Cu-O," *Pismo v ZTF*, Vol. 14, No. 19, 1988, pp. 1770-1772.

[48]Murr, L. E., Hare, A. W., and Eror, N.G., "Shock-Compression fabrication of High-Temperature superconductor/metal Composite monoliths," *Nature*, Vol. 329, No. 6134, Sept. 1987, pp. 37-39.

[49]Murr, L. E., et.al., "Interfacial Phenomena and Microstructural Connectivity in Explosively Fabricated Y-Ba-Cu-O Superconductors," *Journal of Superconductivity*, Vol. 1, No. 1, 1988, pp. 3-19.

[50]Murr, L. E., et al., "Shock-Induced Microstructures in Explosively Fabricated Superconductors," *Journal of Metals*, Vol. 40, No. 1, Jan. 1988, pp. 19-23.

[51]Akashi, T., and Sawaoka, A. B., "Dynamic Compaction of Silicon Nitride Powder," *Journal of Materials Science*, Vol. 22, 1987, pp. 1031-1036.

[52]Kamiya, K., Ikazaki, F., Uchida, K., Goto, A., Kawamura, M., Tanaka, K., and Fujiwara S., "Effect of Shock Pressure on One-Dimensional Explosively Shocked Characteristics of Si_3N_4," *Yogyo-Kyokai-Shi*, Vol. 95, No. 5, 1987, pp. 480-485.

[53]Dremin, A. N., and Breusov, O. N., "Explosion-Chemist," *Nauka i Zizn*, No. 2, 1978, pp. 28-33.

[54]Batsanov, S.S., "Inorganic Chemistry of High Dynamic Pressures," *Uspekhi Khimii*, Vol. 55, Issue 4, 1986, pp. 579-607.

Author Index

PROGRESS IN ASTRONAUTICS AND AERONAUTICS SERIES VOLUMES

*1. **Solid Propellant Rocket Research** (1960)
Martin Summerfield
Princeton University

*2. **Liquid Rockets and Propellants** (1960)
Loren E. Bollinger
Ohio State University
Martin Goldsmith
The Rand Corp.
Alexis W. Lemmon Jr.
Battelle Memorial Institute

*3. **Energy Conversion for Space Power** (1961)
Nathan W. Snyder
Institute for Defense Analyses

*4. **Space Power Systems** (1961)
Nathan W. Snyder
Institute for Defense Analyses

*5. **Electrostatic Propulsion** (1961)
David B. Langmuir
Space Technology Laboratories, Inc.
Ernst Stuhlinger
NASA George C. Marshall Space Flight Center
J.M. Sellen Jr.
Space Technology Laboratories, Inc.

*6. **Detonation and Two-Phase Flow** (1962)
S.S. Penner
California Institute of Technology
F.A. Williams
Harvard University

*7. **Hypersonic Flow Research** (1962)
Frederick R. Riddell
AVCO Corp.

*8. **Guidance and Control** (1962)
Robert E. Roberson, Consultant
James S. Farrior
Lockheed Missiles and Space Co.

*9. **Electric Propulsion Development** (1963)
Ernst Stuhlinger
NASA George C. Marshall Space Flight Center

*10. **Technology of Lunar Exploration** (1963)
Clifford I. Cummings
Harold R. Lawrence
Jet Propulsion Laboratory

*11. **Power Systems for Space Flight** (1963)
Morris A. Zipkin
Russell N. Edwards
General Electric Co.

*12. **Ionization in High-Temperature Gases** (1963)
Kurt E. Shuler, Editor
National Bureau of Standards
John B. Fenn, Associate Editor
Princeton University

*13. **Guidance and Control—II** (1964)
Robert C. Langford
General Precision Inc.
Charles J. Mundo
Institute of Naval Studies

*14. **Celestial Mechanics and Astrodynamics** (1964)
Victor G. Szebehely
Yale University Observatory

*15. **Heterogeneous Combustion** (1964)
Hans G. Wolfhard
Institute for Defense Analyses
Irvin Glassman
Princeton University
Leon Green Jr.
Air Force Systems Command

*16. **Space Power Systems Engineering** (1966)
George C. Szego
Institute for Defense Analyses
J. Edward Taylor
TRW Inc.

*17. **Methods in Astrodynamics and Celestial Mechanics** (1966)
Raynor L. Duncombe
U.S. Naval Observatory
Victor G. Szebehely
Yale University Observatory

*18. **Thermophysics and Temperature Control of Spacecraft and Entry Vehicles** (1966)
Gerhard B. Heller
NASA George C. Marshall Space Flight Center

---
*Out of print.

***19. Communication Satellite Systems Technology** (1966)
Richard B. Marsten
Radio Corporation of America

***20. Thermophysics of Spacecraft and Planetary Bodies: Radiation Properties of Solids and the Electromagnetic Radiation Environment in Space** (1967)
Gerhard B. Heller
NASA George C. Marshall Space Flight Center

***21. Thermal Design Principles of Spacecraft and Entry Bodies** (1969)
Jerry T. Bevans
TRW Systems

***22. Stratospheric Circulation** (1969)
Willis L. Webb
Atmospheric Sciences Laboratory, White Sands, and University of Texas at El Paso

***23. Thermophysics: Applications to Thermal Design of Spacecraft** (1970)
Jerry T. Bevans
TRW Systems

24. Heat Transfer and Spacecraft Thermal Control (1971)
John W. Lucas
Jet Propulsion Laboratory

25. Communication Satellites for the 70's: Technology (1971)
Nathaniel E. Feldman
The Rand Corp.
Charles M. Kelly
The Aerospace Corp.

26. Communication Satellites for the 70's: Systems (1971)
Nathaniel E. Feldman
The Rand Corp.
Charles M. Kelly
The Aerospace Corp.

27. Thermospheric Circulation (1972)
Willis L. Webb
Atmospheric Sciences Laboratory, White Sands, and University of Texas at El Paso

28. Thermal Characteristics of the Moon (1972)
John W. Lucas
Jet Propulsion Laboratory

***29. Fundamentals of Spacecraft Thermal Design** (1972)
John W. Lucas
Jet Propulsion Laboratory

30. Solar Activity Observations and Predictions (1972)
Patrick S. McIntosh
Murray Dryer
Environmental Research Laboratories, National Oceanic and Atmospheric Administration

31. Thermal Control and Radiation (1973)
Chang-Lin Tien
University of California at Berkeley

32. Communications Satellite Systems (1974)
P.L. Bargellini
COMSAT Laboratories

33. Communications Satellite Technology (1974)
P.L. Bargellini
COMSAT Laboratories

***34. Instrumentation for Airbreathing Propulsion** (1974)
Allen E. Fuhs
Naval Postgraduate School
Marshall Kingery
Arnold Engineering Development Center

35. Thermophysics and Spacecraft Thermal Control (1974)
Robert G. Hering
University of Iowa

36. Thermal Pollution Analysis (1975)
Joseph A. Schetz
Virginia Polytechnic Institute
ISBN 0-915928-00-0

37. Aeroacoustics: Jet and Combustion Noise; Duct Acoustics (1975)
Henry T. Nagamatsu, Editor
General Electric Research and Development Center
Jack V. O'Keefe, Associate Editor
The Boeing Co.
Ira R. Schwartz, Associate Editor
NASA Ames Research Center
ISBN 0-915928-01-9

38. Aeroacoustics: Fan, STOL, and Boundary Layer Noise; Sonic Boom; Aeroacoustics Instrumentation (1975)
Henry T. Nagamatsu, Editor
General Electric Research and Development Center
Jack V. O'Keefe, Associate Editor
The Boeing Co.
Ira R. Schwartz, Associate Editor
NASA Ames Research Center
ISBN 0-915928-02-7

39. Heat Transfer with Thermal Control Applications (1975)
M. Michael Yovanovich
University of Waterloo
ISBN 0-915928-03-5

*40. **Aerodynamics of Base Combustion** (1976)
S.N.B. Murthy, Editor
J.R. Osborn,
Associate Editor
Purdue University
A.W. Barrows
J.R. Ward,
Associate Editors
Ballistics Research Laboratories
ISBN 0-915928-04-3

41. Communications Satellite Developments: Systems (1976)
Gilbert E. LaVean
Defense Communications Agency
William G. Schmidt
CML Satellite Corp.
ISBN 0-915928-05-1

42. Communications Satellite Developments: Technology (1976)
William G. Schmidt
CML Satellite Corp.
Gilbert E. LaVean
Defense Communications Agency
ISBN 0-915928-06-X

*43. **Aeroacoustics: Jet Noise, Combustion and Core Engine Noise** (1976)
Ira R. Schwartz, Editor
NASA Ames Research Center
Henry T. Nagamatsu,
Associate Editor
General Electric Research and Development Center
Warren C. Strahle,
Associate Editor
Georgia Institute of Technology
ISBN 0-915928-07-8

*44. **Aeroacoustics: Fan Noise and Control; Duct Acoustics; Rotor Noise** (1976)
Ira R. Schwartz, Editor
NASA Ames Research Center
Henry T. Nagamatsu,
Associate Editor
General Electric Research and Development Center
Warren C. Strahle,
Associate Editor
Georgia Institute of Technology
ISBN 0-915928-08-6

*45. **Aeroacoustics: STOL Noise; Airframe and Airfoil Noise** (1976)
Ira R. Schwartz, Editor
NASA Ames Research Center
Henry T. Nagamatsu,
Associate Editor
General Electric Research and Development Center
Warren C. Strahle,
Associate Editor
Georgia Institute of Technology
ISBN 0-915928-09-4

*46. **Aeroacoustics: Acoustic Wave Propagation; Aircraft Noise Prediction; Aeroacoustic Instrumentation** (1976)
Ira R. Schwartz, Editor
NASA Ames Research Center
Henry T. Nagamatsu,
Associate Editor
General Electric Research and Development Center
Warren C. Strahle,
Associate Editor
Georgia Institute of Technology
ISBN 0-915928-10-8

47. Spacecraft Charging by Magnetospheric Plasmas (1976)
Alan Rosen
TRW Inc.
ISBN 0-915928-11-6

48. Scientific Investigations on the Skylab Satellite (1976)
Marion I. Kent
Ernst Stuhlinger
NASA George C. Marshall Space Flight Center
Shi-Tsan Wu
University of Alabama
ISBN 0-915928-12-4

49. Radiative Transfer and Thermal Control (1976)
Allie M. Smith
ARO Inc.
ISBN 0-915928-13-2

50. Exploration of the Outer Solar System (1976)
Eugene W. Greenstadt
TRW Inc.
Murray Dryer
National Oceanic and Atmospheric Administration
Devrie S. Intriligator
University of Southern California
ISBN 0-915928-14-0

51. Rarefied Gas Dynamics, Parts I and II (two volumes) (1977)
J. Leith Potter
ARO Inc.
ISBN 0-915928-15-9

52. Materials Sciences in Space with Application to Space Processing (1977)
Leo Steg
General Electric Co.
ISBN 0-915928-16-7

53. Experimental Diagnostics in Gas Phase Combustion Systems (1977)
Ben T. Zinn, Editor
Georgia Institute of Technology
Craig T. Bowman, Associate Editor
Stanford University
Daniel L. Hartley, Associate Editor
Sandia Laboratories
Edward W. Price, Associate Editor
Georgia Institute of Technology
James G. Skifstad, Associate Editor
Purdue University
ISBN 0-015928-18-3

54. Satellite Communications: Future Systems (1977)
David Jarett
TRW Inc.
ISBN 0-915928-18-3

55. Satellite Communications: Advanced Technologies (1977)
David Jarett
TRW Inc.
ISBN 0-915928-19-1

56. Thermophysics of Spacecraft and Outer Planet Entry Probes (1977)
Allie M. Smith
ARO Inc.
ISBN 0-915928-20-5

57. Space-Based Manufacturing from Nonterrestrial Materials (1977)
Gerard K. O'Neill, Editor
Brian O'Leary, Assistant Editor
Princeton University
ISBN 0-915928-21-3

58. Turbulent Combustion (1978)
Lawrence A. Kennedy
State University of New York at Buffalo
ISBN 0-915928-22-1

59. Aerodynamic Heating and Thermal Protection Systems (1978)
Leroy S. Fletcher
University of Virginia
ISBN 0-915928-23-X

60. Heat Transfer and Thermal Control Systems (1978)
Leroy S. Fletcher
University of Virginia
ISBN 0-915928-24-8

61. Radiation Energy Conversion in Space (1978)
Kenneth W. Billman
NASA Ames Research Center
ISBN 0-915928-26-4

62. Alternative Hydrocarbon Fuels: Combustion and Chemical Kinetics (1978)
Craig T. Bowman
Stanford University
Jorgen Birkeland
Department of Energy
ISBN 0-915928-25-6

63. Experimental Diagnostics in Combustion of Solids (1978)
Thomas L. Boggs
Naval Weapons Center
Ben T. Zinn
Georgia Institute of Technology
ISBN 0-915928-28-0

64. Outer Planet Entry Heating and Thermal Protection (1979)
Raymond Viskanta
Purdue University
ISBN 0-915928-29-9

65. Thermophysics and Thermal Control (1979)
Raymond Viskanta
Purdue University
ISBN 0-915928-30-2

66. Interior Ballistics of Guns (1979)
Herman Krier
University of Illinois at Urbana-Champaign
Martin Summerfield
New York University
ISBN 0-915928-32-9

***67. Remote Sensing of Earth from Space: Role of "Smart Sensors"**(1979)
Roger A. Breckenridge
NASA Langley Research Center
ISBN 0-915928-33-7

68. Injection and Mixing in Turbulent Flow (1980)
Joseph A. Schetz
Virginia Polytechnic Institute and State University
ISBN 0-915928-35-3

69. Entry Heating and Thermal Protection (1980)
Walter B. Olstad
NASA Headquarters
ISBN 0-915928-38-8

70. Heat Transfer, Thermal Control, and Heat Pipes (1980)
Walter B. Olstad
NASA Headquarters
ISBN 0-915928-39-6

***71. Space Systems and Their Interactions with Earth's Space Environment** (1980)
Henry B. Garrett
Charles P. Pike
Hanscom Air Force Base
ISBN 0-915928-41-8

72. Viscous Flow Drag Reduction (1980)
Gary R. Hough
Vought Advanced Technology Center
ISBN 0-915928-44-2

73. Combustion Experiments in a Zero-Gravity Laboratory (1981)
Thomas H. Cochran
NASA Lewis Research Center
ISBN 0-915928-48-5

74. Rarefied Gas Dynamics, Parts I and II (two volumes) (1981)
Sam S. Fisher
University of Virginia
ISBN 0-915928-51-5

75. Gasdynamics of Detonations and Explosions (1981)
J.R. Bowen
University of Wisconsin at Madison
N. Manson
Université de Poitiers
A.K. Oppenheim
University of California at Berkeley
R.I. Soloukhin
Institute of Heat and Mass Transfer, BSSR Academy of Sciences
ISBN 0-915928-46-9

76. Combustion in Reactive Systems (1981)
J.R. Bowen
University of Wisconsin at Madison
N. Manson
Université de Poitiers
A.K. Oppenheim
University of California at Berkeley
R.I. Soloukhin
Institute of Heat and Mass Transfer, BSSR Academy of Sciences
ISBN 0-915928-47-7

77. Aerothermodynamics and Planetary Entry (1981)
A.L. Crosbie
University of Missouri-Rolla
ISBN 0-915928-52-3

78. Heat Transfer and Thermal Control (1981)
A.L. Crosbie
University of Missouri-Rolla
ISBN 0-915928-53-1

79. Electric Propulsion and Its Applications to Space Missions (1981)
Robert C. Finke
NASA Lewis Research Center
ISBN 0-915928-55-8

80. Aero-Optical Phenomena (1982)
Keith G. Gilbert
Leonard J. Otten
Air Force Weapons Laboratory
ISBN 0-915928-60-4

81. Transonic Aerodynamics (1982)
David Nixon
Nielsen Engineering & Research, Inc.
ISBN 0-915928-65-5

82. Thermophysics of Atmospheric Entry (1982)
T.E. Horton
University of Mississippi
ISBN 0-915928-66-3

83. Spacecraft Radiative Transfer and Temperature Control (1982)
T.E. Horton
University of Mississippi
ISBN 0-915928-67-1

84. Liquid-Metal Flows and Magnetohydrodynamics (1983)
H. Branover
Ben-Gurion University of the Negev
P.S. Lykoudis
Purdue University
A. Yakhot
Ben-Gurion University of the Negev
ISBN 0-915928-70-1

85. Entry Vehicle Heating and Thermal Protection Systems: Space Shuttle, Solar Starprobe, Jupiter Galileo Probe (1983)
Paul E. Bauer
McDonnell Douglas Astronautics Co.
Howard E. Collicott
The Boeing Co.
ISBN 0-915928-74-4

86. Spacecraft Thermal Control, Design, and Operation (1983)
Howard E. Collicott
The Boeing Co.
Paul E. Bauer
McDonnell Douglas Astronautics Co.
ISBN 0-915928-75-2

87. Shock Waves, Explosions, and Detonations (1983)
J.R. Bowen
University of Washington
N. Manson
Université de Poitiers
A.K. Oppenheim
University of California at Berkeley
R.I. Soloukhin
Institute of Heat and Mass Transfer, BSSR Academy of Sciences
ISBN 0-915928-76-0

88. Flames, Lasers, and Reactive Systems (1983)
J.R. Bowen
University of Washington
N. Manson
Université de Poitiers
A.K. Oppenheim
University of California at Berkeley
R.I. Soloukhin
Institute of Heat and Mass Transfer, BSSR Academy of Sciences
ISBN 0-915928-77-9

89. Orbit-Raising and Maneuvering Propulsion: Research Status and Needs (1984)
Leonard H. Caveny
Air Force Office of Scientific Research
ISBN 0-915928-82-5

90. Fundamentals of Solid-Propellant Combustion (1984)
Kenneth K. Kuo
Pennsylvania State University
Martin Summerfield
Princeton Combustion Research Laboratories, Inc.
ISBN 0-915928-84-1

91. Spacecraft Contamination: Sources and Prevention (1984)
J.A. Roux
University of Mississippi
T.D. McCay
NASA Marshall Space Flight Center
ISBN 0-915928-85-X

92. Combustion Diagnostics by Nonintrusive Methods (1984)
T.D. McCay
NASA Marshall Space Flight Center
J.A. Roux
University of Mississippi
ISBN 0-915928-86-8

93. The INTELSAT Global Satellite System (1984)
Joel Alper
COMSAT Corp.
Joseph Pelton
INTELSAT
ISBN 0-915928-90-6

94. Dynamics of Shock Waves, Explosions, and Detonations (1984)
J.R. Bowen
University of Washington
N. Manson
Universitė de Poitiers
A.K. Oppenheim
University of California at Berkely
R.I. Soloukhin
Institute of Heat and Mass Transfer, BSSR Academy of Sciences
ISBN 0-915928-91-4

95. Dynamics of Flames and Reactive Systems (1984)
J.R. Bowen
University of Washington
N. Manson
Universitė de Poitiers
A.K. Oppenheim
University of California at Bereley
R.I. Soloukhin
Institute of Heat and Mass Transfer, BSSR Academy of Sciences
ISBN 0-915928-92-2

96. Thermal Design of Aeroassisted Orbital Transfer Vehicles (1985)
H.F. Nelson
University of Missouri-Rolla
ISBN 0-915928-94-9

97. Monitoring Earth's Ocean, Land, and Atmosphere from Space — Sensors, Systems, and Applications (1985)
Abraham Schnapf
Aerospace Systems Engineering
ISBN 0-915928-98-1

98. Thrust and Drag: Its Prediction and Verification (1985)
Eugene E. Covert
Massachusetts Institute of Technology
C.R. James
Vought Corp.
William F. Kimzey
Sverdrup Technology AEDC Group
George K. Richey
U.S. Air Force
Eugene C. Rooney
U.S. Navy Department of Defense
ISBN 0-930403-00-2

99. Space Stations and Space Platforms — Concepts, Design, Infrastructure, and Uses (1985)
Ivan Bekey
Daniel Herman
NASA Headquarters
ISBN 0-930403-01-0

100. Single- and Multi-Phase Flows in an Electromagnetic Field: Energy, Metallurgical, and Solar Applications (1985)
Herman Branover
Ben-Gurion University of the Negev
Paul S. Lykoudis
Purdue University
Michael Mond
Ben-Gurion University of the Negev
ISBN 0-930403-04-5

101. MHD Energy Conversion: Physiotechnical Problems (1986)
V.A. Kirillin
A.E. Sheyndlin
Soviet Academy of Sciences
ISBN 0-930403-05-3

102. Numerical Methods for Engine-Airframe Integration (1986)
S.N.B. Murthy
Purdue University
Gerald C. Paynter
Boeing Airplane Co.
ISBN 0-930403-09-6

103. Thermophysical Aspects of Re-Entry Flows (1986)
James N. Moss
NASA Langley Research Center
Carl D. Scott
NASA Johnson Space Center
ISBN 0-930403-10-X

104. Tactical Missile Aerodynamics (1986)
M.J. Hemsch
PRC Kentron, Inc.
J.N. Nielsen
NASA Ames Research Center
ISBN 0-930403-13-4

105. Dynamics of Reactive Systems Part I: Flames and Configurations; Part II: Modeling and Heterogeneous Combustion (1986)
J.R. Bowen
University of Washington
J.-C. Leyer
Université de Poitiers
R.I. Soloukhin
Institute of Heat and Mass Transfer, BSSR Academy of Sciences
ISBN 0-930403-14-2

106. Dynamics of Explosions (1986)
J.R. Bowen
University of Washington
J.-C. Leyer
Université de Poitiers
R.I. Soloukhin
Institute of Heat and Mass Transfer, BSSR Academy of Sciences
ISBN 0-930403-15-0

107. Spacecraft Dielectric Material Properties and Spacecraft Charging (1986)
A.R. Frederickson
U.S. Air Force Rome Air Development Center
D.B. Cotts
SRI International
J.A. Wall
U.S. Air Force Rome Air Development Center
F.L. Bouquet
Jet Propulsion Laboratory, California Institute of Technology
ISBN 0-930403-17-7

108. Opportunities for Academic Research in a Low-Gravity Environment (1986)
George A. Hazelrigg
National Science Foundation
Joseph M. Reynolds
Louisiana State University
ISBN 0-930403-18-5

109. Gun Propulsion Technology (1988)
Ludwig Stiefel
U.S. Army Armament Research, Development and Engineering Center
ISBN 0-930403-20-7

110. Commercial Opportunities in Space (1988)
F. Shahrokhi
K.E. Harwell
University of Tennessee Space Institute
C.C. Chao
National Cheng Kung University
ISBN 0-930403-39-8

111. Liquid-Metal Flows: Magnetohydrodynamics and Applications (1988)
Herman Branover, Michael Mond, and Yeshajahu Unger
Ben-Gurion University of the Negev
ISBN 0-930403-43-6

112. Current Trends in Turbulence Research (1988)
Herman Branover, Michael Mond, and Yeshajahu Unger
Ben-Gurion University of the Negev
ISBN 0-930403-44-4

113. Dynamics of Reactive Systems Part I: Flames; Part II: Heterogeneous Combustion and Applications (1988)
A.L. Kuhl
R & D Associates
J.R. Bowen
University of Washington
J.-C. Leyer
Université de Poitiers
A. Borisov
USSR Academy of Sciences
ISBN 0-930403-46-0

114. Dynamics of Explosions (1988)
A.L. Kuhl
R & D Associates
J.R. Bowen
University of Washington
J.-C. Leyer
Université de Poitiers
A. Borisov
USSR Academy of Sciences
ISBN 0-930403-47-9

115. Machine Intelligence and Autonomy for Aerospace (1988)
E. Heer
Heer Associates, Inc.
H. Lum
NASA Ames Research Center
ISBN 0-930403-48-7

116. Rarefied Gas Dynamics: Space-Related Studies (1989)
E.P. Muntz
University of Southern California
D.P. Weaver
U.S. Air Force Astronautics Laboratory (AFSC)
D.H. Campbell
University of Dayton Research Institute
ISBN 0-930403-53-3

117. Rarefied Gas Dynamics: Physical Phenomena (1989)
E.P. Muntz
University of Southern California
D.P. Weaver
U.S. Air Force Astronautics Laboratory (AFSC)
D. Campbell
University of Dayton Research Institute
ISBN 0-930403-54-1

118. Rarefied Gas Dynamics: Theoretical and Computational Techniques (1989)
E.P. Muntz
University of Southern California
D.P. Weaver
U.S. Air Force Astronautics Laboratory (AFSC)
D.H. Campbell
University of Dayton Research Institute
ISBN 0-930403-55-X

119. Test and Evaluation of the Tactical Missile (1989)
Emil J. Eichblatt Jr.
Pacific Missile Test Center
ISBN 0-930403-56-8

120. Unsteady Transonic Aerodynamics (1989)
David Nixon
Nielsen Engineering & Research, Inc.
ISBN 0-930403-52-5

121. Orbital Debris from Upper-Stage Breakup (1989)
Joseph P. Loftus Jr.
NASA Johnson Space Center
ISBN 0-930403-58-4

122. Thermal-Hydraulics for Space Power, Propulsion and Thermal Management System Design (1989)
William J. Krotiuk
General Electric Co.
ISBN 0-930403-64-9

123. Viscous Drag Reduction in Boundary Layers (1990)
Dennis M. Bushnell
Jerry N. Hefner
NASA Langley Research Center
ISBN 0-930403-66-5

124. Tactical and Strategic Missile Guidance (1990)
Paul Zarchan
Charles Stark Draper Laboratory, Inc.
ISBN 0-930403-68-1

125. Applied Computational Aerodynamics (1990)
P.A. Henne
Douglas Aircraft Company
ISBN 0-930403-69-X

126. Space Commercialization: Launch Vehicles and Programs (1990)
F. Shahrokhi
University of Tennessee Space Institute
J.S. Greenberg
Princeton Synergetics Inc.
T. Al-Saud
Ministry of Defense and Aviation Kingdom of Saudi Arabia
ISBN 0-930403-75-4

127. Space Commercialization: Platforms and Processing (1990)
F. Shahrokhi
University of Tennessee Space Institute
G. Hazelrigg
National Science Foundation
R. Bayuzick
Vanderbilt University
ISBN 0-930403-76-2

128. Space Commercialization: Satellite Technology (1990)
F. Shahrokhi
University of Tennessee Space Institute
N. Jasentuliyana
United Nations
N. Tarabzouni
King Abulaziz City for Science and Technology
ISBN 0-930403-77-0

129. Mechanics and Control of Large Flexible Structures (1990)
John L. Junkins
Texas A&M University
ISBN 0-930403-73-8

130. Low-Gravity Fluid Dynamics and Transport Phenomena (1990)
Jean N. Koster
Robert L. Sani
University of Colorado at Boulder
ISBN 0-930403-74-6

131. Dynamics of Deflagrations and Reactive Systems: Flames (1991)
A. L. Kuhl
Lawrence Livermore National Laboratory
J.-C. Leyer
Université de Poitiers
A. A. Borisov
USSR Academy of Sciences
W. A. Sirignano
University of California
ISBN 0-930403-95-9

132. Dynamics of Deflagrations and Reactive Systems: Heterogeneous Combustion (1991)
A. L. Kuhl
Lawrence Livermore National Laboratory
J.-C. Leyer
Université de Poitiers
A. A. Borisov
USSR Academy of Sciences
W. A. Sirignano
University of California
ISBN 0-930403-96-7

133. Dynamics of Detonations and Explosions: Detonations (1991)
A. L. Kuhl
Lawrence Livermore National Laboratory
J.-C. Leyer
Université de Poitiers
A. A. Borisov
USSR Academy of Sciences
W. A. Sirignano
University of California
ISBN 0-930403-97-5

134. Dynamics of Detonations and Explosions: Explosion Phenomena (1991)
A. L. Kuhl
Lawrence Livermore National Laboratory
J.-C. Leyer
Université de Poitiers
A. A. Borisov
USSR Academy of Sciences
W. A. Sirignano
University of California
ISBN 0-930403-98-3

135. Numerical Approaches to Combustion Modeling (1991)
Elaine S. Oran
Jay P. Boris
Naval Research Laboratory
ISBN 1-56347-004-7

136. Aerospace Software Engineering (1991)
Christine Anderson
U.S. Air Force Wright Laboratory
Merlin Dorfman
Lockheed Missiles & Space Company, Inc.
ISBN 1-56346-005-5

137. High-Speed Flight Propulsion Systems (1991)
S. N. B. Murthy
Purdue University
E. T. Curran
Wright Laboratory
ISBN 1-56347-011-X

138. Propagation of Intensive Laser Radiation in Clouds (1992)
O. A. Volkovitsky
Yu. S. Sedunov
L. P. Semenov
Institute of Experimental Meteorology
ISBN 1-56347-020-9

139. Gun Muzzle Blast and Flash (1992)
Günter Klingenberg
Fraunhofer-Institut für Kurzzeitdynamik, Ernst-Mach-Institut (EMI)
Joseph M. Heimerl
U.S. Army Ballistic Research Laboratory (BRL)
ISBN 1-56347-012-8

140. Thermal Structures and Materials for High-Speed Flight (1992)
Earl A. Thornton
University of Virginia
ISBN 1-56347-017-9

141. Tactical Missile Aerodynamics: General Topics (1992)
Michael J. Hemsch
Lockheed Engineering & Sciences Company
ISBN 1-56347-015-2

142. Tactical Missile Aerodynamics: Prediction Methodology (1992)
Michael R. Mendenhall
Nielsen Engineering & Research, Inc.
ISBN 1-56347-016-0

143. Nonsteady Burning and Combustion Stability of Solid Propellants (1992)
Luigi De Luca
Politecnico di Milano
Edward W. Price
Georgia Institute of Technology
Martin Summerfield
Princeton Combustion Research Laboratories, Inc.
ISBN 1-56347-014-4

144. Space Economics (1992)
Joel S. Greenberg
Princeton Synergetics, Inc.
Henry R. Hertzfeld
HRH Associates
ISBN 1-56347-042-X

145. Mars: Past, Present, and Future (1992)
E. Brian Pritchard
NASA Langley Research Center
ISBN 1-56347-043-8

146. Computational Nonlinear Mechanics in Aerospace Engineering (1992)
Satya N. Atluri
Georgia Institute of Technology
ISBN 1-56347-044-6

147. Modern Engineering for Design of Liquid-Propellant Rocket Engines (1992)
Dieter K. Huzel
David H. Huang
ISBN 1-56347-013-6

148. Metallurgical Technologies, Energy Conversion, and Magnetohydrodynamic Flows (1993)
Herman Branover
Yeshajahu Unger
Ben-Gurion University of the Negev
ISBN 1-56347-019-5

149. Advances in Turbulence Studies (1993)
Herman Branover
Yeshajahu Unger
Ben-Gurion University of the Negev
ISBN 1-56347-018-7

150. Structural Optimization: Status and Promise (1993)
Manohar P. Kamat
Georgia Institute of Technology
ISBN 1-56347-056-X

151. Dynamics of Gaseous Combustion (1993)
A. L. Kuhl
Lawrence Livermore National Laboratory
J.-C. Leyer
Université de Poitiers
A. A. Borisov
Russian Academy of Sciences
W. A. Sirignano
University of California
ISBN 1-56347-060-8

152. Dynamics of Heterogeneous Combustion and Reacting Systems (1993)
A. L. Kuhl
Lawrence Livermore National Laboratory
J.-C. Leyer
Université de Poitiers
A. A. Borisov
Russian Academy of Sciences
W. A. Sirignano
University of California
ISBN 1-56347-058-6

153. Dynamic Aspects of Detonations (1993)
A. L. Kuhl
Lawrence Livermore National Laboratory
J.-C. Leyer
Université de Poitiers
A. A. Borisov
Russian Academy of Sciences
W. A. Sirignano
University of California
ISBN 1-56347-057-8

154. Dynamic Aspects of Explosion Phenomena (1993)
A. L. Kuhl
Lawrence Livermore National Laboratory
J.-C. Leyer
Université de Poitiers
A. A. Borisov
Russian Academy of Sciences
W. A. Sirignano
University of California
ISBN 1-56347-059-4

(Other Volumes are planned.)